ADVANCES IN CHEMICAL PHYSICS

VOLUME LXXIV

ADVANCES IN CHEMICAL PHYSICS

EDITED BY

I. PRIGOGINE

University of Brussels
Brussels, Belgium
and
University of Texas
Austin, Texas

AND

STUART A. RICE

Department of Chemistry
and
The James Franck Institute
The University of Chicago
Chicago, Illinois

VOLUME LXXIV

AN INTERSCIENCE® PUBLICATION
JOHN WILEY & SONS
NEW YORK • CHICHESTER • BRISBANE • TORONTO • SINGAPORE

An Interscience® Publication

Library of Congress Cataloging Number: 58-9935

ISBN 0-471-61212-X

Printed in the United States of America

10 9 8 7 6 5 4 3 2 1

CONTRIBUTORS TO VOLUME LXXIV

V. V. Barelko, Institute of Chemical Physics of the Academy of Sciences of the USSR, Chernogolovka Branch, Chernogolovka, Moscow District, USSR

I. M. Barkalov, Institute of Chemical Physics of the Academy of Sciences of the USSR, Chernogolovka Branch, Chernogolovka Moscow District, USSR

V. I. Goldanskii, Institute of Chemical Physics of the Academy of Sciences of the USSR, Chernogolovka Branch, Chernogolovka, Moscow District, USSR

D. P. Kiryukhin, Institute of Chemical Physics of the Academy of Sciences of the USSR, Chernogolovka Branch, Chernogolovka, Moscow District, USSR

Philemon Kottis, Centre de Physique Moleculaire Optique et Hertzienne, Universite de Bordeaux I, Talence, France

Michel Orrit, Centre de Physique Moleculaire Optique et Hertzienne, Universite de Bordeaux I, Talence, France

A. J. C. Varandas, Departamento de Quimica, Universidade de Coimbra, Coimbra, Portugal

A. M. Zanin, Institute of Chemical Physics of the Academy of Sciences of the USSR, Chernogolovka Branch, Chernogolovka, Moscow District, USSR

INTRODUCTION

Few of us can any longer keep up with the flood of scientific literature, even in specialized subfields. Any attempt to do more and be broadly educated with respect to a large domain of science has the appearance of tilting at windmills. Yet the synthesis of ideas drawn from different subjects into new, powerful, general concepts is as valuable as ever, and the desire to remain educated persists in all scientists. This series, *Advances in Chemical Physics*, is devoted to helping the reader obtain general information about a wide variety of topics in chemical physics, which field we interpret very broadly. Our intent is to have experts present comprehensive analyses of subjects of interest and to encourage the expression of individual points of view. We hope that this approach to the presentation of an overview of a subject will both stimulate new research and serve as a personalized learning text for beginners in a field.

I. Prigogine
Stuart A. Rice

CONTENTS

ADVANCES IN CHEMICAL PHYSICS

VOLUME LXXIV

SURFACE AND BULK SPECTROSCOPY OF A MOLECULAR CRYSTAL: EFFECT OF RELAXATION AND THERMAL OR STATIC DISORDER

MICHEL ORRIT AND PHILÉMON KOTTIS

Centre de Physique Moléculaire Optique et Hertzienne
Université de Bordeaux I
Talence, France

CONTENTS

INTRODUCTION

An assembly of molecules, weakly interacting in a condensed phase, has the general features of an oriented gas system, showing spectral properties similar to those of the constitutive molecules, modulated by new collective and cooperative intrinsic phenomena due to the coherent dynamics of the molecular excitations. These phenomena emerge mainly from the resonant interactions of the molecular excitations, which have to obey the lattice symmetry (with edge boundary, dimensionality, internal radiation, and relativistic conditions), with couplings to the phonon field and to the free radiation field.

For the investigation of collective phenomena it is simplest (although not necessary), with regard to translational symmetry and boundary conditions, to deal with a 3D infinite molecular crystal, the positions and the orientations of the constitutive molecules being given with high precision by X-ray spectroscopy. Assuming the structure and the molecular characteristics to be well known and not significantly affected by the crystal field, it is possible, in principle, to calculate the new collective and cooperative properties and to relate them to the topology of the molecular excitations. Various properties may thus be investigated, such as the internal and the external vibrations in the lattice, mechanical and thermodynamical properties, the crystal cohesion energy, the excited electronic states with renormalized oscillator strengths, charge-transfer states, and associated transport properties.

The aim of this chapter is the investigation of the dielectric properties in the optical spectral region of a 3D crystal, with pertinent emphasis on the severe modifications of the dynamics of the electronic excitations when the excitons are confined in a surface or subsurface monolayer to create 2D excitons. This happens in all crystals with layered structure *and* transitions strongly coupled to the radiation field. Recently, many works have discussed the crucial role as doorway states played by the surface states in the exchange of mechanical, chemical, and electromagnetic energy between the substrate crystal and its environment. In this work, we investigate the enhanced radiative properties of the surface excited states of the anthracene single crystal and discuss the way they may allow us to probe the mechanisms of relaxation of the surface states in (1) excitation exchange between surface and substrate by coulombic and retarded interactions, (2) damping of 2D surface excitons, by static, thermal, and photochemical-reaction disorder, and (3) interface perturbations of 2D excitons with a nonresonant substrate, or confined between the substrate and a nonresonant disordered clothing layer.

Our investigation is carried out in the domain of validity of the dipolar exciton. Indeed, the properties of the constitutive molecules, mainly derived by spectroscopy in the gaseous phase, in solutions, or in supersonic jets, are not known with the desired accuracy, especially in the UV part of the spectrum, for which spectroscopy becomes difficult. Also, from the theoretical point of view, electronic wave functions of large molecules are still poorly known in their details. These two circumstances explain why a large class of cooperative properties—involving, for instance, dominantly electronic exchange terms (electron transfer in CT states, or triplet exciton transfer or annihilation)—are poorly understood more than 50 years after the pioneering works on molecular-crystal excitons.[1–13] In contrast, in strong or intermediate dipolar transitions to singlet states, the coulombic interactions between electronic clouds of the molecules largely dominate the exchange terms and decay smoothly with intermolecular distances, so that even a crude knowledge of the electronic wave function may allow a realistic discussion of the collective electronic properties of bulk or surface excitons due to such transitions. These excitons are termed dipolar excitons.

Consequently, in an investigation restricted to molecular crystals, we must emphasize, that only the general outline of the optical spectra can be understood using the Frenkel–Davydov model of excitons in a perfect, infinite, rigid crystal. The purpose of this chapter is also to provide a theoretical study of a general matter–radiation system and to illustrate in the well-known example of the anthracene crystal, representing a large class of crystals, such deviations from the Frenkel–Davydov model as nuclear motion, surfaces and finite volume, amalgamation, or persistent mixed crystals in the substitutional-disorder limit.

The first section recalls the Frenkel–Davydov model in terms of a set of electromagnetically coupled point dipoles. A compact version of Tyablikov's quantum-mechanical solution is displayed and found equivalent to the usual semiclassical theory. The general solution is then applied to a 3D lattice. Ewald summation and nonanalyticity at the zone center are discussed.[14] Separating short and long-range terms in the equations allows us to introduce Coulomb (dipolar) excitons and polaritons.[15,16] Lastly, the finite extent of actual molecules is considered, and consequent modifications of the above theory qualitatively discussed.[14–22]

In Section II we consider in detail the way the nuclear motion affects absorption by and emission from the crystal. After surveying the main coupling mechanisms between excitons and nuclear motion, we separately examine the coupling to high-energy lattice phonons, simplifying the general problem in each case. Vibronic structures of the anthracene crystal are quantitatively discussed using the theory—in particular, the two-particle-state absorption continua and edges, and the broadened vibronic peaks. Current theories of the exciton–phonon coupling are presented and applied to the 0–0 absorption peak (polariton-forbidden transition of anthracene crystal) as obtained from a careful Kramers–Kronig analysis. The role of acoustical phonons in broadening the optical line is emphasized and compared with that of optical phonons. Concerning spontaneous secondary emissions of the crystal, a general perturbative classification is proposed for Raman scattering and for hot and relaxed fluorescence. Fluorescence of anthracene crystals is described and discussed according to current interpretations.

Section III deals with the surface excitations of the anthracene crystal, confined in the first (S_1), second (S_2), and third (S_3) (001) lattice planes. The experimental observations are briefly summarized. A simple model shows how the fast radiative decay arises and how the underlying bulk reflection modulates this "superradiant" emission, as well as why gas condensation on the crystal surface strongly narrows this emission, thus accounting for the observed structures. An intrinsic process is proposed to explain the surface-to-bulk relaxation at low temperatures, observed in spite of the very weak surface-to-bulk coupling for $k \sim 0$ states.

The excitation spectra of surface emission show, besides the a-polarized surface-exciton Davydov component, vibronic surface structures which include intramolecular vibronic peaks and a surface phonon peak with frequency lower than in the bulk. We interpret the asymmetry of the S_1 emission of homogeneous width $\sim 15\ \mathrm{cm}^{-1}$ as due to growing relaxation rates for larger relaxed energy, and we distinguish between Raman and fluorescence processes.

The collected data on surface excitons and surface phonons are confronted in order to estimate the extent of the weak surface reconstruction of the anthracene crystal. A surface destabilization (relative to the bulk) corresponding to a negative pressure (-4 kbar) is inferred and thought to lead to angular reconstruction less than or about $1°$. The observed energy sequence for surface resonances is shown to be compatible only with R^{-7} van der Waals forces, for both mechanisms proposed.

Section IV is devoted to excitons in a disordered lattice. In the first subsection, restricted to the 2D radiant exciton, we study how the coherent emission is hampered by such disorder as thermal fluctuation, static disorder, or surface annihilation by surface-molecule photodimerization. A sharp transition is shown to take place between coherent emission at low temperature (or weak extended disorder) and incoherent emission of small excitonic coherence domains at high temperature (strong extended disorder). Whereas a mean-field theory correctly deals with the long-range forces involved in emission, these approximations are reviewed and tested on a simple model case: the nondipolar triplet naphthalene exciton. The very strong disorder then makes the inclusion of aggregates in the theory compulsory. From all this study, our conclusion is that an effective-medium theory needs an effective interaction as well as an effective potential, as shown by the comparison of our theoretical results with exact numerical calculations, with very satisfactory agreement at all concentrations. Lastly, the 3D case of a dipolar exciton with disorder is discussed qualitatively.

I. DIPOLAR EXCITONS

In the first part of this introductory section, we summarize the main collective phenomena acquired by the dipolar exciton from the lattice-symmetry collectivization of molecular properties. The crystal is considered as an assembly of electrically neutral systems, the molecules, physically separated from each other and in electromagnetic interaction. This N-body problem will be treated quantum-mechanically in the limit of low exciton densities. We redemonstrate the complete equivalence of this treatment with the theories of Lorentz and Ewald, as well as with the semiclassical approximation. In Section I.A, in a more compact but still gradual way, we establish the model of the rigid lattice of dipoles and the general theory of low-exciton-density systems in interaction with the radiation field. Coulombic excitons, photons, and polaritons are analyzed physically in a more "continuous" way than in the direct application of the Tyablikov's method diagonalizing the matter–radiation hamiltonian. In Section I.B, dipole sums are calculated, and the distinction is established between excitons and polaritons. In Section I.C, we show the way these notions of interactions are generalized and applied in aromatic 2D or 3D crystals.

A. Electromagnetic Interactions in an Assembly of N Dipoles

At large distances compared with the size of the molecules, the dipolar transitions may be treated with point dipoles. This approximation is not valid for intermolecular distances met with in condensed phases. However, the point-dipole approximation allows one to discuss the various levels of interaction and proves very useful for the discussion of the general case, as illustrated in Section II. Historically, the point-dipole approximation was the first to be applied to molecular-crystal excitations.[17–20]

1. *The Electric-Dipole Interaction*

a. *The Quasi-boson Approximation.* We consider the transition from the ground electronic state in an assembly of point dipoles with a number of excited states f. Neglecting the dipole–dipole interactions, the energy of the assembly of n point dipoles appears as the sum of point-dipole hamiltonians:

$$H_0 = \sum_{nf} E_{nf} |nf\rangle\langle nf| \tag{1.1}$$

Omitting the ground-state energy, H_0 can be rewritten in terms of the operators $B^\dagger_{nf}$ (B_{nf}) representing the excitation (deactivation) of a dipole from (to) the ground state:

$$H_0 = \sum_{nf} E_{nf} B^\dagger_{nf} B_{nf} \tag{1.2}$$

At this stage, it is convenient to assume a very low density of excited dipoles. In other words, we assume that the exciting external source is sufficiently weak so that at each instant the probability of finding a given dipole in an excited state is very small compared to 1. In this condition, the system satisfies the linear-response approximation. Since the elementary excitations are very dilute (i.e., the occupation numbers are very small), all statistics are equivalent. For the convenience of further calculations (e.g. interaction with photons), the operators $B^\dagger$, B are assumed to obey Bose statistics[21,22]:

$$[B_{nf}, B^\dagger_{mg}] = \delta_{nm}\delta_{fg} \tag{1.3}$$

This Bose approximation amounts to adding to each electronic excited state a sequence of equidistant levels, creating a harmonic oscillator; that is, it is Thomson's classical model of elastically bound electrons, with this difference: that each point dipole presents many eigenfrequencies associated with the transition dipole.[23]

b. *Interactions in the Coulomb Gauge.* The hamiltonian H_0 has to be complemented with the dipole–dipole interactions. In the Coulomb gauge, they are electrostatic instantaneous interactions, providing the matter hamiltonian in the low-excitation-density case:

$$H_M = \sum_{nf} E_{nf} B_{nf}^{\dagger} B_{nf} + \sum_{\substack{nf \\ mg}} J_{nf}^{mg} B_{mg}^{\dagger} B_{nf} \tag{1.4}$$

where J_{nf}^{mg} are the coulombic interactions (approximated by the dipole–dipole term) between charge densities of dipoles nf and mg. Additional interactions of the point dipoles with the transverse (free) electromagnetic field must be included[24]:

$$-\sum_i \frac{e}{m_i} \mathbf{p}_i \cdot \mathbf{A}(\mathbf{r}_i) + \sum_i \frac{e^2}{2m_i} A^2(\mathbf{r}_i) \tag{1.5}$$

where $\mathbf{A}(\mathbf{r}, t)$ is the vector potential of the field at point $\mathbf{r}$, acting on electrons with charge e and momentum $\mathbf{p}_i$. Finally, the matter–radiation energy must be completed with the total energy of the quantized radiation field:

$$H_R = \sum_k \hbar\omega_k (a_k^{\dagger} a_k + \tfrac{1}{2}) \tag{1.6}$$

As usual, the index k includes the wave vector $\mathbf{k}$ and the polarizations $\boldsymbol{\varepsilon}_1$ and $\boldsymbol{\varepsilon}_2$ perpendicular to $\mathbf{k}$, and $a_k^{\dagger}$ (a_k) is the creation (annihilation) operator for photons of energy $\hbar\omega_k = \hbar c|\mathbf{k}|$. We must write in this formalism the operators, vector potential, and electric field which are involved in our calculations:

$$\mathbf{A}(\mathbf{r}) = \sum_k \left(\frac{\hbar}{2\varepsilon_0 \omega_k V} \right)^{1/2} \boldsymbol{\varepsilon}_k (a_k e^{i\mathbf{k}\cdot\mathbf{r}} + a_k^{\dagger} e^{-i\mathbf{k}\cdot\mathbf{r}}) \tag{1.7}$$

$$\mathbf{E}(\mathbf{r}) = \sum_k i \left(\frac{\hbar\omega_k}{2\varepsilon_0 V} \right)^{1/2} \boldsymbol{\varepsilon}_k (a_k e^{i\mathbf{k}\cdot\mathbf{r}} - a_k^{\dagger} e^{-i\mathbf{k}\cdot\mathbf{r}}) \tag{1.8}$$

(with V indicating the quantization volume).

We must remark that $\mathbf{A}$ and $\mathbf{E}$, evolving at frequencies ω_k, cannot represent the electrostatic field created by the dipoles and responsible for the coulombic interactions J in (1.4). **A** and **E** represent the incident waves and the waves radiated by the dipoles.

The substitution of **A** in the expression (1.5) leads to the matter–radiation hamiltonian H, with a quadratic form of Bose operators, which is diagonalized in a straightforward manner using Tyablikov's method.[21] We shall proceed to the same diagonalization with a previous simplification of the hamiltonian H using the electric dipolar gauge.

c. *Interactions in the Electric-Dipole Gauge.* Relativistic causality arguments imply electromagnetic interactions between two systems separated in space cannot be instantaneous, as they appear to be in the Coulomb gauge. Therefore, the terms J in (1.4) are "unphysical."[25] It can be shown that the terms J are canceled by instantaneous interactions issuing from (1.5) and from the transverse field. To eliminate the instantaneous terms J, we make a unitary transformation of the total hamiltonian H with[25]

$$\mathscr{S} = \exp\left(\sum_i \mathbf{r}_i \cdot \mathbf{A}(\mathbf{r}_i)\right)$$

This unitary transformation redefines the variables of the field (a_k, E, A, etc.) and displaces the momentum operator of each charge. Retaining only the electronic dipolar terms, the total hamiltonian becomes

$$H = \sum_{nf} E_{nf} B_{nf}^{\dagger} B_{nf} + H_R - \sum_{nf} \mathbf{D}_{nf} \cdot \mathbf{E}(\mathbf{r}_n) \tag{1.9}$$

H_R is formally of the type (1.6), and **E** of the type (1.8), with new operators $a^{\dagger}(a)$ for the photons. The operator $\mathbf{D}_{nf}$ is the dipole-moment operator associated with dipole n for the excited state f: It may be written as

$$\mathbf{D}_{nf} = \mathbf{d}_{nf}(B_{nf} + B_{nf}^{\dagger}) \tag{1.10}$$

where $\mathbf{d}_{nf}$ is the transition dipole.

The electric-field operator now contains the total field of the system, viz. the free field and the field radiated by the matter. The J dipole–dipole interactions have disappeared: In this formalism all the interactions are retarded, carried by the electric field. We notice the analogy of this transformation with the change from Coulomb to Lorentz gauge, where potentials and fields are retarded. In (1.9) the interaction operator $-\mathbf{D} \cdot \mathbf{E}$ is the analog of the corresponding classical interaction: In the semiclassical approach we are looking for the response of the dipole (or the polarizability) to the total electric field, the interaction energy being $-\mathbf{D} \cdot \mathbf{E}_{\text{total}}$.

2. *Elementary Excitations of the Coupled System*

In this subsection, we present a new formulation of Tyablikov's method[21] to diagonalize a quadratic form for bosons. Since we are operating in the case of fields, the term diagonalization takes on a slightly different sense than for the ordinary Hilbert space. We must transform the hamiltonian (1.9) in the following form:

$$H = \sum_i E_i (P_i^\dagger P_i + \tfrac{1}{2}) \tag{1.11}$$

where E_i and P_i are the energies and operators of the new bosons. We shall analyze the various steps leading to the diagonal form (1.11), starting from the most general quadratic form

$$H = \frac{1}{2}\sum_{\alpha\beta} \omega_{\alpha\beta}(B_\alpha^\dagger B_\beta + B_\beta^\dagger B_\alpha) + \chi_{\alpha\beta}(B_\alpha B_\beta + B_\beta B_\alpha) + \chi_{\beta\alpha}^*(B_\alpha^\dagger B_\beta^\dagger + B_\beta^\dagger B_\alpha^\dagger) \tag{1.12}$$

This symmetric form may always be obtained with the use of the commutation relation

$$[B_\alpha, B_\beta] = 0, \qquad [B_\alpha, B_\beta^\dagger] = \delta_{\alpha\beta} \tag{1.13}$$

with a possible global shift in the origin of the energies. We must notice that this symmetrical form is obtained directly on quantizing the classical fields. Therefore, since the formalism that follows, although concerned with the commutation conditions, never uses them explicitly, we find here the fundamental reason for the equivalence between the classical and quantum-mechanical treatments of interactions with radiation: The operators $B, B^\dagger$ may be formally treated as classical fields.[26]

The expansion (1.12) of H may be written in a compact matrix form using a scalar product in a fictitious space where B_α and $B_\alpha^\dagger$ are components of vectors, with the notation

$$|BB^\dagger) = \text{column matrix} \begin{pmatrix} B_1 \\ B_\alpha \\ B_\alpha^\dagger \end{pmatrix}, \qquad (BB^\dagger|$$
$$= \text{row matrix } (B_1, B_1^\dagger, \ldots, B_\alpha B_\alpha^\dagger)$$

Then H takes the compact form

$$H = (BB^\dagger | h | B^\dagger B) \tag{1.14}$$

where the matrix of the coefficients h decomposes into four blocks:

$$h = \frac{1}{2}\begin{pmatrix} \omega & \chi \\ \chi^* & \omega^* \end{pmatrix} \tag{1.15}$$

The hermiticity of H leads to the hermiticity of ω and to the symmetry of χ. [The inversion of B and $B^\dagger$ in (1.14) is a notational device to bring the coefficients of $B^\dagger B$ to the diagonal of h.] By a linear transformation on the $B^\dagger$, B, we introduce the new operators P and $P^\dagger$, which must yield a symmetric form for the hamiltonian

$$H = \frac{1}{2}\sum_i E_i(P_i P_i^\dagger + P_i^\dagger P_i) \tag{1.16}$$

as may be checked by the interchange of $P^\dagger$ and P and of $B^\dagger$ and B, taking into account the hermiticity of H. The linear relation between B and P is expressed in the following compact form:

$$\begin{aligned} |P^\dagger P) &= S|B^\dagger B) \quad \overset{t}{\Leftrightarrow} \quad (P^\dagger P| = (B^\dagger B|{}^tS \\ |PP^\dagger) &= S^*|BB^\dagger) \quad \overset{t}{\Leftrightarrow} \quad (PP^\dagger| = (BB^\dagger|{}^tS^* \end{aligned} \tag{1.17}$$

(The operation $*$ must be understood as hermitian conjugation on B and P.) In addition, the new operators P, $P^\dagger$ must satisfy the boson commutation relation satisfied by B and $B^\dagger$. The relations (1.13) may also be expressed in a compact form through the use of a matrix φ, defined as

$$\varphi(B, B^\dagger) = |BB^\dagger)(B^\dagger B| - {}^t(|BB^\dagger)(B^\dagger B|)^* \tag{1.18}$$

and take the form

$$\varphi(B, B^\dagger) = K, \qquad \text{or} \quad K = \begin{pmatrix} \mathbb{1} & 0 \\ 0 & -\mathbb{1} \end{pmatrix} \tag{1.19}$$

The boson conditions on the P, $P^\dagger$ become $\phi(P, P^\dagger) = K$, which, using (1.17) and (1.19), leads to the condition on S:

$$K = S^* K\,{}^tS \tag{1.20}$$

The relation (1.16) for the diagonalization of H can be written in a compact form using the unknown diagonal matrix Λ:

$$H = (PP^{\dagger} | \Lambda | P^{\dagger}P) \tag{1.21}$$

leading to the condition

$$h = {}^{t}S^{*} \Lambda S \tag{1.22}$$

From (1.20) and (1.22) we arrive at the system determining S and Λ:

$$\begin{aligned} K\Lambda &= SKhS^{-1} \\ K &= S^{*}K\,{}^{t}S \end{aligned} \tag{1.23}$$

The first of these conditions expresses the diagonalization of the matrix Kh and not that of h. To satisfy the relation (1.16) the eigenvalues of $K\Lambda$ must consist of pairs whose sums are zero (the characteristic polynomial of $K\Lambda$ must be in z^2).

Rewriting (1.16) in the normal form (1.11), we note the global energy shift. In fact, in a real case of radiation interaction with dipoles, the global shift $\frac{1}{2}\sum_i E_i$ or $\frac{1}{2}\sum_{\alpha}\omega_{\alpha\alpha}$ associated with (1.12) diverges owing to the nonrelativistic and nonnormalized form of the matter–radiation interaction. On the contrary, the difference $\frac{1}{2}(\sum_i E_i - \sum_{\alpha\alpha}\omega_{\alpha\alpha})$ remains well defined and represents the cohesion energy of the system of dipoles[19] in the internal radiation, originating from the work of the van der Waals retarded forces and expressing their quantum-mechanical nature.

To complete this method of gradual diagonalization, we write the matrix Kh in a block-diagonal form:

$$Kh = \frac{1}{2}\begin{pmatrix} \omega & \chi \\ -\chi^{*} & -\omega^{*} \end{pmatrix} \tag{1.24}$$

3. *Eigenfrequency Equation and Elementary Excitations of the Matter–Radiation System*

In the case of an assembly of dipoles coupled to radiation and described by the hamiltonian (1.9), we must diagonalize a matrix of the type (1.24) which may be written with explicitly separated matter variables $B, B^{\dagger}$ and radiation variables $a, a^{\dagger}$:

$$H = (B^{\dagger}a^{\dagger}Ba | h | BaB^{\dagger}a^{\dagger}) \quad \text{with} \quad Kh = \begin{pmatrix} \omega_0 & C & 0 & C^{*} \\ C^{*} & \omega & C^{*} & 0 \\ 0 & -C & -\omega_0 & -C^{*} \\ -C & 0 & -C & -\omega \end{pmatrix} \tag{1.25}$$

where ω_0 is a diagonal matrix giving the transition energies of the dipoles, ω is a diagonal matrix of the photon energies, and the matrix C expresses the matter–radiation coupling, with each element C_{kn} expressing the coupling between dipole n and radiation mode k following (1.8):

$$C_{kn} = -i\mathbf{d}_n \cdot \boldsymbol{\varepsilon} \left(\frac{\hbar\omega_k}{2\varepsilon_0 V} \right)^{1/2} e^{i\mathbf{k}\cdot\mathbf{n}} \tag{1.26}$$

Separating matter and photon variables, h may be given the form

$$\begin{pmatrix} \Omega_0 & C \\ C' & \Omega \end{pmatrix} \tag{1.27}$$

from which the photon variables may be eliminated, using the method of Appendix B.1, to give, for each energy z, the diagonalization of the matrix $z - \Omega_0 - C[1/(z - \Omega)]C'$. The inversion of $z - \Omega$ causes no problem, since Ω is diagonal, provided that z does not coincide with any of the eigenvalues of Ω. Since the spectrum of Ω extends from $-\infty$ to $+\infty$, we take $z = \omega + i\varepsilon$ with $\varepsilon \to 0$. Let us put for the shift operator

$$\mathscr{R}(z) = C \frac{1}{z - \Omega} C' \tag{1.28}$$

The matrix elements between the various pairs of matter operators provide the expressions

$$\begin{aligned} (B_m|\mathscr{R}(z)|B_n) &= (B_m|\mathscr{R}(z)|B_n^\dagger) = -(B_m^\dagger|\mathscr{R}(z)|B_n) \\ &= -(B_m^\dagger|\mathscr{R}(z)|B_n^\dagger) = R_{mn}(z) \end{aligned} \tag{1.29}$$

with

$$R_{mn}(z) = \sum_k \frac{C_{km}C_{kn}^*}{z - \omega_k} + \frac{C_{km}^*(-C_{kn})}{z + \omega_k}$$

which reduces, because of the symmetry in $\mathbf{k}$ and $-\mathbf{k}$ of the coefficients ω_k and C_{kn}, to

$$R_{mn}(z) = \sum_k 2\omega_k \frac{C_{km}C_{kn}^*}{z^2 - \omega_k^2} \tag{1.30}$$

Therefore, the new matrix to diagonalize has the form

$$z - \tilde{H} = \begin{pmatrix} z - \omega_0 - R & R \\ -R & z + \omega_0 + R \end{pmatrix} \tag{1.31}$$

Here ω_0 is a diagonal matrix (i.e. without electromagnetic interactions between the dipoles) which is not, in general, proportional to unity, for instance when considering many transitions. The matrices ω_0 and R do not commute in general. Nevertheless, we may simplify the relation $\det(z - \tilde{H}) = 0$ by a passage to the representation defined by

$$\mathscr{P} = \frac{1}{\sqrt{2}}\begin{pmatrix} \mathbb{1} & \mathbb{1} \\ \mathbb{1} & -\mathbb{1} \end{pmatrix} \tag{1.32}$$

The condition $\det[\mathscr{P}^{-1}(z - \tilde{H})\mathscr{P}] = 0$ simplifies to the equation[27]

$$\det[z^2 - \omega_0^2 - 2\omega_0 R(z)] = 0 \tag{1.33}$$

This equation is equivalent to the characteristic equation of the matrix (1.25) and therefore provides, when exactly resolved, all the eigenenergies of the total system in spite of the privileged role played by the matter subsystem in our calculations. We notice that an equation in z^2 is effectively obtained, owing to the equivalent positions devoted to the creation ($B^\dagger$) and annihilation (B) operators in the present theory. More generally, an equation in z^2, with Feynman propagators of the type (1.30), are obtained when problems of boson *fields* are treated, as for instance photons in Section II.A.1.e. On the contrary, equations in z, with propagators of the resolvent type (see Appendix B), appear when "dynamical" problems are treated where the number of particles involved is specified, as for instance in the near vicinity of one resonance, as treated in Section II.B.2.e.

4. *Radiated Fields in the Semiclassical Theory*

We calculate in this subsection the explicit matrix elements $R_{mn}(z)$ of (1.30). The substitution in that equation of the values (1.26) of C_{kn} and $\omega_k = \hbar c|\mathbf{k}|$ gives the expression

$$R_{mn}(z) = \sum_{\mathbf{k},\boldsymbol{\varepsilon}} \frac{(\mathbf{d}_m \cdot \boldsymbol{\varepsilon})(\boldsymbol{\varepsilon} \cdot \mathbf{d}_n)}{2\varepsilon_0 V} \frac{2(\hbar c|\mathbf{k}|)^2}{z^2 - (\hbar c|\mathbf{k}|)^2} e^{i\mathbf{k}\cdot(\mathbf{m}-\mathbf{n})}. \tag{1.34}$$

Since for each pair of dipoles we have many possible directions of transition moments, it is useful to rewrite (1.34) in a tensorial form using the dipole–dipole interaction tensor $\boldsymbol{\phi}(m - n)$:

$$R_{mn}(z) = \mathbf{d}_n \boldsymbol{\phi}(m - n)\mathbf{d}_m, \qquad \boldsymbol{\phi}(r) = \frac{1}{\varepsilon_0 V} \sum_{\mathbf{k},\boldsymbol{\varepsilon}} \frac{|\hbar c\mathbf{k}|^2 e^{i\mathbf{k}\cdot\mathbf{r}}}{z^2 - |\hbar c\mathbf{k}|^2} \boldsymbol{\varepsilon}\boldsymbol{\varepsilon}. \tag{1.35}$$

For each value of $\mathbf{k}$, we have $\sum_{\mathbf{k},\varepsilon} \boldsymbol{\varepsilon\varepsilon} = 1 - \mathbf{kk}/k^2$. Then we proceed to integrate over all $\mathbf{k}$ in (1.35) to find (with the "retarded" boundary condition $z = \omega + i\varepsilon$)

$$\boldsymbol{\phi}(r, \omega) = -\frac{1}{4\pi\varepsilon_0}\left(\frac{\omega^2}{c^2} + \nabla\nabla\right)\frac{e^{i\omega r/c}}{r} \tag{1.36}$$

$$\boldsymbol{\phi}(r, \omega) = -\frac{1}{4\pi\varepsilon_0} e^{i\omega r/c}\left[\frac{\omega^2}{c^2 r}\left(1 - \frac{\mathbf{rr}}{r^2}\right) - \frac{1}{r^3}\left(1 - \frac{i\omega r}{c}\right)\left(1 - \frac{3\mathbf{rr}}{r^2}\right)\right] \tag{1.37}$$

It is easy to recognize in $\boldsymbol{\phi}(r)$ the tensorial expression for the retarded field created at the point $\mathbf{r}$ by a dipole located at the origin $\mathbf{r} = \mathbf{0}$. The instantaneous term in $1/r^3$ coincides, in the approximation of the retarded effects, with the electrostatic field and reestablishes the coulombic dipole–dipole interaction. We shall investigate $\boldsymbol{\phi}(r)$ at distances short compared to λ, which allows us to expand (1.37) in powers of $\omega r/c$:

$$\begin{aligned}\boldsymbol{\phi}(r, \omega) = -\frac{1}{4\pi\varepsilon_0}\Bigg[&-\frac{1}{r^3}\left(1 - \frac{3\mathbf{rr}}{r^2}\right) \\ &+ \frac{1}{r}\frac{\omega^2}{2c^2}\left(1 + \frac{\mathbf{rr}}{r^2}\right) + \tfrac{2}{3} i \frac{\omega^3}{c^3} + O\left(\left(\frac{\omega r}{c}\right)^4\right)\Bigg]\end{aligned} \tag{1.38}$$

First, this expansion shows that the difference between the instantaneous electrostatic term and the retarded field appears only in second order. Secondly, we find in the third order a pure imaginary scalar (nontensorial) although the imaginary part introduced via z is infinitely small. This pure imaginary scalar originates from the continuous summation over $\mathbf{k}$ in (1.35) (a discrete summation introduces only infinitely small imaginary terms) and describes the damping of the dipolar field owing to radiation. To see this, we may consider the action on itself of the dipole located at the origin $\mathbf{0}$, i.e. $\boldsymbol{\phi}(\mathbf{r} \to \mathbf{0}, \omega)$: The real part of $\boldsymbol{\phi}$ diverges, in a way that even depends on the way $\mathbf{r}$ tends to zero. These two singularities are due in part to the point character of the dipole and to the divergence of its electrostatic energy, and in part to the inadequacy of this treatment for large wave vectors, for which the relativistic description of matter is necessary. In contrast, the imaginary part of $\boldsymbol{\phi}$ remains finite and well defined when it is contracted with the dipole $\mathbf{d}_0$; we obtain

$$\frac{i\gamma_0}{2} = \tfrac{2}{3} i \frac{\omega^3}{c^3}\frac{d_0^2}{4\pi\varepsilon_0} \tag{1.39}$$

where γ_0 is the radiative width due to the emission of the isolated dipole,[28] and

$i\gamma_0/2$ is a complex energy describing an exponential evolution. It may appear surprising to find complex energies when looking for the real eigenvalues of an isolated system. This is only an artifact created by the separation of the whole system into matter and radiation subsystems, which are not independent. In reality, solutions of (1.33) are all real; they acquire an imaginary part only when the z dependence of $R(z)$ is neglected.

Coming back briefly to the semiclassical theory,[15,16] the local field acting on a dipole $\mathbf{d}_n$ is defined as a sum of the field of the sources, external to the system, and the field created by all the dipoles except the considered dipole $\mathbf{d}_n$:

$$\mathbf{E}_{\text{loc}}(n) = \mathbf{E}_{\text{ext}} + \mathbf{E}_{\text{dipoles}} \qquad (m \neq n) \tag{1.40}$$

$\mathbf{d}_n$ is assumed to respond linearly (cf. Appendix A) to the local field, the latter being, at the limit of low densities of excitation, as small as one wishes it to be. In temporal Fourier transform, the response of $\mathbf{d}_n$ to the perturbation $-\mathbf{d}_n \cdot \mathbf{E}_{\text{loc}}(n)$ is determined by the polarizability tensor $\boldsymbol{\alpha}_n(z)$

$$\mathbf{d}_n = \boldsymbol{\alpha}_n \mathbf{E}_{\text{loc}}(n) \tag{1.41}$$

with

$$\boldsymbol{\alpha}_n(\hbar\omega) = \sum_f \frac{2\omega_{nf}}{\omega_{nf}^2 - \omega^2} \mathbf{d}_{nf}\mathbf{d}_{nf} \tag{1.42}$$

where $\mathbf{d}_{nf}$ is the transition dipole to the molecular state f, and ω_{nf} its frequency. Denoting by $\mathscr{d}$ the column vector of the electric dipoles and by $\mathscr{E}$ that of the local fields

$$\mathscr{d} = \begin{bmatrix} \mathbf{d}_1 \\ \vdots \\ \mathbf{d}_n \\ \vdots \end{bmatrix}, \qquad \mathscr{E}_{\text{loc}} = \begin{bmatrix} \mathbf{E}_{\text{loc}}(1) \\ \vdots \\ \mathbf{E}_{\text{loc}}(n) \\ \vdots \end{bmatrix}, \qquad \text{etc.} \tag{1.43}$$

we may write (1.41) in a matrix form

$$\mathscr{d} = \boldsymbol{\alpha}\mathscr{E}_{\text{loc}} \tag{1.44}$$

where $\boldsymbol{\alpha}$ is a diagonal matrix of polarizability tensors. The field created at $\mathbf{d}_n$ by all the other dipoles is expressed as a function of all the dipoles ($m \neq n$) with the use of $\boldsymbol{\phi}(m, n)$ of (1.35):

$$\mathbf{E}_{\text{dipoles}}(m \neq n) = \sum_{m \neq n} \boldsymbol{\phi}(m, n)\mathbf{d}_m \tag{1.45}$$

with a matrix form

$$\mathscr{E} = \boldsymbol{\phi} \mathscr{d} \tag{1.46}$$

The column of dipole vectors is then determined from that of the external field using the relations (1.40), (1.44), (1.46):

$$\mathscr{d} = \boldsymbol{\alpha}(\mathscr{E}_{\text{ext}} + \boldsymbol{\phi} \mathscr{d}) \tag{1.47}$$

or

$$\mathscr{d} = (1 - \boldsymbol{\alpha}\boldsymbol{\phi})^{-1} \boldsymbol{\alpha} \mathscr{E}_{\text{ext}} \tag{1.48}$$

which defines a generalized polarizability, including the local field, $\boldsymbol{\alpha}_e$:

$$\boldsymbol{\alpha}_e = (1 - \boldsymbol{\alpha}\boldsymbol{\phi})^{-1} \boldsymbol{\alpha} = (\boldsymbol{\alpha}^{-1} - \boldsymbol{\phi})^{-1} \tag{1.49}$$

Thus, $\boldsymbol{\alpha}_e$ provides the general response of the system of dipoles to an external field of any form.

Most interesting solutions of (1.48) are those in the absence of an external field ($\mathscr{E}_{\text{ext}} = 0$). Then, the nontrivial solutions are given by the poles of det $\boldsymbol{\alpha}_e$:

$$\det(\boldsymbol{\alpha}^{-1} - \boldsymbol{\phi}) = 0 \tag{1.50}$$

This compatibility equation is exactly equivalent to (1.33) when we take into account the expression (1.35) for $R(z)$ and its dependence on $\mathbf{d}_{nf}$ and on $\boldsymbol{\phi}$. We find that there is a complete correspondence between the quantum-mechanical and classical theories as to the equation determining the elementary excitations.

This approach clearly distinguishes two ranges of interaction: At the scale $r < \lambda$, where the electrostatic interaction dominates, and at the scale $r > \lambda$, where the retardation effects dominate. This scale property justifies the separation, implicit in the Coulomb gauge, between instantaneous terms and retarded terms. However, the electric-dipole gauge shows that these two distinct aspects of the electromagnetic interaction are physically undissociable, even though it is possible in many problems to omit retardation effects.

Finally, we must underline that the above formalism applies to any system of dipoles; the only limitation is that of low excitation densities.

B. The Crystal of Dipoles

In an infinite lattice with point dipoles placed at the nodes—a first model of a molecular crystal—the translational symmetry allows us to simplify equation

(1.33) greatly and to look for the polarization wave in the subspace of a given wave vector **K**.

1. *Dipole Sums*

Let us first consider an infinite 3D lattice (2D and 1D infinite and truncated lattices are considered in Sections III and IV) with one dipole or many dipoles per unit cell. The matrix of $R(z)$ in (1.33) is diagonalized by a transformation to Bloch states of wave vector **K**, defined, for instance, by the creation operator

$$B^{\dagger}_{\alpha f}(\mathbf{K}) = \frac{1}{\sqrt{N}} \sum_{\mathbf{m}} e^{i\mathbf{K}\cdot\mathbf{m}} B^{\dagger}_{\mathbf{m}\alpha f} \tag{1.51}$$

with the cell index α. Then we obtain for the matrix elements of R

$$\langle \mathbf{K}\alpha f | R(z) | \mathbf{K}\beta g \rangle = \mathbf{d}^{f}_{\alpha} \sum_{p} \boldsymbol{\Phi}_{\alpha\beta}(p, \omega) e^{i\mathbf{K}\cdot\mathbf{p}} \mathbf{d}^{g}_{\beta} \tag{1.52}$$

where we use the fact that the transition moment $\mathbf{d}^{f}_{n\alpha}$ does not depend on the unit cell n. The solution of (1.33) implies the calculation of the elements of R of the type (1.52), with many possible directions and moduli of the transition moments, according to the electronic states of the excited molecules in the unit cell. On the contrary, the sums $\boldsymbol{\Phi}$, not contracted with the dipoles, depend only on the wave vector **K**; they may be calculated, once for all, in the cases of equivalent ($\alpha = \beta$) and nonequivalent ($\alpha \neq \beta$) molecules. Hence, the fundamental quantities are the dipole sums

$$\boldsymbol{\Phi}_{\alpha\beta}(\mathbf{K}, \omega) = \sum_{\mathbf{p}} \boldsymbol{\Phi}_{\alpha\beta}(\mathbf{p}, \omega) e^{i\mathbf{K}\cdot\mathbf{p}} \tag{1.53}$$

with the $\boldsymbol{\Phi}(p, \omega)$ given by (1.37). A naive calculation of (1.53) for $\mathbf{K} = \mathbf{0}$, for the description of optical transitions, leads to the summation of decaying terms in $1/R^1$, $1/R^2$, $1/R^3$, with a number $4\pi R^2\, dR$ for a 3D lattice. All three series diverge, or converge conditionally: their limit depends on the order of summation, and even when the summation is performed on a finite sample, it is strongly dependent on the shape of the sample. These results, which provoked tremendous controversies in the early years of the exciton theory,[28] are a consequence of the infinite range of the electromagnetic interactions: The energy of the optical transition is not determined by the sole condition $\mathbf{K} = \mathbf{0}$, or, in other words, the excitonic band is not analytic at $\mathbf{K} = \mathbf{0}$. However, the divergence of (1.53) at $\mathbf{K} = \mathbf{0}$ does not mean that $\boldsymbol{\Phi}(K, \omega)$ is not well defined: Indeed, the oscillating exponential term in (1.53) leads to the convergence of the series for any value of $\mathbf{K} \neq \mathbf{0}$, as can be seen through a summation by wave planes.[29] Nevertheless, the sum of the series depends on the direction of **K**.

Therefore, the selection rule for an optical transition must take into account the direction of the incident photon (hence, that of the created polarization wave) and must be stated as

$$\mathbf{K} = |\mathbf{K}|\hat{\mathbf{K}} \qquad \text{with} \quad \lim |\mathbf{K}| = 0 \tag{1.54}$$

2. *The Method of Ewald*

The summation method for $\phi(K)$, which we briefly summarize, was proposed by Ewald in 1915.[30] Since then, more efficient methods have been utilized[29] for the summation of dipolar interactions. However, the Ewald method has the advantage, for our purposes, of expressing analytically the separation between short and long ranges, resulting in coulombic and retarded interactions. The long-range interactions are responsible for the singularity at $\mathbf{K} = \mathbf{0}$. The philosophy of this method is the following: The convergence of the series (1.53) being very slow in the direct space, especially near $\mathbf{K} \to \mathbf{0}$, it is desirable to make a passage to the reciprocal space in order to carry out the summation. Unfortunately, because of the singularity of a point dipole, the series diverges in the reciprocal space. To overcome this difficulty, Ewald used the following artifice: He replaced the distribution of point dipoles by a distribution of gaussians centered around each dipole. This distribution creates a field whose expansion in the reciprocal space converges rapidly. Then, Ewald expressed in the real space the difference between the fields of the two distributions; this difference vanishes far from the gaussian distributions, and the series in the real space converges equally rapidly. This decomposition is illustrated in Fig. 1.1.

The following comments may be made on the fields of the two distributions.

1. The field of the gaussian distribution is summed in the reciprocal space.

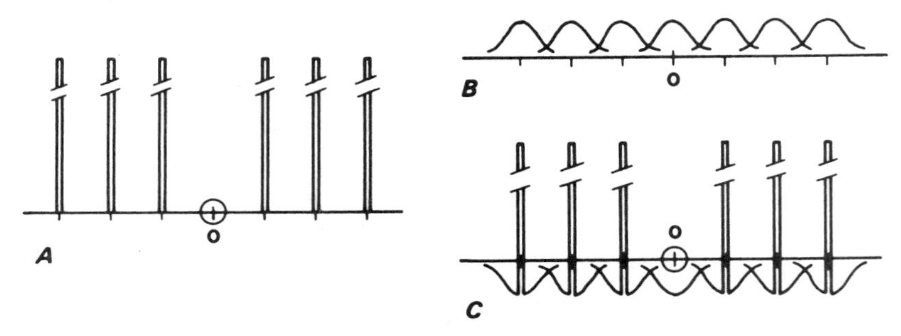

Figure 1.1. Scheme of dipole-moment densities utilized in the method of Ewald: (A) Distribution of point dipoles whose field is calculated at the origin **0**. (B) distribution of gaussians centered at each point, point **0** included. (C) distribution of dipoles plus gaussians: $A = B + C$.

The dipole-moment density of the gaussian centered at each point $\mathbf{x}$ is

$$\rho(\mathbf{x}) = \pi^{-3/2}\eta^3 e^{-\eta^2 x^2} \tag{1.55}$$

The width η^{-1} of the gaussian is chosen to be of the order of the lattice step, so that the two series, in the real and reciprocal spaces, converge rapidly. For the whole lattice, including the origin $\mathbf{0}$, and for a wave vector $\mathbf{K}$, the dipole-moment gaussian density is

$$\rho_g(\mathbf{r}) = \sum_{\mathbf{n}} \rho(\mathbf{r} - \mathbf{n}) e^{i\mathbf{K}\cdot\mathbf{n}} \tag{1.56}$$

$\rho_g(\mathbf{r})$ may be represented by its expansion in Fourier series in the reciprocal space:

$$\rho_g(\mathbf{r}) = \frac{1}{V_0} \sum_{\mathbf{G}} \tilde{\rho}(\mathbf{K} + \mathbf{G}) e^{i(\mathbf{K}+\mathbf{G})\cdot\mathbf{r}} \tag{1.57}$$

$$\tilde{\rho}(k) = e^{-k^2/4\eta^2} \tag{1.58}$$

where $\tilde{\rho}(K)$ is the Fourier transform of $\rho(\mathbf{r})$, and V_0 is the volume of the unit cell. Moreover, in the theory of Green's functions[31] in classical electrodynamics, the field radiated by a distribution of dipoles may be expressed using the Green's function G_g of the distribution,

$$\boldsymbol{\phi}_g(\mathbf{r}, \omega) = \frac{1}{4\pi\varepsilon_0}\left(\frac{\omega^2}{c^2} + \boldsymbol{\nabla\nabla}\right) G_g(\mathbf{r}, \omega) \tag{1.59}$$

and

$$\left(\Delta + \frac{\omega^2}{c^2}\right) G_g(\mathbf{r}, \omega) = -4\pi\rho_g(\mathbf{r}) \tag{1.60}$$

Using relations (1.59)–(1.60), $\boldsymbol{\phi}_g$ becomes

$$\boldsymbol{\phi}_g(\mathbf{K}, \omega) = \frac{1}{\varepsilon_0 V_0} \sum_{G} \frac{\omega^2/c^2 - (\mathbf{K}+\mathbf{G})(\mathbf{K}+\mathbf{G})}{\omega^2/c^2 - (\mathbf{K}+\mathbf{G})^2} \exp\left[-\frac{(\mathbf{K}+\mathbf{G})^2}{4\eta^2}\right] \tag{1.61}$$

(The tensorial form of the numerator must be contracted with the transition dipoles whose interaction is calculated.)

2. The field of the gaussian–dipole distribution is calculated from the Green's functions of the two distributions, as a sum in the direct space. (In this calculation the gaussian placed at $\mathbf{0}$ must be taken into account; it plays a

special role, since its counterpart, the point dipole at **0**, is not included in the dipole distribution.) In a few lattice steps, the electrostatic fields of the two distributions compensate exactly (Gauss's theorem); in fact, retardation effects introduce a slight discrepancy, of the order of $(\omega/c\eta)^2$, following expansion of (1.38). We do not reproduce the calculation of these terms, which may be found in ref. 28.

We conclude that the dipole sum $\boldsymbol{\phi}(K, \omega)$ decomposes into

$$\boldsymbol{\phi}(\mathbf{K}, \omega) = \frac{1}{\varepsilon_0 V_0} \sum_{G} \frac{\omega^2/c^2 - (\mathbf{K}+\mathbf{G})(\mathbf{K}+\mathbf{G})}{\omega^2/c^2 - (\mathbf{K}+\mathbf{G})^2} \exp\left(\frac{\omega^2/c^2 - (\mathbf{K}+\mathbf{G})^2}{4\eta^2}\right) + e^{(\omega/2c\eta)^2} \sum_{n \neq 0} e^{i\mathbf{K}\cdot\mathbf{n}} F(\mathbf{n}) + F(0) \tag{1.62}$$

$F(\mathbf{n})$ is the field resulting from the difference between point dipole and gaussian at the point **n**, and $F(\mathbf{0})$ is the gaussian term centered at **0**. The latter exists only in sums of dipolar interactions between equivalent molecules.

We may expand (1.62) in powers of $\omega/c\eta$, which is a small parameter of the order of 10^{-3} to 10^{-2}. The terms $F(\mathbf{n})$ converge at a distance of the order of η^{-1}; retardation effects contribute, according to (1.38), by multiplicative terms in $(\omega/c\eta)^2$. Thus, the sum (1.62) may be approximated by its instantaneous value at $\omega = 0$: $t_1(\mathbf{K}) + O((\omega/c\eta)^2)$, where $t_1(\mathbf{K})$ is a tensor analytic at $\mathbf{K} = \mathbf{0}$.

When **K** lies in the first Brillouin zone, only the term $\mathbf{G} = \mathbf{0}$ is important for retardation effects: the other terms, with $\mathbf{G} \neq \mathbf{0}$, have retardation effects still smaller, of the order of $\omega^2/c^2G^2 \ll 1$. On the contrary, in the term with $\mathbf{G} = \mathbf{0}$ the dependence on ω is very strong in the vicinity of $K = \omega/c$. Finally, we may compare $\boldsymbol{\phi}(\mathbf{K}, \omega)$ with the static dipole sum $\boldsymbol{\phi}(\mathbf{K}, 0)$, which following (1.62) may be written

$$\boldsymbol{\phi}(\mathbf{K}, 0) = \frac{1}{\varepsilon_0 V_0} \hat{\mathbf{K}}\hat{\mathbf{K}} + \mathbf{t}(\mathbf{K}) \tag{1.63}$$

with separation of the analytic term $\mathbf{t}(\mathbf{K})$ and the singular term $(1/\varepsilon_0 V_0)\hat{\mathbf{K}}\hat{\mathbf{K}}$. Therefore, we obtain

$$\boldsymbol{\phi}(\mathbf{K}, \omega) = \frac{1}{\varepsilon_0 V_0} \hat{\mathbf{K}}\hat{\mathbf{K}} + \mathbf{t}(\mathbf{K}) + \frac{1}{\varepsilon_0 V_0} \frac{\omega^2}{c^2} \frac{(1 - \hat{\mathbf{K}}\hat{\mathbf{K}})}{(\omega^2/c^2 - K^2)} \tag{1.64}$$

where only the last term contains retardation effects. The relation (1.64) is correct up to terms in $(\omega a/c)^2$, with a indicating a typical lattice distance. $\hat{\mathbf{K}}$ indicates the direction of the wave vector. It is clear from (1.64) that the nonanalyticity at $\mathbf{K} = \mathbf{0}$ extends to both coulombic and retarded terms.

To conclude this discussion on short- and long-range interactions, let us calculate the dielectric permittivity tensor of the dipoles of the crystal. The wave vector **K** being fixed by the external field, we may write with the notation (1.43)

$$\mathscr{E}_{\text{ext}} = \mathbf{E}_{\text{ext}\mathbf{K}}\sqrt{N}\,|\mathbf{K}\rangle \tag{1.65}$$

where $|\mathbf{K}\rangle$ is the column vector $(\cdots\ (1/\sqrt{N})e^{i\mathbf{K}\cdot\mathbf{r}}\ \cdots)$. The vector of induced dipoles is

$$\mathscr{d} = \boldsymbol{\alpha}_e\mathscr{E}_{\text{ext}} = \sqrt{N}\boldsymbol{\alpha}_\varepsilon|K\rangle\mathbf{E}_{\text{ext}\mathbf{K}} \tag{1.66}$$

As a consequence of the translational invariance of the matrix $\boldsymbol{\alpha}_e$, $|\mathbf{K}\rangle$ is an eigenstate, so that the induced dipoles build up a plane wave $|\mathbf{K}\rangle$. The macroscopic polarization-density vector is given by

$$\mathbf{P}_K = \frac{1}{V}\sum_{\mathbf{n}} e^{-i\mathbf{K}\cdot\mathbf{n}}\mathbf{d}_{\mathbf{n}} = \frac{\sqrt{N}}{V}\langle\mathbf{K}|\mathscr{d}\rangle \tag{1.67}$$

using the definitions $\mathbf{D}_\mathbf{K} = \varepsilon_0\boldsymbol{\varepsilon}(\mathbf{K},\omega)\mathbf{E}_{\text{ext}\mathbf{K}} = \varepsilon_0\mathbf{E}_{\text{ext}\mathbf{K}} + \mathbf{P}_{\mathbf{K}'}$, we obtain

$$\boldsymbol{\varepsilon}(\mathbf{K},\omega) = 1 + \frac{1}{\varepsilon_0 V_0}\langle K|\boldsymbol{\alpha}_e|K\rangle \tag{1.68}$$

or

$$\boldsymbol{\varepsilon}(\mathbf{K},\omega) = 1 + \frac{1}{\varepsilon_0 V_0}\frac{1}{(\boldsymbol{\alpha}_0^{-1} - \boldsymbol{\phi}(\mathbf{K},\omega))} \tag{1.69}$$

Here $\boldsymbol{\alpha}_0^{-1}$ denotes the inverse polarizability tensor of the unit cell. For example, in a crystal with two molecules per unit cell, $\boldsymbol{\alpha}_0^{-1}$ is a 2×2 matrix of tensors $[(\boldsymbol{\alpha}_0^{-1})_{12} = (\boldsymbol{\alpha}_0^{-1})_{21} = 0, \boldsymbol{\alpha}_0^{-1} =$ inverse polarizability of molecule $i]$ and ϕ is a 2×2 matrix of the dipolar sums.

The expression (1.69) for $\boldsymbol{\varepsilon}$ is quite general in the sense that it gives the response of the crystal to an external field of any wave vector. In particular the poles of $\boldsymbol{\varepsilon}(\mathbf{K},\omega)$ provide, over the whole Brillouin zone, the dispersion curves of the new elementary excitations built up by coulombic and retarded interactions.

3. *Excitons and Polaritons*

Equation (1.64) clearly indicates the separation, according to the wave vector, of the phenomena, retarded and instantaneous, caused by the electric-dipole interactions. First, we consider the coulombic interactions alone for wave vectors $K \gg \omega/c$, which allow us to introduce the dipolar exciton. Then, the retarded interactions will be taken into account for wave vectors near the center of the Brillouin zone, which is the domain of the polaritons.

a. *Coulombic Interactions: The Molecular-Crystal Exciton.* For crystal wave vectors $K \gg \omega/c$, the last term of (1.64) is negligible compared to $\boldsymbol{\phi}(\mathbf{K}, 0)$. This condition is effective almost over the whole Brillouin zone, the exception is a small area around its center, outside of which the search of the elementary excitations may validly be restricted to $\boldsymbol{\phi}(\mathbf{K}, 0)$.

For a maximum of simplification we restrict ourselves to a crystal with two dipoles per unit cell and to two transitions, characterized by the transition dipoles $\mathbf{d}_\alpha^f$ (α is the cell index, while f indicates the transition). Then, the determinant of (1.33), with the explicit matter variables, factorizes, in each subspace of momentum $\mathbf{K}$, into determinants of order 4:

$$\begin{vmatrix} \omega^2 - \omega_1^2 - 2\omega_1 \mathbf{d}_1^1 \cdot \boldsymbol{\phi}_{11} \cdot \mathbf{d}_1^1 & -2\omega_1 \mathbf{d}_1^1 \cdot \boldsymbol{\phi}_{11} \cdot \mathbf{d}_1^2 & -2\omega_1 \mathbf{d}_1^1 \cdot \boldsymbol{\phi}_{12} \cdot \mathbf{d}_2^1 & -2\omega_1 \mathbf{d}_1^1 \cdot \boldsymbol{\phi}_{12} \cdot \mathbf{d}_2^2 \\ -2\omega_2 \mathbf{d}_1^2 \cdot \boldsymbol{\phi}_{11} \cdot \mathbf{d}_1^1 & \omega^2 - \omega_2^2 - 2\omega_2 \mathbf{d}_1^2 \cdot \boldsymbol{\phi}_{11} \cdot \mathbf{d}_1^2 & -2\omega_1 \mathbf{d}_1^2 \cdot \boldsymbol{\phi}_{12} \cdot \mathbf{d}_2^1 & -2\omega_2 \mathbf{d}_1^2 \cdot \boldsymbol{\phi}_{12} \cdot \mathbf{d}_2^2 \\ -2\omega_1 \mathbf{d}_2^1 \cdot \boldsymbol{\phi}_{21} \cdot \mathbf{d}_1^1 & -2\omega_1 \mathbf{d}_2^1 \cdot \boldsymbol{\phi}_{21} \cdot \mathbf{d}_1^2 & \omega^2 - \omega_1^2 - 2\omega_1 \mathbf{d}_2^1 \cdot \boldsymbol{\phi}_{22} \cdot \mathbf{d}_2^1 & -2\omega_1 \mathbf{d}_1^2 \cdot \boldsymbol{\phi}_{22} \cdot \mathbf{d}_2^2 \\ -2\omega_2 \mathbf{d}_2^2 \cdot \boldsymbol{\phi}_{21} \cdot \mathbf{d}_1^1 & -2\omega_2 \mathbf{d}_2^2 \cdot \boldsymbol{\phi}_{21} \cdot \mathbf{d}_1^2 & -2\omega_2 \mathbf{d}_2^2 \cdot \boldsymbol{\phi}_{22} \cdot \mathbf{d}_2^1 & \omega^2 - \omega_2^2 - 2\omega_2 \mathbf{d}_2^2 \cdot \boldsymbol{\phi}_{22} \cdot \mathbf{d}_2^2 \end{vmatrix} = 0 \qquad (1.70)$$

When the two dipoles occupy symmetrical positions in the cell, as for instance in the case of the anthracene crystal, (1.70) may be further simplified by introducing symmetric and antisymmetric states for all directions of $\mathbf{K}$, with respect to the assumed symmetry. Then (1.70) reduces to 2×2 determinants for the two (symmetric and antisymmetric) transitions. The solution of (1.70) leads to four values of ω for each wave vector $\mathbf{K}$, i.e. to four excitonic branches. In general, the crystal field is assumed weak compared to intramolecular forces, so that coupling between excitonic branches may be neglected. To a first approximation, each of the excitonic branches, symmetric and antisymmetric, is given by the equation

$$\omega^2 - \omega_1^2 - 2\omega_1 \mathbf{d}_1 \cdot \boldsymbol{\phi}_{11} \cdot \mathbf{d}_1 \pm 2\omega_1 \mathbf{d}_1 \cdot \boldsymbol{\phi}_{12} \cdot \mathbf{d}_2 = 0 \qquad (1.71)$$

For $|\mathbf{K}| \to 0$, the two energies [solutions of (1.71)] are those of the Davydov components of the dipolar exciton, their difference being the Davydov

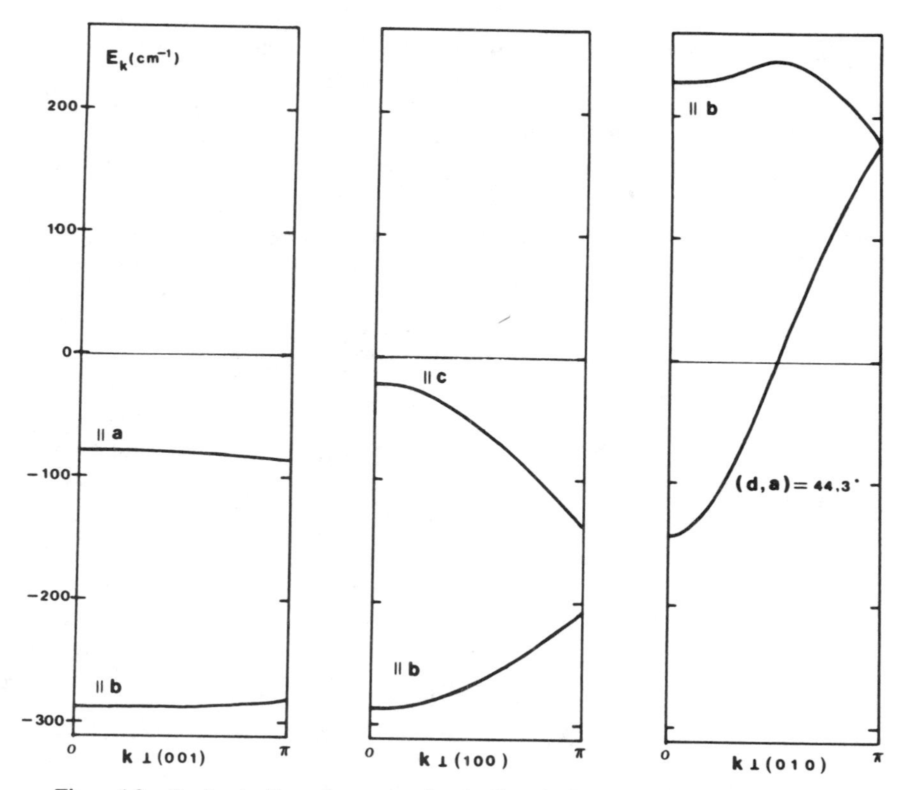

Figure 1.2. Excitonic dispersion curves for the first singlet state of the anthracene crystal. These curves are calculated in the point-dipole approximation, the transitions to upper states being accounted for by a constant dielectric permittivity.[13,37]

splitting. All these quantities depend, through the term $(1/\varepsilon_0)\hat{\mathbf{K}}\hat{\mathbf{K}}$ of (1.64), on the direction of the excitonic wave vector **K**. This dependence, illustrated in Fig. 1.2, has been analyzed by model calculations and experiments on the anthracene crystal.[32]

REMARK: The approximation $\boldsymbol{\phi}(\mathbf{K}, \omega) \sim \boldsymbol{\phi}(\mathbf{K}, 0)$ has broken the implicit symmetry embodied in the equations between matter bosons (excitons) and radiation bosons (photons), in spite of the mathematical elimination of the photons in equation (1.33). This is the reason why equations (1.70)–(1.71) give only the exciton energies, contrary to (1.33), which gives the eigenenergies of the whole matter–radiation system. Consideration of $\boldsymbol{\phi}(\mathbf{K}, \omega)$ will allow us to recover the photons—at least those with large wavelengths compared to the lattice step.

b. *Retarded Interactions: The Molecular-Crystal Polariton.* The last term of (1.64) is important in the area of a few times ω/c around the center $\mathbf{K} = \mathbf{0}$ of the Brillouin zone. Then we may replace $t(\mathbf{K})$ by $t(\mathbf{0})$ for the investigation of this critical area, since $t(\mathbf{K})$ is analytic around the center $\mathbf{K} = 0$.

Let us assume that the coulombic branch is resolved, for $|\mathbf{K}| = 0$ and for a given direction $\hat{\mathbf{K}}$, by the diagonalization of (1.70). This means that we know, for each direction $\hat{\mathbf{K}}$, the eigenenergies $\omega_e(\mathbf{K})$ for each excitonic mode, as well as the eigenvectors $|e\rangle$, which are linear functions by the transformation (1.32), (1.51) of the creation and annihilation operators of molecular states. Furthermore, let us define the excitonic dipolar moment by the same linear transformation on the molecular dipoles,

$$\mathbf{d}_e = \sum_{f\alpha} \langle f\alpha | e \rangle \mathbf{d}_\alpha^f \tag{1.72}$$

Then we check that the complete matrix (of the whole matter–radiation system) to be diagonalized, (1.33), may be written as a sum of (1.70) and the retarded term

$$\sum_{ee'} \mathbf{d}_e \boldsymbol{\psi} \mathbf{d}_{e'} |e\rangle\langle e'| \tag{1.73}$$

where $\boldsymbol{\psi}$ is a tensor of purely retarded interaction,

$$\boldsymbol{\psi} = \frac{1}{\varepsilon_0 V_0} \frac{(1 - \hat{\mathbf{K}}\hat{\mathbf{K}})}{(1 - c^2K^2/\omega^2)} \tag{1.74}$$

Thus, diagonalization of (1.33) is transformed to a diagonalization of

$$\sum_e \frac{1}{2\omega_e}(\omega^2 - \omega_e^2)|e\rangle\langle e| + \sum_{ee'} \mathbf{d}_e \cdot \boldsymbol{\psi} \cdot \mathbf{d}_{e'} |e\rangle\langle e'| \tag{1.75}$$

The expression (1.75) is simplified if we note that $\boldsymbol{\psi}$ acts only in the subspace of wave vectors orthogonal to $\hat{\mathbf{K}}$: $\boldsymbol{\psi}$ is *purely transverse*. Thus, let us define for each wave vector its *transverse component* $\mathbf{d}_e^\perp$ by

$$\mathbf{d}_e^\perp = \mathbf{d}_e - \hat{\mathbf{K}} \cdot \mathbf{d}_e \tag{1.76}$$

The retarded terms of (1.75) may be written as a scalar product of transverse vectors

$$\frac{1}{\varepsilon_0 V_0} \frac{1}{(1 - c^2K^2/\omega^2)} \left(\sum_e \mathbf{d}_e^\perp |e\rangle \right) \cdot \left(\sum_{e'} \mathbf{d}_e^\perp \langle e'| \right) \tag{1.77}$$

The "perturbation" term of (1.75) may be given the form $V_a|a\rangle\langle a|$, so that (see Appendix B.2) the eigenvalue equation may be put in the form $1/V_a = \langle a|G_0|a\rangle$, which, in our case, may be expressed in tensorial representation:

$$\left| 1 - \frac{c^2K^2}{\omega^2} - \sum_e \frac{\mathbf{d}_e^\perp \mathbf{d}_e^\perp}{(\omega^2 - \omega_e^2)} \frac{2\omega_e}{\varepsilon_0 V_0} \right| = 0 \tag{1.78}$$

This equation in ω^2, which is exact if we consider the dependence $\hat{\mathbf{K}}$ of ω_e and of $\mathbf{d}_e$, determines all the eigenenergies of the exciton–photon coupled system. However, it is preferable to put it in the classical form, which privileges the photon subspace, by introducing the transverse dielectric tensor[12,13]

$$\varepsilon^\perp(\mathbf{K}, \omega) = 1 - \frac{1}{\varepsilon_0 V_0} \sum_e \frac{\mathbf{d}_e^\perp \mathbf{d}_e^\perp}{\omega^2 - \omega_e^2} 2\omega_e \tag{1.79}$$

This tensor is less general than the dielectric tensor of classical electrodynamics (1.69), since it contains the interaction with only the retarded transverse fields. For each wave vector $\mathbf{K}$, (1.78) provides two solutions whose eigenpolarizations are orthogonal. The principal dielectric constants are obtained by the evaluation of the 2×2 determinants of (1.78) ($i = 1, 2$):

$$\begin{aligned} \varepsilon_{ii}^\perp = 1 &- \frac{1}{\varepsilon_0 V_0} \sum_e \frac{|\mathbf{d}_e^\perp|^2}{\omega^2 - \omega_e^2} \omega_e \\ &\pm \frac{1}{\varepsilon_0 V_0} \left[\left(\sum_e \frac{\mathbf{d}_{ex}^{\perp 2} - \mathbf{d}_{ey}^{\perp 2}}{\omega^2 - \omega_e^2} \omega_e \right)^2 + 4 \left(\sum_e \frac{\mathbf{d}_{ex}^\perp \cdot \mathbf{d}_{ey}^\perp}{\omega^2 - \omega_e^2} \omega_e \right)^2 \right]^{1/2} \end{aligned} \tag{1.80}$$

with x and y two arbitrary directions perpendicular to $\hat{K}$. The dispersion equation for each polarization is written

$$\frac{c^2K^2}{\omega_i^2} = \varepsilon_{ii}^\perp(\omega_i, \mathbf{K}) \tag{1.81}$$

The form of the solution (1.81) for a given direction $\mathbf{K}$ is illustrated in Fig. 1.3 for a single excitonic branch. Far from the area $K \sim \omega_e/c$, the energy solutions of (1.81) are those of a coulombic free exciton in one part, and, in the other part, of a free or quasi-free photon with a given nonresonant refractive index. In the vicinity of the crossing point ($K = \omega_e/c$), strong coupling mixes excitons and photons of same wave vector, forming a stable exciton–photon quasi-particle with variable amplitude, the polariton,[33] which obeys boson

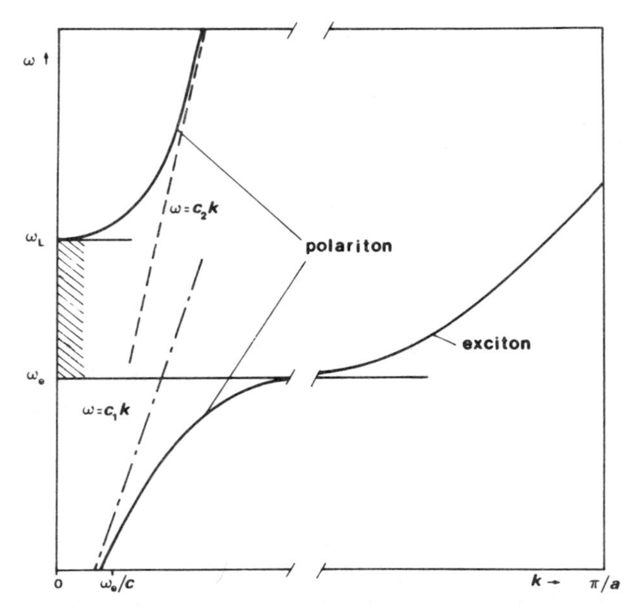

Figure 1.3. Sketch of the polariton dispersion for a given direction **K** (notice the scale change to cover the entire Brillouin zone). The broken straight lines indicate the dispersion of the electromagnetic waves in the crystal far from the excitonic b transition. In the stopping band (hatched), only excitonic states with large wave vectors may be created, and the crystal reflection is "quasi-metallic".

statistics. In the energy region $\omega_e < \omega < \omega_L$, the only possible propagation mode for a plane wave is a pure coulombic exciton with very large wave vector compared to that of the photons, hence inaccessible to photons and resulting in a reflectivity very near to 1 for the energy band. In fact, other elementary couplings (phonons, defects, etc.) modify strongly the reflectivity along the stopping band (see Section II).

Compared to the dispersion equation $c^2K^2 = \varepsilon\omega^2$, (1.81) introduces the dependence of $\varepsilon(\omega)$ on **K**: This is the spatial dispersion phenomenon studied first by Pekar.[34] For molecular crystals spatial dispersion may be ignored, since the slope of the excitonic branch is often negligible near $K \sim \omega/c$. This is not the case for several ionic crystals, for which the spatial dispersion is strong enough for a new wave, corresponding to the excitonic branch, to be excited. New boundary conditions must then be found for the study of the reflectivity of the exciton–polariton system.[35]

When a transition is investigated near the resonance, the tensor $\boldsymbol{\varepsilon}$ must be written as a sum of one resonant contribution and one nonresonant contribution including all nonresonant levels:

$$\boldsymbol{\varepsilon}^{\perp}(\omega) = \boldsymbol{\varepsilon}_{\mathrm{nr}}^{\perp}(\omega) - \frac{1}{\varepsilon_0 V_0} \frac{\mathbf{d}_e^{\perp} \mathbf{d}_e^{\perp} 2\omega_e}{\omega^2 - \omega_e^2} \tag{1.82}$$

The dependence of $\boldsymbol{\varepsilon}_{\mathrm{nr}}$ on ω may be omitted.

To summarize, the retarded interactions are important only for small wave vectors, of the order of that of the photons. For larger wave vectors the retarded interactions are uncoupled, in the sense that they do not contribute to the local field which describes the interaction between dipoles. This property allows us to understand why in global effects (cohesion energy, dispersion, etc.) retarded interactions make very small contributions, although for small **K**, the retarded interactions may show very strong effects (such as the quasi-metallic reflection of certain dyes,[15] or of the second singlet of the anthracene crystal). In particular, in all phenomena that involve interactions between excitons and free radiation, the retarded effects are by no means essential.

We close this summary with two remarks.

1. Umklapp process: In the interaction of a continuous wave (photon, electron, etc.) with the lattice, the quasi-momentum of the wave is conserved, modulo a vector in the reciprocal lattice. The introduction of these quanta of momentum leads to the Umklapp process. In many macroscopic treatments the matter is treated as a continuous medium and Umklapp processes are neglected. In our treatment, Umklapp processes are included in the coulombic interactions (calculation of the local field), but implicitly omitted in the retarded interactions, since we dropped the term $(\omega a/c)^2$ in (1.64).

2. Imaginary terms: The imaginary terms introduced by the complex variable $z = \omega + i0$ in (1.33) must vanish in a *exact* calculation of the eigenvalues of the coupled system. This is the case in (1.81), which is an exact expression. For systems with weak matter–radiation coupling, we are able to make in (1.33) the following approximation:

$$R(z) \sim R(\omega_0 + i0) \tag{1.83}$$

Then the approximate solutions of (1.33) are generally complex. However, in the case of the polariton, these solutions are still real, owing to the geometry of the system: A photon emitted by the infinite 3D lattice is necessarily reabsorbed; in other words, the exciton is not coupled to a continuum of photons (in which it is irreversibly diluted), but it is coupled to discrete photons with well-defined wave vectors, with subsequent undamped oscillations, which are the essence of the polariton. The situation is dramatically different when emission occurs in a crystal with finite extension (real crystal): The exciton decays into a continuum of photons provided its wave vector is conserved (cf. Section III.A).

C. Influence of the Extended Molecular Electronic Clouds

It is clear that the point-dipole lattice model should be abandoned for real crystals where the size of the interacting molecules may become comparable to the intermolecular distances, especially for highly excited $\pi\pi^*$ electronic states. Historically, various nondipolar contributions have been successively considered to obtain reasonable agreement with experimental[17] data. The most important effect has been found to be mixing of molecular states by the crystal field, strongly perturbing the dipolar interactions of weak or moderate transitions. In addition, inclusion of nondipolar interactions has the effect of diminishing dipolar interactions of strong transitions (Davydov splitting, exciton band width, and associated quantities such as exciton transport or exciton annihilation rates) and of leading to better agreement with experiment.[27]

1. *Lattice of Realistic Molecules*

We should consider only the effects of extended electronic clouds that could modify the point-dipole–lattice interactions investigated in Section I.B. We assume the isolated-molecule wave function to be valid, neglecting its perturbation by inclusion in the crystal lattice. For all our calculations we neglect the overlap of neighboring molecules' wave functions, which is a valid approximation for low-lying excited states, but fails for states near the crystal-lowered ionization threshold for which CT states, or conduction states, appear and invalidate the use of isolated-molecule wave functions.

Using the electric-dipole gauge (Section I.A.1.c), the intermolecular interactions have the additive form on the molecular charges given by

$$-\sum_i e\mathbf{r}_i \cdot \mathbf{E}(\mathbf{r}_i) \tag{1.84}$$

where $\mathbf{E}(\mathbf{r}_i)$ is the field created by the molecules except the one located at $\mathbf{r}_i$, whose field is assumed to be included in its molecular hamiltonian. Then, we may introduce the exact shape of the molecular wave function for the appropriate interactions with the neighboring molecules. Obviously, at longer distances (1.84) reduces to an electric dipole interaction where $\mathbf{E}(\mathbf{r}_i)$ varies more and more smoothly over a molecular size. Thus, the difference between real and dipolar interactions will decay rapidly with distance, so that the exact structure of the molecular wave function will be necessary only within a radius of a few lattice steps. Therefore, it is possible, inside this radius, to neglect retarded effects $(\omega a/c)^2$ and to correct the point-dipole interactions with nondipolar coulombic interactions. Outside the radius, long-range coulombic and retarded interactions will preserve the validity of the dipolar approximation calculated in Section I.B.

REMARK. For short-range interactions, a multipole expansion of the molecular potential has been proposed.[17] In fact, at short ranges, the expansion converges very poorly. More, it diverges over distances smaller than the molecular diameter. Furthermore, the calculation of the multipole coefficients requires that of the wave functions of the molecule, which is not a simplification of the problem.

The coulombic interactions between transitions f and g of molecules $n\sigma$ and $m\rho$ are written for short distances as

$$J_{n\sigma f}^{m\rho g} = \langle m\rho{:}g;\, n\sigma{:}0 | V_{\text{coul}} | m\rho{:}0;\, n\sigma{:}f \rangle \tag{1.85}$$

They occur mainly between "excited electrons" in the transition (i for $m\rho$ and j for $n\sigma$)

$$V_{\text{coul}} = \sum_{ij} \frac{e^2}{r_{ij}} \tag{1.86}$$

The integrals (1.86) allow us to consider only singly excited states of the crystal. Multiply excited states are coupled to singly excited states by V_{coul}. They will simply be dropped; that corresponds to the quasi-boson approximation (low excitation densities) implicit in all our calculations. Schlosser and Philpott have also considered[27,36] a series of vibronic states. However, to avoid any inconsistency related with the assumption of the single-excitation character of vibronic states, we assume all states, 0, f, $g, \ldots,$ to be purely electronic. Vibronic coupling and its effects on the nature of elementary excitations will be considered in detail in Section II. The expression (1.85) generalizes the dipolar interaction (1.35) at short distances, but at long distances it must be replaced by (1.35) to recover the retardation effects. The transition dipoles in (1.35) are calculated from the wave functions

$$\mathbf{d}_{n\sigma}^{f} = \langle n\sigma{:}f | \sum_i e_i \mathbf{r}_i | n\sigma{:}0 \rangle \tag{1.87}$$

Then, for the exact interactions $\tilde{J}$, we obtain

$$\tilde{J}_{n\sigma f}^{m\rho g}(z) = R_{n\sigma f}^{m\rho g}(z) + I_{n\sigma f}^{m\rho g} \tag{1.88}$$

where I contains the purely nondipolar interactions between molecules at short distances.

In a Fourier transformation, the sums of the terms $I(R)$, with a very rapid decay (at least as $1/R^4$), are analytical functions of $\mathbf{K}$, in particular for $\mathbf{K} = \mathbf{0}$. The nondipolar terms contribute simply to the term $t(\mathbf{K})$ of (1.64), to give the general form of $\tilde{J}^{\rho g}_{\sigma f}(\mathbf{K}, z)$:

$$\tilde{J}^{\rho g}_{\sigma f}(\mathbf{K}, z) = \frac{1}{\sqrt{N}} \sum_m \tilde{J}^{m\rho g}_{0\sigma f}(z) e^{i\mathbf{K}\cdot\mathbf{m}} \tag{1.89}$$

$$\tilde{J}^{\rho g}_{\sigma f}(\mathbf{K}, z) = [I^{\rho g}_{\sigma f}(\mathbf{K}) + \mathbf{d}^f_\sigma t(\mathbf{K}) \mathbf{d}^g_\rho] + \frac{1}{\varepsilon_0 V_0} \hat{\mathbf{K}}\hat{\mathbf{K}} + \frac{1}{\varepsilon_0 V_0} \frac{\omega^2}{c^2} \frac{1 - \hat{\mathbf{K}}\hat{\mathbf{K}}}{(\omega^2/c^2 - K^2)} \tag{1.90}$$

the term in brackets being an analytic function of $\mathbf{K}$.

2. *The Anthracene Crystal Lattice*

The first calculations on the anthracene crystal, including the nonanalyticity at $\mathbf{K} = \mathbf{0}$ of the excitonic band, were proposed by Davydov and Sheka in ref. 18. In these calculations, the vibronic aspect was ignored, while the upper electronic levels were included by means of an effective dielectric constant, the term $(1/\varepsilon_0)\mathbf{dd}$ being replaced by $(1/\varepsilon_0\varepsilon_r)\mathbf{dd}$. In 1977, Honma[37] corrected a few points in these calculations. Still in the dipolar approximation, Philpott has since calculated[20] the excitonic energies of the main aromatic crystals, taking into account explicitly many excited electronic states and series of vibronic levels. The orders of magnitude of the Davydov splitting and of the polarization ratios were found in good agreement with the experimental data.

D. Concluding Remarks

In this introductory section on the basic theories used, we have summarized the notion of exciton in a rigid lattice together with the specific characteristics introduced by long-range electromagnetic interactions. First, we examined the general case of a system of point dipoles coupled to the electromagnetic field. The electric dipolar gauge is shown to be well adapted to these problems: It allows us to include both the coulombic and the retarded interactions in an electric field acting on the dipoles. Thus, we have derived an interaction form which is strictly equivalent to the semiclassical form for the interactions between dipoles. The diagonalization of the polarization-field hamiltonian and that of the interacting electric field have been performed by a compact version of the Tyablikov method of diagonalization of a quadratic form of bosons. Treating matter and radiation elementary excitations in a symmetric form, we have shown the gradual transformation of exciton bosons and photons into polariton bosons, assuming low densities of matter excitations (linear-response approximation).

The formalism derived has been applied to an infinite 3D lattice of dipoles. A brief discussion of the method of Ewald allows us to distinguish, in the dipolar sums, the dispersion due to short-range forces (coulombic interactions between neighboring molecules) and the nonanalytic dispersion due to long-range forces originating both from long-range coulombic interactions and from retarded interactions. The structure at $\mathbf{K} = \mathbf{0}$ of the elementary excitations, depending on the direction of the wave vector (nonanalyticity), is thus shown to be the signature of the infinite range of the dipole–dipole interactions.

Outside of a small region around the center of the Brillouin zone, (the optical region), the retarded interactions are very small. Thus the concept of coulombic exciton may be used, as well the important notions of mixure of molecular states by the crystal field and of Davydov splitting when the unit cell contains many dipoles. On the basis of coulombic excitons, we studied retarded effects in the optical region $\mathbf{K} \sim \mathbf{0}$, introducing the polariton, the mixed exciton–photon quasi-particle, and the transverse dielectric tensor. This allows a quantitative study of the polariton from the properties of the coulombic exciton.

When the point-dipole approximation is no longer valid, the exact distribution of transition charges on the molecule is introduced. The difference between this distribution and that of the point dipoles is important only in short-range interactions and modifies only the analytic part of the dispersion. In particular, the retarded interactions (and the associated properties) are not modified.

II. COUPLING OF THE BULK EXCITONS TO THE MOLECULAR VIBRATIONS AND TO THE LATTICE PHONONS

In a previous publication[1] we reported high-resolution luminescence and reflection spectra of the (001) face of the anthracene crystal. We analyzed the bulk optical response (reflection and transmission spectra) and the surface and subsurface signatures. Exciton and polariton theories were used, and the parametrization of their coupling to the phonons allowed almost perfect simulations of the experimental data. The recording of the surface and subsurface 2D coherent fluorescence, checked by the surface-clothing technique, provided the final proof of the existence of surface-confined 2D excitons and allowed experimental and theoretical investigations—for the deeper comprehension of the phenomena (surface quantum percolation,[38,39] nonlinear optics[40]) and for applications (surface-selective photochemistry[41])—of the specific radiative dynamics of the surface excitations. The existence of surface monolayer exciton has been contested for years on the grounds that coupling of the surface states to lower bulk states (generally true for condensed matter made of polyatomic molecules) or scattering by the disorder or surface trapping (true for the bulk coherent emission) would surely quench the coherent surface emission.

In this section we describe the absorption of light by a crystal, creating coherent or localized excitons, and relaxation by coupling to vibrations and ultimately to lattice phonons. We provide a very detailed description of the doorway states of the crystal and of their relaxation to the emitting states. A very detailed Kramers–Kronig analysis allows us to understand the nonradiative damping of the bulk excitons and the reason why surface states are so weakly coupled to the bulk (even states of the second subsurface layer, which is almost degenerate with the bulk layers), with the result that their emission yield is quite high.

In Section I we analyzed the properties of a rigid lattice of rigid molecules in the framework of the Born–Oppenheimer approximation. The nuclei were assumed to be fixed in their equilibrium position, which is a very crude approximation, since the nuclei vibrate around the equilibrium positions with the vibration potential depending on the electronic state. Conversely, the electronic state is affected by the nuclear vibration. This interdependence is the source of the coupling between electronic and nuclear motions.

In the isolated molecule, this "nonadiabatic" coupling leads to phenomena such as internal conversion, where an excited electronic state is coupled to a highly excited isoenergetic vibronic state belonging to a lower electronic state with a large energy gap. The rates of internal conversion are relatively low for the low-lying electronic states (of the order of a nanosecond) and may be neglected in the investigation of crystal states, for which typical evolution times are much shorter. However, in the crystal the electronic states of the excitonic bands have very small separations compared to the isolated-molecule levels. Thus, the mixing of the electronic states with the vibrations becomes essential for the comprehension of the optical properties of the crystal. In particular, the dissipation of the light energy by irreversible decay of the electronic states, etc., can only be accounted for by the excitation of the crystal vibrations. For instance, the theory in Section I leads to an optically forbidden square-shaped reflection band which is very far from the observed shape. Many other phenomena, such as crystal luminescence or those related specifically to vibration (e.g. Raman resonant scattering or vibron fission), necessitate the explicit treatment of the molecular and lattice vibrations. In compact molecular crystals, such as anthracene, it is possible to distinguish lattice modes, with block motion of many molecules as a solid, from intramolecular modes, the vibrational excitons,[13] whose motion is essentially intramolecular but in phase. In what follows, we call them phonons and vibrations, respectively. In the study of the coupling to the lattice, we have to consider an initial thermal population of phonons, while for our usual temperatures ($kT \ll 300\,\text{cm}^{-1}$) the vibration will be considered initially at the zero-point level.

In this section we study the interaction of a branch of electronic excitons with a mode of vibrations and phonons. The parameters in this model are the dispersion width $2B$ of the excitonic band, the average energy quantum $\hbar\Omega_0$ of the vibration, the coupling intensity, and the temperature. According to the ordering of these parameters, the system shows very different behavior, whose general treatment is beyond the scope of this section. We restrict ourselves to the usual cases that are relevant to the first singlet exciton of the anthracene crystal and to its absorption and emission mechanisms.

A. Mechanisms of Coupling to the Vibrations

In this subsection we comment on the origin of the interdependence of electronic and vibrational excitations. This will allow us to define our notation, starting with the vibrations in the ground electronic state.

1. *Phonons and Vibrations*

Let us consider a crystal of N unit cells, each one containing σ nuclei. In the expansion of the electronic energy in powers of nuclear displacements around their equilibrium positions at $T = 0$, the linear term vanishes. It is usual to make the harmonic approximation, keeping only the quadratic terms. Then the crystal hamiltonian is expressed as a function of the momentum $P_{n\alpha}$, of the mass $m_{n\alpha}$, and of the position $r_{n\alpha}$ of each nucleus $n\alpha$ (n indexes the cell and α the coordinate):

$$H = \sum_{n\alpha} \frac{p_{n\alpha}^2}{2m_\alpha} + \frac{1}{2}\sum_{\substack{n\alpha \\ m\beta}} V_{n-m}^{\alpha\beta} r_{n\alpha} r_{m\beta} \tag{2.1}$$

Making use of the translational symmetry, we define new conjugate variables:

$$\rho_{k\alpha} = \sqrt{\frac{m_\alpha}{N}} \sum_n e^{i\mathbf{k}\cdot\mathbf{n}} r_{n\alpha}, \qquad \pi_{k\alpha} = \sqrt{\frac{1}{Nm_\alpha}} \sum_n e^{i\mathbf{k}\cdot\mathbf{n}} p_{n\alpha} \tag{2.2}$$

Then, we obtain

$$H = \frac{1}{2}\sum_k \left(\sum_\alpha \pi_{k\alpha}\pi_{-k\alpha} + \sum_{\alpha\beta} \mathscr{V}_{\alpha\beta}(k)\rho_{k\alpha}\rho_{-k\beta} \right) \tag{2.3}$$

with

$$\mathscr{V}_{\alpha\beta}(\mathbf{k}) = \sum_m (m_\alpha m_\beta)^{-1/2} e^{i\mathbf{k}\cdot\mathbf{m}} V_m^{\alpha\beta} \tag{2.4}$$

Thus, we have to treat 3σ variables coupled in each subspace of the wave vector $\mathbf{k}$. To uncouple (2.3) we look for solutions of the form

$$\pi_{k\alpha} = -i\omega\rho_{k\alpha} \tag{2.5}$$

The diagonalization of each H_k reduces to that of $\omega^2 - \mathscr{V}_{\alpha\beta}(\mathbf{k})$. Let $\Omega^2_{\mathbf{k}s}$ be the eigenvalues, real and positive, of $\mathscr{V}_{\alpha\beta}(\mathbf{k})$ associated with the orthonormal eigenvectors $(e^\alpha_{\mathbf{k}s})$, which must satisfy

$$\Omega^2_{\mathbf{k}s} e^\alpha_{\mathbf{k}s} = \sum_\beta \mathscr{V}_{\alpha\beta}(k) e^\beta_{\mathbf{k}s} \tag{2.6}$$

With the new set of variables $\hat{\rho}_{\mathbf{k}s} = \sum_\alpha e^\alpha_{\mathbf{k}s}\rho_{\mathbf{k}\alpha}$, $H_\mathbf{k}$ is expressed in a diagonal form which may be quantized with creation and annihilation operators for each harmonic mode $\mathbf{k}s$:

$$\begin{aligned} \hat{\rho}_{\mathbf{k}s} &= \sqrt{\frac{\hbar}{2\Omega_{\mathbf{k}s}}}(b_{\mathbf{k}s} + b^\dagger_{-\mathbf{k}s}) \\ \hat{\pi}_{\mathbf{k}s} &= -i\sqrt{\frac{\hbar\Omega_{\mathbf{k}s}}{2}}(b_{\mathbf{k}s} - b^\dagger_{-\mathbf{k}s}) \end{aligned} \tag{2.7}$$

to get finally

$$H = \sum_{\mathbf{k}s} \hbar\Omega_{\mathbf{k}s}(b^\dagger_{\mathbf{k}s}b_{\mathbf{k}s} + \tfrac{1}{2}), \quad [b_{\mathbf{k}s}, b^\dagger_{\mathbf{k}'s'}] = \delta_{\mathbf{k}\mathbf{k}'}\delta_{ss'} \tag{2.8}$$

The nucleus position operator may be expressed in terms of normal modes of phonons:

$$r_{n\alpha} = \sum_{\mathbf{k}s} e^\alpha_{\mathbf{k}s}\sqrt{\frac{\hbar}{2Nm_\alpha\Omega_{\mathbf{k}s}}}\, e^{i\mathbf{k}\cdot\mathbf{n}}(b_{\mathbf{k}s} + b^\dagger_{-\mathbf{k}s}) \tag{2.9}$$

When the harmonic approximation is dropped, potential-energy terms in r^3, r^4,... couple the phonon modes. These terms are responsible for processes such as (at low temperatures) phonon fission or vibration fission into phonons. To take account of the variation of the frequencies and of the equilibrium positions with temperature, the phenomenological quasi-harmonic approach is often used, in which the eigenfrequencies $\Omega_{\mathbf{k}s}$ are functions of the crystal volume.[42]

In the strong-binding approximation, valid for an aromatic molecular crystal, the intramolecular degrees of freedom are practically uncoupled in

$\mathcal{V}_{\alpha\beta}(\mathbf{k})$ of (2.4), so that it is possible to treat the crystal field as a perturbation of the intramolecular potential (an invalid approximation for soft molecules in general[42]). Vibrational interactions between molecules lead to collective vibrations or to vibrational excitons with *vibration Davydov components* at $\mathbf{k}=\mathbf{0}$ when the unit cell contains many molecules. Raman scattering at low temperature allows one to measure Davydov splittings, which for the anthracene crystal do not exceed a few reciprocal centimeters and are negligible for the most active modes.[43] Another effect of the crystal field is a slight shift, generally to higher values, of the energies of the intramolecular vibrations relative to the vapor-phase values.[44] This effect is connected with the slight variation of the interatomic distance from the vapor to the crystal phase.[45]

For the investigation of crystal phonons, we may in a first approximation consider rigid molecules. Then we obtain $6s$ branches of phonons in a crystal with s molecules per unit cell. In these $6s$ branches, $6s$-3 are optical phonon branches and 3 acoustical phonon branches; for the latter $\Omega_s(\mathbf{k}) \underset{k \to 0}{\sim} c_s|\mathbf{k}|$ (c_s being the sound velocity in branch s). For directions of $\mathbf{k}$ of high symmetry the acoustical branches are divided into one longitudinal and two transverse branches. In general the optical branches correspond to mixed motions of translation and rotation. Nevertheless, for centrosymmetric crystals where the center of the molecule is center of symmetry, for $\mathbf{k}=\mathbf{0}$, the translation optical modes are of u symmetry, while the rotation modes are of g symmetry and are called librations. IR absorption spectroscopy allows one to observe the u modes at $\mathbf{k}=\mathbf{0}$, while Raman scattering allows one to observe g modes at $\mathbf{k}=\mathbf{0}$ in a centrosymmetric crystal. Inelastic neutron scattering allows one to trace the dispersion branches of all the phonons[46]: see Fig. 2.1.

In the anthracene crystal, the center of the molecule is the center of symmetry. Since we have two molecules per unit cell, Raman spectroscopy allows us to observe six libration modes, three symmetric (A_g) and three antisymmetric (B_g), with respect to the b axis of the crystal.[47] The Raman spectra obtained in our laboratory will be discussed in Section II.D.2. We may estimate, with very good approximation, that the librations occur around the three inertial principal axes of the molecule, with the following ordering: 49.4 and 56.9 cm^{-1} around the normal (N) axis, 70.7 and 81.4 cm^{-1} around the mean (M) axis, and 131.7 and 140.1 cm^{-1} around the long axis (with weight between 60 and 70% for each of these axes[47]).

For the intramolecular modes, Raman scattering allows us to observe many modes of g symmetry,[47] among which the most intense are the completely symmetric modes of the aromatic skeleton: 390 cm^{-1} for bending of C–C–C angles, and 1400 cm^{-1} for stretching of the C–C bonds (breathing mode).

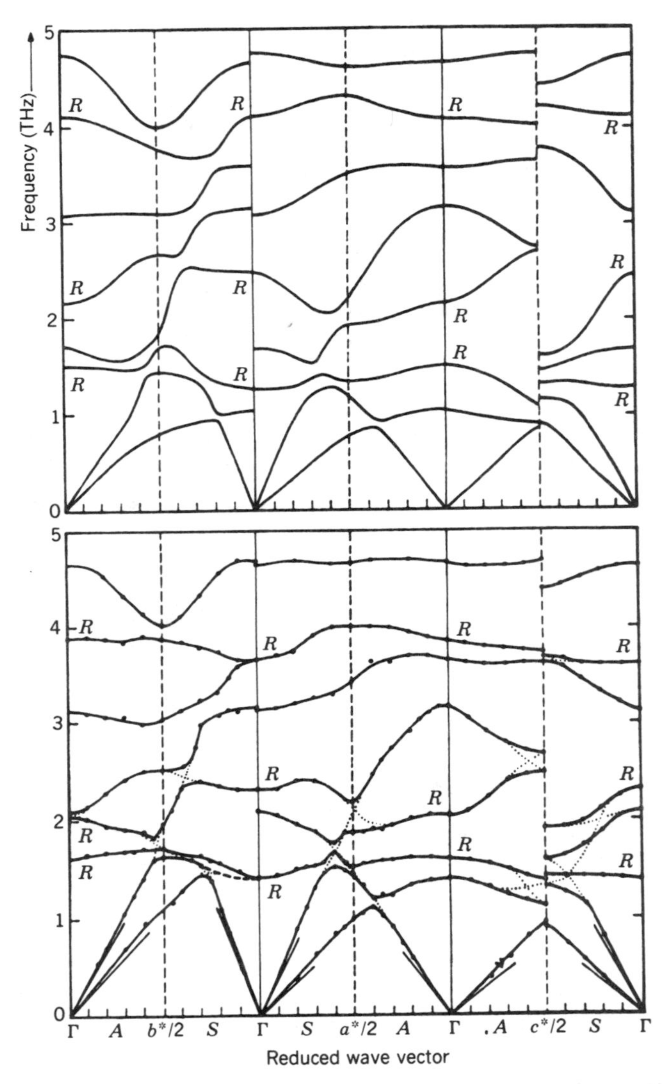

Figure 2.1. Dispersion of phonons in a deuterated anthracene crystal (S. L. Chaplot et al., 1982). Upper part: calculated spectra. Lower part: dispersion measured by neutron scattering. The Raman-active modes at $\mathbf{k} = \mathbf{0}(\Gamma)$ are marked R, and have symmetry $A_g(S)$ or $B_g(A)$. Note the weak dispersion of the lower R mode along c^*.

2. *Coupling to the Electronic Excitations*

In the Born–Oppenheimer approximation[13] the electronic energy $E^e(R)$ and the wave function $e(R)$ at the nuclear configuration of the isolated molecule allow one to calculate the potential in which the nuclei move, the electrons following adabatically in the state $e(R)$. In the crystal, the presence of a vibration coupled to the electronic exciton will modify the motion of the latter; thus the electronic state $e(R)$ and the Born–Oppenheimer approximation must be abandoned for the description of the crystal states. Conversely, the electronic state is able to strongly modify the vibration state. Consequently, the exciton–vibration coupling is a very complex problem where average factorization (separability) of degrees of freedom is a priori excluded. The complexity persists even for the simplest mechanisms of coupling: Indeed, the translational symmetry applies only to the total (exciton plus vibration) system. In what follows, we analyze the effects of the translational symmetry on various simple mechanisms of interaction encountered in molecular crystals.

a. *The Intramolecular Coupling.* Let us consider the vibrational mode and its normal coordinate Q. In the ground electronic state, the potential $E^0(Q)$ acting on this mode is given in Fig. 2.2 in the harmonic approximation, the

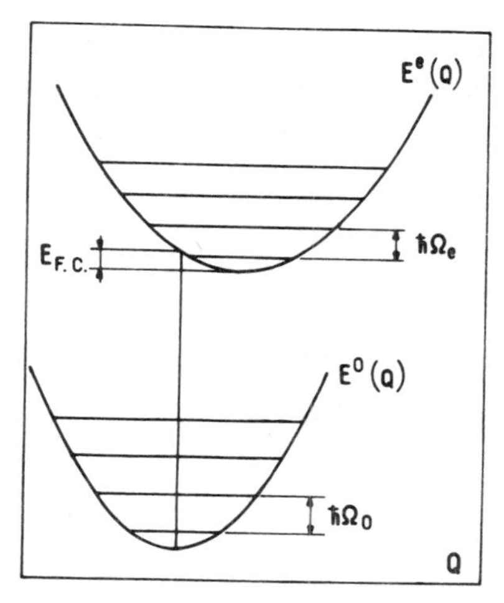

Figure 2.2. Scheme of nuclear potentials in the ground electronic state $E^0(Q)$ and the excited electronic state $E^e(Q)$. In the excited state, the frequency changes ($\Omega_0 \rightarrow \Omega_e$) and the equilibrium point is shifted. The classical relaxation energy to the new nuclear configuration in the excited state is the Franck–Condon energy E_{FC} and characterizes the linear exciton–vibration coupling.

equilibrium point in the ground state is taken as origin, and the vibration frequency is Ω_0:

$$E^0(Q) = E^0(0) + \tfrac{1}{2}m\Omega_0^2 Q^2 \tag{2.10}$$

The associated hamiltonian is

$$h^0 = \hbar\Omega_0(b^\dagger b + \tfrac{1}{2}) \tag{2.11}$$

where the vibration boson operators are related to Q by

$$Q = \sqrt{\frac{\hbar}{2m\Omega_0}}(b + b^\dagger) \tag{2.12}$$

When the molecule is raised to the electronic excited state, the normal mode Q preserves the same decomposition with respect to the nuclear displacements, except for Jahn–Teller effects, which we exclude in this work. In these conditions, still in the harmonic approximation, two parameters, the equilibrium point and the potential curvature, will change in the excited state (in the particular case of non-totally-symmetric vibrations, the equilibrium point does not change). In the excited state e, the nuclear potential becomes

$$E^e(Q) = E^e(Q_e) + \tfrac{1}{2}m\Omega_e^2(Q - Q_e)^2 \tag{2.13}$$

The kinetic energy of the mode Q being invariant, we may write for the molecular hamiltonian, using the excitation boson operators B, $B^\dagger$,

$$h = h^0(Q) + B^\dagger B[E^e(Q) - E^0(Q)] \tag{2.14}$$

or, explicitly in terms of b, $b^\dagger$,

$$h = \hbar\Omega_0(b^\dagger b + \tfrac{1}{2}) + B^\dagger B\left[\hbar\omega_e + \frac{\hbar\Delta}{2}(b + b^\dagger)^2 + \xi\hbar\Omega_0(b + b^\dagger)\right] \tag{2.15}$$

with the new notation

$$\hbar\omega_e = E^e_{(0)} - E^0_{(0)} + \tfrac{1}{2}m\Omega_e^2 Q_e^2, \qquad \Delta = \frac{\Omega_e^2 - \Omega_0^2}{2\Omega_0}, \qquad \xi = \frac{\Omega_e}{\Omega_0}\sqrt{\frac{E_{FC}}{\hbar\Omega_0}} \tag{2.16}$$

where the Franck–Condon energy E_{FC} represents the potential energy in the excited state of the equilibrium configuration of the ground state (Fig. 2.2):

$$E_{\mathrm{FC}} = \tfrac{1}{2} m \Omega_e^2 Q_e^2 \tag{2.17}$$

If $\Omega_e = \Omega_0$, then E_{FC} represents also the stabilization in the ground state between the ground-state equilibrium configuration and the excited-state equilibrium configuration.

Thus the hamiltonian (2.15) couples the electronic excitations to the vibrations by linear terms in ξ and by quadratic terms in Δ. The molecular eigenstates of (2.15) are the vibronic states; they are different from tensorial products of electronic excitations and "undressed" vibrations. Even for this simple intramolecular effect, we cannot, when moving to the crystal, consider excitonic and vibrational motions as independent.

b. *The Coupling to Phonons.* We consider here the way coupling to phonons may affect the electronic motion of the crystal states. Let us write the rigid-crystal hamiltonian of Section I for a single excitonic branch:

$$H = \sum_n (\hbar\omega_e^M + D_n) B_n^\dagger B_n + \sum_{nm} J_{nm} B_n^\dagger B_m \tag{2.18}$$

The terms D_n represent the energy stabilization due to van der Waals forces for one excitation localized on molecule n and due to neighboring molecules in the ground state. In general, these forces are much larger than the interactions J, which are neglected in a zero-order approximation. The localized electronic excitations are then eigenstates of H; we may investigate the nuclear potential in these electronic states. Let (ne) be the electronic state of an excited molecule n, the other $m \neq n$ molecules being in their ground state. The energy of the state $E^{(ne)}(R)$, as well as that of $E^{(0)}(R)$, depends on the positions R of all the nuclei. On the assumption that the crystal energy is a sum of bimolecular terms $\mathscr{D}_{mn}(R_m, R_n)$, the energy difference between excited (ne) and ground (0) states becomes

$$E^{(ne)}(R) - E^{(0)}(R) = \sum_m (\mathscr{D}_{nm}^{(ne)} - \mathscr{D}_{nm}^{(0)}) + \hbar\omega_e^M \tag{2.19}$$

Summation over m gives the stabilization term D_n. In general, the R dependences of $\mathscr{D}^{(ne)}$ and $\mathscr{D}^{(0)}$ are different. If we assume this difference to be small, or if the crystal is rigid enough, we may estimate the equilibrium configuration of state (ne) to be close to that of the ground state, $R^{(0)}$. So a linear expansion of the D_{nm}'s is significant:

$$D_{nm}^e(R) = \mathscr{D}_{nm}^{(ne)} - \mathscr{D}_{nm}^{(0)} = D_{nm}^e(R^{(0)}) + (R - R^{(0)}) \left.\frac{\partial D_{nm}^e}{\partial R}\right|_{R^{(0)}} \tag{2.20}$$

Using the R dependence of the D_n's in (2.20), we may write for the hamiltonian of the linear exciton–phonon interaction

$$H_{\text{int}}^{(1)} = \sum_n B_n^\dagger B_n \sum_m (R - R^{(0)}) \left. \frac{\partial D_{nm}^e}{\partial R} \right|_{R^{(0)}} \tag{2.21}$$

On the assumption of bimolecular interactions of rigid molecules, D_{nm}^e depends only on the translational and rotational degrees of freedom of molecules n and m. Let x_n^i be the six coordinates of molecule α in unit cell n. Using (2.9) with appropriate coefficients $e_{qs}^{\alpha i}$, $H_{\text{int}}^{(1)}$ may be expressed in terms of phonon operators:

$$H_{\text{int}}^{(1)} = \sum_{\substack{n\alpha \\ qs}} \frac{e^{i\mathbf{q}\cdot\mathbf{n}}}{\sqrt{N}} B_{n\alpha}^\dagger B_{n\alpha} (b_{qs} + b_{-qs}^\dagger) \chi_{s\alpha}(q) \tag{2.22}$$

with

$$\chi_{s\alpha}(q) = \sum_{im\beta} \sqrt{\frac{\hbar}{2m_i \Omega_{qs}}} \left(e_{qs}^{\alpha i} \frac{\partial D_{n\alpha m\beta}}{\partial x_{n\alpha}^i} + e^{i\mathbf{q}\cdot(\mathbf{m}-\mathbf{n})} e_{qs}^{\beta i} \frac{\partial D_{n\alpha m\beta}}{\partial x_{m\beta}^i} \right) \tag{2.23}$$

By the same reasoning, the R dependence of terms J_{nm} induces an additional exciton–phonon interaction

$$H_{\text{int}}^{(2)} = \sum_{\substack{n\alpha m\beta \\ qs}} \frac{e^{i\mathbf{q}\cdot\mathbf{n}}}{\sqrt{N}} \tilde{F}_{\alpha\beta}^{qs}(n-m) B_{n\alpha}^\dagger B_{m\beta} (b_{qs} + b_{-qs}^\dagger) \tag{2.24}$$

with

$$\tilde{F}_{\alpha\beta}^{qs}(n-m) = \sum_i \sqrt{\frac{\hbar}{2m_i \Omega_{ks}}} \left(e_{qs}^{\alpha i} \frac{\partial J_{n\alpha m\beta}}{\partial x_{n\alpha}^i} + e^{i\mathbf{q}\cdot(\mathbf{m}-\mathbf{n})} \frac{\partial J_{n\alpha m\beta}}{\partial x_{m\beta}^i} e_{qs}^{\beta i} \right) \tag{2.25}$$

Using the exciton-wave representation, the exciton–phonon interaction hamiltonians (2.22), (2.24) fuse to a single expression:

$$H_{\text{int}} = \sum_{k,\mathbf{q}s} B_{\mathbf{k}+\mathbf{q}}^\dagger B_{\mathbf{k}} (b_{\mathbf{q}s} + b_{-\mathbf{q}s}^\dagger) \frac{1}{\sqrt{N}} f_s(\mathbf{k}, \mathbf{q}) \tag{2.26}$$

with

$$\frac{1}{\sqrt{N}} f_s(k, q) = \frac{1}{\sqrt{N}} [\chi_s(q) + F(k, q)] \tag{2.27}$$

where $F(\mathbf{k}, \mathbf{q})$ is the Fourier transform of $\tilde{F}_{\mathbf{q}}(\mathbf{n} - \mathbf{m})$.

In most molecular crystals, the D terms are much larger than the J terms. Therefore, it is reasonable to neglect the contribution of the latter to the exciton–phonon coupling. In the following, we shall neglect the $\mathbf{k}$ dependence of $f_s(\mathbf{k}, \mathbf{q})$. We must notice that (2.26) is valid for any strength of the exciton–phonon coupling, and that the expression (2.26) is invariant under translation, conserving the total wave vector, contrary to certain other forms of hamiltonian[48,49] supposed to describe correctly the strong-interaction case.

c. *Exciton–Photon–Phonon Coupling.* The motion of the molecules modifies not only the intermolecular interactions, but also the interaction with an external field. When the exciton–photon coupling is strong, the nuclear motion is coupled to the radiation via the electronic motion, so there is exciton–photon–phonon coupling. When we take into account the retarded part of the intermolecular interactions J, the exciton–photon–phonon interaction is implicitly included in $\partial J/\partial R$. Since in what follows J will describe only the coulombic interaction, we must study the exciton–photon–phonon interaction explicitly.

In the exciton–photon interaction, the translational molecular motions have negligible effects owing to the small amplitude of the translation compared to the optical wavelength. In contrast, the molecular rotations may cause an important variation of the transition dipole: the librations may be strongly coupled to the incident photon via its coupling to the exciton. If $D_\alpha(R)$ is the transition dipole of an α molecule in a unit cell, the first-order expansion in the libration coordinate θ around the $\mathbf{u}$ axis will give

$$\mathbf{D}_\alpha(R) = \mathbf{D}_\alpha^0 + \theta \mathbf{D}_\alpha^0 \times \mathbf{u} \tag{2.28}$$

θ is expressed in terms of libration operators of the type (2.9); the interaction hamiltonian is deduced in a straightforward manner by expressing $\mathbf{D}_\alpha^0$ in terms of exciton operators following (1.8.10):

$$H_{ep\gamma} = -\sum_{\substack{n\alpha \\ \mathbf{q}s}} (\mathbf{d}_\alpha^0 \times \mathbf{u})\psi_{\mathbf{q}s}(B_{n\alpha} + B_{n\alpha}^\dagger)(b_{\mathbf{q}s} + b_{\mathbf{q}s}^\dagger)(a_\mathbf{k} - a_{-\mathbf{k}}^\dagger) \tag{2.29}$$

which contains the resonant processes in $B^\dagger a(b + b^\dagger)$ and $Ba^\dagger(b + b^\dagger)$ represented by the diagrams

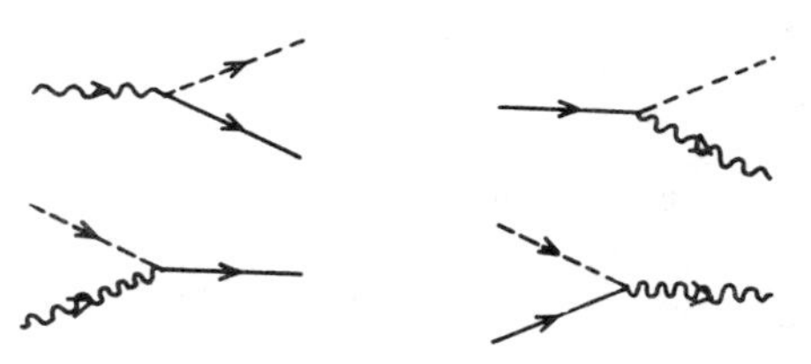

with

〰〰➤〰〰 = photon propagator

——➤—— = exciton propagator

---➤--- = phonon propagator

We must remark that the amplitude of these processes is generally weak compared to the direct exciton–photon amplitude, owing to the small libration amplitudes (of the order of 1°) at low temperatures. It is still smaller when the incident light polarization is parallel to the molecular transition dipole. For instance, in anthracene-crystal excitation, we expect the exciton–photon–phonon contribution to be more important for the *a* than for the *b* polarization. On the contrary, these processes become much more important in nonresonant excitations, in Raman scattering for instance (cf. Section II.D).

3. *General Aspects of the Exciton–Vibration Coupling*

This subsection summarizes what is known about exciton–vibration coupling in general. Thereafter, specific cases will be treated. We restrict our discussion to a "minimal" model of coupling characterized by the following approximations[50]:

One mode of vibration is considered, in the harmonic approximation.

Dispersion of vibrations (i.e. of phonons) is neglected. This approximation is well adapted to the intramolecular vibrations and, to a less degree, to the libration modes, particularly to optical modes.

The exciton–vibration coupling is assumed to be linear, given by (2.26) or by (2.15), when the frequency change upon excitation is neglected.

This coupling is assumed *local*, i.e., creation or absorption of vibrations occurs without exciton transfer, as in (2.15). This approximation amounts to considering only the R_n dependence of the local energy D_{nm}. We find that the χ_s do not depend on $\mathbf{q}$.

The crystal hamiltonian embodying the above approximations is written in second quantization:

$$H = \sum_{nm} V_{nm} B_n^\dagger B_m + \sum_n \hbar\Omega_0 b_n^\dagger b_n + \sum_n \xi\hbar\Omega_0 B_n^\dagger B_n (b_n + b_n^\dagger) \tag{2.30}$$

The $B(b)$ are operators of excitons (vibrations); the V_{nm} are intermolecular interactions; ξ, a dimensionless parameter, characterizes the linear coupling. H is qualitatively characterized by the following parameters:

B = half width of the excitonic dispersion band;
$\hbar\Omega_0$ = energy quantum of the vibration;
$E_{FC} = \xi^2\hbar\Omega_0$ = Franck–Condon energy, an energy stabilization of the localized exciton in the absence of transfer.

The detailed description of H, for which there is no analytical expression, needs the knowledge of other parameters: the structure and dimensionality of the lattice, and the nature of the interactions V_{nm}. For a qualitative description, we restrict ourselves to the above three parameters and consider first the low-temperature case in limit situations:

1. The weak-coupling case $E_{FC} \ll B$: Tensorial products of phonons and free excitons are practically eigenstates. The dispersion diagram for these excitations is sketched in Fig. 2.3*a*, *c*. In the case $\hbar\Omega_0 > 2B$ (Fig. 2.3*a*), the subspaces with different numbers of vibrations are energetically separated, and the coupling induces only a level shift: The electronic exciton is shifted downwards. For $\hbar\Omega_0 < 2B$, the free exciton undergoes real scattering in the exciton band with absorption and emission of one phonon (Fig. 2.3*b*). In

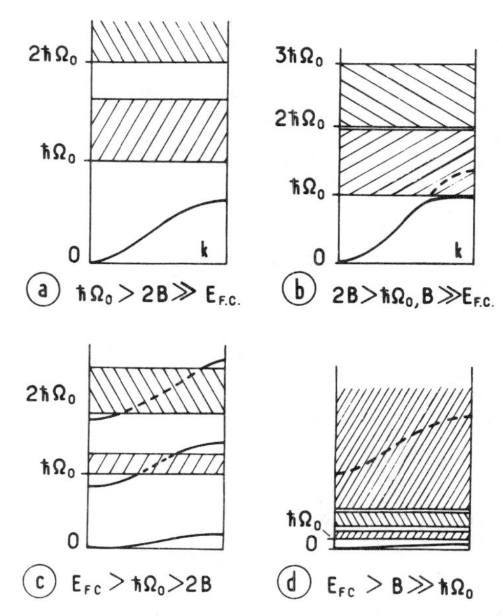

Figure 2.3. Scheme, in the various types of coupling, of the dispersion of the exciton–vibration system as a function of the total wave vector. The energy origin is the lower excitonic state. For weak coupling, the free-exciton scattering on the vibrations provides a good picture (*a*, *b*). For strong coupling the localized exciton provides the most appropriate picture: molecular vibronic states (*c*) or self-trapped exciton (*d*). The hatched areas represent continua of states where the excitonic wave vector is arbitrary.

particular, the free excitons, with energies larger than the vibration energy quantum $\hbar\Omega_0$, become very unstable to the spontaneous creation of one phonon on intersecting the two-particle continuum (see Section II.B.2), that is, when the inequality

$$J(\mathbf{k}) > J(\mathbf{k}-\mathbf{q}) + \hbar\Omega_{\mathbf{q}} = J(\mathbf{k}-\mathbf{q}) + \hbar\Omega_0 \tag{2.31}$$

holds. Here $J(\mathbf{k}')$ and $J(\mathbf{k}'-\mathbf{q})$ indicate the initial and final exciton states with creation of one phonon $\hbar\Omega_{\mathbf{q}}$ of wave vector $\mathbf{q}$. The coupling of exciton states $J(\mathbf{k}')$ is accompanied by a downward detachment of a continuum state (a general phenomenon analyzed in Section IV.A.2) and by the reduction of the dispersion curve of the pure exciton (without phonon) to a width of the order of $\hbar\Omega_0$.[13]

2. The case $B \ll \hbar\Omega_0$ is one of the most important for molecular crystals. One expects to recover the vibronic molecular states satisfying the translational symmetry of the crystal (see Fig. 3*a, c*). In Section II.B, we shall examine their perturbation due to the presence of two-particle degenerate states.

3. The last interesting case is $\hbar\Omega_0 \ll B, E_{FC}$: The lattice stabilization around an excited molecule is strong enough to create many vibrational quanta, with a subsequent deformation of the lattice around the localized exciton. This coupling leads to two limit situations according to the value g of the ratio E_{FC}/B, relative to a critical value g_c (with $g_c \sim 1$). For $E_{FC} < g_c B$, there is no trapping; the transfer forces, although reduced, dominate the exciton dynamics and the lattice is not distorted. The exciton scatters on the vibrations, resulting in a scheme similar to Fig. 3*b*. For $E_{FC} > g_c B$ (Fig. 3*d*), the exciton is trapped in the lattice deformation. This self-trapped state (as for the small polaron in the electron–phonon dynamics) causes large shifts in the nuclear motion, so that its dispersion and its oscillator strength are very weak. Thus it is practically localized. From the spectral point of view, its emission is red-shifted, weak, and diffusive like that in an excimer.[51] We must remark that, at zero temperature, light absorption proceeds from the purely excitonic state, without any distortion: This follows from the assumption that $\hbar\Omega_0 \ll B, E_{FC}$, which amounts to a rigid-lattice (or, at zero temperature, a quasi-rigid-lattice) approximation. Eventually, on a longer time scale, the pure excitonic energy relaxes either to the bottom of the excitonic band or to the self-trapped state, according to the ordering of the couplings $E_{FC} \lessgtr g_c B$.

The temperature effects on cases 1–3 above are as follows: We may consider a temperature increase to a finite value, creating an initial distribution of phonons. Consider case 3, with $\hbar\Omega_0 \ll B$, E_{FC}. The nuclei move slowly compared to the excitons. Thus, we make the static-lattice approximation,[52] with a thermal distribution of the nuclear coordinates $\mathbf{q}_n$ on the different sites n. In the case of $kT \gg \hbar\Omega_0$ we have a Maxwell (gaussian) distribution. The resonance energy of site n undergoes a linear shift V_n proportional to q_n, with [cf. (2.30)]

$$V_n = \xi\hbar\Omega_0(b_n + b_n^\dagger) \tag{2.32}$$

where $b + b^\dagger$ is given its classical significance. Thus, the Boltzman probability of observing oscillator n with a shift $b_n + b_n^\dagger$ provides the probability of energy V_n,

$$p(V_n) = (4\pi E_{\mathrm{FC}}kT)^{1/2} \exp\left(-\frac{V_n^2}{4E_{\mathrm{FC}}kT}\right) \tag{2.33}$$

This problem is that of diagonal disorder (see Section IV). With a gaussian distribution of the energies V_n, we calculate a width $\Delta(T)$:

$$2\Delta(T) = 4\sqrt{\log 2}\,(E_{\mathrm{FC}}kT)^{1/2} \tag{2.34}$$

Applying the results of Section IV, we find the two well-known limit results: At high temperature, $\Delta \gg B$, the excitonic coherence is thermalized; the lattice is an assembly of incoherent molecules with a response consisting of localized excitations with gaussian shape. At low temperature, $\Delta \ll B$, the excitonic coherence is nearly unaffected; the lattice is a virtual crystal with scattering of excitons on the disorder as, for instance, investigated in Section IV by the CPA and related methods.[52] Numerical simulations of this problem, reported recently,[53] have allowed the interpretation of the Urbach law (Section II.C) in terms of the absorption of excitonic states, momentarily trapped in a lattice thermal fluctuation, when local interactions are not strong enough to create self-trapped excitons.

From the above brief discussion we see that although the coupling mechanisms are analogous, the coupled systems (exciton–vibration and exciton–phonon) lead to radically different situations: In the first case we have $B \lesssim \hbar\Omega_0$, while in the second case we have $B > \hbar\Omega_0$ (for the first singlet exciton of anthracene). In these conditions, the approximations, models, or pictures will be also very different. In the vibration case, the molecular vibronic states will always be a very good basis for the vibronic coupling discussed in Section II.B. For the lower vibronic component of the crystal (purely electronic exciton) only the coupling to phonons has to be considered; the discussion of coupling to phonons is presented in Section II.C, from experimental and theoretical points of view, with the introduction of a new parameter: the temperature.

B. The Vibronic Coupling

In this section, we reconsider the intramolecular vibronic coupling (cf. Section II.A.2.a). In general the intramolecular vibrations satisfy the condition $\hbar\Omega_0 > 2B$, so that the approximation that the number of vibrational quanta is a constant of the motion[54] is valid. This approximation simplifies the vibronic problem of the crystal and allows to obtain analytical results. The utility and the difficulty of applying this model to real cases will thus appear.

1. *The Vibron Model*

a. *Molecular Vibronic States.* In Fig. 2.2, the nuclear potentials for the ground and the excited states are different. For the normal coordinate Q, let us denote by $\chi_n^{(0)}$ and $\chi_m^{(e)}$ the vibrational states in the ground and the excited electronic states, with n and m quanta of vibration, respectively. We may write for the corresponding Franck–Condon factor $(F_n^m)^2$, with

$$F_n^m = \langle \chi_n^{(0)} | \chi_m^{(e)} \rangle = \int \chi_n^{(0)*}(Q)\chi_m^{(e)}(Q)\, dQ \tag{2.35}$$

In the crude Born–Oppenheimer approximations, the oscillator strength of the 0–n vibronic transition is proportional to $(F_0^n)^2$. Furthermore, the Franck–Condon factor is analytically calculated in the harmonic approximation. From the hamiltonian (2.15), it is clear that the exciton coupling to the field of vibrations finds its origin in the fact that we use the same vibration operators in the ground and the excited electronic states. By a new definition of the operators, it becomes possible to eliminate the terms $B^\dagger B(b + b^\dagger)$, $B^\dagger B(b + b^\dagger)^2$. For that, we apply to the operators the following canonical transformation:

$$e^S = \exp\{\tau[(b^2 - b^{\dagger 2}) + \theta(b - b^\dagger)]\} \tag{2.36}$$

$$\tau = B^\dagger B \tfrac{1}{4} \log \frac{\Omega_e}{\Omega_0} \tag{2.37a}$$

$$\theta = 2\sqrt{\frac{E_{FC}}{\hbar\Omega_0}} \frac{\Omega_0/\Omega_e}{(1 - \sqrt{\Omega_0/\Omega_e})} \tag{2.37b}$$

(cf. Appendix C). The old operators are expressed using the new operators ($\tilde{B} = e^S B e^{-S}$) as follows:

$$B = \tilde{B} \exp\{\tau[\tilde{b}^2 - \tilde{b}^{\dagger 2} + \theta(\tilde{b} - \tilde{b}^\dagger)]\} \tag{2.38}$$

$$b = \tilde{b} \cosh 2\tau - \tilde{b}^+ \sinh 2\tau - \frac{\theta}{2}(1 - e^{-2\tau}) \tag{2.39}$$

So the hamiltonian (2.15) is given an uncoupled form:

$$h = \hbar\Omega_0 \left(\frac{\Omega_e}{\Omega_0}\right)^{\tilde{B}^\dagger \tilde{B}} \tilde{b}^\dagger \tilde{b} + \tilde{B}^\dagger \tilde{B}\left(\hbar\omega_0 + \frac{\hbar\Omega_e}{2} - \frac{\hbar\Omega_0}{2} - E_{FC}\right) \tag{2.40}$$

From (2.40) we obtain the ground state with $\tilde{B}^\dagger \tilde{B} = 0$, and the excited state with $\tilde{B}^\dagger \tilde{B} = 1$, with its new frequency Ω_e, and with a shift of the fundamental vibration, due in part to the change of the zero-point energy $(\hbar\Omega_e/2 - \hbar\Omega_0/2)$ and in part to the Franck–Condon shift giving the energy stabilization $-E_{FC}$.

In the new representation, the state $(n!)^{-1/2}\tilde{b}^{\dagger n}|\tilde{0}\rangle$ represents the state $\chi_n^{(0)}$ when $B^\dagger B = 0$, and the state $\chi_n^{(e)}$ when $B^\dagger B = 1$. The Franck–Condon factor becomes

$$F_m^n = \langle m|e^S|n\rangle \tag{2.41}$$

Following Appendix C, the operator e^S is written as a product:

$$\exp\{\tau[b^2 - b^{\dagger 2} + \theta(b - b^\dagger)]\} = e^{f_1 b^{\dagger 2} + g_1 b^\dagger} e^{h b^\dagger b} e^{f_2 b^2 + g_2 b} e^{\psi} \tag{2.42}$$

where

$$\begin{aligned} f_1(\tau) &= -\tfrac{1}{2}\tanh 2\tau = f_2(-\tau) \\ g_1(\theta, \tau) &= \frac{\theta(1 - e^{2\tau})}{2\cosh 2\tau} = g_2(\theta, -\tau) \\ h(\tau) &= -\log\cosh 2\tau \\ \psi(\theta, \tau) &= \frac{\theta^2}{4}\left(\frac{1}{\cosh 2\tau} - 1\right) - \tfrac{1}{2}\log\cosh 2\tau \end{aligned} \tag{2.43}$$

The relations (2.41)–(2.42) allow us to calculate all Franck–Condon factors, for example

$$\begin{aligned} & F_0^0 = e^\psi, \qquad F_0^1 = e^\psi g_2, \qquad F_1^0 = e^\psi g_1, \\ & F_1^1 = e^\psi g_1 g_2 + e^\psi e^h \\ & F_0^2 = e^\psi\left(\frac{g_2^2}{2} + f_2\right)\sqrt{2}, \qquad F_2^0 = e^\psi\left(\frac{g_1^2}{2} + f_1\right)\sqrt{2}, \\ & F_1^2 = e^\psi\left(\frac{g_2^2}{2} + f_2\right)\sqrt{2}g_1 + e^\psi e^h g_2 \\ & F_2^1 = e^\psi g_2\left(\frac{g_1^2}{2} + f_1\right)\sqrt{2} + e^\psi e^h g_1, \\ & F_2^2 = e^\psi 2\left(\frac{g_2^2}{2} + f_2\right)\left(\frac{g_1^2}{2} + f_1\right) + e^\psi g_1 g_2 e^h + e^\psi e^{2h} \end{aligned} \tag{2.44}$$

The factorization (2.42) allows us to express the undressed electronic operators (2.38) as combinations of the renormalized excitation operators.

b. *The Vibronic Exciton.* The hamiltonian is written as a sum of molecular hamiltonians h_n of the form (2.40), with purely electronic intermolecular interactions J_{nm} like those studied in Section I. The dependence of J_{nm} on the nuclear motion and the dispersion of the molecular vibrations are neglected, leading to the crystal hamiltonian

$$H = \sum_n h_n + \sum_{nm} J_{nm} B_n^\dagger B_m \tag{2.45}$$

The eigenstates of h_n are molecular vibronic states; B and J are respectively operators and undressed excitonic interactions, purely electronic.

The vibronic exciton approximation restricts H to a subspace corresponding to a given vibronic molecular state. In this subspace the degeneracy of the localized vibronic states is lifted by the interactions $J_{nm} B_n^\dagger B_m$. Using the translational invariance, the eigenstates of the crystal are seen to be the vibronic excitons, or vibrons:

$$|k{:}e\chi_\nu^{(e)}\rangle = \frac{1}{\sqrt{N}} \sum_n e^{i\mathbf{k}\cdot\mathbf{n}} |n{:}e\chi_\nu^{(e)}\rangle \tag{2.46}$$

The vibronic exciton is a collective oscillation of the crystal where vibronic molecular states (e.g. electronic and molecular excitations) stay on the same site. To the first order of the perturbations J_{nm} the vibron dispersion is given by

$$\langle k{:}e\chi_\nu^{(e)} | \sum_{nm} J_{nm} B_n^\dagger B_m | k{:}e\chi_\nu^{(e)} \rangle = (F_0^\nu)^2 J(k) \tag{2.47}$$

The above approximation neglects, in particular, the states of the type $|m{:}0\chi_\mu^{(0)}; n{:}e\chi_{\nu'}^{(e)}\rangle$, which are quasi-degenerate with the vibronic states $|n{:}e\chi_\nu^{(e)}\rangle$ when $\mu + \nu' = \nu$. In the next subsection we shall study the influence of these "two-particle" states, which build up a continuum of states coupled to the vibrons.

The importance of the vibron model lies in the fact that it is the doorway state of the optical absorption. (The two-particle states have no oscillator strength, since we cannot create vibrations in the ground state.) Indeed, the creation of vibrations is subsequent to an electronic excitation of a molecule, in the approximation where the absorbing dipole is the sum of molecular dipoles (with the appropriate phase), which is the approximation of weak intermolecular forces.

This allows us to understand the reason why the zero- and first-order moments, the oscillator strength and center of gravity of the absorption band, are correctly calculated in the vibron model.[54] In contrast, the shape of the absorption band is very sensitive to the interaction with the two-particle states; see Section II.B.2 below.

c. *Collective Coupling.* When the excitonic bandwidth approaches the vibrational quantum ($B \sim \hbar\Omega_0$), interaction between many vibrons—the so-called collective coupling[55]—is possible in the following way. Let us write the vibronic hamiltonian for a given wave vector **k**:

$$H_{\mathbf{k}} = \sum_{\nu} \nu\hbar\Omega_0 |\nu\rangle\langle\nu| + \sum_{\mu\nu} J_{\mathbf{k}} F_0^{\mu} F_0^{\nu} |\nu\rangle\langle\mu| \tag{2.48}$$

where $|\nu\rangle$ represents the vibron $|\mathbf{k}, \nu\rangle$ of (2.46).

Equation (2.48) contains all the interactions between vibrons. Since $J_{\mathbf{k}}$ is a purely electronic interaction, the second term of (2.48) may be expressed using the state with nuclei in the state $\nu = 0$ and the ground electronic state,

$$\sum_{\mu\nu} J_{\mathbf{k}} F_0^{\mu} F_0^{\nu} |\nu\rangle\langle\mu| = J_{\mathbf{k}} |\varphi\rangle\langle\varphi| \tag{2.49}$$

where

$$|\varphi\rangle = \sum_{\nu} F_0^{\nu} |\nu\rangle = |k{:}e\chi_0^{(0)}\rangle \tag{2.50}$$

The form of the perturbation (2.49) allows us (see Appendix B) to express the resolvent of H_k:

$$G_k = (z - H_k)^{-1} = G_0 + G_0 |\varphi\rangle t \langle\varphi| G_0 \tag{2.51}$$

where

$$G_0 = \sum_{\nu} (z - \nu\hbar\Omega_0)^{-1} |\nu\rangle\langle\nu| \tag{2.52}$$

and

$$t = (J_k^{-1} - \langle\varphi|G_0|\varphi\rangle)^{-1} \tag{2.53}$$

The poles of G_k, which are the eigenenergies of the coupled system, are also those of t; they are the solution in z of the equation:

$$g(z) \doteqdot \sum_{\mu} \frac{(F_0^{\mu})^2}{z - \mu\hbar\Omega_0} = J_k^{-1} \tag{2.54}$$

These solutions are given in Fig. 2.4 as functions of J_k for a linear vibronic

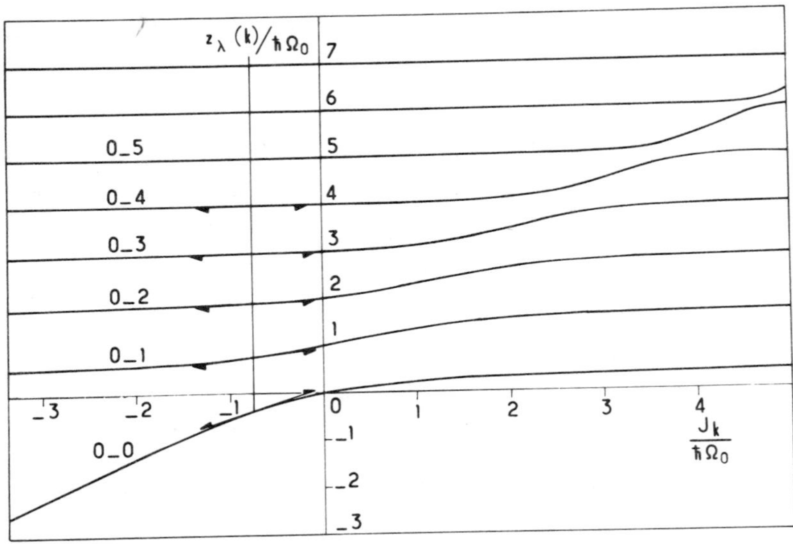

Figure 2.4. Variation vs J_k of the solutions z_λ of (2.54) for the eigenenergies of the system of coupled vibrons. The curve labeled 0–*n* represents the *n*th vibronic transition. The oscillator strengths of each state are given by the slopes of the curves. In particular, for J_k negative and strong, only the 0–0 vibron (purely electronic) is optically accessible.

coupling with $\xi = 1$ [then $(F_0^\mu)^2 = \xi^{2\mu}/\mu! e^{-\xi^2}$]. Let the $|\lambda\rangle$, with energies z_λ given by solutions of (2.54), be the eigenstates of (2.48). The oscillator strength of a state $|\lambda\rangle$ is given by the resolvent G_k (2.51). Indeed, as the operator of the purely electronic electric dipole moment is proportional to $B + B^\dagger$, the absorption is proportional to the imaginary part of $\langle\phi|G_k|\phi\rangle$, with $k = 0$. Thus, the oscillator strength of each $|\lambda\rangle$ is proportional to the residue of $\langle\varphi|G_k|\varphi\rangle$ in z_λ, since

$$\langle\varphi|G_k|\varphi\rangle = (g^{-1} - J_k)^{-1} \tag{2.55}$$

and

$$f_\lambda = \mathrm{Res}\langle\varphi|G_k|\varphi\rangle = -\frac{1}{J_k^2}\left\{\frac{\partial g}{\partial z}\right\}_{z_\lambda}^{-1} = \frac{\partial z_\lambda}{\partial J_k} \tag{2.56}$$

We may read directly from Fig. 2.4 the oscillator strength of each vibron $|\lambda\rangle$, which is proportional to the slope in J ($k = 0$).

Now we can discuss the influence of the collective coupling. For $J_k \ll \hbar\Omega_0$, the crystal vibrons are deduced from the molecular states by means of the dispersion $(F_0^\mu)^2 J_k$, as in Section II.B.1.b above, with the molecular oscillator strengths $(F_0^{\mu})^2$ for the states $k=0$. When J_k becomes comparable to $\hbar\Omega_0$, there is redistribution of displacements and oscillator strengths: For $J_k<0$, this redistribution favors the lowest state $\lambda=0$, which gradually takes on the character of a pure electronic state. The oscillator strengths of the crystal vibronic states are different from the $(F_0^{\mu})^2$; they appear to have new Franck–Condon factors $(\tilde{F}_0^{\lambda})^2$ depending on J_k. For larger $J_k>\hbar\Omega_0$, E_{FC}, we find the pure electronic state $|\varphi\rangle$ and vibronic states with neither oscillator strength nor dispersion. However, we must note that in this case the vibron model is insufficient and that the influence of the two-particle states, describing the exciton scattering on the vibrations, becomes predominant and must be included in the crystal hamiltonian.

2. *Inclusion of Two-Particle Mechanisms*

a. *The Two-Particle-State Continuum.* Let us consider the vibron (2.46) with a single vibration quantum ($v=1$). Following (2.40), its energy is $\tilde{\hbar}\omega_0+\hbar\Omega_0+(F_0^1)^2 J_k$. In cases where Ω_e is not very different from Ω_0, the vibron is coupled to a continuum of $N(N-1)$ crystal states, in which the electronic and vibrational excitations are on different sites, with energies $\tilde{\hbar}\omega_0+(F_0^0)^2 J_{k'}+\hbar\Omega_0$. Each vibron, with wave vector $\mathbf{k}$, is coupled to scattering states (of the Lippmann–Schwinger type) characterized by the exciton wave vector $\mathbf{k}'$ and by the vibration wave vector $\mathbf{k}-\mathbf{k}'$, with k' sweeping the whole Brillouin zone. Thus, the vibron is coupled to a continuum of two-particle states into which it dilutes irreversibly (see Fig. 2.5) in a so-called fission process. In specific cases this fission is not possible; then the model of a bound exciton–vibration state will be used to describe the vibron.

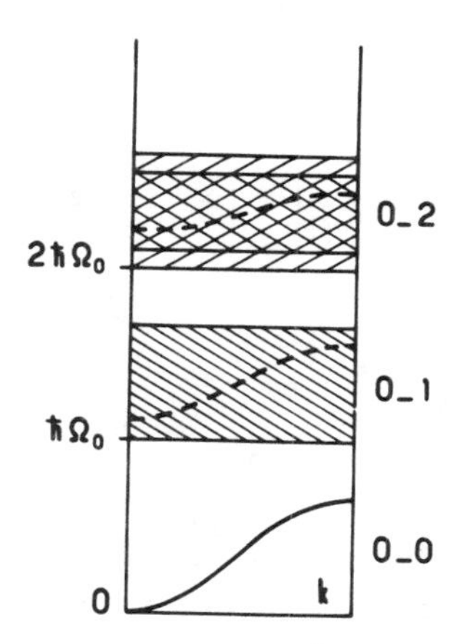

Figure 2.5. Dispersion diagram of the system of vibron-multiparticle states. The vibrons coupled to two-particle states become unstable (broken curves).

For cases where $\hbar\Omega_0 \gg J_k$, the field problem (2.30) may be reduced to a dynamical problem[54] where the interactions conserve the number of particles and allow one to distinguish between real processes and virtual processes. The latter are always neglected for $\hbar\Omega_0 \gg J_k$, and for $\hbar\Omega_0 \gtrsim J_k$ they are eliminated by an appropriate renormalization as in Section II.B.1.c, for instance.

b. *The Rashba Hamiltonian.* We consider the dynamical problem in the above case $\hbar\Omega_0 \gg J_k$, so that the coupling $J^2/\hbar\Omega_0$ between various subspaces of constant vibrational quanta can be neglected. In these conditions, the most general expression for the intermolecular interactions can be obtained using the renormalized excitations of Section II.B.1.a. In the local representation, we denote by $|nm\rangle$ the state with an electronic excitation on site n and one vibrational excitation on site m (the case $n = m$ gives back the vibron $v = 1$, since the excitations are renormalized). Assuming only bimolecular interactions, for processes on a single particle, with $p \neq m, n$ and $m \neq n$, we obtain the following terms illustrated diagrammatically:

p
n
m

$$D_{nm}|np\rangle\langle np| \quad M_{nm}|np\rangle\langle mp| \quad \mu_{nm}|pn\rangle\langle pm| \quad \Delta_{nm}|pn\rangle\langle pn|, \tag{2.57}$$

and for processes involving two particles $m \neq n$, we have the following diagrams:

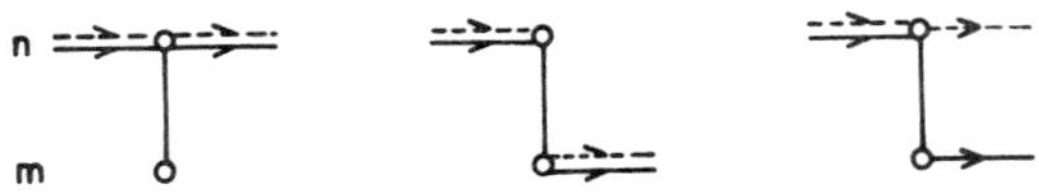

n
m

$$D^{(1)}_{nm}|nn\rangle\langle nn| \quad M^{(1)}_{nm}|nn\rangle\langle mm| \quad P_{nm}|nn\rangle\langle mn|$$

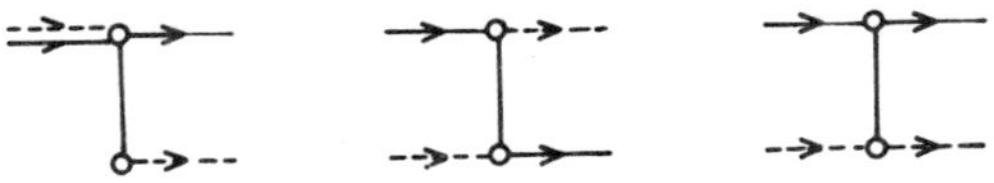

$$Q_{nm}|nn\rangle\langle nm| \quad K_{nm}|nm\rangle\langle mn| \quad D^{(2)}_{nm}|nm\rangle\langle nm| \tag{2.58}$$

Here D and M denote the exciton energy stabilization and its transfer amplitude respectively; Δ and μ are analogs for the vibration, and $D^{(1)}$ and $M^{(1)}$ for the vibron. P and Q are vibron–two-particle-state interactions accompanied, respectively, by exciton and vibration transfer; $D^{(2)}$ and K are respectively the two-particle-state energy stabilization and the amplitude of two-particle-state transfer.

In what follows, we treat first the case without interaction between excitons and vibrationa, which allows us to suppress the restriction $p \neq m, n$ in the one-particle processes. For that, it suffices to appropriately modify the two-particle processes issuing from the one-particle processes with $p = m$ or n:

$$\hat{D}^{(1)} = D^{(1)} - D - \Delta, \qquad \hat{P} = P - M,$$
$$\hat{Q} = Q - \mu, \qquad \hat{D}^{(2)} = D^{(2)} - D - \Delta. \tag{2.59}$$

The quantities (2.59) may be expressed using complementary assumptions. For instance, for purely electronic intermolecular interactions J_{nm}, we obtain the following relations:

$$\begin{aligned} M_{mn} &= J_{mn}(F_0^0)^2, & \mu_{mn} &= 0, & \Delta_{mn} &= 0 \\ M_{mn}^{(1)} &= J_{mn}(F_0^1)^2, & P_{mn} &= J_{mn}F_1^1 F_0^0, & Q_{mn} &= 0 \\ K_{mn} &= J_{mn}(F_1^0)^2, & D_{mn}^{(2)} &= D_{mn} \end{aligned} \tag{2.60}$$

with furthermore

$$\hat{P}_{mn} = J_{mn}[F_1^1 F_0^0 - (F_0^0)^2], \qquad \hat{Q} = 0 \tag{2.61}$$

The same results can be obtained from (2.45) by replacing $B_n^\dagger$ and B_m according to (2.38) and expressing the exponential according to (2.42): The zero- and first-order terms in $b_n^\dagger b_m$ describe propagation in subspaces with a single vibration.

Finally, the Rashba hamiltonian in the general case is obtained by the addition of the interactions (2.57) and (2.58) to the molecular hamiltonian (2.40):

$$\begin{aligned} H = &\sum_{mn} (\hbar\tilde{\omega}_0 + \hbar\Omega_0)|mn\rangle\langle mn| + \sum_n (\hbar\Omega_e - \hbar\Omega_0)|nn\rangle\langle nn| \\ &+ \sum_{\substack{mnp \\ m \neq n}} M_{mn}|mp\rangle\langle np| + \mu_{mn}|pm\rangle\langle pn| + \sum_{n,m} \hat{D}_{mn}^{(1)}|nn\rangle\langle nn| \\ &+ \sum_{\substack{mn \\ m \neq n}} (M_{mn}^{(1)}|mm\rangle\langle nn| + \hat{P}_{mn}(|nn\rangle\langle mn| + |mn\rangle\langle nn|) \\ &+ \hat{Q}_{mn}(|nn\rangle\langle nm| + |nm\rangle\langle nn|) \\ &+ K_{mn}|mn\rangle\langle nm| + \hat{D}_{mn}^{(2)}|mn\rangle\langle mn|) \end{aligned} \tag{2.62}$$

The excitation stabilizations D and Δ have been included in $\hbar\tilde{\omega}_0$ and $\hbar\Omega_0$. In the expression (2.62) we take as energy origin $\hbar\tilde{\omega}_0 + \hbar\Omega_0$ and we restrict our investigation to states with total wave vector $\mathbf{q}$ equal to that of the incident photon, the resulting hamiltonian H remaining translationally invariant. Then it is appropriate to use two new basis sets:

The tensorial-product basis $\{|kk'\rangle\}$ with

$$|kk'\rangle = \frac{1}{N}\sum_{nm} e^{i\mathbf{k}\cdot\mathbf{n} + i\mathbf{k}'\cdot\mathbf{m}}|nm\rangle \tag{2.63}$$

from which we retain only vectors of the type $|k, q-k\rangle$.

The basis of vibron and two-particle states, $\{|k\rangle; |kR\rangle\}$, with

$$|k\rangle = \frac{1}{\sqrt{N}}\sum_{n} e^{i\mathbf{k}\cdot\mathbf{n}}|nn\rangle,$$
$$|kR\rangle = \frac{1}{\sqrt{N}}\sum_{n} e^{i\mathbf{k}\cdot\mathbf{n}}|n, n+R\rangle, \qquad R \neq 0 \tag{2.64}$$

from which we retain only the vectors $|q\rangle$ and $|q, R\rangle$ (with $R \neq 0$).

In the new basis sets, the effective hamiltonian $H(q)$ decomposes into three terms:

1. A term without interactions between exciton and vibration:

$$H_0(q) = \sum_{k} (M_k + \mu_{q-k})|k, q-k\rangle\langle k, q-k| \tag{2.65}$$

with

$$M_k = \sum_{R} M_{0R} e^{i\mathbf{k}\cdot\mathbf{R}}, \qquad \mu_k = \sum_{R} e^{i\mathbf{k}\cdot\mathbf{R}} \mu_{0R} \tag{2.66}$$

2. A term containing vibron interactions:

$$V(q) = (\Delta E + M_q^{(1)})|q\rangle\langle q| + \mathscr{V}(|q\rangle\langle\sigma_q| + |\sigma_q\rangle\langle q|) \tag{2.67}$$

with

$$\Delta E = \hbar(\Omega_e - \Omega_0) + \hat{D}^{(1)}, \qquad M_q^{(1)} = \sum_{R} M_{0R}^{(1)} e^{i\mathbf{q}\cdot\mathbf{R}}$$

$$|\sigma_q\rangle = \frac{1}{\mathscr{V}}\sum_{R} (\hat{Q}_{0R} + e^{i\mathbf{q}\cdot\mathbf{R}}\hat{P}_{R0})|qR\rangle \tag{2.68}$$

$$\mathscr{V} = \left(\sum_{R} |\hat{Q}_{0R} + e^{i\mathbf{d}\cdot\mathbf{R}}\hat{P}_{0R}|^2\right)^{1/2}$$

3. A term with interactions between two-particle states:

$$W(q) = \sum_{R} \hat{D}^{(2)}_{0R} |qR\rangle\langle qR| + K_{0R} e^{-i\mathbf{q}\cdot\mathbf{R}} |q, -R\rangle\langle q, R| \tag{2.69}$$

c. *Exact Derivation of the Optical Absorption.* Following Section II.B.1.b,c, the optical absorption is related to the creation of one vibron by the incident photon and therefore is proportional to $\mathrm{Im}\,\langle\varphi|G_q|\varphi\rangle$. This gives for the absorption in the 0–1 region at energy z

$$\mathscr{A}(z) \propto (F_0^1)^2\, \mathrm{Im}\,\langle q|G(z)|q\rangle \tag{2.70}$$

with

$$\langle q|G(z)|q\rangle = \langle q|[z - H(q)]^{-1}|q\rangle \tag{2.71}$$

In what follows, we give the principal stages of the calculation of (2.71).

The first remark is that the term $V(q)$ of the hamiltonian operates only in a subspace of dimension 2, spanned by $\{|q\rangle$ and $\sigma(q)\}$. Thus the resolvent $\langle q|G|q\rangle$ may be given (see Appendix B) a simple form using the projector P:

$$P = |q\rangle\langle q| + |\sigma_q\rangle\langle\sigma_q| \tag{2.72}$$

$$PGP = [(P\mathscr{G}P)^{-1} - PVP]^{-1} \tag{2.73}$$

This allows us to calculate $\langle q|G|q\rangle$ with

$$\mathscr{G}(z) = [z - H_0(q) - W(q)]^{-1} \tag{2.74}$$

The calculation of $\mathscr{G}$ cannot be carried out analytically[54] unless we use a truncated form for $W(q)$. However, when W embodies only short-range interactions, the calculation (2.74) amounts to the inversion of a small matrix.[56] In the case of (2.60), where the intermolecular interactions are purely electronic, this approximation amounts to limiting the range of J to the near neighbors. For a domain $\mathscr{D}$ of action of W around a given site, W acts in the finite subspace

$$W = P_1 W P_1, \qquad P_1 = \sum_{R \in \mathscr{S}} |qR\rangle\langle qR| \tag{2.75}$$

which leads to the expression

$$\mathscr{G} = G_0 + G_0 P_1 T P_1 G_0 \tag{2.76}$$

$$G_0 = (z - H_0)^{-1}, \qquad P_1 T P_1 = [(P_1 W P_1)^{-1} - P_1 G_0 P_1]^{-1} \tag{2.77}$$

Equations (2.73), (2.76), (2.77) allow a straightforward calculation of $\langle \mathbf{q}|G|\mathbf{q}\rangle$.

For particular lattices with two molecules per unit cell, with simple structure and nearest-neighbor interactions (linear alternating chains, square 2D lattices, or simple cubic lattices of the NaCl type), the absorption occurs for two values of q: $q=0$ (the b component) and $q=Q$ (the a component); see Section II.B.1.b. In these simple lattices the environment of the site is symmetric and the complete inversion in P_1TP_1 (2.77) is not necessary, because, in the coupling of P_1 to P, only the completely symmetric state $|S\rangle$ intervenes:

$$|S\rangle = \frac{1}{\sqrt{v}} \sum_{R \in \mathscr{S}} |qR\rangle \tag{2.78}$$

with v indicating the number of the neighbor sites. We have $P_1TP_1|S\rangle = T_S|S\rangle$, with

$$T_S = [(\hat{D}^{(2)} \pm K_{OR})^{-1} - g_S]^{-1} \tag{2.79}$$

where the $+$ and $-$ go with the cases $q=0$ and $q=Q$, respectively, and

$$g_s = \langle S|P_1G_0P_1|S\rangle = \sum_{R' \in \mathscr{S}} \langle qR|G_0|qR'\rangle \tag{2.80}$$

Finally, we may write the result obtained for the optical absorption when the intermolecular complings are purely electronic [following (2.60)]:

$$\mathscr{A}_{\pm}(z) \propto \operatorname{Im} \frac{(F_0^1)^2}{z - \hbar(\Omega_e - \Omega_0) - (F_0^1)^2 J_{q\pm} - (z - f_{\pm}^{-1})(F_1^2/F_0^0)^2} \tag{2.81}$$

where

$$\begin{gathered} f_{\pm} = g_0 + v g_1^2 T_S^{\pm} \\ g_0 = \frac{1}{N}\sum_k \frac{1}{z - (F_0^0)^2 J_k}, \qquad g_1 = \frac{1}{N}\sum_k \frac{e^{i\mathbf{k}\cdot\mathbf{R}_1}}{z - (F_0^0)^2 J_k} \end{gathered} \tag{2.82}$$

R_1 being the vector and J the interaction between two nearest neighbors, and

$$T_S^{\pm} = [\pm (F_1^0)^2 J^{-1} - g_S]^{-1}$$

From (2.80), if $g_2 = (1/N)\sum_k e^{2i\mathbf{k}\cdot\mathbf{R}_1}/[z - (F_0^0)^2 J_k]$, we have for the linear chain

$g_s = g_0 + g_2$, for the square lattice $g_s = g_0 + g_2 + 2g_1$, and for the cubic lattice $g_s = g_0 + g_2 + 4g_1$. The sign $\pm$ of $T_S^{\pm}$ corresponds to the incident photon polarization and to the sign of $J_{q\pm} = \pm 2vJ$.

d. *Discussion of the Results.* Full numerical calculations of (2.82) have been carried out for the three dimensionalities considered above.[56] The results are similar in the sense that dimensionality is not important in the vibron–two-particle-state coupling. For that reason, we discuss the absorption spectra relative to the square lattice, with two molecules per unit cell and nonequivalent interactions between nearest neighbors (cf. the lattice in Section IV.B.1 with $V_{11} = 0$). Since the structure of the lattice appears secondary, this model will be used for a qualitative discussion of the vibronic absorption of the anthracene crystal (cf. Section II.B.3 below).

The denominator of (2.81) presents an imaginary part in two distinct cases: Either (1) g_0 (and g_1) are complex and the absorbing state is diluted in the two-particle-state continuum, or (2) the real part of the denominator vanishes, $f_{\pm}$ being real, for discrete values of z, and we have absorption by a discrete state. The calculated absorption spectra are presented in Fig. 2.6 for various values of the linear coupling (ξ) and the quadratic coupling ($\Delta\Omega = \Omega_e - \Omega_0$); the corresponding Franck–Condon factors are given by (2.44).

The lineshape of the vibronic 0–1 absorption is dramatically modified. We give the critical shapes in Fig. 2.6. For weak quadratic $\Delta\Omega$ coupling (*a*–*d*), the absorption shape is a continuum, variously shaped for $\xi^2 < 1$. The case $\xi^2 \ll 1$ shows a very weak absorption which reproduces the density of states of the 0–0 excitonic band with smoothed van Hove points. This is the case of the most unstable vibron (all the two-particle states are equally vibron-contaminated), the width of the 0–1 resonance being that of the dispersion, a few hundred of reciprocal centimeters for the dipolar exciton (Fig. 2.6*a*). For increasing linear coupling, the vibronic resonance emerges near the vibronic energy $\hbar\Delta\Omega + (F_0^1)^2 J_q$ (Fig. 2.6*b*). The famous case[58] $\xi^2 = 1$ and $\Delta\Omega = 0$ shows the discrete unperturbed vibronic resonance, uncoupled from the two-particle-state continuum because $F_1^1 = 0$. For stronger linear coupling ($\xi^2 > 1$), two discrete states emerge from both sides of the very weak two-particle continuum. These two states are practically pure vibrons, dressed with and repelled by the two-particle continuum (Fig. 2.6*e*). (As indicated in Section II.B.1.b, the vibron is at the center of gravity of the absorption band.)

As for the effect of the quadratic coupling, we consider here the usual case $\Delta\Omega < 0$. As $|\Delta\Omega|$ increases while the two-particle continuum remains fixed (at least for values $|\Delta\Omega| \ll \Omega_0$), the vibron states are stabilized by $\Delta\Omega$ (Fig. 2.6*e*). The emergence of the discrete state below the band is favored, while for sufficiently large values of $\Delta\Omega$ (and large linear couplings $\xi^2 > 1$) the two

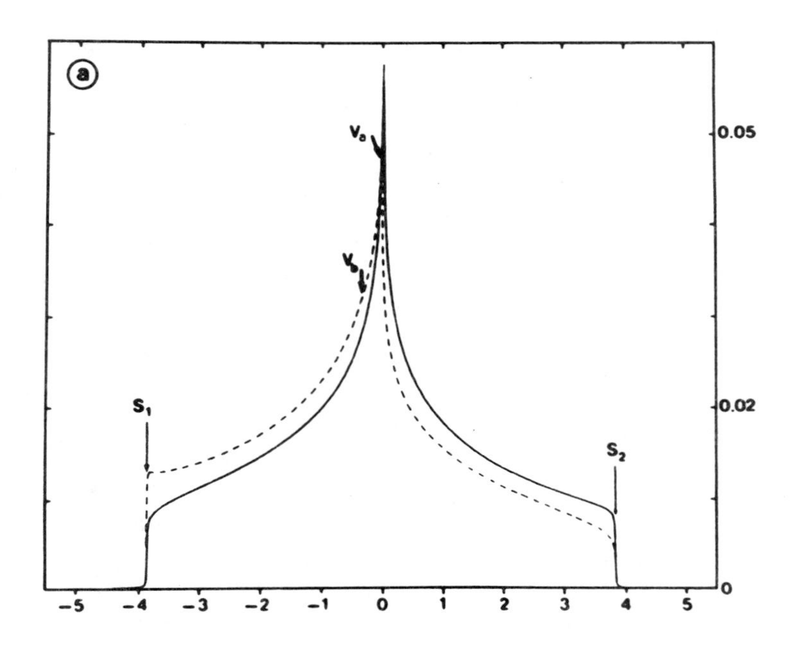
a
V_a
V_b
S_1
S_2
0.05
0.02
0
−5
−4
−3
−2
−1
0
1
2
3
4
5

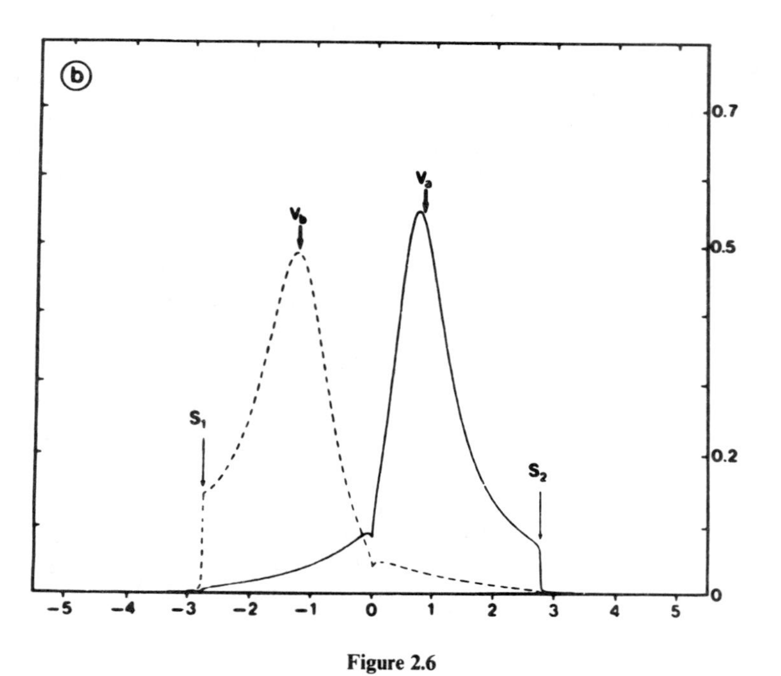
b
V_a
V_b
S_1
S_2
0.7
0.5
0.2
0
−5
−4
−3
−2
−1
0
1
2
3
4
5

Figure 2.6

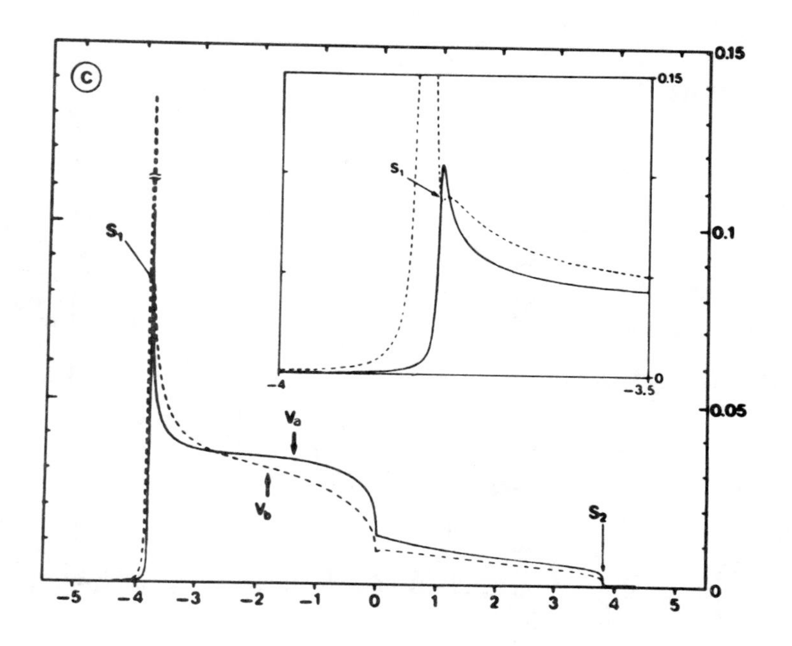
c
0.15
0.1
0.05
0
S1
Va
Vb
S2
-5
-4
-3
-2
-1
0
1
2
3
4
5
0.15
-4
0
-3.5

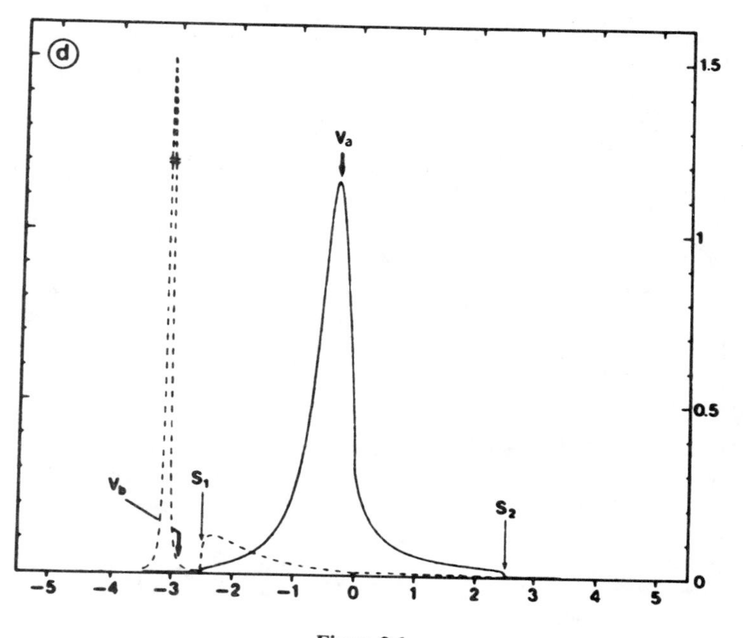
d
1.5
1
0.5
0
Va
Vb
S1
S2
-5
-4
-3
-2
-1
0
1
2
3
4
5

Figure 2.6

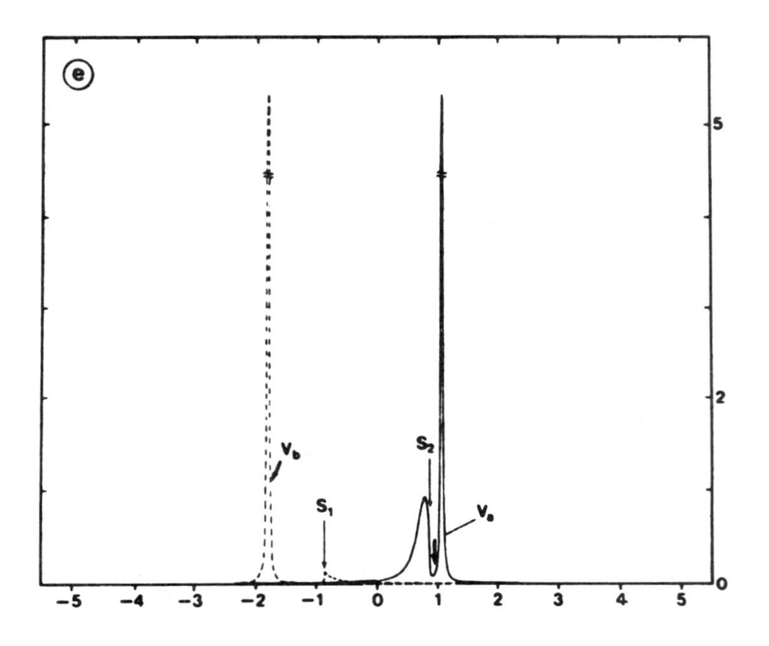

e
V_b
S_1
S_2
V_a
5
2
0
−5
−4
−3
−2
−1
0
1
2
3
4
5

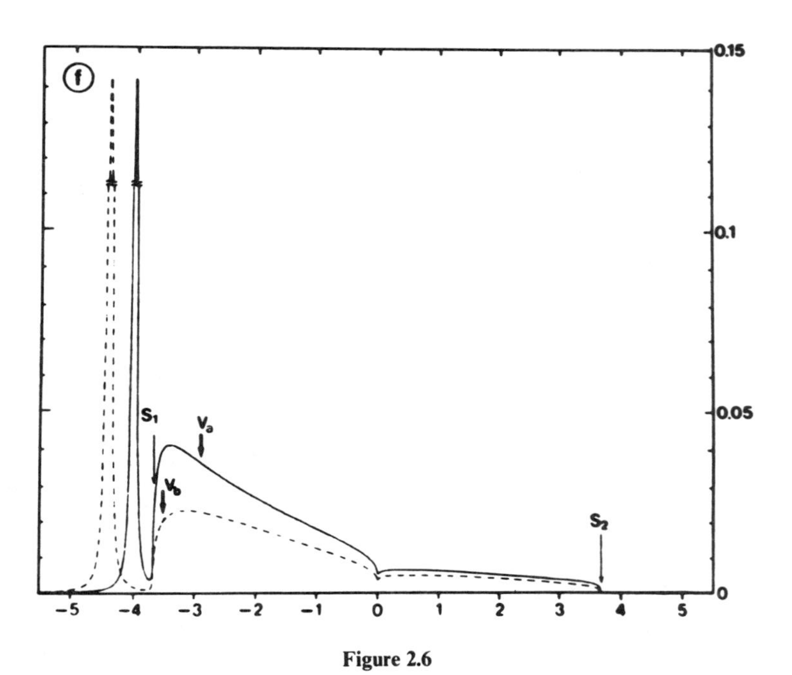

f
S_1
V_a
V_b
S_2
0.15
0.1
0.05
0
−5
−4
−3
−2
−1
0
1
2
3
4
5

Figure 2.6

discrete resonances rejoin below the continuum: see Fig. 2.6f (the upper resonance has moved back across the continuum under the influence of increasing quadratic coupling).

The various types of critical absorption lineshapes are illustrated schematically in Fig. 2.7. In the region I of weak couplings, the lineshape is that of large and weak continua for both polarizations of the crystal. Region II is that of strong linear couplings ($\xi^2 > 1$) and weak quadratic couplings: The lineshape is that of two peaks emerging from both sides of a weak continuum. Region III is characterized by equally strong quadratic coupling, with the upper peak passing back into the continuum: The lineshape is that of a discrete resonance well detached below a broad band. Still stronger quadratic couplings characterize region IV: The lineshape is that of two discrete strong peaks below a very weak broad band. A few borderline situations may be discussed: For $\Delta\Omega = 0$, regions I and II rejoin at $\xi^2 = 1$; for $\xi^2 = 0$, there is neither vibronic absorption nor Davydov splitting, but a discrete state is detached from the continuum, roughly when $\Delta\Omega$ crosses half the value of the two-particle bandwidth. (This is the analog of an impurity isolated in the lattice; see Section IV.B.2.) Rigorously speaking, the above discussion should not hold for 1D and 2D lattices, where the divergence of g_0 (2.82), at the van Hove points, determines, in all kinds of coupling, discrete states which are very often near the two-particle continuum band. However, for the situations of Fig. 2.7, where only continuum bands exist, the oscillator strength of 1D and 2D discrete states is very weak, and the above discussion, summarized in Fig. 2.7, holds qualitatively.

This discussion has shown the importance of the two-particle states in critically modulating the lineshape of the vibronic absorption, which appears very broad, generally nonlorentzian, and with shifts in the vibronic resonances

Figure 2.6. Calculated 0–1 vibronic absorption spectra for a 2D lattice with two molecules per cell (broken lines for a polarization and solid lines for b polarization). V_a and V_b indicate the positions of the vibrons, calculated from (2.47), whereas S_1 and S_2 indicate the positions, respectively, of the low- and high-energy thresholds of the two-particle-state band. The parameters corresponding to each spectrum are the following (in units of $\hbar\Omega_0$): (a) $E_{FC} = 0.04$, $\Delta\Omega = -0.02$; (b) $E_{FC} = 0.36$, $\Delta\Omega = -0.02$; (c) $E_{FC} = 0.04$, $\Delta\Omega = -0.16$; (d) $E_{FC} = 0.36$, $\Delta\Omega = -0.16$; (e) $E_{FC} = 1.44$, $\Delta\Omega = -0.04$; (f) $E_{FC} = 0.04$, $\Delta\Omega = -0.32$. The axes of abscissae (energies) are graduated in J units ($J = \hbar\Omega_0/10$); the absorption units are arbitrary, but consistent in the various spectra. The spectra (a) and (b) correspond to a weak quadratic coupling: in (a) we find the density-of-states distribution; then, in (b), broadened vibronic resonances, which eventually (e) shift out of the band as discrete peaks with increasing linear coupling. For an intermediate quadratic coupling (c, d), the shape of the absorption varies strongly, showing eventually (d) a discrete peak for one polarization and one broad band for the other polarization. Last, in strong quadratic coupling (f) a discrete state appears, below the band, for both polarizations.

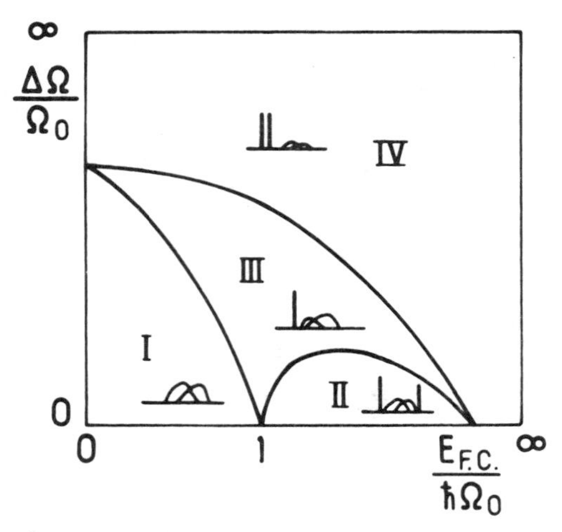

Figure 2.7. The various responses of the 0–1 vibron–two-particle states described by the hamiltonian (2.65)–(2.69) vs the two parameters E_{FC} and $\Delta\Omega$, for a 2D lattice with two molecules per cell. The scheme of the absorption is given in each delimited area. Apart from the well-known case $E_{FC} = \hbar\Omega_0$ ($\xi^2 = 1$), the boundaries of the areas depend on the excitonic dispersion of the lattice. Rigorously speaking, this scheme is valid only for a 3D lattice.

when the latter exist. The most restrictive approximation used in applying the Rashba hamiltonian to the dynamical problem ($B \ll \hbar\Omega_0$, the so-called dynamical restriction) amounts to ignoring the collective coupling. Philpott[55,57] has carefully considered the collective coupling, including approximately the upper vibronic transitions. However, the calculations and results are difficult to interpret. We think that, at least in the case of energetically separated vibronic subspaces, our treatment with effective Franck–Condon factors (Section II.B.1.c) is a reasonable approach to the collective coupling.

3. *Vibronic Absorption Structures of the Anthracene Crystal*

The principal modes of vibration of the anthracene molecule, active in absorption and in fluorescence, have been given in ref. 58. This subsection is intended to provide a consistent analysis of the whole structure of the crystal absorption (reflectivity) and to discuss its shape in terms of vibron–two-particle states and collective couplings and of the spectral and dynamical (relaxation) effects to which they give rise. These active vibration modes are here summarized in decreased order of strength:

1. The vibrations at 1406, 1397 cm^{-1} in the first excited state, very strong, with a well-developed vibronic series and the associated Franck–Condon factor[59]

$$(F_0^0)^2 = 0.324, \quad (F_0^1)^2 = 0.316, \quad (F_0^2)^2 = 0.218,$$
$$(F_0^3)^2 = 0.092, \quad (F_0^4)^2 = 0.050 \tag{2.83}$$

(The sum of these factors equals 1, so that the vibronic series is assimilated to the first five vibronic levels.) The quadratic couplings are negligible, so that the above Franck–Condon factors correspond roughly to a linear coupling with $\xi^2 \sim 1$.

2. The vibrations at 392, 390 cm^{-1} in the excited state, strong, whose molecular spectrum shows the first and the second harmonics.[58] From the oscillator strength of the vibronic 0–392-cm^{-1} transition, we may roughly derive a purely linear coupling with $\xi^2 \sim 0.14$. Each one of the principal 1400-cm^{-1} vibronic transitions is flanked with a vibration at 390 cm^{-1}, building the sequence $1400 + 392$, $2 \times 1400 + 392, \ldots$ cm^{-1}.
3. Other modes and combinations of modes, with weak intensities, are also observed, mainly the vibronic transitions at 1162 cm^{-1} (1159 excited) and 1564 cm^{-1} (1550 excited), which will be commented on below.

a. *Vibronic Reflectivity Spectra.* These spectra were first resolved and discussed by Philpott and Turlet[60] for the two polarizations, a and b, of the crystal. We follow the principal vibronic structures assigned by Philpott and Turlet, and we summarize them below, with the notation of these authors, using spectra of Fig. 2.8:

I. THE 0–0 LINE. The analysis of this structure will be made in Section II.C below. The b component at $E_{vb}^{00} = 25\,095$ cm^{-1} (line 1b in Fig. 2.8) is very narrow at low temperatures, while the a component, at about $E_{va}^{00} = 25\,319$ cm^{-1} (line 1a in Fig. 2.8) is very broad (~ 80 cm^{-1} at low temperatures).

II. THE VIBRONIC SYSTEM AT 392 cm^{-1}. We note first, almost at the same energy for both polarizations, the structures 2b at 25 487 cm^{-1} and 2a at 25 485 cm^{-1} assigned to the two-particle 0–392-cm^{-1} vibronic absorption threshold. The bottom of the 0–0 excitonic band being the fluorescent b component, these threshold structures are located at $\hbar\Omega_0 = 392$ cm^{-1} above the b component *for both polarizations*, the 392-cm^{-1} vibrational exciton dispersion of the order of 1 cm^{-1}[61] being negligible.[43] The shape of the threshold of these structures is presumably related to that of the density of states of the excitonic band, which is practically of 2D character (see Section II.D below). Thus, we find the low-energy threshold shape of the vibron–two-particle system calculated in Section II.B.2 (cf. Fig. 2.6).

The next structures are 3b at 25 665 cm^{-1} and 3a at 25 672 cm^{-1}, which are broad bumps attributed to vibronic resonances greatly diluted in the two-

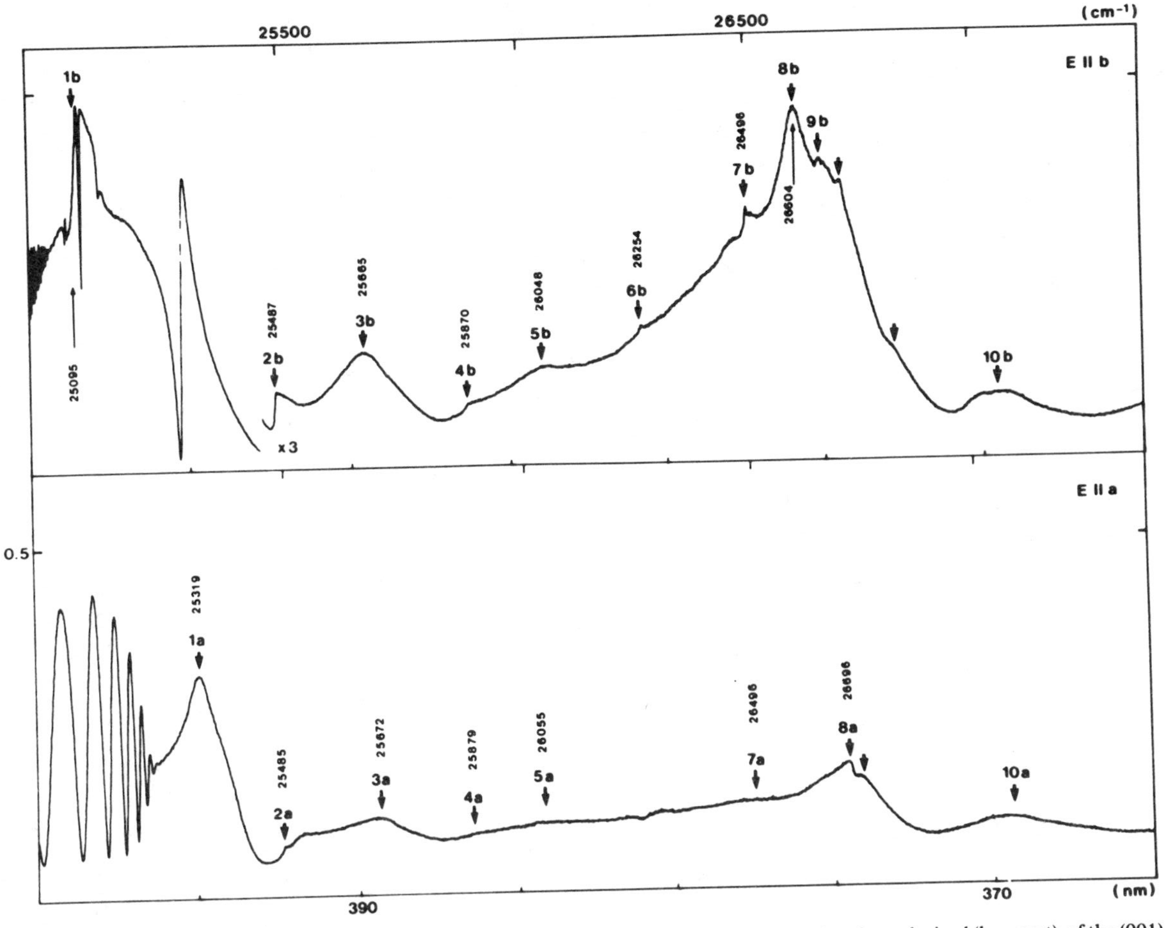

Figure 2.8. Reflectivity spectra in the various vibronic regions, *b*-polarized (upper part) and *a*-polarized (low part), of the (001) face of the anthracene crystal at 5 K. The energies of the main structures are indicated in reciprocal centimeters. (The interference structures are discussed in Section III.)

particle continuum (Fig. 2.6*b*). The intensity of this vibronic resonance does not appear consistent with the weak linear coupling ($\xi^2 \sim 0.14$), but we must remark that on one hand the strong inequality $\hbar\Omega_0 = 392\,\text{cm}^{-1} \gg J$ is not satisfied, and on the other hand the shape of the density of excitonic states is not well known.

The following structures are feeble replicas of the 0–392 vibronic structures, assigned to the 0–2 vibronic transition (0–2) × 392: We notice first that the two-particle thresholds, 4*b* at 25 870 cm^{-1} and 4*a* at 25 879 cm^{-1} (very weak), are followed by large structures assigned to (0–2) × 392 vibrons, 5*b* at 26 048 cm^{-1} and 5*a* at 26 055 cm^{-1}. The fact that the energy threshold, from the bottom of the excitonic band (0–0) + 2 × 392, is predominant is an indication that the direct vibron fission, creating two vibrations and one 0–0 electronic exciton, overwhelms the cascade process of creation of one vibration and one 0–392 vibron and so on. This fission ordering is consistent with the values of the Franck–Condon factors (2.44) for the two processes controlled respectively by $F_2^2 F_0^0$ and $F_1^2 F_0^1$ with

$$\begin{aligned} F_2^2 F_0^0 &= e^{-\xi^2}\left(1 - \xi^2 + \frac{\xi^4}{2}\right) \\ F_0^1 F_1^2 &= e^{-\xi^2}\xi^2\left(1 - \frac{\xi^2}{\sqrt{2}}\right) \end{aligned} \tag{2.84}$$

A linear coupling, $\xi^2 \sim 0.14$, leads to $F_2^2 F_0^0 \gg F_1^2 F_0^1$.

III. THE VIBRONIC SYSTEM AT 1400 cm^{-1}. The molecular vibronic 1400-cm^{-1} progression is found in the reflectivity spectra, where the 0–0, 0–1, 0–2 transitions are well resolved, the 0–3 transition poorly. Comparing the intensities of the various bands, it is found that the *b*-polarized lines are considerably modified relative to the molecular analogs while the *a*-polarized lines remain unchanged[62]; see Table I. The most striking aspect is the unusual

TABLE I

	Intensity		
		Crystal	
Transition	Molecule	‖*b*	‖*a*
0–0	0.38	0.60	0.38
0–1	0.37	0.31	0.38
0–2	0.25	0.09	0.24

large width of the vibronic 0–1 and 0–2 bands. This broadening has been explained by the mixing of nearby modes and combinations of modes, borrowing intensity from the mode at 1400 cm^{-1}; this assumption has been confirmed by the analysis of reflectivity[60] and fluorescence[58] spectra (see also Section II.D below).

In what follows, we analyze only the 0–1 vibronic transition. The corresponding structures are well resolved (Fig. 2.8): The two-particle threshold (7*b*)[60] $E_{vb}^{00} + 1400\,\text{cm}^{-1}$ at 26 462 cm^{-1}; the attribution of 7*a* is less certain. Two well-resolved maxima (8*b* and 8*a*) are assigned to 0–1 vibrons at 26 604 and at 26 096 cm^{-1}. These peaks are very broad, consistent with their position relative to the threshold; they are located at the center of the two-particle continuum. The *b* component is compatible with the model case calculated for a linear coupling $\xi^2 \sim 0.5$; see Fig. 2.6*b*. The *a* component cannot enter the model, because this state lies at the center of the excitonic band and undergoes strong intraband broadening processes; see below.

Many other structures may be observed in the 0–1 reflectivity envelope: 6*b* at 26 254 cm^{-1} is attributed[60] to the threshold of the two-particle states at $E_{vb}^{00} + 1162\,\text{cm}^{-1}$, which is a vibration visible in fluorescence and in absorption; 9*b* is the analog for the vibration at 1564 cm^{-1}. Finally, the transitions 0–1 and 0–2 have their vibronic counterparts associated with the vibration at 392 cm^{-1} (10*b* and 10*a*) for the 0–1 transition.

b. *Positions and Intensities of the Vibronic Lines.* As stated in Section II.B.1.c, the intensity redistribution by the collective coupling is determined by the purely electronic interaction J_k, which cannot be derived directly from experiments. To determine J, involved in the values of the *a* and *b* Davydov components (or, which is the same, in the center of gravity of the excitonic band), we make a zero-order approximation, ignoring the collective coupling. Using the model in Section II.B.1.b, the centers of gravity of 0–0 and of 0–1 must be separated by 1400 cm^{-1}, and connected in the same linear manner to the *a* and *b* components of 0–0 and 0–1 (see Fig. 3.19):

$$E_{\text{CG}}^{00} = \lambda E_b^{00} + (1-\lambda)E_a^{00} \tag{2.85}$$

$$E_{\text{CG}}^{01} = E_{\text{CG}}^{00} + 1400 = \lambda E_b^{01} + (1-\lambda)E_a^{01}$$

The positions of the vibronic line, reported in Section II.B.3.a above, allow us to situate the 0–0 center of gravity at $E_{\text{CG}}^{00} = 25\,279\,\text{cm}^{-1}$, which is 40 cm^{-1} below the *a* component; the "392" 0–1 vibron provides the same value, $E_{\text{CG}}^{00} \sim 25\,280\,\text{cm}^{-1}$. This remarkable coincidence reinforces our confidence in the vibron model of Section II.B.1.b.

However, while the line positions are consistent with the absence of significant collective coupling, the intensity of the vibronic structures (Table I) indicates, on the contrary, a large redistribution of the oscillator strength on the b components. This contradiction caused us to reconsider more closely the collective coupling in anthracene. The principal investigations on the matter were carried out by Philpott et al.[59,63] Their model consists in a diagonalization, for each wave vector **k**, of a set of five vibronic levels (0–0, 0–1,...,0–4) and other electronic states coupled by dipolar and nondipolar interactions. The principal restriction results from the monoexcitation assumption, which excludes all two-particle states, although in fact their role is very important, as demonstrated below. Indeed, let us consider the van der Waals D term stabilizing the electronic state. Using second-order perturbation theory[61] on purely electronic functions, we write

$$D_{S_1} = \sum_{m,S_i} \frac{\langle nS_1|V|mS_i\rangle\langle mS_i|V|nS_1\rangle}{E_{S_1} - E_{S_i}} \tag{2.86}$$

Consider now the D contribution between two vibronic states, ν and μ of S_1. We have, for instance, for the calculation of the level ν

$$\langle \nu|D_{S_1}|\mu\rangle = \sum_{m,S_i,\lambda} \frac{\langle nS_1\nu|V|mS_in\lambda\rangle\langle mS_in\lambda|V|nS_1\mu\rangle}{E_{S_1} - E_{S_i} + (\nu - \lambda)\hbar\Omega_0} \tag{2.87}$$

where the summation on λ includes the vibrational levels of the ground state n. Thus, the intermediate states in (2.87) are two-particle states. In the Born–Oppenheimer approximation, neglecting $\hbar\Omega_0$ compared to $E_{S_1} - E_{S_i}$, we obtain

$$\langle \nu|D_{S_1}|\mu\rangle = D_{S_1}\delta_{\mu\nu} \tag{2.88}$$

The corrective terms are of the order $D_{S_1}\cdot\hbar\Omega_0/(E_{S_1} - E_{S_i})$ and effectively couple the vibrons $S_1\mu$ and $S_1\nu$. For anthracene, these terms are of the order of $D/10$, and we leave them aside in a first analysis. Thus, we are led to a global shift (D_{S_1}) of all the vibronic levels, without coupling. On the contrary, the monoexcitation condition leads to the condition $\lambda = 0$ in (2.87), with the following result:

$$\langle \nu|D_{S_1}|\mu\rangle \overset{?}{=} D_{S_1}F_0^{\mu}F_0^{\nu} \tag{2.89}$$

This expression means both a differential shift and coupling of the vibronic states. In dipolar excitons the D term is very important, and (2.89) explains why the calculations reported in refs.[59,63] clearly overestimate the collective coupling effect: Their D term is comparable to J (500 cm^{-1}) and adds to it.

Although the Davydov splittings of the vibronic excitons are in good agreement with the observations (0–0: 949 cm^{-1}; 0–1: 106 cm^{-1}; 0–2: 56 cm^{-1}), the calculated energy differences between the structures 0–1 and 0–0 are too large:

polarization a: 1479 cm^{-1} instead of 1377 cm^{-1},
polarization b: 1622 cm^{-1} instead of 1507 cm^{-1},

and the calculated intensity ratios favor too much the 0–0 transition for the two polarizations, in contradiction with the observations by Sceats[62] (Table I). In addition, the calculated stabilization of the upper vibronic levels reported by these authors is too weak (40 cm^{-1} for the 0–4) according to (2.89). In view of these three discrepancies, we cannot apply the results concerning the collective coupling reported by Philpott et al.[59,63]

To evaluate the collective coupling effects, we make the assumption that the only efficient mechanism of coupling is that introduced by J_k of Section II.B.1.c. Thus, we neglect the terms discarded in order to pass from (2.87) to (2.88), which result from the interactions with the other electronic states (in particular S_2), and which may introduce differential stabilization of the vibronic levels.

First, we consider the positions of the b vibrons relative to the center of gravity. Starting from the position of the 0–0 vibron (determined above with $z_0 = -180$ cm^{-1}), using the factors (2.83) and a slightly different form for (2.84), we find in units of $\hbar\Omega_0$

$$z_\nu = \nu + J_b \frac{(F_0^\nu)^2}{1 - J_b \Sigma_\nu} \tag{2.90}$$

with

$$\Sigma_\nu = \sum_{\mu \neq \nu} \frac{(F_0^\mu)^2}{z - \mu} \tag{2.91}$$

We obtain for J_b, the purely excitonic (undressed) dispersion of the b

TABLE II

Transition	Displacement (cm^{-1})
0–0	−180
0–1	−146
0–2	−97
0–3	−38
0–4	−15

component, the value $-0.340\hbar\Omega_0$, or $-476\,\text{cm}^{-1}$. Then, we determine the dispersion origin displacements of the various b vibrons as shown in Table II. Secondly, we consider the intensity of the various vibronic bands evaluated by using (2.56), which can be rewritten in units of $\hbar\Omega_0$:

$$f_\nu = \frac{(F_0^\nu)^2}{(1 - J_b\Sigma_\nu)^2 + J_b^2(F_0^\nu)^2\Pi_\nu} \tag{2.92}$$

with

$$\Pi_\nu = \sum_{\mu \neq \nu} \frac{(F_0^\mu)^2}{(z-\mu)^2} \tag{2.93}$$

It is clear that, to the first order in J, the renormalizations of the oscillator strengths lead to results twice as important, in relative values, as those of the displacements. The exact calculation of f_ν provides the values given in Table III. For the a vibrons, the much smaller 0–0 displacement relative to the center of gravity ($+40\,\text{cm}^{-1}$) leads to very weak relative variations in the positions (less than $+4\%$) as well as in the intensities (less than $+8\%$).

To summarize, the following conclusions can be drawn:

1. The collective coupling of the vibronic excitons by the exchange energy term J allows us to understand qualitatively the intensity ratios measured by Sceat for both polarizations (Table I), if we assume the 0–0 center of gravity to be, according to our analysis,[61] at $40\,\text{cm}^{-1}$ below the a component. The resulting value $J_b = -476\,\text{cm}^{-1}$ is close to the values generally accepted ($-500\,\text{cm}^{-1}$).[62]

2. More quantitatively, the value (0.66) that we obtain for the intensity ratio (0–1)/(0–0) of the b lines disagrees with the measured value (0.52). [The calculations by Philpott et al. Provide, on the contrary, too weak a value (0.37)]. In addition to the lack of accuracy inherent in methods of measuring oscillator strengths of strong transitions, it appears probable that inclusion of upper electronic states in (2.87) would, by the induction of a complementary collective coupling, improve the accord with the measured values.

TABLE III

Transition	f_ν
0–0	0.436
0–1	0.288
0–2	0.185
0–3	0.070
0–4	0.021

3. Measured positions, relative to the centers of gravity, for the b vibrons, 0–0 ($-180\,\text{cm}^{-1}$) and 0–1 ($-75\,\text{cm}^{-1}$), are in severe disagreement with the values of Table II. This disagreement is easy to understand when we know that the maxima associated with the upper vibronic bands are not the centers of gravity of these bands. As noticed above in Section II.B.3.a.iii, the coupling to other nearby modes borrows oscillator strength, and hence Davydov splitting from the vibron 1400-cm^{-1} mode. Consequently, the displacements of b and a vibrons are appreciably reduced. According to (2.90), this cannot possibly be explained by the sole effect of the collective coupling: the 0–1 vibron displacement associated with the intensity ratio (0–1)/(0–0) would be about $127\,\text{cm}^{-1}$.

C. Optical Absorption in the Field of Phonons

We have indicated in Section I that the optical properties of the crystal are characterized by the transverse dielectric tensor $\boldsymbol{\varepsilon}^{\perp}(\mathbf{k}, \omega)$ (1.79). The real and imaginary parts of this tensor being related by the Kramers–Kronig relations resulting from the linearity, $\boldsymbol{\varepsilon}^{\perp}(\mathbf{k}, \omega)$ is itself determined by its imaginary part. In what follows, we assume that an eigendirection of $\boldsymbol{\varepsilon}^{\perp}$ is excited, and we consider $\boldsymbol{\varepsilon}''(\mathbf{k}, \omega)$ and the optical conductivity $\omega\boldsymbol{\varepsilon}''(\mathbf{k}, \omega)$ under the common denomination of "optical absorption". In fact, it is the conductivity that determines the absorption by the crystal of the energy of the plane wave (see Appendix A).

In Section I, the spectra of $\boldsymbol{\varepsilon}''(\omega)$ consist of Dirac δ peaks (1.79). In a real crystal these peaks are broadened by static disorder, thermal fluctuations, and excitation–relaxation processes. Discarding for the moment the static disorder, we focus our attention on broadening processes due to lattice phonons, which may be described alternatively in terms of fluctuations of the local energies of the sites, or in terms of exciton relaxation by emission and absorption of phonons. These two complementary aspects of the fluctuation–dissipation theorem[64] will allow us to treat the exciton–phonon coupling in the so-called strong and weak cases. The extraordinary (polariton) 0–0 transition of the anthracene crystal will be analyzed on the basis of these theoretical considerations and the semiexperimental data of the Kramers–Kronig analysis.

1. *Theoretical Models of Optical Absorption*

For the sake of an orderly discussion of the absorption mechanisms, we consider only coupling to the field of phonons via the D terms: see (2.22). Usually, the D terms provide the dominant coupling terms ($D \sim 2500\,\text{cm}^{-1}$ vs $400\,\text{cm}^{-1}$ for J for the 0–0 transition). In addition the $1/R^6$ variation for D terms is stronger than the $1/R^3$ variation for J terms. Thus the hamiltonian of the system coupled to the field of the phonons is given the form

$$H = \sum_{\mathbf{k}} \hbar\omega_{\mathbf{k}} B_{\mathbf{k}}^{\dagger} B_{\mathbf{k}} + \sum_{s\mathbf{q}} \hbar\Omega_s(q) b_{\mathbf{q}s}^{\dagger} b_{\mathbf{q}s} + \sum_{ns\mathbf{q}} B_n^{\dagger} B_n (b_{\mathbf{q}s}^{\dagger} + b_{-\mathbf{q}s}) \chi_s(q) \qquad (2.94)$$

The exciton–phonon coupling strength may be characterized by the stabilization[53] S obtained when the lattice is left to relax around an exciton localized on site n. If we consider the independent variables $(b_{-qs} + b_{+qs}^{\dagger})$ as continuous, we find, with the use of the last two terms of (2.44),

$$S = \sum_{s\mathbf{q}} \frac{|\chi_s(\mathbf{q})|^2}{\hbar\Omega_s(\mathbf{q})} \qquad (2.95)$$

This stabilization ($-S$ is always negative) is the lattice analog of the energy E_{FC} of molecular excitation (2.17). For a given exciton–phonon coupling, the "softer" the $s\mathbf{q}$ modes are (i.e., the smaller $\hbar\Omega_0$), the stronger the corresponding stabilization is. Furthermore, the $s\mathbf{q}$ modes being independent, the stabilization terms are cumulative.

According to the value of S relative to the exciton bandwidth B, the relaxed form of the exciton will be delocalized or localized for $-B < -S$ or $-S < -B$, respectively. Adopting Toyozawa's terminology,[53] we can say that for $S < g_c B$ the exciton–phonon coupling is weak and for $S > g_c B$ it is strong, where g_c is a characteristic parameter of the lattice structure of the order of unity ($g_c = 0.87$ for the square lattice[53]).

At the usual temperatures, the phonon modes are populated, so that the initial populations must enter the calculation of the optical absorption. It can be shown (see Appendix A, where E indicates the complex energy $\hbar\omega + i0$) that the absorption is proportional to the following function:

$$I(E) = \sum_{\alpha} p_{\alpha} \operatorname{Im} \langle \mathbf{k}\alpha | \frac{1}{E + E_{\alpha} - H + i0} | \mathbf{k}\alpha \rangle \qquad (2.96)$$

with α indicating the lattice state of energy E_α and of population P_α at the temperature T $[\beta = (kT)^{-1}]$:

$$P_{\alpha} = \left(\sum_{\alpha'} e^{-\beta E_{\alpha'}} \right)^{-1} e^{-\beta E_{\alpha}} \qquad (2.97)$$

(Here $\mathbf{k}$ is the wave vector of the exciton created by the incident photon; in the following we shall take $\mathbf{k} = \mathbf{0}$.)

We must make a remark on the exciton–phonon coupling: Rigorously speaking the hamiltonian H (2.96) should contain the retarded interactions, and the problem of coupling should be that of polariton–phonon coupling. In reality, the exciton–photon interaction is important only during the absorption of the incident photon, and such a mechanism is embodied in the expression for $\boldsymbol{\varepsilon}^{\perp}(\omega)$: In the subsequent evolution, as in Section II.C.2, the coupling to photons does not contribute to the absorption in the linear-response approximation, because the phonon interactions populate the whole Brillouin excitonic zone, where retarded effects are in average negligible. This peculiarity is due to the local nature of the exciton–phonon interactions in the absorption of molecular crystals. On the contrary, in radiative relaxation phenomena, such as fluorescence, the explicit polariton–phonon coupling cannot be avoided (cf. Sections II.D and IV.A).

The excitonic lineshape broadening is described quantitatively by the width of the line, its shape (mainly its asymmetry), the existence of sidebands, the absorption wing at low energies, etc. An exact solution of (2.94), (2.96) being impossible, we consider below the two limit cases: (1) the exciton–phonon coupling is weak ($S < B$), and the quasi-free exciton model is adequate; (2) the exciton–phonon coupling is strong ($S > B$) and the quasi-localized exciton is more appropriate. In both cases, effects of the other parameters, $\hbar\Omega_0$ and kT, will be specified.

a. *Weak Exciton–Phonon Coupling.* A perturbative approach to free-exciton states is adequate. Consider the self-energy $\Sigma_\alpha(E)$, corresponding to each state α of lattice phonons, defined by

$$\langle \mathbf{k}\alpha | \frac{1}{E + E_\alpha - H} | \mathbf{k}\alpha \rangle = \frac{1}{E - \Sigma_\alpha(E) - \hbar\omega_{\mathbf{k}}} \tag{2.98}$$

To the second order of perturbation in the exciton–phonon coupling, we obtain for the self-energy

$$\Sigma_\alpha^{(2)}(E) = \sum_{s\mathbf{q}} \frac{|\chi_s(\mathbf{q})|^2 n_s(\mathbf{q})}{E - \hbar\omega_{\mathbf{k}+\mathbf{q}} + \hbar\Omega_s(\mathbf{q})} + \frac{|\chi_s(\mathbf{q})|^2 [1 + n_s(\mathbf{q})]}{E - \hbar\omega_{\mathbf{k}+\mathbf{q}} - \hbar\Omega_s(\mathbf{q})} \tag{2.99}$$

where $n_s(\mathbf{q})$ is the number of phonons s, $\mathbf{q}$ in the α lattice state. To the equation (2.99) correspond the diagrams

(2.100) [diagram] + [diagram]

(where the solid and the dashed lines indicate the free-exciton and the phonon propagator). By convention the first term describes an emission in $1 + n_s(q)$

and the second an absorption in $n_s(q)$. The optical response is given by the average of (2.98) over the various states α. An expansion of (2.98) in powers of Σ_α shows that the average can be taken directly on Σ_α: Indeed, the averages of the type $\langle n_s(q) n_{s'}(q') \rangle$ can be uncoupled and replaced by $\langle n_s(q) \rangle \langle n_{s'}(q') \rangle$; the error introduced by the terms $\langle n_s(q)^2 \rangle$ is of the order of $1/N$.[65] The uncoupling is quite justified in the quasi-free-exciton approximation. Thus, we obtain

$$I(E) = [E - \hbar\omega_k - \Sigma^{(2)}(E)]^{-1}$$

where

$$\Sigma^{(2)}(E) = \sum_{s\mathbf{q}} \frac{|\chi_s(\mathbf{q})|^2 \langle n_s(\mathbf{q}) \rangle}{E - \hbar\omega_{\mathbf{k}+\mathbf{q}} + \hbar\Omega_s(\mathbf{q})} + \frac{|\chi_s(q)|^2 (1 + \langle n_s(q) \rangle)}{E - \hbar\omega_{\mathbf{k}+\mathbf{q}} - \hbar\Omega_s(\mathbf{q})} \tag{2.101}$$

The first term of $\Sigma^{(2)}$ accounts for the broadening and the shift caused by phonon absorption [in $\langle n_s(q) \rangle$], and the second term for those caused by spontaneous and induced emissions $[1 + \langle n_s(q) \rangle]$. The Bose–Einstein population $\langle n_s(q) \rangle$ has the form

$$\langle n_s(q) \rangle = (e^{\beta\hbar\Omega_s(q)} - 1)^{-1} \tag{2.102}$$

1. For an optical transition to the bottom of the excitonic band, only phonon-absorption broadening occurs, necessitating an activation temperature of the order of $\hbar\Omega_s(q)/k$. For an optical phonon branch, if the absorption broadening process dominates, then the imaginary part of Σ varies slowly in the vicinity of the excitonic transition, and one expects a Lorentzian lineshape at least as long as its width is smaller than $\hbar\Omega_s$. Then this width must vary as $\langle n_s \rangle = \{\exp(\beta\hbar\Omega_s) - 1\}^{-1}$ if Ω_s is the average frequency of the optical branch. The high-temperature limit of (2.101) is reached when the width is larger than the energies of the coupled phonons. Then, various structures of the phonons fuse to give a large width, absorption and emission processes play an equivalent role, and the width of the optical absorption takes the approximate form

$$\Gamma(E) = 2\,\mathrm{Im}\,\Sigma^{(2)}(E) \sim 2\pi \cdot 2kT \sum_{s\mathbf{q}} \frac{|\chi_s(\mathbf{q})|^2}{\hbar\Omega_s(\mathbf{q})} \delta(E - \hbar\omega_{\mathbf{k}+\mathbf{q}}) \tag{2.102'}$$

Neglecting the variation of χ_s with q, and using the density

$$n(E) = \frac{1}{N\pi} \mathrm{Im} \sum_k \frac{1}{E - \hbar\omega_k + i0} \tag{2.103}$$

of states of the excitonic band, we have

$$\Gamma(E) \sim 2\pi\, 2SkTn(E) \tag{2.104}$$

As a consequence of the rapid variation of $n(E)$ around the bottom of the excitonic band, one expects a stronger broadening in the upper energies, i.e. an asymetric lineshape. We must remark that (2.104) shows that the perturbations theory to second order does not give the high-temperature limit in $T^{1/2}$ of (2.33), obtained using the localized-exciton model, which is better adapted for high-temperature cases.

2. For an optical transition to the middle of the excitonic band, the low-temperature limit width is dominated by phonon spontaneous emission with a lorentzian lineshape, more or less distorted by the density of excitonic states at the final energy $E - \hbar\Omega_s$. With increasing temperature, the line broadens and reaches the high-temperature limit (2.104).

We now examine more carefully the lineshape in the low-energy part far from the 0–0 transition value.[52,53] From a perturbational point of view, this absorption may be visualized as a successive absorption of many phonon energy quanta up to the excitonic band, where a density of excitonic final states exists. In this scheme, the low-energy absorption wing exists only at finite temperature. If ν phonons are necessary to reach the excitonic band, the dominant term in Σ originates from the consecutive absorption of ν phonons following the diagram

(2.105)

The form of (2.105) suggests the use of the approximation of Migdal,[52] which consists in "dressing" the free-exciton propagator with Σ itself in (2.100). The self-consistent equation to solve, in diagrammatic form, is

(2.106)

or as a functional equation with the unknown function $\Sigma(E)$:

$$\begin{aligned}\Sigma(k, E) = \sum_{s,q} & \frac{|\chi_s(q)|^2 \langle n_s(q)\rangle}{E - \hbar\omega_{kq} + \hbar\Omega_s(q) - \Sigma(k+q, E + \hbar\Omega_s(q))} \\ & + \frac{|\chi_s(q)|^2 (1 + \langle n_s(q)\rangle)}{E - \hbar\omega_{k+q} - \hbar\Omega_s(q) - \Sigma(k+q, E - \hbar\Omega_s(q))}\end{aligned} \tag{2.107}$$

This equation is an approximation in the sense that high-order diagrams are missing. The first neglected diagrams, in fourth order in the exciton–phonon interaction, are of the form

To find the low-energy absorption, we iterate the expression (2.107) retaining only the phonon absorption term in $\langle n \rangle$, and stopping at the second-order term for the last real process, which leads to the excitonic band at the energy $E + \nu\hbar\Omega_s(q)$ and provides an imaginary part for $\Sigma^{(2)}\ (E + \nu\hbar\Omega_s(q))$. We may evaluate the optical absorption using the simplification $[\chi_s(q) = \chi_s, \hbar\Omega_s(q) = \hbar\Omega_0, \Sigma$ independent of $k]$

$$\operatorname{Im} \Sigma(E) = (\chi^2 \langle n \rangle)^{\nu-1} \mathscr{G}(E + \hbar\Omega_0) \cdots (E + (\nu - 1)\hbar\Omega_0) \times \operatorname{Im} \Sigma^{(2)}(E + \nu\hbar\Omega_0) \qquad (2.108)$$

where

$$\mathscr{G}(E) = \frac{1}{N} \sum_k \frac{1}{(E - \hbar\omega_k)^2}$$

Furthermore, assuming a constant density of excitonic states and a large excitonic bandwidth $B \gg \hbar\Omega_0$, we arrive at the simple expression

$$\operatorname{Im} \Sigma(E) \sim \frac{\hbar\Omega_0}{(\nu - 1)!} \left(\frac{\chi^2 \langle n \rangle}{2B\hbar\Omega_0} \right)^{\nu} \qquad (2.109)$$

for low enough temperature. The expression (2.109) varies exponentially because of the lack of resonance at the optical transition energy E_0, which is the well-known Urbach law[53]

$$I(E) \propto \exp\left(-\sigma(T) \frac{E_0 - E}{kT} \right) \qquad (2.110)$$

where $\sigma(T)$ is the steepness index and varies smoothly with T. With the simple model (2.109) we obtain approximately, for low temperatures,

$$\sigma(T) \sim 1 - \frac{kT}{\hbar\Omega_0} \log \frac{S}{2B} \qquad (2.111)$$

In this model of weak exciton–phonon coupling ($S < B$), $\sigma(T)$ is larger than unity. In addition, the exponential wing disappears for $T \to 0$, which is consistent with the absence of phonons to assist the optical absorption.

b. *Strong Exciton–Phonon Coupling.* For strong coupling ($S > B$) the picture of free-exciton scattering is obviously less well adapted than the diffusion of localized excitons. As in Section II.B for the vibron, a canonical transformation allows us to eliminate the linear term $b + b^{\dagger}$ at the cost of introducing higher-order terms proportional to B. The treatment of these terms in the lower orders recalls that of the collective coupling of the vibronic states in Section II.B.1.c and the interaction with two-particle states[55] in Section II.B.2. This approach allowed Cho and Toyozawa[66] to show the transition, as S increases, from the stable free-exciton regime to the stable self-trapped-exciton regime, the other regime in each case being metastable. In each case, the metastable regime has been detected experimentally in molecular crystals by Matsui et al.[51]

As shown in Section II.A.3, the localized exciton in a lattice can be treated as an impurity in the lattice. The methods of treating the disorder, particularly the CPA (see Section IV.B), are then applicable for the quasi-static case, i.e. when the nuclear motion is slow relative to the other dynamic parameters (or $\hbar\Omega_0 \ll B, S$). The classical exciton–phonon coupling is then treated exactly, and the interaction with the disordered lattice is treated either approximately by uncoupling the local scattering[52,50] or exactly using numerical simulations.[53] The latter work, by Toyozawa et al., shows that the picture of the localized exciton allows one to develop a unifying theory of the exciton–phonon coupling where the quasi-free exciton is treated equally well. Furthermore, numerical simulations have shown that the Urbach law is remarkably accurate in various cases of coupling ($S > B$); in the case $S > g_c B$, an absorption wing persists even at zero temperature, due to relaxation to the self-trapped exciton.

The work by Sumi and Toyozawa[52] has been extended by Sumi[50] to the cases where phonons must be treated quantum-mechanically (large values of $\hbar\Omega_0$). This very general formalism, called the dynamical CPA (DCPA) method, allows a unified description of the exciton–vibration coupling (Section II.A.3) and the interpolation between various limit cases of the vibron, of the free exciton, and of the self-trapped exciton. This formalism has been extended to collective coupling including two-particle states in all cases of exciton–phonon couplings, linear and quadratic.[56]

2. *Absorption Spectra as Kramers–Kronig-Transformed Reflection Spectra*

One of the finest tools to investigate the theoretical approach (2.101), (2.111) to absorption, without using numerous ad hoc parameters, is the use of the

Kramers–Kronig (KK) transformation of the reflection spectra. This provides the "optical absorption" $\varepsilon''(\omega)$ semiexperimentally and allows a thorough analysis of the various relaxation mechanisms creating the absorption lineshape (2.102), (2.111) of an ideal finite crystal in its phonon bath. This method is currently used. However, two major difficulties often obscure the credibility of the results:

1. the existence of uncontrolled crystal defects or impurities, which may cause fortuitous structures;
2. the arbitrariness of the choice of the two boundary parameters, which may strongly affect the derived spectra.

To avoid these two difficulties for our analysis, we used reflectivity spectra of thin, high-quality crystals, with surface structures subtracted out to recover the bulk reflectivity of a perfect semiinfinite crystal. Since this problem and its inherent difficulties are often ignored, we describe below our approach, up to the final application to UV spectroscopy, in order to point out the source of discrepancies originating either from the sample spectra or from the theoretical approximations.

a. *The Ideal Crystal in Its Phonon Bath.* The ideal sample is a thin monocrystalline layer with parallel (001) faces, homogeneous except for surface effects. The incident light is assumed normal to the front face, polarized along a principal direction, a or b. With $n(\omega) = [\varepsilon(\omega)]^{1/2}$ denoting the complex refractive index in this direction, the well-known expressions for the amplitudes of reflection and transmission for a crystal of thickness e are the following:

$$r = \frac{1 - n^2}{1 + n^2 + 2in \cot(n\omega e/c)} \tag{2.112}$$

$$t = \frac{2in}{\sin(n\omega e/c)} \frac{1}{(1 + n^2 + 2in \cot(n\omega e/c))} \tag{2.113}$$

The behavior of the functions $r(\omega)$ and $t(\omega)$ is complicated by a series of interference oscillations due to the complex cotangent. For a transparent layer [$n(\omega)$ real], the oscillations have zero minima for $r(\omega)$ at $n(\omega)\omega e/\pi c =$ integers. In the case of weak absorption by the phonon continuum ($n = \nu + i\kappa$, $\kappa \neq 0$), the back-face reflection component, for a sufficiently thick sample, is absorbed, and we obtain

$$|t|^2 \sim \frac{16|n|^2}{|1 + n|^4} e^{-2\kappa\omega e/c} \tag{2.114}$$

For a well-chosen thickness e, the transition intensity directly provides K and hence $\varepsilon''(\omega)$, the real part of n being derived from inspection of $|r(\omega)|^2$. However, for real crystals and for moderate or strong transitions near resonance, $\kappa(\omega)$ becomes so large that no transmission is detectable (even for $e \sim 1\ \mu m$ for the first singlet transition of the anthracene crystal). Then the reflection power takes the simple form

$$|r|^2 = \left|\frac{1-n}{1+n}\right|^2 \tag{2.115}$$

Nevertheless, the mere knowledge of $|r(\omega)|^2$ is not sufficient to provide the two parameters $\nu(\omega)$ and $\kappa(\omega)$: both reflection and transmission spectra are needed. This is the main reason that the KK transformation is used for reflectivity spectra, as we summarize below in application to our problem of UV spectroscopy.

b. *The Real Crystal.* Real crystals have characteristics which are not so simple as those of an ideal crystal. For instance, even if the back face is neglected, any defect of planarity of the front face causes a loss of the reflectivity yield which varies smoothly with the wavelength. In contrast, on investigating a transparent region, the oscillations (2.112), (2.113) will be strongly affected by the lack of parallelism of the front and back faces: In particular, the oscillations will be damped, or even suppressed, if the thickness varies by more than one wavelength in the irradiated area. In this case, we collect the sum of the intensities of the various multiple reflections (usually twice that of one face, when direct reflection dominates). Therefore, in the absence of oscillations and of absorption affecting the back-face reflection, the total reflectivity is as follows:

$$R_c = \frac{2R}{1+R}, \qquad R = \left|\frac{1-n}{1+n}\right|^2 \tag{2.116}$$

These few remarks show that the quality of the crystal must be carefully checked before any attribution of structures of the reflectivity spectrum to absorption mechanisms is made at the observed frequencies. Conversely, observation of unexpected strong reflectivity, possibly related to the presence of various thin dips[68] in the low-energy wing of the reflectivity, will appear strange if we do not consider the back-face contribution with the dips being signatures of various types of bulk absorption such as impurities. Another, very characteristic dip, with width proportional to kT [hence very sensitive to the temperature; see Fig. 2.9 (arrow)], is attributed to phonon-assisted

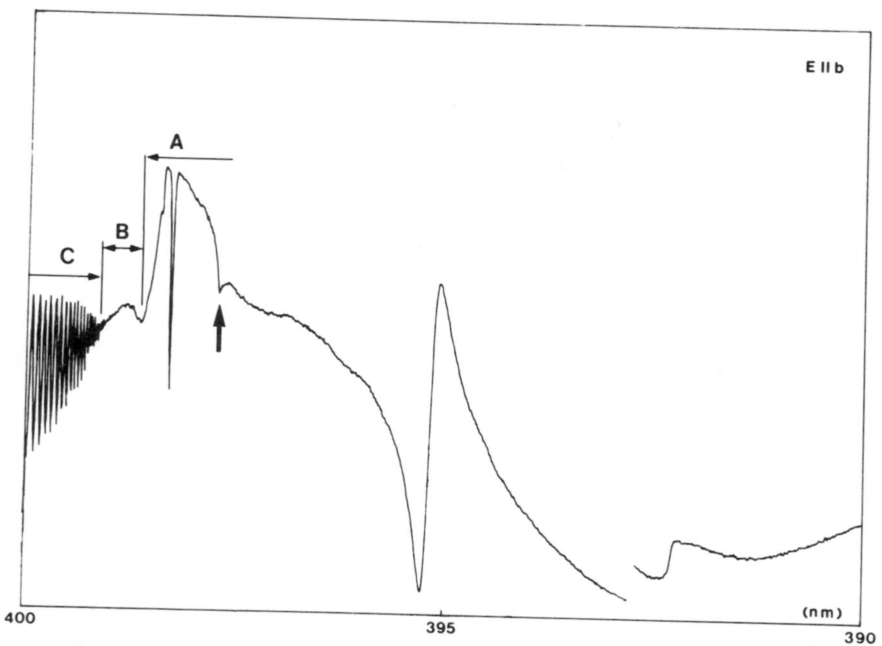

Figure 2.9. Detail of the 0–0, *b*-polarized reflectivity at 5 K (cf. Fig. 2.8). The arrow indicates the threshold of creation of "46-cm^{-1}" phonons. Part A is due to reflection from the front face alone, part B to the total reflectivity of incoherent contributions from front and back faces, and part C to the reflectivity resulting from coherent superposition of front and back faces (oscillations).

absorption in the low-energy wing, according an Urbach-law scheme analyzed in Section II.C.1.a. The analysis of the transmission spectra confirmed this interpretation.

This brief discussion, summarized in Fig. 2.9, indicates that, before any KK transformation, reflectivity spectra must be handled with great care in distinguishing three regions:

1. In region *A*, the bulk absorption is large enough to cancel any contribution from the back face, the total reflectivity is $R(\omega)$, and region *A* may be interpreted, in the surface-structure approximation, as the reflectivity of a semiinfinite perfect crystal.
2. In region *B*, back reflectivity contributes to the total reflectivity, but because of macroscopic defects, crystal shape, finite spectral resolution, etc., the front- and back-face reflection components are incoherent, and the reflectivities merely add to give the observed reflectivity. In the

absence of any bulk absorption, the reflectivity in region B amounts to $2R/(1 + R)$ and accounts for the very often observed very slow decay of the reflectivity in region B.

3. In region C, front- and back-face reflection components interfere to give rise to the well-known oscillations, or channeled spectrum of a slab.

To conclude, regions B and C may show absorption-induced structures, especially thermally activated absorptions (hot bands). The diminution of this activated absorption causes the transition from region A to B in Fig. 2.9. Region $B + C$ is a region of impurity, X-trap, or other spurious absorptions;[41] it is unusable for quantitative analysis of the exciton–phonon or polariton–phonon intrinsic relaxation mechanisms we investigate below. Therefore, our analysis will be concerned only with region A of the b- and a-polarized reflection spectra as the best candidates of a KK analysis.

c. *Two Candidates for a* KK *Analysis.* We discuss the absorption in the vicinity of the 0–0 transition in a- and b-polarized spectra, illustrated in

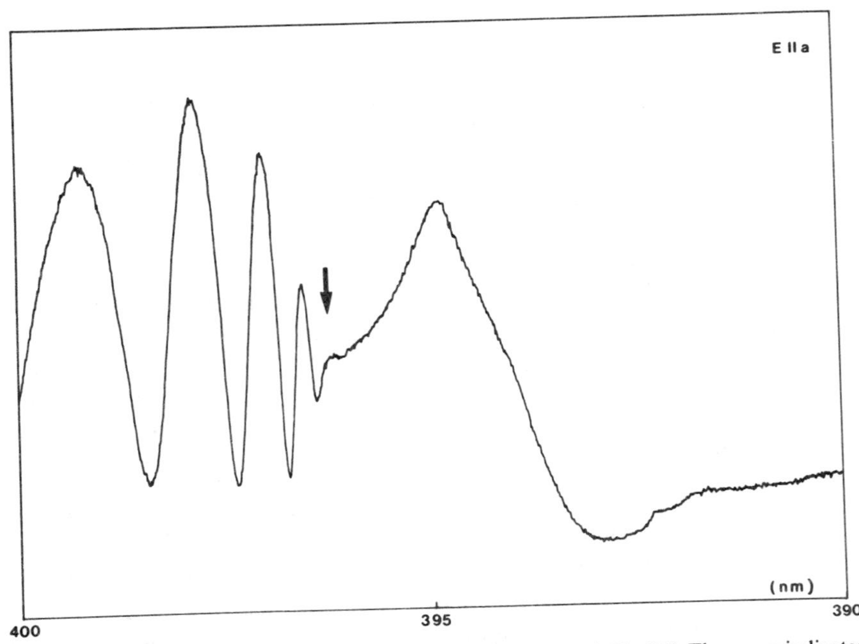

Figure 2.10. Detail of the 0–0 a-polarized reflectivity at 5 K (cf. Fig. 2.8). The arrow indicates the setup of coherent superposition of reflectivity amplitudes from front and back faces. This structure is located at 140 cm^{-1} above the bottom of the excitonic band; it indicates the threshold of $a \rightarrow b$ relaxation with creation of "140 cm^{-1}" B_g phonons and b excitons.

Figs. 2.9, 2.10, recorded at very low temperatures and in high resolution. For a crystal of very high quality, the broadening of these transitions is mainly due to phonons. We briefly analyze their influence on the reflectivity in order to assure that no spurious structures are considered. These spectra have been recorded for the first time and analyzed by Turlet et al.[1,67]. We simply summarize a few points necessary to test the KK transformation, to point out the specificity of the intrinsic relaxation mechanisms related to exciton–phonon couplings, and to evaluate quantitatively the corresponding coupling parameters.

b polarization: Comparison of the recorded spectrum (region A) of Fig. 2.9 with the square stopping band of the polariton of width 350 to 400 cm^{-1}[67] (see also Section I) suffices to show the importance and the variety of relaxation processes, involving strong acoustical phonon relaxation of the 0–0 transition and other mechanisms, including two-particle interactions (cascades of exciton and vibron fissions) as the excitation energy sweeps upwards the region A. Conversely, the fast transition of the reflectivity from $R(\omega)$ to $2R(\omega)/[1 + R(\omega)]$ reflects the very rapid decay of the phonon-assisted absorption coefficient when the excitation energy sweeps downwards from region A to B, obeying the Urbach law (2.110); see Fig. 2.9. As a general rule, the overall reflectivity broadens with increasing temperature, with a correlative decay of its maximum. However, the evolution of the 0–0 transition peak is negligible up to 15 K; then it accelerates suddenly above this temperature.

a polarization: This spectrum, illustrated in Fig. 2.10 shows a broad and weak reflectivity, which is the consequence of both the weak *a*-polarized transition and the final state of the 0–0 transition (located in the middle of the excitonic band), undergoing strong downwards relaxation, even at zero temperature. This relaxation is responsible for the observed large width, about 80 cm^{-1}, comparable to the value derived from absorption spectra.[62] Region A shows no significant structures, except a two-particle threshold related to the fission of an *a* vibron into one 392-cm^{-1} vibration and one *b* exciton. In contrast, the low-energy part (region $B + C$) shows a surprising sudden buildup of the oscillations far below the 0–0 transition, at the energy corresponding to $E_{vb}^{00} + 140\,cm^{-1}$ (Fig. 2.10, arrow). The interpretation of this kind of $B \rightarrow C$ transition is a fission threshold of the *a* exciton, the latter relaxing predominantly to the *b* exciton with creation of a 140-cm^{-1} B_g libration. Transmission spectra confirm this interpretation. As to the thermal behavior of the *a* reflectivity spectrum, it is comparable to that for the *b* polarization, compatible with the existence with an intrinsic instability of the *a* exciton.

d. *Phonon Broadening of the 0–0 Transition.* In order to discuss phonon broadening of the 0–0 (polariton-forbidden) transition in a quantitative way for the above two candidates, the complete complex function $\varepsilon(\omega)$,[69] instead of the reflection power, has to be determined. Moreover, $\varepsilon(\omega)$ allows a correct analysis of surface exciton structures.[91,92] However, experimental determination of $\varepsilon(\omega)$ is very difficult, since no light is transmitted, even by micrometer-thick crystals, in the 0–0 energy region, and since a direct measurement of the reflection dephasing angle (e.g. by ellipsometry) is rather involved in the anisotropic anthracene crystal. Thus, we were led to perform the following KK transformation on the reflection spectra, to get the phase shift upon reflection at temperatures between 1.6 and 100 K.

e. *The Principle of the* KK *Transformation.* It is well known[69] that the real and imaginary parts of the dielectric permittivity satisfy the KK relations, as direct consequences of causality and of the linearity of the dielectric response to a weak field. Analogous relations exist between the modulus and the phase of any linear susceptibility. Using the fact that the energy absorption by a physical system in a *positive-defined* function of the excitation frequency ω, it can be shown[69] that the susceptibility has neither zeros nor poles in the complex upper half plane of ω: see Fig. 2.11. If this susceptibility is, for instance, the reflectivity amplitude, then its logarithm, $\log r(\omega)$, is well defined, without cuts, in the entire half plane. Using the integration countour C in Fig. 2.11, one obtains

$$\oint_c \frac{\log r(\omega)}{\omega - \omega_0} d\omega = \int_\gamma \frac{\log r(\omega)}{\omega - \omega_0} d\omega + \fint_{-\infty}^{+\infty} \frac{\log r(\omega)}{\omega - \omega_0} d\omega - i\pi \log r(\omega_0) = 0 \tag{2.117a}$$

$$\oint_c \frac{\log r(\omega)}{\omega + \omega_0} d\omega = \int_\gamma \frac{\log r(\omega)}{\omega + \omega_0} d\omega + \fint_{-\infty}^{+\infty} \frac{\log r(\omega)}{\omega + \omega_0} d\omega - i\pi \log r(-\omega_0) = 0 \tag{2.117b}$$

Subtraction of the relations (2.117) eliminates the integration over γ (Jordan's lemma). Then, according the general relation[69] $r^*(-\omega) = r(\omega^*)$, with

$$r(\omega) = \sqrt{R(\omega)}\, e^{i\theta(\omega)} \tag{2.118}$$

we obtain

$$\theta(\omega_0) = \frac{-\omega_0}{\pi} \fint_0^{+\infty} \frac{\log R(\omega)}{\omega^2 - \omega_0^2} d\omega \tag{2.119}$$

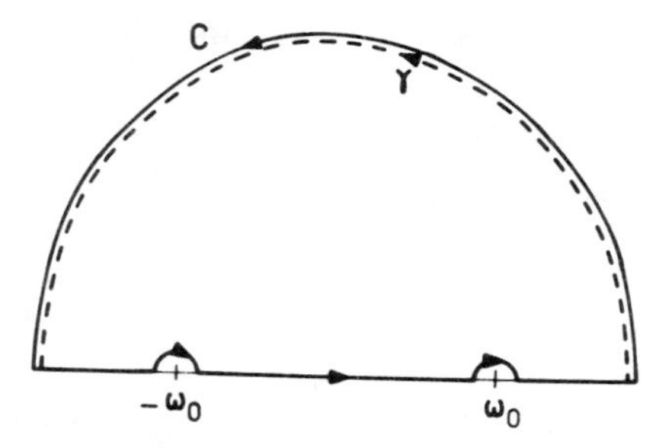

Figure 2.11. The integration contour in the complex plane allowing one to generalize the Kramers–Kronig relations to the modulus and phase of the reflectivity amplitude.

Switching to variables λ and suppressing the principal value in λ_0 with the use of relations $⨍_0^\infty d\lambda\,(\lambda^2 - \lambda_0^2) = 0$, we obtain

$$\theta(\lambda_0) = \frac{\lambda_0}{\pi} \int_0^{+\infty} \frac{\log[R(\lambda)/R(\lambda_0)]}{\lambda^2 - \lambda_0^2}\, d\lambda \tag{2.120}$$

f. *The Application of the* KK *Transformation to the* UV *Domain.* Writing $r(\lambda) = \sqrt{R(\lambda)}\, e^{i\theta(\lambda)}$, $r(\lambda)$ is completely determined from the knowledge of the whole reflectivity spectrum $R(\lambda)$.

To get the dielectric functions $\varepsilon(\lambda)$, since the geometrical shape of the crystal is not perfectly known, we ought to select experimental conditions so as to have the simplest possible relation between $\varepsilon(\lambda)$ and $r(\lambda)$, namely, the Fresnel formula for the normal reflection amplitude of a semi-infinite dielectric:

$$r(\lambda) = \frac{1 - |\varepsilon(\lambda)|^{1/2}}{1 + |\varepsilon(\lambda)|^{1/2}} \tag{2.121}$$

[in particular, spatial dispersion is neglected in (2.121)]. Melt-grown crystals are thick enough to allow the semiinfinite model to be applied. However, those crystals generally are of poor crystallographic quality[93] and give somewhat broadened reflectivity structures, even at low temperatures, which exclude a fine investigation of exciton–phonon coupling mechanisms. Conversely, vapor-grown crystals often are very clear, but are so thin (a few micrometers), that the back-face reflection is involved in the spectra whenever the crystal becomes transparent, thus hindering the use of (2.121). Thus, we chose to use a vapor-grown crystal in region A of the spectrum (Fig. 2.9), where only the front face contributes to R for $\lambda < \lambda_A$. Here λ_A depends on temperature and was taken as $\lambda_A = 3990$ Å at $T = 5$ K. The long-wavelength part of the spectrum (from 4000 to 5000 Å) was replaced by an analytical fit, using a very crude model for $\varepsilon(\lambda)$:

$$\varepsilon(\omega) = \varepsilon_0 - \frac{A}{\omega - \omega_0} \tag{2.122}$$

The parameters ε_0, A, λ_0 were adjusted to retain continuity with the actual spectrum used, and to fit the refractive-index variations in the visible region published by other authors.[94] At low temperature, we had

$$\begin{aligned} A &= 1170\,\mathrm{cm}^{-1} \\ \varepsilon_0 &= 3.46 \end{aligned} \tag{2.122'}$$

at $\omega_0 = 25\,095\,\mathrm{cm}^{-1}$. The corresponding polariton stopping band is 340 cm^{-1} wide.

Proceeding to the evaluation of the integral (2.120), we now look at further requirements on the raw spectra. First, from equation (2.120) we see that the most extended available spectrum should be used. Our reflectivity data ranged from 3000 to 4000 Å, so they encompassed the main vibronic bands[91] (0–0, 0–1, 0–2) of the first singlet (S_1) system. Unfortunately, experimental limitations[61] made it impossible to include in the spectrum the very strong second singlet S_2 (around 25000 Å). In the spectral region of interest, where $\varepsilon(\lambda)$ was sought, and where the sharpest structures lay, we recorded[61] high-resolution (0.1 Å) spectra, typically from 3500 to 4000 Å. The rest of the data, at higher energies (from 3000 to 3500 Å) were recorded with 1-Å resolution to save computation time. The reflectivity was then normalized, using a numerically smoothed mirror spectrum as reference; thereby, only the absolute reflectivity at one wavelength (i.e. at the maximum) remained to be determined (see discussion below).

Next, the relation (2.121) only pertains to a homogeneous bulk dielectric. Therefore, the surface structures of the anthracene crystal, though leaving (2.120) valid, are an obstacle to the calculation of the bulk permittivity.[91] As suggested in ref. 91, surface structures I, II, and (whenever it appeared) III were eliminated. The ersatz spectrum, in the intervals surrounding structures I and II (+ III), consisted of a cubic function adjusted so as to preserve the continuity of the spectrum and of its first derivative at both ends of the interval. Such a procedure is justified, first by the theoretical simulation of surface structures,[91,92] and second because such a structureless shape of the bulk reflectivity manifests itself when the surface structures are either shifted (e.g. under gas coating) or destroyed (by photodimerization, or in samples lacking surface structure).[91,120] However, as structures II and III are so close to the resonance, the final transformed spectrum in the resonance region sometimes depended strongly on the particular interpolation we used. We shall come back to this point in the discussion of the results.

The KK transformation itself was carried out in two steps. First, the integral in (2.120) was computed using the available data (the trapezoidal rule was adopted, since it assigns the same weight to all points, thus minimizing the noise). The contributions from the missing parts of the spectrum are taken into account through the extrapolation function $E(\lambda)$, so that

$$\theta(\lambda_0) = \frac{\lambda_0}{\pi} \int_{\text{data}} \frac{\log[R(\lambda)/R(\lambda_0)]}{\lambda^2 - \lambda_0^2} d\lambda + E(\lambda_0) \tag{2.123}$$

The extrapolations $E(\lambda)$, originating from remote spectral regions, should vary slowly in the narrow interval around the resonance, where $\theta(\lambda)$ is sought. Rather than attempting to evaluate $E(\lambda)$ analytically on the grounds of the asymptotic variations of $R(\lambda)$ for $\lambda \to 0$ or $\lambda \to \infty$, we chose to parametrize $E(\lambda)$ itself in the form

$$E(\lambda) = A\lambda + B + \frac{C}{\lambda} \tag{2.124}$$

which is asymptotically valid for a very extended spectrum of data. The parameters A, B, C were adjusted by least-squares fitting, so as to cancel the complete phase shift $\theta(\lambda)$ in a transparency region of the crystal, *where θ is known to be zero* (typically between 4100 and 4300 Å at low temperatures, and in a more red-shifted interval at higher temperatures). Finally, in the second step of the transformation, θ was calculated in the region studied (here from 3900 to 4000 Å) using equation (2.124) for $E(\lambda)$. This method was tested using analytical reflectivity spectra: even in the presence of strong bands outside the data spectrum, discrepancies from the expected permittivity did not exceed 10% in relative value.[70]

Once the complex reflectivity amplitude is known, the relation (2.121) determines $\varepsilon(\omega)$. Nevertheless, the absolute reflectivity, unimportant as regards $\theta(\omega)$ [see (2.120)] becomes essential for the permittivity. Earlier work on a number of crystal samples showed us that the maximum reflectivity at 4 K was greater than 90%.[1] However, due to free sample mounting, the front face of the crystal is not perfectly planar, and accurate direct measurements of the absolute reflectivity are impossible. Fortunately, surface structures II and III allow probing the bulk reflectivity around 3982 Å: High-resolution spectra ($0.3\ \text{cm}^{-1}$) of structure II (cf. Section III) show the absence of any constructive intereference. This, together with numerical simulations,[121] indicates that the bulk reflectivity should be very close to 100% (within 2%) at the maximum.

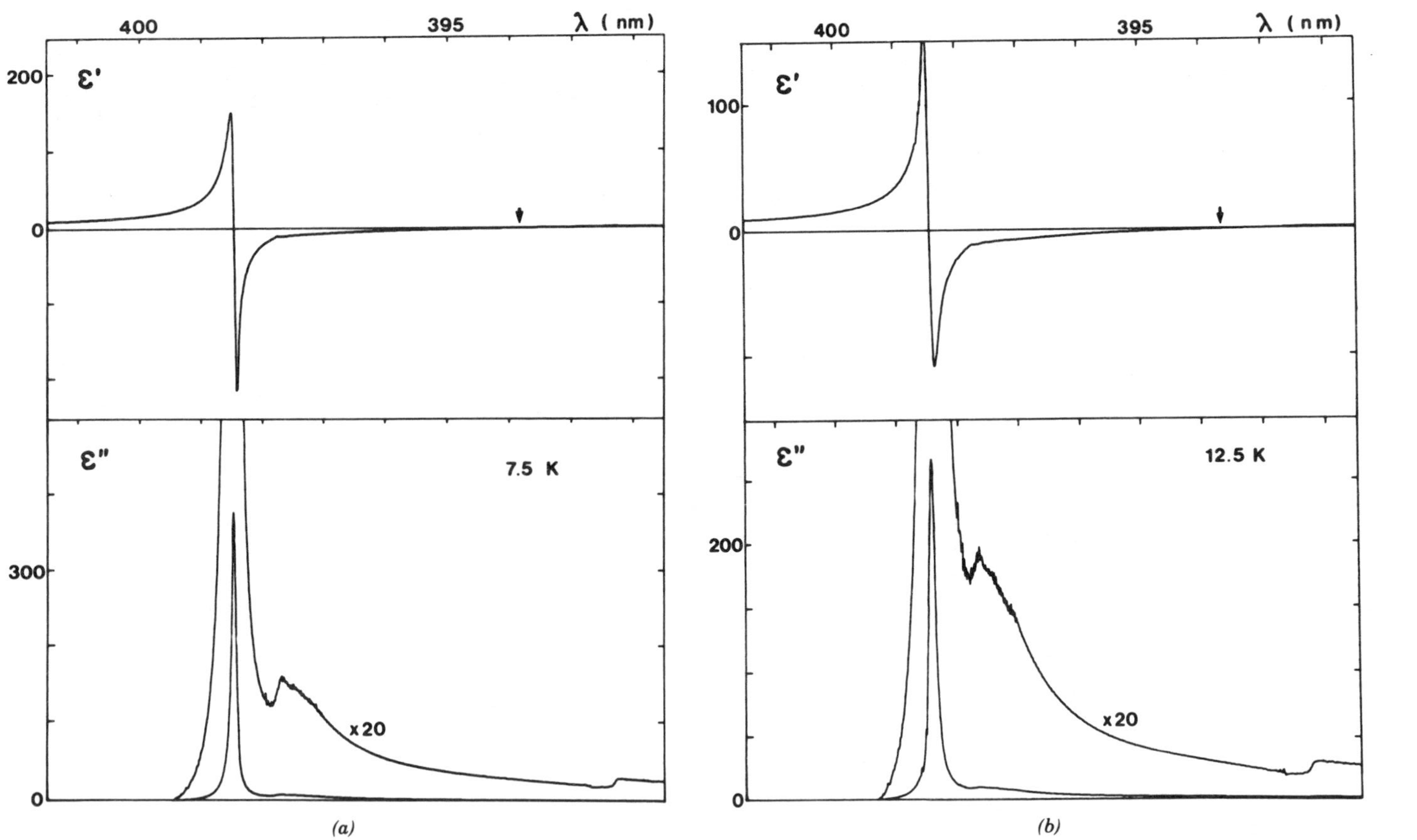

Figure 2.12. Real and imaginary parts of the dielectric permittivity $\varepsilon(\omega)$ around the b-polarized 0–0 transition, obtained from Kramers–Kronig analysis of reflectivity spectra at temperature ranging from 7 to 77 K. The arrow on the ε' curves indicates the point where $\varepsilon' = 0$. We note the stepwise threshold of the "46-cm^{-1}" phonon sideband (a). At higher temperatures (b–h) it broadens to give the smooth asymmetrical absorption curve at 50 K (g).

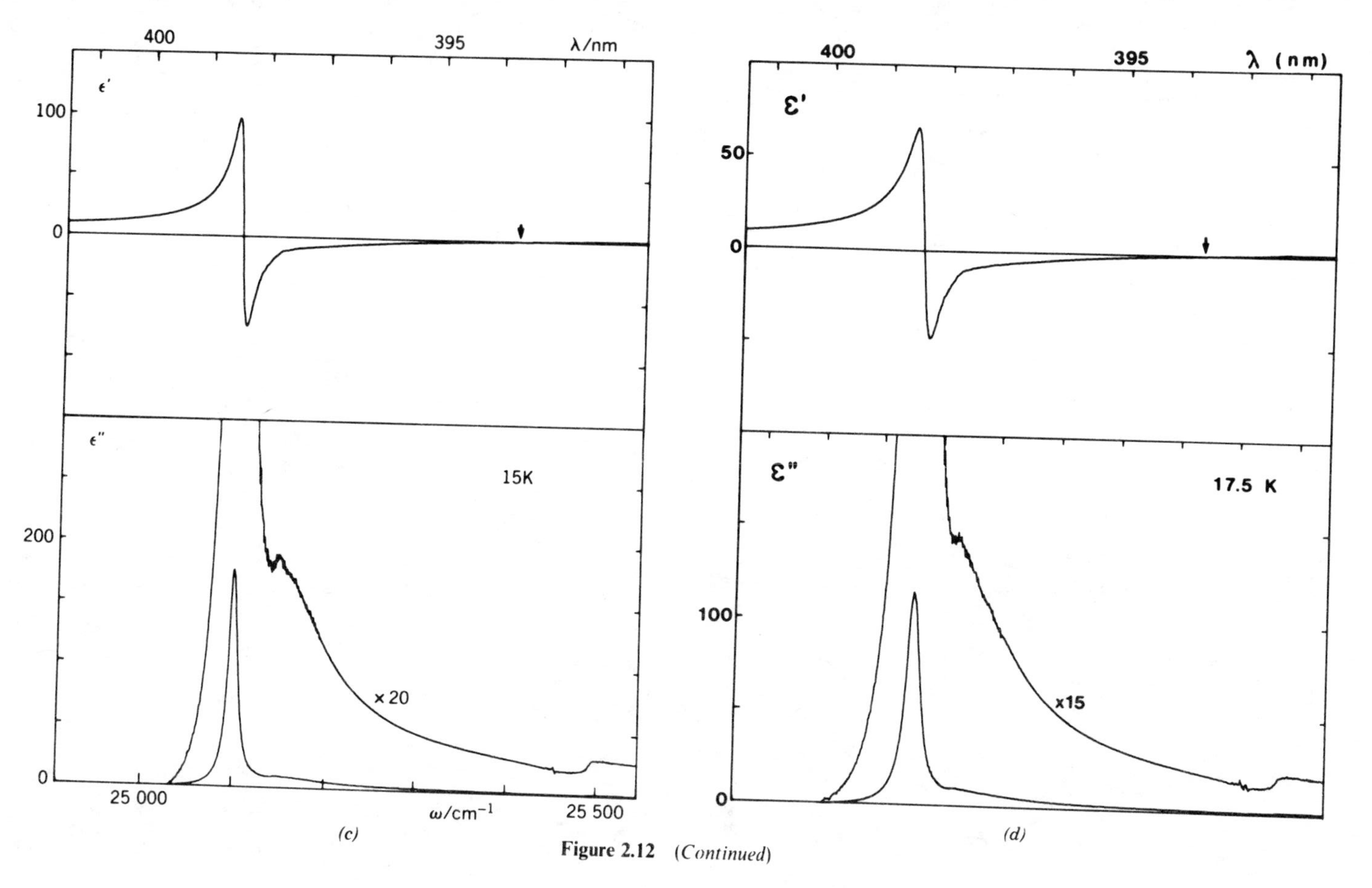

(c) (d)

Figure 2.12 (Continued)

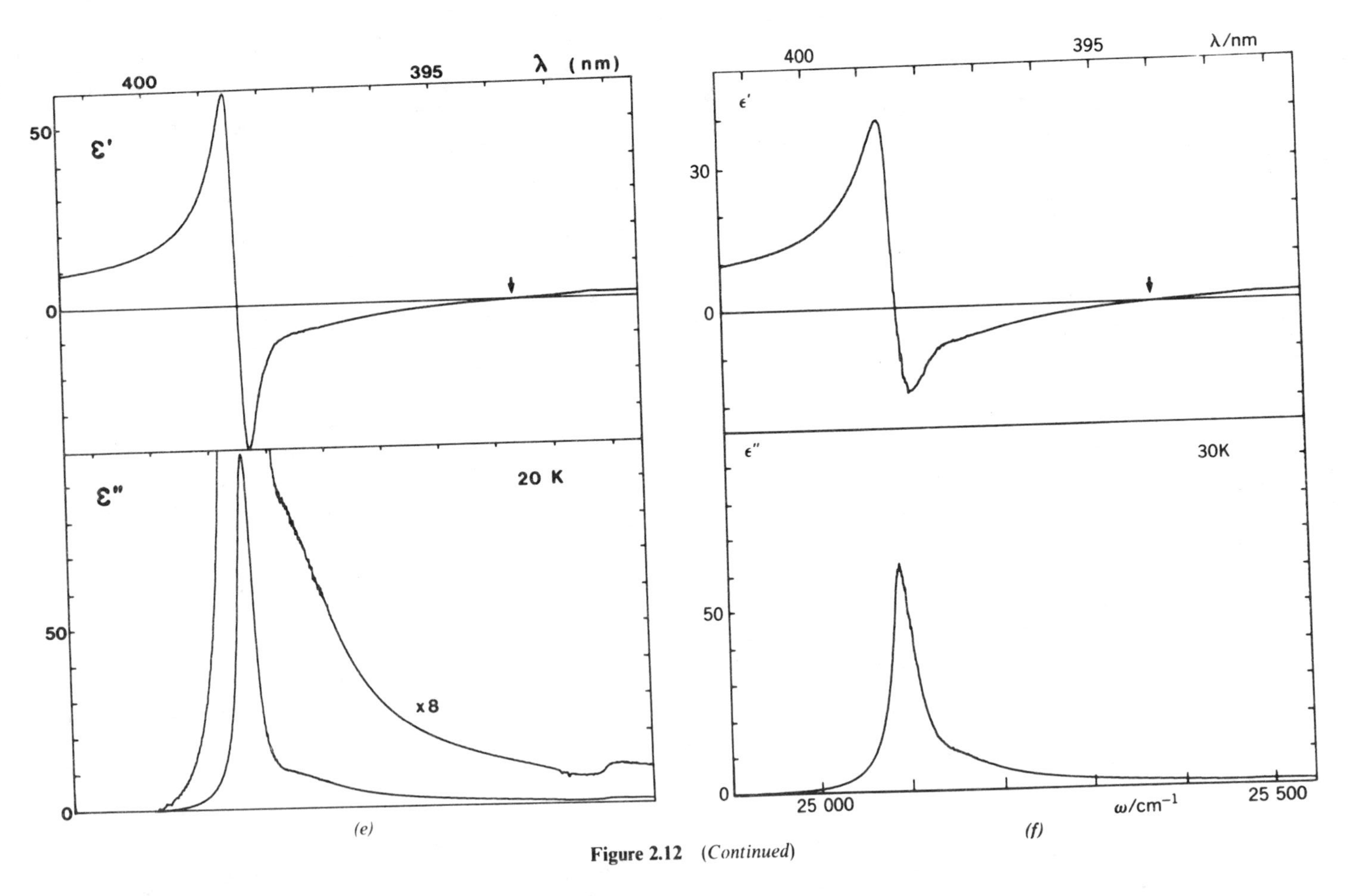

Figure 2.12 *(Continued)*

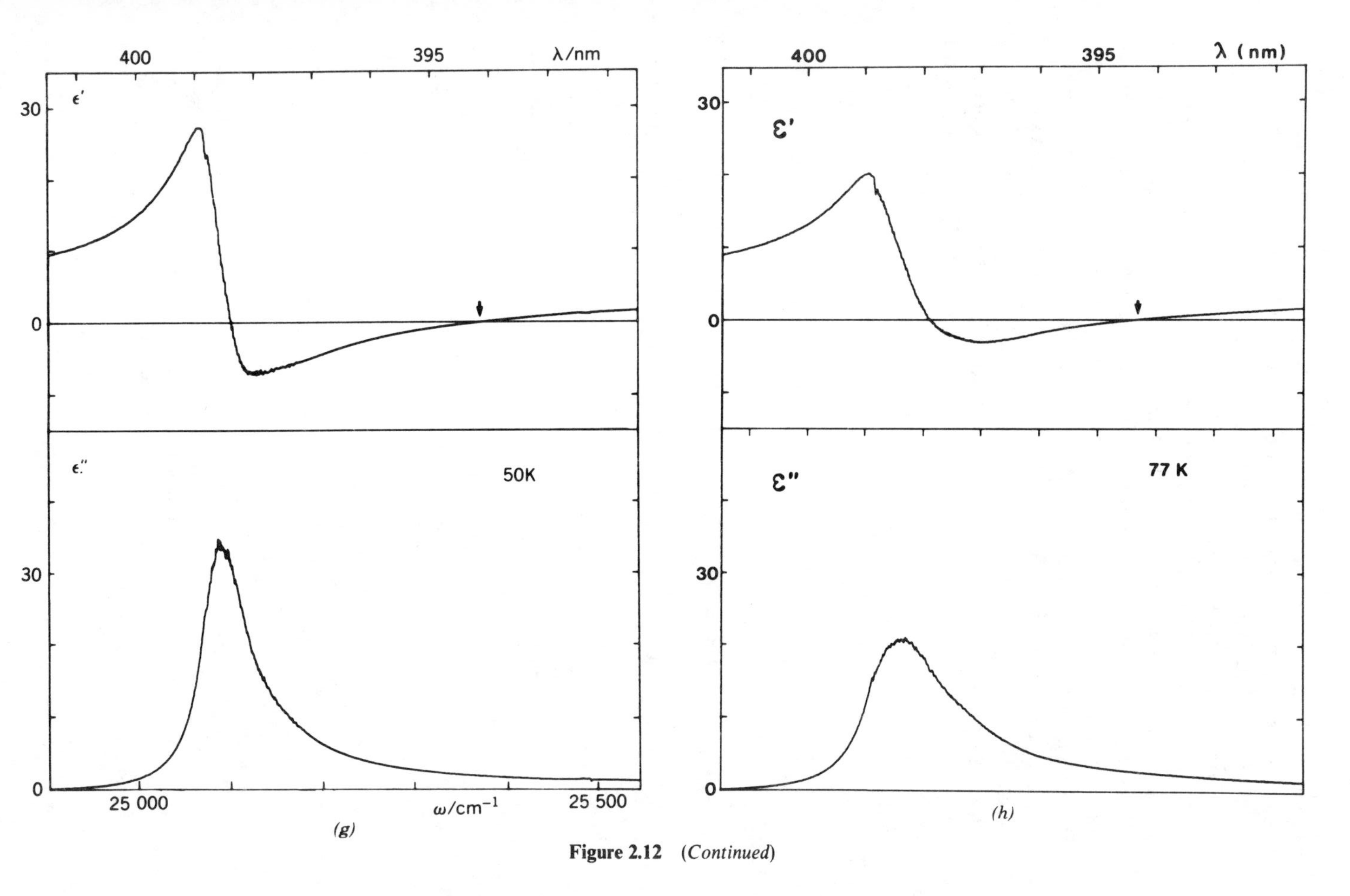

Figure 2.12 (*Continued*)

Thus, we assigned to the maximum reflectivity at 1.7 K a value of 98%, and scaled other spectra accordingly at different temperatures.

Figure 2.12 presents the real and imaginary parts of $\varepsilon(\lambda) = \varepsilon'(\lambda) + i\varepsilon''(\lambda)$ at typical temperatures $5 < T < 77$ K. The main structure is the 0–0 absorption (ε'') peak (3985 Å at 5 K). Below 10 K, the half-maximum residual width of the peak, $\gamma_r \sim 5\ \mathrm{cm}^{-1}$, seems to be chiefly fixed by the experimental ($1\ \mathrm{cm}^{-1}$) and "numerical" (1–$2\ \mathrm{cm}^{-1}$) resolution. Thus we were not able to reach the ultimate width of the 0–0 transition, $1\ \mathrm{cm}^{-1}$ (broadened, according to Galanin et al.[85], by the relaxation to polariton states). The asymmetrical shape of the peak appears in the magnified spectrum, with a pronounced phonon sideband $46\ \mathrm{cm}^{-1}$ above the main peak. At still higher energies the absorption decreases, though slower than in a lorentzian wing, to the 390-cm^{-1} intramolecular-vibration threshold. Making use of the simple model (2.122), we introduce a

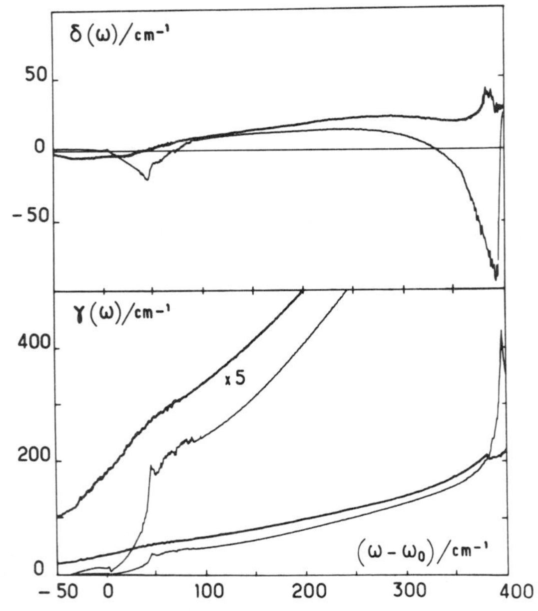

Figure 2.13. Shift function $\delta(\omega)$ and damping function $\gamma(\omega)$ of the b exciton at two limit temperatures: 5 K (light lines) and 50 K (bold lines), calculated from (2.125) and $\varepsilon(\omega)$ KK spectra. At 5 K, $\gamma(\omega)$ grows smoothly in the acoustical-phonon region (between ω_0 and the "46-cm^{-1}" threshold), then presents a sudden increase at the threshold. The parameters of the model (2.127)–(2.130) were fitted using this function $\gamma(\omega)$. At higher temperatures ($T > 50$ K), a smoothing of the phonon structures in $\gamma(\omega)$ and $\delta(\omega)$ is observed, together with a general growth of $\gamma(\omega)$ across the investigated spectral region.

renormalized expression for the excitonic energy and set

$$\varepsilon(\omega) = \varepsilon_0 - \frac{A}{\omega - \omega_0 - \delta(\omega) - \frac{1}{2}i\gamma(\omega)} \tag{2.125}$$

$\Sigma(\omega) = \delta(\omega) + \frac{1}{2}i\gamma(\omega)$ is the excitonic self-energy and satisfies KK relations; ε_0 and A assume the previous values. The derived variations of $\delta(\omega)$ and $\gamma(\omega)$ are presented in Fig. 2.13. We notice that $\delta(\omega)$, though fairly weak throughout the 0–0 region, reaches values of $\sim 20\,\mathrm{cm}^{-1}$ around the phonon sideband, and should therefore be included in quantitative estimations of $\gamma(\omega)$.[95] We find that $\gamma(\omega)$ increases between ω_0 and $\omega_0 + 46\,\mathrm{cm}^{-1}$, crosses a stepwise threshold at $\omega_0 + 46\,\mathrm{cm}^{-1}$, and then steadily grows to $\omega_0 + 394\,\mathrm{cm}^{-1}$, where a vibronic study[56] would be better suited than the parametrization (2.125).

At higher temperatures, all structures rapidly broaden. The 0–0 absorption peak becomes markedly asymmetric above 20 K, with its high-energy wing being nearly twice as broad as the low-energy wing at 77 K. In addition, the 46-cm^{-1} phonon sideband readily spreads into the high-energy wing, contributing to its broadening. The half-maximum width γ_0 is determined by the variations of $\gamma(\omega)$ and $\delta(\omega)$ in the region around the optical resonance, and cannot be simply related to the function $\gamma(\omega)$. The width γ_0 is plotted in Fig. 2.14 against temperature, together with high-temperature widths from Morris and Sceats's work.[96,62] Whereas the general trend is a linear variation with T, above 200 K, a sublinear character appears when the width becomes appreciable with respect to the excitonic bandwidth,[70] and we may suspect a change to a $T^{1/2}$ law. In the temperature range $0 < T < 150\,\mathrm{K}$, the experimental points roughly fit a phonon-average-number law, namely

$$\gamma_0(T) = \gamma_r + \gamma_1[\exp(\hbar\Omega/k_B T) - 1]^{-1} \tag{2.126}$$

with effective phonon energy $\hbar\Omega = 27\,\mathrm{cm}^{-1}$, $\gamma_1 = 72\,\mathrm{cm}^{-1}$, and $\gamma_r \sim 5\,\mathrm{cm}^{-1}$ for the residual width discussed above. The rather low value of the effective phonon energy stresses the importance of low-frequency acoustic phonons ($\hbar\Omega < 46\,\mathrm{cm}^{-1}$) in the broadening of the 0–0 peak. Two points at 17.5 and 20 K lie well above the phenomenological $\gamma_0(T)$ curve, outside the expected inaccuracy limits ($\pm 15\%$). These aberrant points are presumably due to an ill-adapted replacement of the broadened surface structure II, resulting in an artificial broadening of the 0–0 transition. Lastly, another important parameter of the absorption curve is the peak position, but since its evolution with temperature[96] is complicated by the thermal expansion of the crystal, it will be disregarded hereafter.

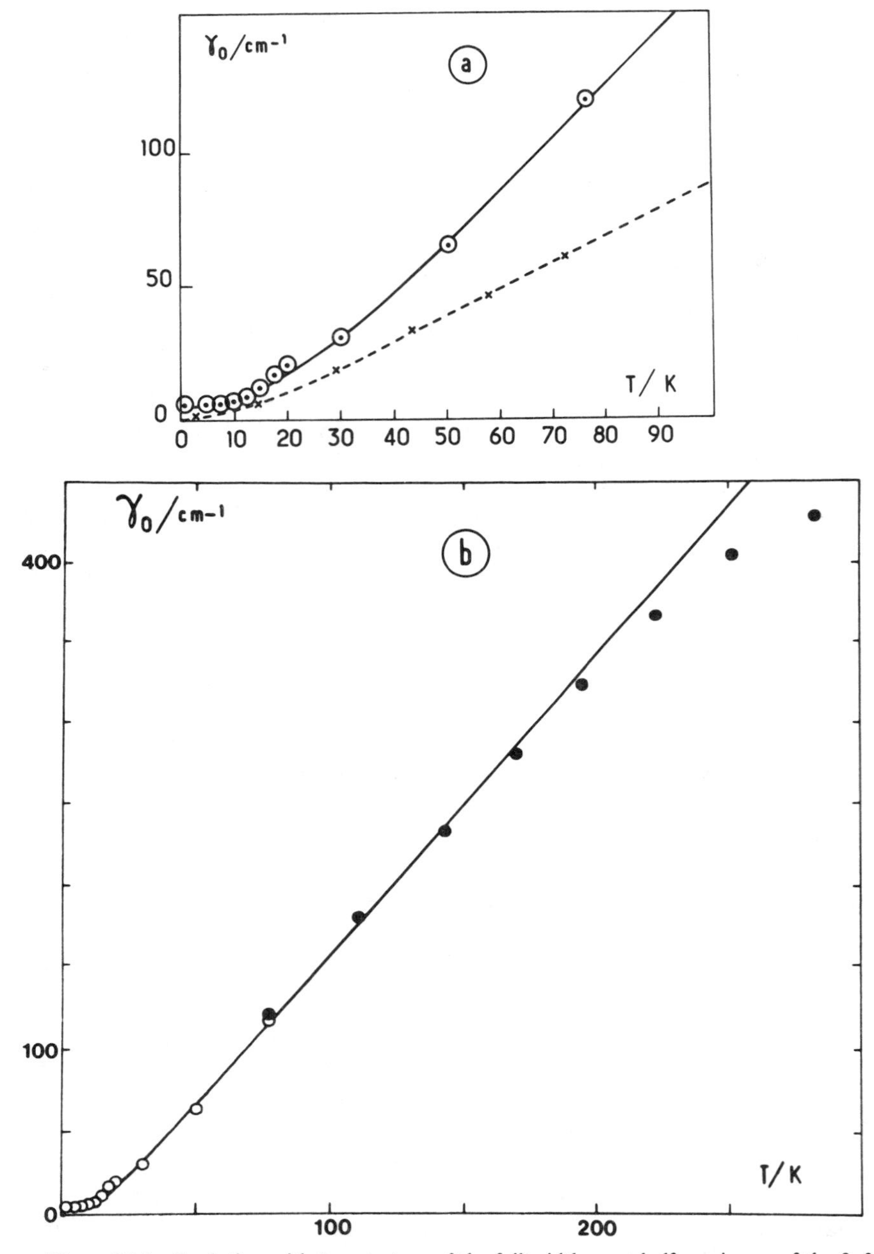

Figure 2.14. Evolution with temperature of the full width γ_0 at half maximum of the 0–0 absorption peak. Hollow circles represent our results from Kramers–Kronig analysis. (*a*) Evolution between 0 and 77 K. The solid line was drawn using equation (2.126) and adjusted parameters $\gamma_1 = 72\,\mathrm{cm}^{-1}$, $h\Omega = 27\,\mathrm{cm}^{-1}$. The dashed line connects the results of our model (2.127)–(2.130) for six different temperatures. (*b*) Evolution between 0 and 300 K. The full circles are taken from ref. 62. This summary of the experimental results shows the linear behavior between 30 and 50 K, and the sublinear curvature at temperatures above 200 K.

3. *Theory and Discussion*

The basic problem of coupling an exciton with one or several phonon branches is summarized in the following second-quantized hamiltonian:

$$H = \sum_{\mathbf{k}} \hbar\omega_{\mathbf{k}} B_{\mathbf{k}}^{\dagger} B_{\mathbf{k}} + \sum_{s\mathbf{q}} \hbar\Omega_s(\mathbf{q}) b_{\mathbf{q}s}^{\dagger} b_{\mathbf{q}s} + \sum_{ns\mathbf{q}} \chi_s(q) B_n^{\dagger} B_n (b_{\mathbf{q}s}^{\dagger} + b_{\mathbf{q}s}) \qquad (2.127)$$

where s is the phonon branch index, and $\mathbf{k}$ ($\mathbf{q}$) and $B_{\mathbf{k}}^{\dagger}$ ($b_{\mathbf{q}}^{\dagger}$) are the exciton (phonon) wave vector and creation operator. The exciton–phonon interaction is assumed to be linear and local, with strength $\chi_s(\mathbf{q})$ at every site n. We may limit the coupling to the linear term, since at low temperatures molecular displacements are weak. The reliability of the local-coupling approximation is less clear; it may prove inadequate, for instance in transport problems. Nevertheless, local coupling is a convenient first approximation to exciton–phonon coupling. It originates mainly from the dependence of the excitonic D term on phonon coordinates.[97] Therefore, the order of magnitude of $\chi_s(q)$ is $N^{-1/2}(\partial D/\partial R)\Delta R$, where ΔR is the rms amplitude of the mode.

The dynamics underlying the hamiltonian (2.127) is quite complex, because of the competition between the localization- and delocalization-favoring parts of H, according to temperature. The dynamics of the system is roughly characterized by a few energy parameters:

1. B, the excitonic half bandwidth;
2. S, the stabilization of a localized excitation under lattice relaxation;[66,52]

$$S = \sum_{s\mathbf{q}} \frac{|\chi_s(\mathbf{q})|^2}{\hbar\Omega_s(\mathbf{q})} \qquad (2.128)$$

3. $\hbar\Omega_s$, a typical energy of the phonons, determining the extent of their quantum behavior;
4. $\Delta = (2Sk_BT)^{1/2}$, which measures[53] at high temperatures ($k_BT \gg \hbar\Omega_s$) the energy fluctuation of a localized exciton due to thermal fluctuations of the phonon bath.

As regards dynamics at long times, and for $\hbar\Omega_s \ll B, S$, the most stable state of the system behaves unequivocally either like a free exciton (for $B > S$) or like a self-trapped exciton (for $B < S$).[98] This transition smooths when $\hbar\Omega_s$ becomes comparable with S.[99] Furthermore, the photon absorption, with which we are dealing here, undergoes a continuous change, as pointed out by Toyozawa and coworkers,[52,66,98] from a lorentzian lineshape (weak scattering, $B > \Delta$) to a gaussian lineshape when molecules absorb individually (strong scattering, $B < \Delta$). For $\hbar\Omega_s \sim B$, S, vibrations behave quantum-mechanically, producing vibronic structures in absorption spectra. This is the case of intramolecular vibrations at 390 and 1400 cm^{-1} in the anthracene crystal, where the 0–0 transition is far enough from other vibronic structures to be studied separately. The absence of a Stokes shift in fluorescence spectra shows that the 0–0 anthracene exciton is free.[1] Moreover, the 0–0 exciton half bandwidth is $\sim 200\,cm^{-1}$, and the temperature at which the 0–0 broadening becomes of the order of B is 200 K. Therefore, in our study between 0 and 77 K, we restrict ourselves to weak exciton scattering as well as weak exciton–phonon coupling.[98]

a. *Contribution of Acoustical and Optical Phonons.* Since the adiabatic, or quasi-classical, model of phonons[52,98] cannot account for the phonon sidebands at low temperatures (Fig. 2.12), we tentatively used perturbation theory on the hamiltonian (2.127). The free-exciton picture leads to the excitonic self-energy $\Sigma(k, E)$. By renormalization of the exciton propagator only, a self-consistent functional relation is found[12] for Σ:

$$\begin{aligned}\Sigma(k, E) = \sum_{s,q} \{ & |\chi_s(q)|^2 n_s(q) \\ & \times [E - E_{k+q} + \hbar\Omega_s(q) - \Sigma(k+q, E + \hbar\Omega_s(q))]^{-1} \\ & + |\chi_s(q)|^2 [1 + n_s(q)] \\ & \times [E - E_{k+q} - \hbar\Omega_s(q) - \Sigma(k+q, E - \hbar\Omega_s(q))]^{-1} \} \end{aligned} \qquad (2.129)$$

Equation (2.129) sums an infinite number of diagrams, but still misses fourth-order terms. To solve it numerically we adopted the following model, which is a simplified version of that of Davydov and Myasnikov[100]:

1. Exciton dispersion: As the 0–0 exciton dispersion in anthracene is somewhat complex,[101] we assumed a bidimensional parabolic dispersion (effective-mass approximation). This assumption is justified by the absence of interplane coupling for **k** perpendicular to plane (001) planes[29]—the case of *b*-polarized exciton under study—and it is aimed to give back the observed stepwise thresholds bound to the bottom of the band. The model most likely fails for the exciton band as a whole, but anyhow, we could not avoid it without lengthening computation times excessively.

2. Phonons: At least two phonon branches are involved in the observed absorption: the acoustic phonons and the optical 46-cm^{-1} branch. Our model includes a single acoustic branch [with cutoff frequency Ω_{max}, and isotropic Debye dispersion $\hbar\Omega_{ac}(q) = \hbar\Omega_{max} q/q_{max}$] and an optical dispersionless branch (Einstein's model, with frequency Ω_{op}).

3. Exciton–phonon coupling: The wave-vector dependence of $\chi_{op}(q)$ was

neglected, whereas $\chi_{ac}(q)$ was taken to be proportional to $q^{1/2}$ across the acoustic branch:

$$\chi_{op}(q) = N^{-1/2}\chi_{op}$$
$$\chi_{ac}(q) = N^{-1/2}\chi_{ac}(q/q_{max})^{1/2} \tag{2.130a}$$

When solving equation (2.129) we assumed, like Davydov and Myasnikov,[100] the self-energy Σ to be independent of k; but unlike those authors, we included the complete self-energy $\Sigma_{op} + \Sigma_{ac}$ in the iteration process. The calculation was carried out on a spectrum of 1500 complex Σ values (resolution 1 cm^{-1}), which was recalculated at each step. Satisfactory convergence was reached within ten iterations or less. The spectrum of $\gamma(\omega) = 2\,\mathrm{Im}\,|\Sigma(\omega)|$ is presented in Fig. 2.15 with the corresponding absorption spectra (2.125) at different temperatures. The model was completely determined by fitting the parameters χ_{ac} and χ_{op} to the lowest-temperature spectrum (3 K). Then we obtained

$$\chi_{ac} = 70\,\mathrm{cm}^{-1}, \qquad \chi_{op} = 50\,\mathrm{cm}^{-1} \tag{2.130b}$$

From the roughness of the model and fit, we estimate the inaccuracy of these values to be $\sim 20\%$.

On the other hand, the values (2.130b) may be roughly estimated for translational and rotational modes using (2.23). If R and θ are the characteristic coordinates for the molecular displacements and rotations, then the derivatives of the local interactions $D(R, \theta)$, with respect to R and to θ, are of the order

$$\frac{\partial D}{\partial R} \sim 6\frac{D}{R}, \qquad \frac{\partial D}{\partial \theta} \sim (\tan\theta)D \tag{2.130c}$$

(since D varies as R^{-6} and on angles of the order of $\pi/2$). The coupling parameter χ is related to the zero-point fluctuations of the molecular positions and angles, R and θ, which are given by Pawley[75]:

$$\Delta R \sim 3 \times 10^{-2}\,\text{Å}, \qquad \Delta\theta \sim 2 \times 10^{-2} \tag{2.130d}$$

This leads, with $D \sim 2500\,\mathrm{cm}^{-1}$ for the first singlet state, to

$$\bar{\chi}_{translation} \sim \frac{\partial D}{\partial R}\Delta R \sim 75\,\mathrm{cm}^{-1}$$
$$\bar{\chi}_{rotation} \sim \frac{\partial D}{\partial \theta}\Delta\theta \sim 50\,\mathrm{cm}^{-1} \tag{2.130e}$$

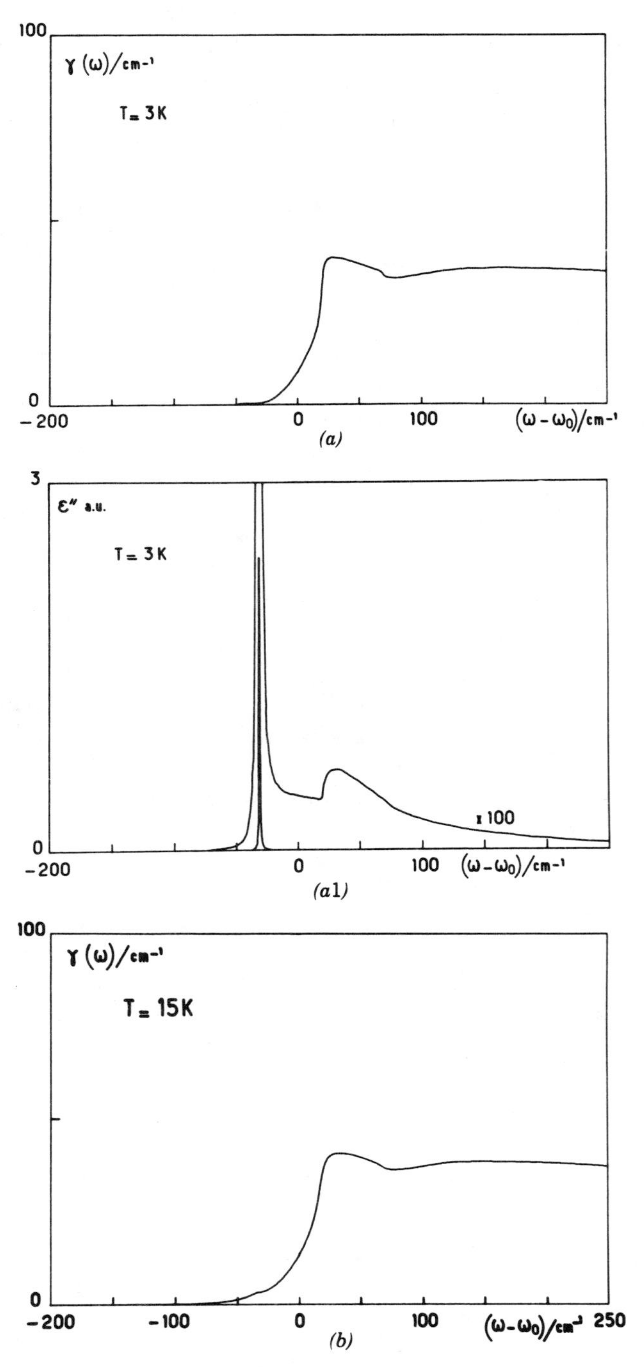
100
γ (ω)/cm⁻¹
T = 3K
0
−200
0
100
(ω − ω₀)/cm⁻¹
(a)
3
ε″ a.u.
T = 3K
x 100
0
−200
0
100
(ω − ω₀)/cm⁻¹
(a1)
100
γ (ω)/cm⁻¹
T = 15K
0
−200
−100
0
100
(ω − ω₀)/cm⁻¹
250
(b)

Figure 2.15

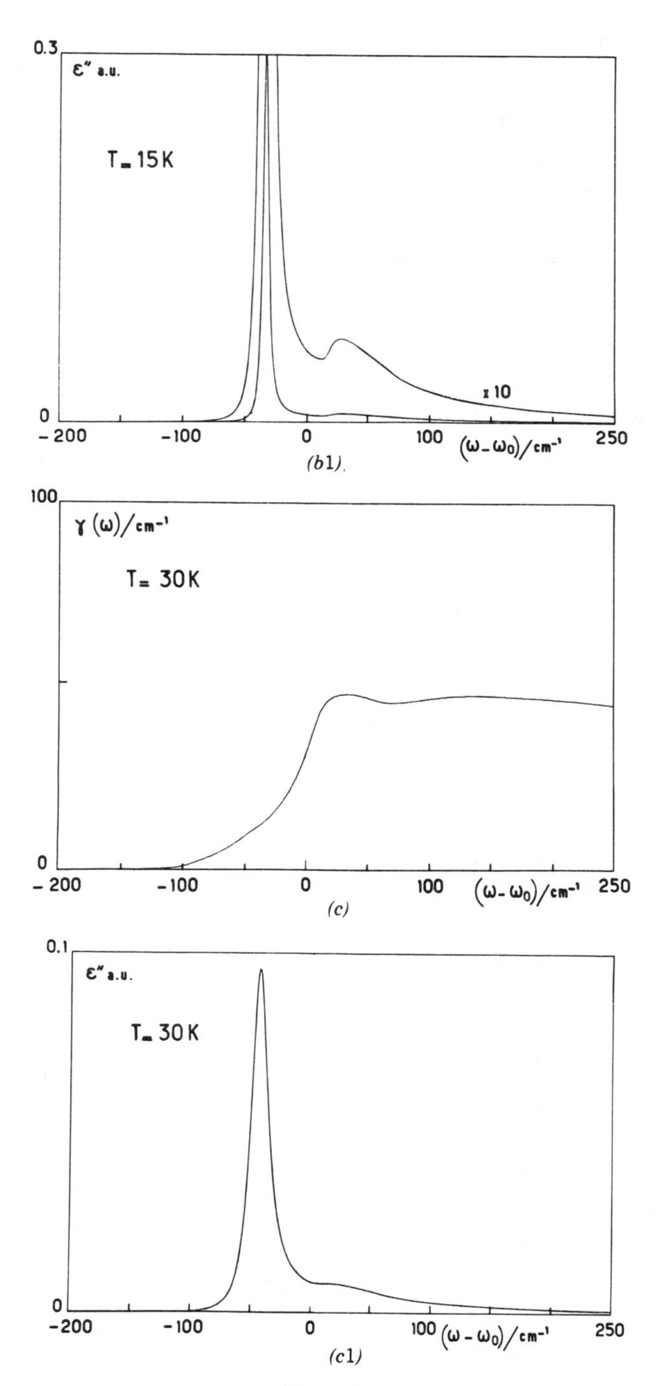
0.3
ε" a.u.
T= 15K
x 10
0
- 200
-100
0
100
(ω - ω₀)/cm⁻¹
250
(b1)
100
γ (ω)/cm⁻¹
T= 30K
0
- 200
-100
0
100
(ω - ω₀)/cm⁻¹
250
(c)
0.1
ε" a.u.
T= 30K
0
-200
-100
0
100
(ω - ω₀)/cm⁻¹
250
(c1)

Figure 2.15

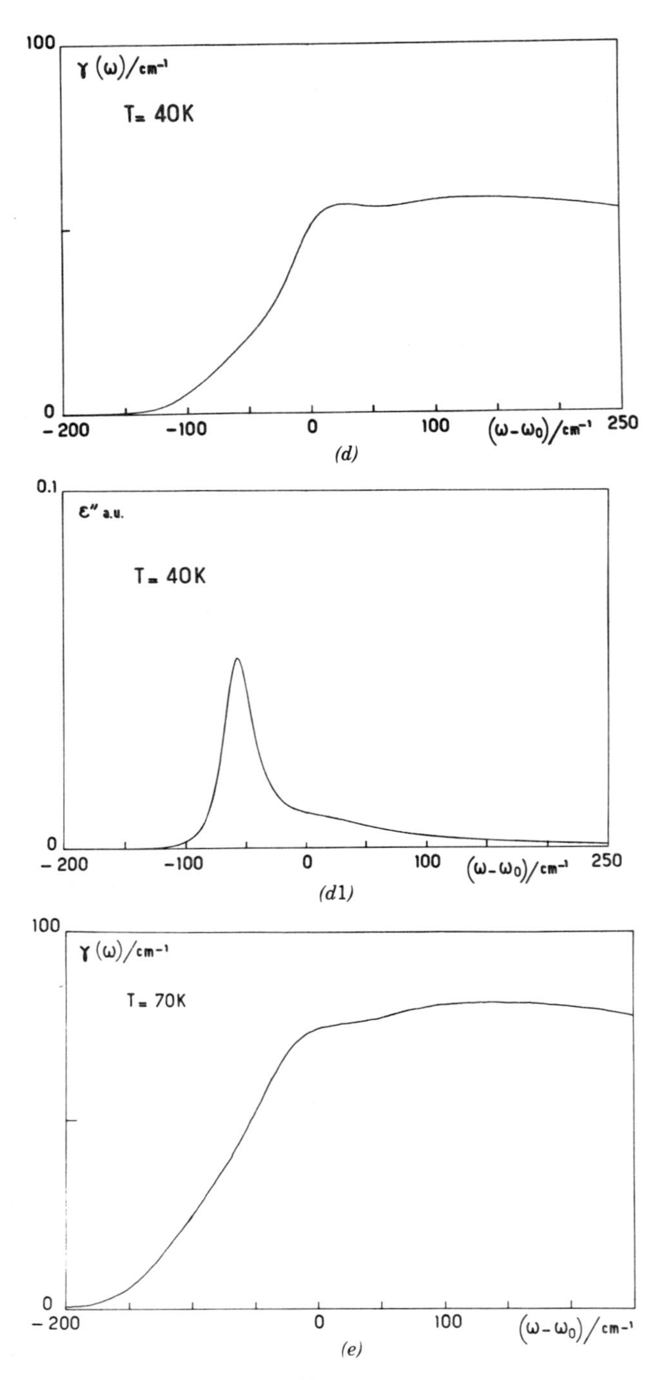
100
γ(ω)/cm⁻¹
T= 40K
0
- 200
-100
0
100
(ω-ω0)/cm⁻¹
250
(d)
0.1
ε″ a.u.
T= 40K
0
- 200
-100
0
100
(ω-ω0)/cm⁻¹
250
(d1)
100
γ(ω)/cm⁻¹
T= 70K
0
- 200
0
100
(ω-ω0)/cm⁻¹
(e)

Figure 2.15

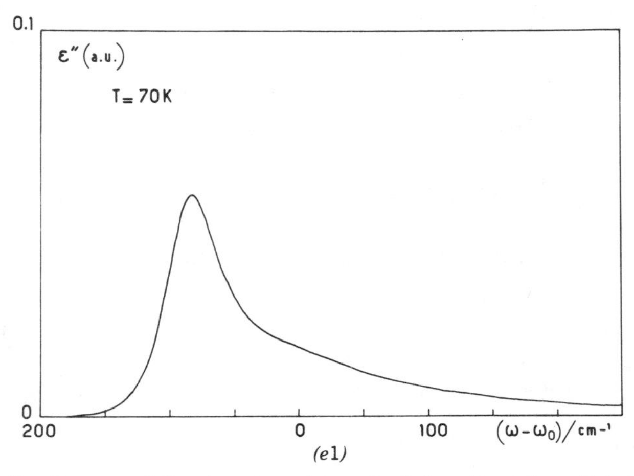

Figure 2.15. Spectra of the damping $\gamma(\omega)$ and of the absorption $\varepsilon''(\omega)$ (with arbitrary absorption units) derived from our model (2.127)–(2.130), at temperatures ranging between 3 and 70 K. At low temperatures we may distinguish in the spectrum pure acoustical effects below the threshold, and combined effects (acoustical + optical) above the threshold. At high temperatures, both branches contribute to yield the broad and asymmetrical lineshape. The energy origin has been chosen at the unperturbed exciton band botton. So the absorption spectra show a red shift even at low temperatures, which should be considered when comparing the model with the experimental spectra of Figs. 2.12–13.

We note particularly the values of $\bar{\chi}_r$ and $\bar{\chi}_t$ for the optical phonons and for the acoustical phonons on the boundary of the zone are of the same order [in contrast, for the acoustical phonons with large wavelengths, (2.23) shows that $\chi_{ac}(q) \sim_{q\to 0} q^{1/2}$]. In addition, the estimated values (2.130e) are in very good agreement with the experimental values (2.130b). Qualitative agreement is also found with the results of Ref. 102 on the (201) crystal face.

The model shows some satisfactory features: The smooth increase in $\gamma(\omega)$ between ω_0 and $\omega_0 + 50\,\text{cm}^{-1}$ agrees with experimental data and strengthens the interpretation of this scattering as due to acoustical phonons.[127] The general shape of the absorption curve and the T dependence of the peak width (cf. Fig. 2.12) are well reproduced, and the exponential drop of the absorption on the low-energy wing reminds us of the Urbach–Martienssen rule.[52,53] The phonon threshold at $\omega_0 + 50\,\text{cm}^{-1}$, though sharp at low temperatures, broadens and even disappears above 50 K (in contrast with the sideband in ref. 100). Yet, serious shortcomings remain:

1. The optical-phonon threshold at low temperature is not as sharp as experimental spectra show. The bidimensional exciton model does not account for the specific dispersions of excitons and phonons in anthracene.

2. Further increase of $\gamma(\omega)$ above $\omega_0 + 50\,\text{cm}^{-1}$ is a characteristic feature of the anthracene absorption that the model fails to explain—the wide and asymetrical a-polarized absorption peak confirms the importance of $\gamma(\omega)$ at higher energies. We expect that higher-frequency phonons (and two-phonon combinations), though ineffective in the low-temperature broadening of the exciton peak, should manifest themselves through a strong relaxation at energies where their creation becomes allowed.

3. The half-maximum width predicted by our model is compared in Fig. 2.14a (dashed line) with experimental widths. The theoretical calculations underestimate the width by a factor close to 2 at high temperatures. It ought to be stressed that no parameter of our model could be adjusted to make up for this difference. Otherwise, coupling to higher phonon modes would not change the width at low temperatures ($T < 50\,\text{K}$). We are thus led to question our perturbative approach. Schreiber and Toyozawa[53] have presented an exact numerical calculation of the adiabatic model of exciton–phonon coupling in one-, two-, and three-dimensional lattices. They show that the renormalized perturbation theory of a two-dimensional exciton underestimates the half-maximum width of the absorption peak by a factor of ~ 1.7, irrespective of the temperature (cf. Fig. 4 of ref. 53). Although the results of Ref. 53 are valid only when the linewidth is larger than the phonon energy (adiabatic model), they show that the lineshape problem is not correctly dealt with by the "renormalized" perturbation theory. This occurs because the basic requirement of a perturbation development, namely that the perturbation be smaller than the energy difference between the mixed levels, is not met at the energy scale of the linewidth. On the contrary, at high energies and low temperatures, the energy difference is greater than the linewidth, and the second-order perturbation theory is found to be a good approximation to the renormalized perturbation theory (2.129). Thus, we expect our values (2.130b) for the coupling strengths to be still correct. Indeed, the factor of ~ 1.8 between experimental and theoretical widths in Fig. 2.14a could correspond to the underestimation factor 1.7 of the perturbation theory for a two-dimensional exciton. (On the other hand, we note from the results of ref. 53 that only a two-dimensional exciton is compatible with the linear behavior of the width γ_0 vs T.)

Last, we may estimate from (2. 130b) the stabilization S of the localized exciton, $S = 150\,\text{cm}^{-1}$. This value does not include the contribution from higher-energy phonons [which should not be very important—see (2.128)], but is still well consistent with the absence of self-trapping of the anthracene exciton: $S < B \sim 200\text{–}250\,\text{cm}^{-1}$.

In conclusion, on phonon-assisted absorption, the investigation of the phonon modes involved in the broadening of an excitonic transition is greatly simplified at low temperatures, where perceptible effects on the lineshape are produced by phonon creation only. High-resolution reflection spectra of the (001) face of vapor-grown crystals are shown to contain the requisite information for such an investigation.

The bulk dielectric function $\varepsilon(\omega)$ has been evaluated through a Kramers–Kronig transformation of reflection spectra. To this purpose, the surface structures[60] had to be carefully analyzed and eliminated. The absorption spectra at low temperatures (1.7–77 K) complement earlier work by Morris and Sceats.[96] When interpreted in terms of the damping function (or exciton self-energy), including both damping and energy shift, these spectra clearly show an optical-phonon theshold (46 cm^{-1} above the zero-phonon line) and a smoothly growing background imputable to acoustical phonons. At higher temperatures, all structures broaden to yield a strong exciton damping. A very coarse model allowed us to roughly estimate the exciton–phonon coupling strengths.

b. *Discussion of the Theoretical Approach.* We would like to stress the following results of this study:

1. Acoustical phonons are important in the thermal broadening of the *b*-polarized exciton of anthracene.
2. The estimated coupling strengths $\chi_{ac} \sim 70\,\text{cm}^{-1}$ for acoustical phonons at the zone boundary and $\chi_{op} \sim 50\,\text{cm}^{-1}$ for the 46-cm^{-1} optical mode, are consistent with weak exciton–phonon coupling.
3. Higher-energy optical phonons may explain the exciton relaxation well above the exciton-band bottom ($\omega > \omega_0 + 50\,\text{cm}^{-1}$); for instance, the 140-cm^{-1} B_g lattice mode dominates the *a*-polarized exciton relaxation. However, acoustical and lower-energy (around 50 cm^{-1}) optical phonons, according to our model, account by themselves for the observed broadening at temperatures below 77 K.
4. We found a correct account of the broadening magnitude and increase with temperature, out of reach of the renormalized perturbation theory. Exact numerical calculations by Schreiber and Toyozawa[53] agree with our data and confirm our values of the exciton–phonon coupling strengths.

D. The Finite-Crystal Secondary Emissions

A sample irradiated with monochromatic light of a given direction **k** gives rise to reemissions at various frequencies, and directions different from **k**. In the linear-response approximation all these reemissions are usually called spontaneous secondary emissions. Among them, emissions at the excitation frequency ω_i may originate either by diffraction on static structures (surfaces, inhomogeneities defects, etc.) or by scattering on the thermodynamical fluctuations of the sample. In the first case, the emitted light is perfectly monochromatic—this is the case, for instance, of reflection and transmission of a slab as in Sections II.A, B, C. In the second case, the emission is diffusive in accordance with the temporal evolution of the refractive-index thermodynamical fluctuation on the scale of one wavelength, leading to the Rayleigh-scattering central peak and the Brillouin-scattering sidebands related to the acoustical phonons. Near these scattering structures, without significant frequency changes, appear scattering structures with frequency changes of a few reciprocal centimeters to a few thousand reciprocal centimeters. These emissions, corresponding, in the case of transparent media, to excitations of nuclear degrees of freedom (rotations, vibrations, etc.), are termed Raman spontaneous scattering.

In absorbing media (e.g. where the exciting light resonates with electronic transitions), two amplitudes, interfering and competing, of secondary emission may appear, according to the energy and phase relaxation mechanisms acting on the incident light. These are the Raman resonant scattering component and the hot luminescence component with its subdivision into ZPL and hot and thermalized emission channels (other secondary emissions, such as phosphorescence and delayed fluorescence, may be neglected for the singlet anthracene crystal). The separation of these resonant secondary emissions is quite involved; it has recently been discussed successfully for a solution of molecules.[104] The investigation of the intrinsic fluorescence of the bulk crystal and its surface, on the basis of the polariton states, will show the connection of these three channels, termed[77] RRS, HL, and RF, in terms of elastic (pure dephasing) and inelastic (phonon creation–annihilation) interaction events during the lifetime of the excitation.

1. *General Aspects of the Crystal Secondary Emissions*

Let us consider a scattering event by the crystal: An incident photon $(\omega_1, \mathbf{k}_1)$ is scattered to a photon $(\omega_2, \mathbf{k}_2)$, the energies $\hbar\omega_1$ and $\hbar\omega_2$ being well defined in a frequency spectroscopy experiment.

First, we envisage the weak exciton–photon coupling (which allows an intuitive description of the phonon effects on the nature of the secondary emissions). Therefore we write the hamiltonian of the total system as sums of free photons (H_γ), free excitons (H_e), and free phonons (H_p), with the appropriate interactions $H_{e\gamma}$ (Section I) and H_{ep} (see Sections II, A, B, C.), including intramolecular vibrations too.

To second order in $H_{e\gamma}$, the transition probability of the scattering event $\gamma(\mathbf{k}_1) \rightarrow \gamma(\mathbf{k}_2)$—the phonon system passing from state R [energy E_R and population $p(R)$] to state R', *while conserving the energy of the total system is*

$$P(\gamma(\mathbf{k}_1)\rightarrow\gamma(\mathbf{k}_2))=\frac{2\pi}{\hbar}\sum_{RR'}P(R)\left|\sum_{ee'}\langle\gamma(\mathbf{k}_1)|H_{e\gamma}|e\rangle\langle e,R|G|e',R'\rangle\right.$$

$$\left.\times\langle e'|H_{e\gamma}|\gamma(\mathbf{k}_2)\rangle\right|^2\delta(\hbar\omega_1+E_R-\hbar\omega_2-E_{R'}) \qquad (2.131)$$

where G denotes the resolvent associated with the exciton–phonon subsystem; cf. Sections II.B and C. The probability P may be calculated with a suitable model, such as the DCPA method, for G.[77] However, to point out qualitatively the effects of H_{ep}, we restrict ourselves here to a perturbative approach in powers of H_{ep}. The successive orders *at zero temperature* correspond to the following diagrams:

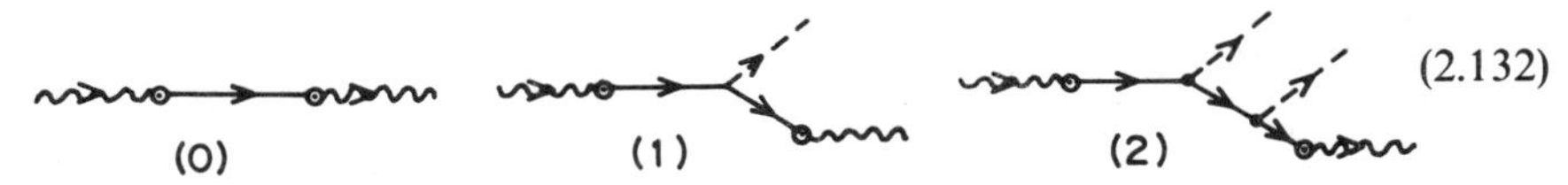

(2.132)

with the zero-order term describing the Rayleigh scattering, while the first-order term describes a direct, peak-to-peak relation between the absorbed and the emitted photons:

$$\hbar\omega_1=\hbar\omega_2\pm E_p \qquad (2.133)$$

giving the Raman scattering component to the lowest order in H_{ep}. Second- and higher-order terms describe the energy conservation with more complex paths, where the energy conservation between $\hbar\omega_1$ and $\hbar\omega_2$ occurs on frequency bands and not on peaks as in (2.133). For that reason, we classify scattering involving relations more complex than (2.133) as luminescence probability.

If the exciton–phonon interaction H_{ep} is strong compared to the emission probability, high-order terms in H_{ep} contribute to P (2.131), providing strong luminescence at the expense of the one-phonon (Raman) process. In contrast, if the emission probability dominates the phonon creation probability, the peak (2.133) dominates the secondary emission at the expense of the luminescence.[77] Examples of this competition will be discussed for the surface-state secondary emission, where the picosecond emission of the surface states, and its possible modulation, allow very illustrating insights into the competition of the various channels modulated by static or thermal disorder, or by interface effects.

Let us now consider the case of strong exciton–photon coupling, which is that in the singlet state of the anthracene crystal. As shown in Section I, we have to consider as zero-order hamiltonian H that of the polaritons, and the transition between polaritons will be induced by the exciton–phonon coupling H_{ep}. Let us denote by $\mathscr{H}$ and $\mathscr{G}$ the hamiltonian of the total system and of its resolvent, respectively, and write for the zero-order hamiltonian

$$H = H_e + H_\gamma + H_{e\gamma} \tag{2.134}$$

with its resolvent G for an infinite crystal in the absence of phonons. The transition probability amplitudes between two polaritons $(\omega_1, \mathbf{k}_1)$ and $(\omega_2, \mathbf{k}_2)$ are given by the matrix elements of the matrix

$$\mathscr{S} = \frac{1}{2i\pi} \lim_{t \to \infty} \int dz\, e^{-iz \cdot t/\hbar}\, \mathscr{G}(z) \tag{2.135}$$

Introducing t matrices for scattering of polaritons on phonons (with the lattice state passing from R to R'), we obtain

$$\mathscr{G}(z) = G(z + E_R) + G(z + E_R) T_{ep} G(z + E_{R'}), \tag{2.136}$$

which allows a straightforward calculation of (2.135) similar to that of $\langle e, R | G | e', R' \rangle$ in (2.131).

However, in real crystals we observe transitions between free photons (the incident and the emerging photons, and not the photons trapped in polariton states), so that we cannot consider the crystal as an infinite 3D system, and there arises a dimensionality problem (with one-to-one correspondence between photons and excitons). Thus, the resolvent $G(z)$ has to refer to an excitonic finite system coupled to a continuum of photons.[78] As T_{ep} does not operate on photons, we find for $\mathscr{S}$ the following exact relation:

$$\langle \mathbf{k}_2 | \mathscr{S} | \mathbf{k}_1 \rangle = \frac{1}{2i\pi} \lim_{t \to \infty} \int dz\, e^{-iz \cdot t\, \hbar} \sum_{\substack{\mathbf{k},\mathbf{k}' \\ e,e' \\ R,R'}} \langle \mathbf{k}_2 | G | \mathbf{k} \rangle \frac{\langle \mathbf{k} | H_{e\gamma} | e \rangle}{z - E_e}$$

$$\times \langle eR | T_{ep} | e'R' \rangle p(R') \frac{\langle e' | H_{e\gamma} | \mathbf{k}' \rangle}{z - E_{e'}} \langle \mathbf{k}' | G | \mathbf{k}_1 \rangle \tag{2.137}$$

with $|\mathbf{k}\rangle$ indicating free photons and $\langle \mathbf{k} | G | \mathbf{k}' \rangle$ describing a phononless propagation of photon $\mathbf{k}$ to photon $\mathbf{k}'$ after scattering on the matter system [in (2.137) the phononless process has been omitted]. At the $t \to +\infty$ limit, only the real poles contribute in (2.137), with energy conservation on all the real intermediate states. If, in T_{ep}, coupling to free photons is neglected, the only real poles are those introduced by $\langle \mathbf{k} | G | \mathbf{k}' \rangle$ for $\hbar\omega_k = \hbar\omega_{k'}$. Therefore, the transition probability $P(k_1 \to k_2)$ contains intrinsically amplitudes of reflection $r(\omega)$ and of transmission $t(\omega)$,[78] "dressing" the incident and emerging photons. For a thin slab, the functions $r(\omega)$ and $t(\omega)$ are given in (2.112), (2.113), and we obtain a transition probability of simple form with well-separated elementary processes:

$$P(\mathbf{k}_1 \to \mathbf{k}_2) = \frac{2\pi}{\hbar} \sum_{RR'} p(R) \left| \sum_{\substack{\varepsilon_1\varepsilon_2 \\ ee'}} s_{\varepsilon_1}(\omega_1) \frac{\langle \varepsilon_1 \mathbf{k}_1 | H_{e\gamma} | e \rangle}{\hbar\omega_1 - E_e} \langle eR | T_{ep} | e'R' \rangle \right.$$
$$\left. \times \frac{\langle e' | H_{e\gamma} | \varepsilon_2 \mathbf{k}_2 \rangle}{\hbar\omega_2 - E_{e'}} s_{\varepsilon_2}(\omega_2) \right|^2 \delta(\hbar\omega_1 + E_R - \hbar\omega_2 - E_{R'}) \quad (2.138)$$

where $\varepsilon_1, \varepsilon_2 = \pm$, and $s_+ = t(\omega)$, $s_- = r(\omega)$. The factors $r(\omega_2)$ and $t(\omega_2)$ in (2.138) allow us to understand why and to what extent the emission of a slab will be modulated by interference of its front and back faces, which may be considered as a dimensionality effect. For a thick crystal, $t(\omega_2) \to 0$, these effects disappear, but we encounter modulation of the surface-state emission, even quenching of emission, by interference with the bulk reflectivity (see Section III).

2. *The Raman Scattering of the Anthracene Crystal*

Since we treat resonant UV spectroscopy, we summarize this emission in relation to the resonant Raman limit, which we investigate for bulk and surface resonant spectroscopy.

The Raman response is investigated through the polarizability tensor $\boldsymbol{\alpha}(\omega)$

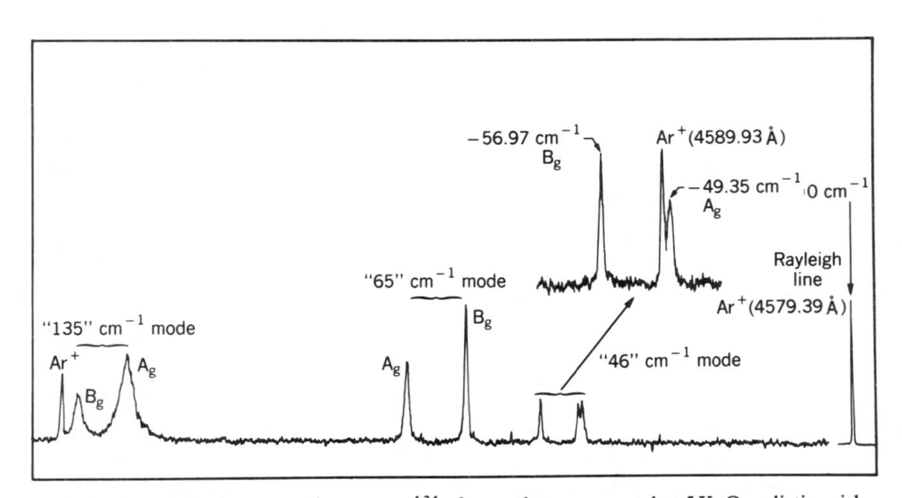

Figure 2.16. Raman-scattering spectra[131] of an anthracene crystal at 5 K. One distinguishes the three frequency doublets, corresponding roughly to the three main molecular inertial axes and to the two components (A_g and B_g) of the unit cell. In this spectrum, only the frequency of the lower A_g mode (49.35 cm^{-1}) is determined with the required accuracy ($\pm 0.10\,\text{cm}^{-1}$).

TABLE IV

		Energy (cm^{-1})		
		Ref. 47	Ref. 79	
Axis	Mode	6 K	5 K	120 K
N	A_g	49	49.37 ± 0.10	44.7
	B_g	56	56.95 ± 0.10	51.5
M	B_g	70	70.53 ± 0.10	67.3
	A_g	82	81.40 ± 0.15	76.7
L	A_g	132	131.68 ± 0.25	126.2
	B_g	140	140.18 ± 0.25	133.6

(Section I) and nuclear derivatives, which treat globally all the excited electronic transitions. In this subsection, we consider only Raman spectroscopy at low temperatures,[47,79] as illustrated in Fig. 2.16.

Since the anthracene crystal is centrosymmetric (see Section II.A.1°), among the various phonon modes at $k = 0$, only the libration pair (g) modes are Raman-active. The three principal axes of libration are roughly taken[47,80] to be the molecular axes, L, M, N. For each axis, there are two modes, symmetric and antisymmetric (A_g and B_g of the unit cell). The energy of these modes are given in Table IV at 5 K and at 120 K to show their temperature sensitivity. The error is probably underestimated, except for the mode 49 cm^{-1}, which lies near the reference Ar^+ line.[79]

The sensitivity of the phonon frequencies to temperature shows quite clearly the importance of their anharmonicity.[42] The width of the Raman peaks, very small at low temperature (~ 1 cm^{-1}), evolves in parallel with the frequency shift with temperature, which is still a consequence of the phonon–phonon interactions due to the anharmonicity. The fundamental reason for this strong anharmonicity, as well as the importance of the equilibrium-position shifts between 4 and 300 K,[45] resides in the weakness of the van der Waals cohesive forces in the molecular crystal.

The most Raman-active intramolecular modes are the two modes (a_g) encountered in the absorption spectra of Fig. 2.8: 397 and 1404 cm^{-1}. With a resolution of 1 cm^{-1}, it is difficult to observe any splitting in these modes; in addition, their frequencies show very little sensitivity to temperature. All these observations are in good agreement with the fact that the intramolecular modes are little affected by the crystal field. In contrast, the weakest vibrations (torsions, flexions, and internal distortions, such as butterfly and twisting motions) contribute significantly to the determination of the intermolecular modes, securing the thermalization of all the molecular degrees of freedom.[45]

3. *Intrinsic Fluorescence of the Anthracene Crystal*

The fluorescence of the anthracene crystal is strong and quite striking. Usually in crystals with weak transitions and poor quality, the emission is due essentially to the traps, shallow or deep, the trapping time being of the order of nanoseconds. It is well known that although the anthracene crystal has no singlet shallow traps because of its wide excitonic bandwidth, deep traps exist, particularly β-methylanthracene.[71] Therefore, the spectrum of the fluorescence of a crystal consists, at low temperatures, of the narrow 0–0 line of the impurity, followed by its vibronic structure. We ignore trap emissions, since our aim is the analysis of the intrinsic emission following the theory developed in Section I and II. Intrinsic emission largely dominates the emission spectrum in anthracene crystals of high quality, sublimation-grown in our laboratory. As shown in Fig. 2.17, the fluorescence spectrum is resolved, at low temperatures, in a series of narrow lines and narrow bands investigated by many authors.[81–87] We summarize below the principal experimental features and then proceed to their interpretation on the basis of a consistent model developed in Sections I and II.

a. *The Experimental Observations*. The experimental investigations are of two types: frequency spectroscopy in various conditions, and time-resolved emission on the nanosecond and picosecond scales.[85,87] Aside from the surface-state emission, investigated in Sections III and IV, the crystal fluorescence has a *fluorescence origin* at 25 100 cm^{-1} above which no crystal emission is observed. Below this origin, a first series of lines, between 25 100 and 24 900 cm^{-1}, are assigned to activation of lattice phonon modes. The remaining emissions are related to the origin by subtraction of intramolecular vibration quanta: the main ones are 396 and 1404 cm^{-1}.

Figure 2.17 shows the excited intrinsic fluorescence obtained in our laboratory by excitation at 3917 Å of a sublimation-grown crystal a few micrometers thick,[71] at 5 K. Results of other authors with sublimation-grown samples[81–84,86] confirm the general lineshape of the fluorescence, but, owing to the various experimental conditions, variations are observed, which can be summarized as follows:

1. Excitation conditions: At about 10 cm^{-1} and farther above the 0–0 transition, the excitation wavelength does not seem to influence the fluorescence spectrum. This observation points to a rapid thermalization of the exciton population.[87] In contrast, the excitation power has an important nonlinear effect: Above a threshold stimulated emission is observed on the 0–1400 line, at 23 692 cm^{-1}.[88] The crystal behaves as an optical resonator, with laser effects, which explains the sharpness and the intensity of this line in Fig. 2.17.[89]

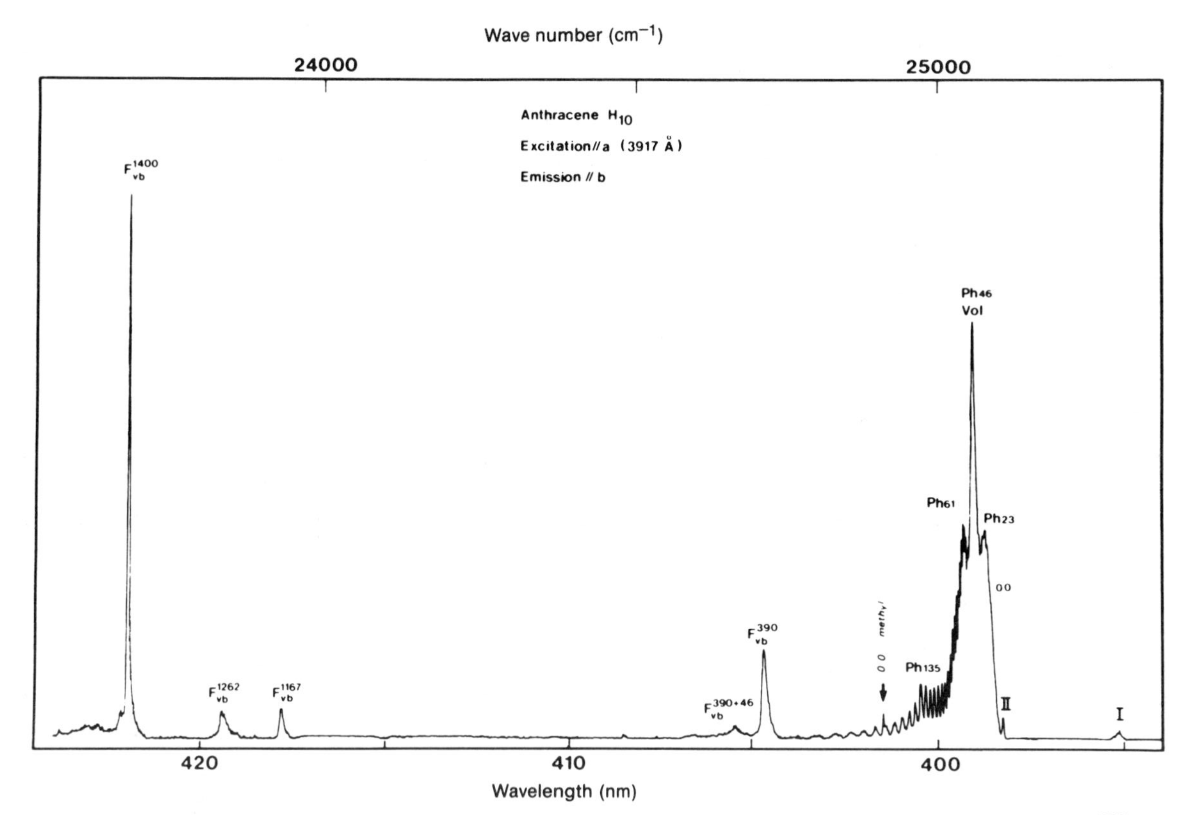

Figure 2.17. Fluorescence spectra of the anthracene crystal at 5 K, observed from the front (001) face of a thin sample.[120] The main lines cited in the text are as follows: Ph 23 = 25 082 cm^{-1}, Ph 46 = 25 052 cm^{-1}, Ph 61 = 25 037 cm^{-1}, $F_{vb}^{390} = 24\,703$ cm^{-1}, $F_{vb}^{390+46} = 25\,052$ cm^{-1} − 390 cm^{-1}, $F_{vb}^{1400} = 23\,692$ cm^{-1}. We note the very narrow and strong line (laser emission) of the 0–1400 fluorescence, as well as the fluorescence oscillations around the structure Ph 135.

2. The thickness effect: The usual thickness is of the order of 10 μm. Brodin et al.[84] analyzed the variation of the spectrum as the thickness is varied from 1.2 to 20 μm. They noticed an intensity redistribution in favor of the wider bands (25 082 and 25 037 cm^{-1}). A thickness effect is also observed on the emission oscillatory modulation visible in Fig. 2.17, which is the counterpart of the reflectivity oscillations in Fig. 2.9.

3. The temperature effect: The strong band between 25 100 and 24 900 cm^{-1} is resolved only below 15 K. Most of the spectra have been recorded at $T = 4.2$ K and $T = 1.7$ K. Increasing T causes broadening of the high-energy part of the narrow lines (25 052, 24 703, 23 692 cm^{-1}) and smoothing of the high-energy wing of the wide bands (25 082, 25 037, ... cm^{-1}). This trend is confirmed at very low temperatures[86]: at $T = 0.4$ K the width of the narrow lines stabilizes at about 5 cm^{-1}, while the asymmetry of the lines persists. Galanin et al.[85] have recorded significantly different spectra with thick crystals ($\sim$ 100 μm). They noticed an important amplification of the wide bands, which is compatible with observations on thick sublimation-grown crystals.

4. Time-resolved emission: Galanin et al.[85] and Aaviksoo et al.[87] have resolved the emission in time, with increasing resolution. Their investigations have revealed two types of emission regimes: one for the narrow lines (25 025, 24 703, and 23 692 cm^{-1}) with a decay rate $\tau = 0.9$ ns, and another for the wide bands (25 082 and 25 037 cm^{-1}) with a slower decay rate $\tau = 1.4$ ns. The first wide band (25 082 cm^{-1}) was investigated in more detail by Aaviksoo (second ref. 87); it shows a distribution of rise and decay times depending on the emission energy inside the band. The emissions nearest to the origin involve shorter dynamics, from 350 to 1300 ps for the decay rates.

b. *Model Relaxation Mechanisms for the Decay of Polaritons.* The fluorescence quantum yield is about 100%, which means that each exciting photon finally creates one photon in the fluorescence band. There are two main reasons for this: (1) the strong transition dipole makes the radiative channel dominant over the nonradiative channel; (2) the *b*-created state lies at the bottom of the exciton band. Thus, intraband thermalization of excitons leads, at low temperature, to the vicinity of the emissive state. As indicated in Sections II.B and C, relaxation of prepared states, with creation of vibrations and phonons, is very fast, and corresponds to homogeneous widths of the order of 50 to 100 cm^{-1}, which involve times shorter than 1 ps. These relaxation times become longer in the vicinity of the bottom of the excitonic band, at low temperature, because only acoustical phonons of long wavelengths, with small densities and weak couplings, may interact with the excitons. We shall assume the creation, a few picoseconds after the excitation, of a quasithermalized distribution of excitons, which will lead to the first characteristic of the anthracene-crystal emission: the fluorescence occurs over very long durations.

The second characteristic is the absence of the 0–0 transition, which is polariton-forbidden.[90] Thus, the renormalization of the strong exciton–photon coupling to obtain new quasi-stationary states, the polariton states (real energies) perturbed only by phonon scattering, is essential for the comprehension of the resulting fluorescence spectrum.[90] Thus, polaritons may either be scattered, on vibrations or phonons, to other polaritons (without creating light), or escape the crystal (surface effect), creating light and phonons. Therefore, the fluorescence, its quantum yield, and its spectral distribution may be understood only when the probabilities of the polariton dynamics are evaluated.

I. DENSITY OF POLARITON STATES. First of all, the shape of the lower part of the Coulomb-exciton band has to be investigated. Taking into account the nonanalyticity (Section I.B.2) and ignoring nondipolar interactions between

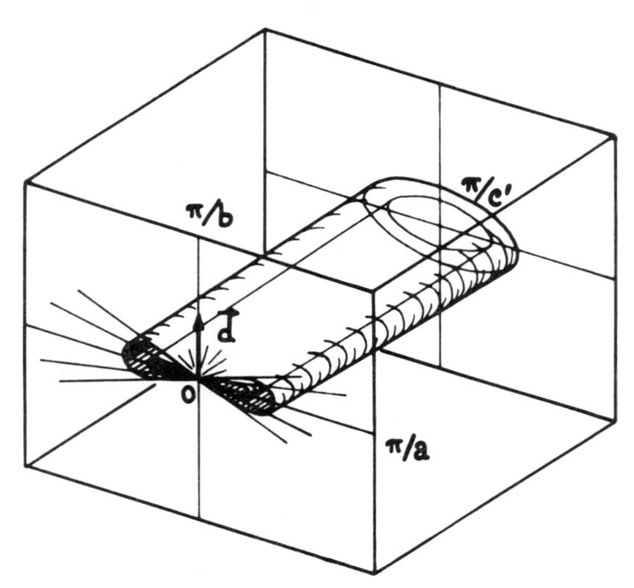

Figure 2.18. Profile of isoenergetic surfaces of the excitonic dispersion in the vicinity of the bottom of the band in the model (2.139) of an orthorhombic crystal. We note the lengthening of the surfaces along c'. Also, the wave vectors tend to orient perpendicular to $\mathbf{d}$ (or the b axis) in the vicinity of the point $\mathbf{K} = \mathbf{0}$.

(a, b) planes, the excitonic dispersion shape[85] in the vicinity of $k = 0$ is

$$E(\mathbf{k}) = E_0 + \Delta \cos^2 \theta + \frac{\hbar^2}{2m_\perp} k_\perp^2 \tag{2.139}$$

where $\Delta = d^2/\varepsilon_0 V_0$ is the polariton stopping bandwidth, and θ the angle between the b axis and $\mathbf{k}$, with $k_\perp$ its component in the (a, b) plane. We have chosen an isotropic model in the (a, b) plane. In the first Brillouin zone, the isoenergetic surfaces have complex shapes which are included in a cylinder about the c' axis, pinched at the point $k = 0$ (Fig. 2.18). At low temperatures, the exciton distribution tends to point the wave vectors along the c' axis, with a maximum of population at the border of the zone (π/c'). The corresponding density of states, zero for $E < E_0$, is discontinuous with a threshold shape at E_0, analogous to that of a 2D system. Actually, retardation effects modify this shape in the vicinity of $k = 0$, leading according to Section I to a corrected dispersion:

$$\hbar\omega - \hbar\omega_0 = \Delta \frac{(\omega/c)^2 - k^2 \cos^2 \theta}{(\omega/c)^2 - k^2} + \frac{\hbar^2}{2m_\perp} k_\perp^2 \tag{2.140}$$

where $E(k) = \hbar\omega$ and $E_0 = \hbar\omega_0$ and an effective mass $m_\perp$ was introduced for the in-plane motion of the exciton. Following (2.140), the isoenergetic surfaces change in the vicinity $E \lesssim E_0$ as illustrated in Fig. 2.19.

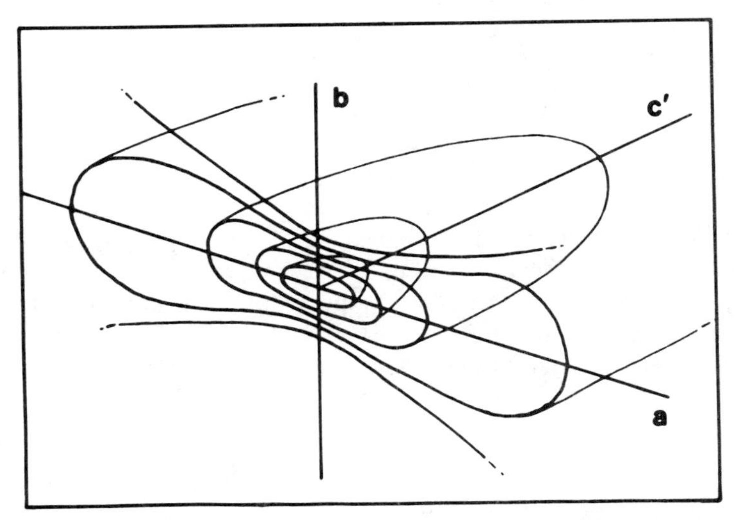

Figure 2.19. Profile of isoenergetic surfaces of the polariton states slightly below the bottom of the excitonic band, in the crystal model (2.139) including the modification (2.140).

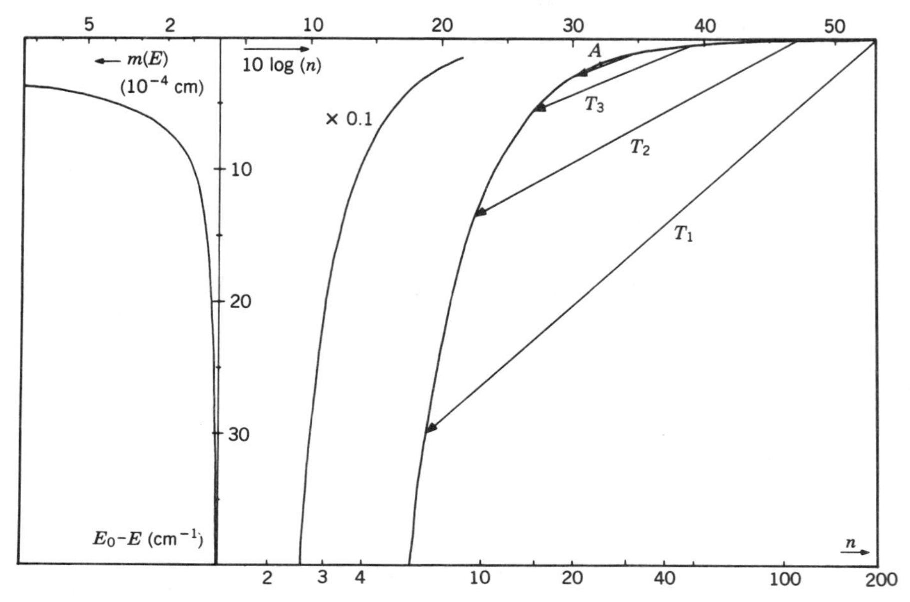

Figure 2.20. Right part: The polariton dispersion at a few tens of reciprocal centimeters below the bottom of the excitonic band, vs the wave vector, or the refractive index $n = ck/\omega$ (notice the logarithmic scale). The arrows indicate transitions with creation of one acoustical phonon, with linear dispersion in k (with a sound velocity of 2000 m/s). For the transitions T_1, T_2, T_3 the final momentum is negligible compared to the initial momentum, and the unidimensional picture suffices. For the transitions between T_3 and the point A, the direction of the final wave vectors should be taken into account. Left part: The density of states $m(E)$ (2.141) of the polaritons in the same energy region. This diagram explains why the transitions T_1 will be much slower than the transitions around T_3 and the point A. The very rapid increase of $m(E)$ at a few reciprocal centimeters below E_0 shows the effect of the thermal barrier.

For a polariton whose photonic component is predominant, at a few thousands reciprocal centimeters below E_0, the isoenergetic surfaces are roughly spherical. Then, for $E \lesssim E_0$, they change to ellipsoids compressed along the b axis, the excitonic component of the polariton becoming predominant. Thus, from a fraction of a reciprocal centimeter below E_0, the polariton density of states shows a variation as $(E - E_0)^{-2}$:

$$m(E) \sim m_0(E)\left(1 + \frac{1}{8}\frac{\hbar\omega\cdot\Delta}{(\hbar\omega_0 - \hbar\omega)^2}\right) \tag{2.141}$$

where $m_0(E) = (2\pi^2 E)^{-1}\ (\omega a/c)^3$ is the density of free photons. $m(E)$ is illustrated in Fig. 2.20 (left part); it decays very rapidly below E_0.

II. POLARITON-TO-POLARITON SCATTERING. The polariton may scatter on vibrations or phonons, through its excitonic component, to another polariton. For the case of anthracene, even for the lowest energy we consider ($E_0 - 1400\,\mathrm{cm}^{-1}$), the photon component is only about 0.1, so that for this scattering we may treat the polaritons as pure excitons and calculate the resulting transition probabilities from polariton E_i to polariton E_f, which may be written[85] as follows:

$$P(E_i \rightarrow E_f) = \frac{2\pi}{\hbar} |\chi_s|^2 \rho(E_f) \begin{Bmatrix} \langle n_s(E_f - E_i) \rangle \\ 1 + \langle n_s(E_i - E_f) \rangle \end{Bmatrix} \tag{2.142}$$

according as $E_f \gtrless E_i$, where $\rho(E_f)$ is practically the density of polariton states, if the phonon dispersion is neglected. It is clear from (2.142) that, at low temperature, the relaxation becomes faster as the final state approaches the bottom of the excitonic band. For vibrations and phonon energies above $50\,\mathrm{cm}^{-1}$, the variation of $\rho(E)$ is slow. In contrast, for relaxation of excitons which are in the vicinity of E_0, assisted by creation of acoustical phonons, the strong variation of $\rho(E_f)$ contributes selectively: Fig. 2.20 illustrates possible relaxation paths of excitons with wave vectors parallel to the c' axis, assisted by creation of acoustical phonons $\Omega = c_s k$, with $c_s = 2 \times 10^3\,\mathrm{ms}^{-1}$. Relaxation of excitons at the zone boundary is slow because the density of final states is very small. In contrast, the excitons near the zone center, 1 to $2\,\mathrm{cm}^{-1}$ below E_0, will undergo fast relaxation, causing polariton transfer in a region below the point A (see Fig. 2.20), at many reciprocal centimeters below E_0, where further relaxation by acoustical phonons is negligible. This is the polariton bottleneck phenomenon.[84,85] Thus, the exciton states in the middle of the zone behave as "macroscopic" traps in the reciprocal space. In particular, as for the ordinary traps, an increase of temperature raises the population to higher states, leading to a thermal barrier in the emission process.

III. RADIATIVE DECAY OF THE POLARITON. Up to now, we have considered an infinite 3D crystal with the polaritons of Section I as quasi-stationary states. In fact, when the polariton encounters the crystal boundaries, there is a probability of transmission in the form of free photon. For sufficiently thick crystals, so long as the transmission probability is low, the emission may be treated as a perturbation in the scheme of infinite polaritons, so that the emission per unit time is the transmission probability divided by the duration between two successive collisions on the crystal surface[90]:

$$\gamma_r(\theta_i) = (1 - R) \frac{\cos\theta_i}{e} \frac{d\omega}{dk}\bigg|_k \tag{2.143}$$

where e indicates the thickness of the crystal and θ the angle of incidence. Furthermore, $\gamma_r(\theta)$ must be averaged on all states of a given energy. If the wave-vector distribution is assumed isotropic in the plane perpendicular to the b axis, one finds an approximate expression which is valid in the vicinity of E_0:

$$\gamma_r(\omega) = \frac{2c}{en^2(\omega)} \frac{1}{(n(\omega) + \omega\, dn/d\omega)} \tag{2.144}$$

Therefore, $\gamma_r(\omega)$ decays rapidly near ω_0, as $(\omega_0 - \omega)^{5/2}$. On the contrary, where the photon component of the polariton becomes important, the emission becomes too strong to be treated as a perturbation for crystal a few micrometers thick. In addition, coherent reflection on the two faces creates interference modulation of the emission, visible in Fig. 2.17 and formally included in (2.138). In a complementary way, phonon relaxation in the vicinity of E_0 dominates the photon relaxation. Then, as the energy decays below E_0, $\gamma_r(\omega)$ increases and $P(E_i \rightarrow E_f)$ decreases very rapidly, the radiative rate $\gamma_r(\omega)$ overwhelming the polariton relaxation below an energy threshold which depends on the crystal thickness and on the temperature.[90]

c. *Interpretation of the Fluorescence Spectra.* The spectra of all crystals point to a first thermalized exciton distribution at the bottom of the excitonic band, just above the energy E_0. For this distribution, relaxation by acoustical phonons and by emission being negligible, the excitons will relax only by the creation of optical phonons and of intramolecular vibrations, to polaritons of energy $E_0 - \hbar\Omega_0$. In their turn, these new polaritons may either undergo further relaxations if the crystal is thick, or emit one photon if the crystal is thin, relaxation of polaritons being more probable for a polariton distribution with larger energies. These competing processes account for the emission pattern of Fig. 2.21. Emissions at $E_0 - 45\,\mathrm{cm}^{-1}$, $E_0 - 390\,\mathrm{cm}^{-1}$, and $E_0 - 1404\,\mathrm{cm}^{-1}$ give the narrow lines at 25 052, 24 705, and 23 692 cm^{-1}. These emissions allow one to locate the energy of the exciton distribution D_1 at about 25 096 cm^{-1}, which agrees with the generally accepted value[60,71] 25 093 $\pm 2\,\mathrm{cm}^{-1}$ when the distribution width $kT \sim 3\,\mathrm{cm}^{-1}$ is taken into account. [We notice that the spectra of refs. 81, 84–87, recorded at very low temperature (0.4 K), lead to the same value, provided one includes the 7-cm^{-1} correction for the refractive index of air.] Note that the value (45 cm^{-1}) for the optical phonon involved in the first narrow emission (Fig. 2.21) is different from that (49 cm^{-1}) found in Raman spectroscopy (at $\mathbf{k} = \mathbf{0}$). This difference is consistent with phonon creation with $\mathbf{k} \neq \mathbf{0}$ pointing in the c'-axis direction,[46] since the wave vector of the final polariton is negligible. In addition, the phonon dispersion along the c' axis may provide the reason why this line (25 052 cm^{-1}) continues to be broader (6 cm^{-1}) than the two other lines (4 cm^{-1}) at very low

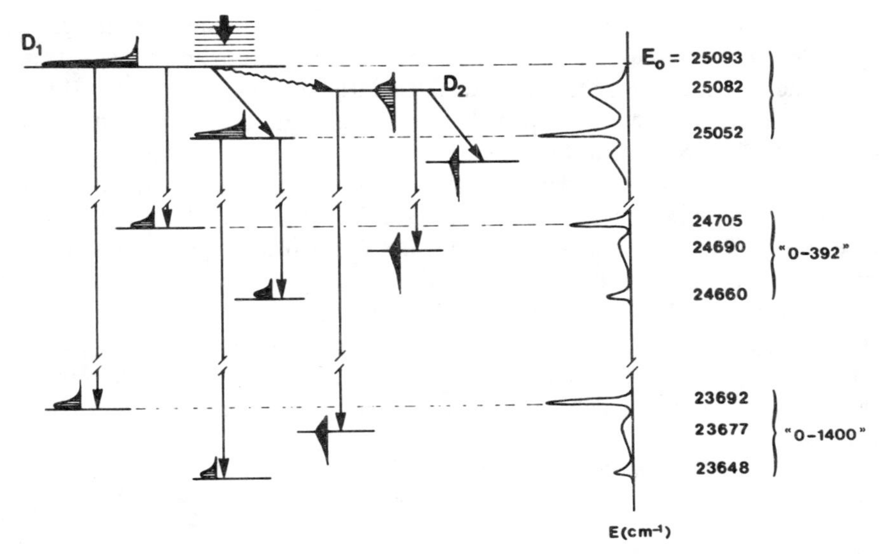

Figure 2.21. Scheme of the various distributions D_1 and D_2 of the polaritons leading to the observed bulk fluorescence. The model of two main distributions accounts for the narrow lines, the satellite broad bands, and their relative intensities. The energy of the main fluorescence lines is given in reciprocal centimeters. The bold arrow represents the relaxation in the excitonic band to states above E_0. The primary distribution of excitons (D_1) relaxes by the creation of acoustical phonons (wavy arrow) to the secondary distribution of polaritons (D_2) below E_0 as well as to other vibrations in the ground state as given by the spectral model[85] or the dynamical model (second ref. 87).

temperature (0.4 K). Last, the polariton distribution at 25 052 cm^{-1} may relax radiatively to give additional (satellite) lines at 25 052 − 390 and 25 052 − 1404 cm^{-1}, which are narrow, with intensities strongly depending on the thickness of the crystal (Fig. 2.17).[84]

Apart from the principal distribution D_1, which is the origin of the narrow emissions, we must invoke a second, broader distribution D_2 of polaritons with energies below E_0, with a maximum at 25 082 cm^{-1} (Fig. 2.21). Relaxation from this distribution gives broad emissions at 25 033 cm^{-1} (49 cm^{-1} below) and 23 677 cm^{-1} (1404 cm^{-1} below),[86] characterized by a high energy threshold at very low temperatures. This broad distribution may be assigned to the relaxation near the middle of the zone by acoustical phonons.

The dependence of the spectra upon the thickness of the crystal may be accounted for in terms of the discussion in Section II.D.3.b.iii. For a thin crystal, the radiative relaxation dominates in the broad (25 082 cm^{-1}) and narrow (25 052 cm^{-1}) distributions, so that the satellite lines at "0" − 390 and "0" − 1404 cm^{-1} are very weak. For the thick crystals, the satellite lines are stronger. The spectra reported by Galanin et al.[85] seem to show strengthening of the broad distribution D_2. However, rather than a thickness effect, this strengthening could well be assigned to the large density of defects in thick crystals, populating the broad distribution D_2 through the breakdown of the translational invariance of the crystal.

The temperature dependence of the observed emission is equally well interpreted in the scheme of Section II.D.3.b: The lines emitted from the main distribution D_1 are asymmetric, with a width of the order of kT for the high-energy part. At very low temperature (0.4 K)[86] the main distribution approaches a width of about 4 cm^{-1}, which indicates that the thermalization regime is slower than the radiative relaxation rate, of the order of 1 ns. In addition, the shape of the second distribution D_2, at 25 082 cm^{-1}, sharpens as the thermal barrier,[85] which inhibits the relaxation very near the middle of the zone, weakens at very low temperature.

Finally, the time-dependent regimes of the emission[85] are consistent with this model. The decay time of the narrow lines reflects the lifetime of the main distribution D_1: $\tau = 0.9$ ns for the three main lines at 25 052, 24 700, and 23 692 cm^{-1}. The satellite lines at 25 052 cm^{-1} have slower decay times, $\tau_2 = 2$ ns; they probably reflect the emission duration. The rise and decay times from the second distribution D_2 indicate[87] that the relaxation becomes faster as the exciton is created nearer the energy E_0 (with a population time of a few picoseconds), in agreement with the increase in the polariton density of states.

To conclude this discussion on the intrinsic emission of a 3D crystal, we must insist on the extreme complexity of the emission dynamics, very recently investigated with high-resolution time-resolved spectroscopy.[87] Many regimes coexist, including "relaxed fluorescence" originating from the main distribution D_1, but also hot luminescence from evolving polaritons. Most of the relaxation mechanisms are one-phonon processes, so that we may as well consider these emissions as preresonant Raman scattering of the various created polaritons (second ref. 87). Experimental studies could help to check this model, by creating the invoked polaritons directly and by subtracting their luminescence from the fluorescence spectrum.

III. SURFACE-EXCITON PHOTODYNAMICS AND INTRINSIC RELAXATION MECHANISMS

In solid-state chemical physics, surface phenomena and surface characterization have long been of considerable interest to scientists working in physics, chemistry, thermodynamics, biology, astrophysics, and other fields.

As the interface of two media, the surface is a confined domain which may catalyze intense activity—physicochemical reactions, charge-carrier generation, nonlinear optics—owing to the transition between the two phases and to the breaking of bulk symmetries.[40] Furthermore, the surface behaves as a necessary doorway channel for elastic[105] and inelastic exchanges between the bulk and its environment, depending critically on its macroscopic and microscopic characteristics.[106] The essential difficulty in a selective investigation and definition of the surface resides in the small number of surface molecules, whose properties are obscured by those of the substrate molecules. Various techniques are currently used for investigating metal and semiconductor surfaces, such as slow-electron scattering,[107] ellipsometry optics,[108] nonlinear surface optics,[40,106] supersonic jets of atoms and small molecules,[105] polarization modulation,[109] and the very recent STM method.[110] Molecular crystals, however, are rarely amenable to these techniques.

In this section we analyze the surface investigation of molecular crystals by the technique of UV spectroscopy, in the linear-response limit of Section I, which allows a selective and sharp definition of the surface excited states as 2D excitons confined in the first monolayer of intrinsic surfaces (surface and subsurfaces) of a molecular crystal of layered structure. The (001) face of the anthracene crystal is the typical sample investigated in this chapter.

A theory of 2D excitons and polaritons is presented for this type of surfaces, with continuity conditions matching 2D states their 3D counterparts in the bulk substrate, investigated in Sections I and II. This leads to a satisfactory description of the excitations (polaritons, excitons, phonons) and their theoretical interactions in a general type of real finite crystals: A crystal of layered structure (easy cleavage) with strong dipolar transitions (triplet states do not build up long-lived polaritons).

In this theory, the dynamics of the intrinsic-surface-confined excitons account surprisingly well—in a natural way, without introducing ad hoc parameters—for the surface emissive properties, and they allow, *a contrario*, a very sensitive probing of various types of surface disorders, whether residual, accidental, or induced. The disorder may be thermal, substitutional, chaotic owing to surface chemistry, or mechanical owing to interface compression. It may be analyzed as a specific perturbation of the surface exciton's coherence and of its enhanced emissive properties.

In contrast with other materials (metals, semiconductors, ionic crystals), where disruption of strong surface bonds leads to a complete reorganization of the surface, with extended disorder and strong perturbation of the intrinsic surface and bulk states, the molecular crystals, held together by weak van der Waals bonding, produce surfaces where disruption of bonds causes only weak perturbation of the intrinsic surface and bulk states. Furthermore, if the surface is an easy cleavage plane, this leads to very weak, nondipolar interplane interactions, so that the surface plane (whose energy degeneracy with the bulk planes is lifted by the missing interactions[1] constitutes a nearly perfect 2D system, creating a new, well-resolved band of 2D coherent crystal states, coupled to the bulk unperturbed states via the radiation field. (This explains the failure of slow-electron scattering in selective investigation of surface molecules and contrasting success of UV spectroscopy, which at low temperatures is very sensitive (1) to the microscopic environment of the molecules (short-range coulombic interactions) by resonant frequency selection, and (2) to long-range coulombic and retarded 2D interactions, leading to intrinsic coherent surface states, by selective coupling to the giant transition dipoles of the surface states). Indeed, the most specific feature of the 2D surface states is the giant transition, which is not polariton-forbidden.[1] This provides the well-known enhanced surface radiative channel, with radiative constants of 1 ps and less, explaining the absence of shallow traps and trapping effects on the surface states, the trapping rates being usually on the nanosecond scale.

Therefore, at low temperatures, the surface is characterized by huge absorption and emission powers, selective at the surface resonance (and unperturbed by the nearby very weak 0–0 bulk transition: see Section II), following a single picosecond emission regime from the state $\mathbf{k} = \mathbf{0}$, in contrast with the very complex and slow emission regime of the bulk polariton, discussed in Section II. This picture has to be complemented by the substrate effect: by coherent coupling the substrate may modulate the surface emission rate from picosecond to forbidden emission, while surface-to-bulk nonradiative decay rates, $\mathbf{k}$-dependent, may reach femtosecond values. This intrinsic scheme of energy exchange allows us to picture the intrinsic surface as a doorway state pumping radiative energy (or other energies capable of exciting the surface states) and transferring it to the bulk with a $\mathbf{k}$-dependent selectivity, except for the $\mathbf{k} = \mathbf{0}$ state, for which the radiative channel dominates. We may single out a few consequences of the enhanced radiative channel of the (001) surface states, investigated in this section:

1. Reflectivity spectroscopy: The quasi-metallic reflection of the bulk near the 0–0 transition is sharply modulated by well-resolved signatures of the surface and of the subsurfaces. The positions, the intensity, and the shape of the signature allow one to investigate the surface–bulk interactions.[121]
2. Excitation spectroscopy: Monitoring of the surface emission allows one to discriminate the upper excited surface states and their relaxation dynamics. Problems such as surface reconstruction, or quantum percolation of surface excitons upon thermal and static disorder, are connected with high accuracy to changes of the exciton spectra.[61,118,119,121]
3. Surface photochemistry: Monitoring the reflectivity of high-quality crystals (defects minimized) allows one to follow in time slow photoreactions, especially autocatalytic photoreactions.[120]
4. True 2D systems: Recent progress in the engineering of monolayer solid

films, or monolayers on water surfaces, has renewed interest in the investigation of 2D systems in building perfect monolayers of very complex, strongly absorbing molecules. In addition, more recent progress has allowed obtaining planned microorganizations (*J* aggregates). These two systems, in connection with the investigation of biological membranes,[106] have led to new and general interest in 2D systems, ordered and disordered, such as are investigated in the present work.[111]

These surface excitation phenomena are investigated in this section, on excitons in the intrinsic-surface–bulk system, and the next section, on disordered 2D excitons.

The concept of a surface-confined exciton in the (001) face of the anthracene crystal was introduced by the Kiev group[112] in order to account for the very peculiar structures observed by Brodin, Dudinski, and Marisova[113] in the reflectivity spectra, and those observed by Glockner and Wolf[82] in the excited fluorescence in a similar experiment. Since then, the interpretation of the reflectivity spectra in terms of monolayer surface effects, in the pioneering work of the IBM group in San José,[114] has definitively established the existence of surface 2D electronic excitons in the anthracene crystal. The Bordeaux group has reported and analyzed the zero-phonon fluorescence of the first three surface (001) monolayers[117]; the Chicago group, using its elaborate ATR technique, has explored and analyzed the dispersion of surface polariton modes and their interactions with bulk states.[115,116] Recently, the Bordeaux group has devised a theory of 2D disordered dipolar excitons accounting for the surprising stability of the surface coherent picosecond emission against disorder[118] and against autocatalytic photodimerization of the surface molecules,[41] and predicting the threshold collapse of the surface emission under the combined effects of disorder and the substrate.[118,119]

In this section we assume the existence of intrinsic surface excitons, and we restrict ourselves to a succinct summary of the basic experimental data, whose detailed and original description has been given elsewhere.[1,67,120]

In Section III.A we examine the optical response in reflectivity of the surface, particularly its coupling to the radiation field, when the matter system changes dimensionality from 3D to 2D, drastically modifying the radiative stability of the system. In Section III.B, we analyze the emission of the surface layer coupled to the bulk substrate and the excitation spectra of this emission. Relaxation mechanisms, involving mainly fission and Raman-like processes in exciton–vibration and exciton–phonon couplings (cf. Section II) are proposed to account for the peculiar intrasurface relaxation to the emitting surface states. This allows us to probe relaxation in 2D lattices, at low temperature, involving intramolecular vibrations and optical lattice phonons, by preresonant and resonant Raman scattering,[77] in conjunction with multiphonon creation mechanisms providing fluorescence. While in Sections III.A and B we do not examine the physical processes which allow the existence of surface states, Section III.C is entirely devoted to a discussion, by no means exhaustive, of these processes and to their compatibility with the information presented in Section III.A and B.

A. The Optical Response of the Surface Exciton

The optical response of a monomolecular layer consists of scattered waves at the frequency of the incident wave. Since the surface model is a perfect infinite layer, the scattered waves are reflected and transmitted plane waves. In the case of a 3D crystal, we have defined (Section I.B.2) a dielectric permittivity tensor providing a complete description of the optical response of the 3D crystal. This approach, which embodies the concept of propagation of dressed photons in the 3D matter space, cannot be applied in the 2D matter system, since the photons continue propagating in the 3D space. Therefore, the problem of the 2D exciton must be tackled directly from the general theory of the matter–radiation interaction presented in Section I.

Before doing so, we summarize the basic experimental features of the surface exciton states.

1. *Experimental Observations*

a. *The Reflectivity of the Crystal Surface.* The most pertinent features of the surface states are the three interference structures (I, II, III) indicated in Fig. 3.1. This spectrum is typical of the (001) face of a crystal flake, sublimation-grown from very pure anthracene.[67,120]

As illustrated in Fig. 2.8 of Section II, the general reflectivity lineshape shows: (1) a sharp rise of the bulk 0–0 reflectivity (Section II.B.C) at E_{vb}^{00}, corresponding to the b coulombic exciton with a wave vector perpendicular to the (001) face; (2) a dip, corresponding to the fission in the surface of a bulk polariton into one "46"-cm^{-1} phonon and one b exciton at E_{vb}^{00} + "46" cm^{-1}; (3) two vibrons E_{vb}^{390} and E_{vb}^{1400} immersed in their two-particle-state continua with sharp low-energy thresholds. On this relatively smooth bulk reflectivity lineshape are superimposed sharp and narrow surface 0–0 transition structures whose observation requires the following:

1. Very good chemical purity: Very low concentration of deep traps causing fast trapping (on the order of picoseconds) and extended disorder. The very low level of chemical impurity required ($< 10^{-6}\, M$[120]) is in fact too severe.
2. Very good physical quality: Minimization of step-shaped surface and of

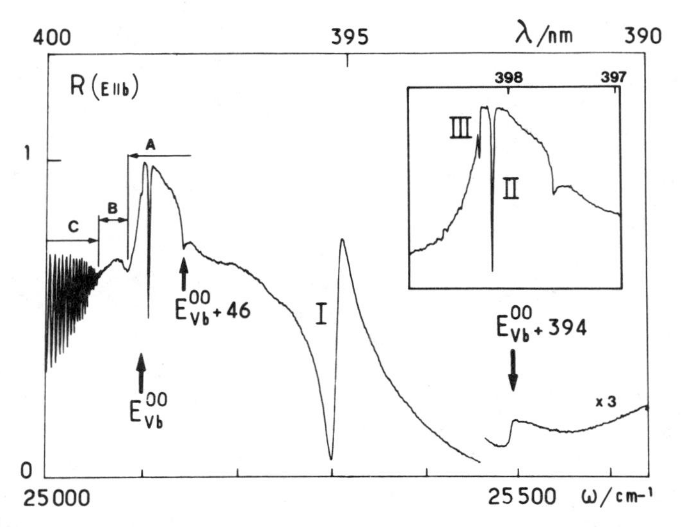

Figure 3.1. The *b*-polarized 0–0 reflectivity of a (001) face at 1.7 K. We note the surface structures, I, II, III, as sharp interference signatures on the strong 0–0 bulk reflectivity; see J. M. Turlet et al.[1] and (3.25)–(3.26).

dislocation defects contributing to extended disorder.[122] A constant thickness ($\sim \lambda$) must be obtained, as confirmed by a uniform color in the polarizing microscope.

3. Absence of chemical and photochemical pollution on the crystal surface, since crystalline anthracene is very sensitive to foreign bodies and to photoreactions.
4. Observation of the spectra at low temperatures, with an upper limit of 80 K. As discussed in Section II.C, the surface structures follow the same thermal broadening as the bulk 0–0 structures. This provides three critical temperatures—$T_{\rm I} < 77\,$K, $T_{\rm II} < 30\,$K, and $T_{\rm III} < 5\,$K—as the energy gap between the surface and the bulk structures decays from 200 to a few reciprocal centimeters: see Fig. 3.1.

As to the homogeneous broadening related to the natural lifetime of the coherent surface states, this will be investigated in this section in connection with the relaxing effect of the substrate bulk, as well as in Section IV in connection with the dephasing surface disorders examined.

It is intuitively obvious, although not yet explained, that the nearer the surface structure lies to the bulk resonance, the more severe the above empirical conditions become, especially those on the physical quality of the crystal. The presence of structure III is the sign of a high-quality crystal, while the presence of a broad structure located between structures II and I (which are strongly attenuated or even quenched) is the sign of poor quality.[67,120] The latter connection has been recently analyzed by UV-controlled destruction of the crystal surface.[41] When the above empirical prerequisites are satisfied, the experimental characteristics of the surface structures are as follows:

Structure I, a quasi-lorentzian line, at 25 301 cm^{-1}, of width 13 cm^{-1} at $T = 1.7$ K;

Structure II, a sharp dip, at 25 102 cm^{-1}, of width 0.8 cm^{-1} at $T = 1.7$ K;

Structure III, a sharp dip, at 25 094 cm^{-1}, of width 0.4 cm^{-1} at $T = 1.7$ K.

The details of structures II and III, near the bulk 0–0 band, are highly resolved in Fig. 3.2.

The most important proof that the origin of the observed structures is the surface monomolecular layer has been provided by Philpott and Turlet[114] using their technique of surface gas coating at low temperatures. They observed, upon coating, reversible shifts of the surface structures, to lower energies and proportional to the new van der Waals forces of the condensed

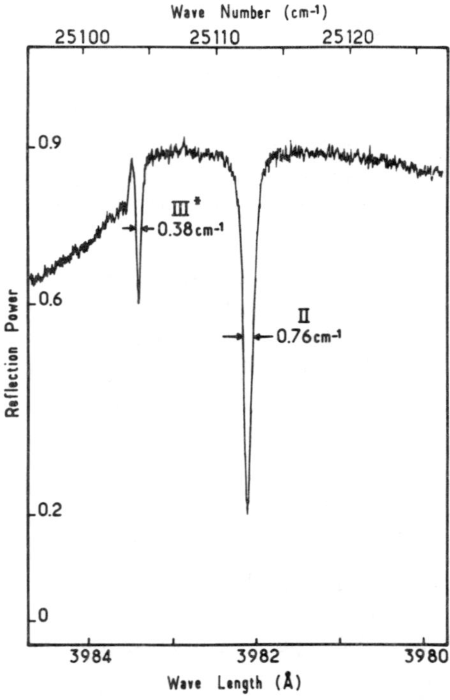

Figure 3.2. High-resolution (0.3 cm^{-1}) detail of structures II and III of Fig. 3.1. We note that the two structures get much deeper, although not yet perfectly resolved.

gas, acting regressively on the three first surface layers.[67,114] However, what is less obvious is that the shifts are accompanied by a significant proportional narrowing (of the order of 50%) of the structures and a shape change due to van der Waals adiabatic interactions originating from the disordered condensed-gas layer, which should rather produce an inhomogeneous additional width in the shifted structure. The analysis of this narrowing will yield significant insight into the combined effects of the condensed-gas–surface interface and surface–substrate-bulk interface; see Fig. 3.3.

As for the surface structures related to vibronic bulk structure, we must first remark that the *b* 0–0 bulk line has been analyzed and found to be extremely narrow (less than 1 cm^{-1} as analyzed by KK transformation in Section II). This very narrow line, related to the strong polariton character of the transition, is the reason why nearby surface structures are observed with very good resolution, as shown in Figs. 3.1–2.

In contrast, surface vibronic structures cannot be clearly observed in reflectivity, because the bulk vibronic structures are intrinsically very broad and so are the surface vibronic structures. However, we have detected, on the bulk vibronic structures, surface substructures which are very sensitive to the surface-gas-coating test: for instance, on the *b* vibronic structure "1400"-cm^{-1} a small bump that shifted reversibly, and on the *a*-polarized broad reflectivity, substructures sensitive to gas coating. In both cases, it is difficult to draw quantitative conclusions on surface vibronic states, as it was for the surface structures I, II, and III on the *b* 0–0 bulk reflectivity spectrum.

This challenge has been met by a new technique based on the excitation of the surface I fluorescence, initiated in our laboratory, which allowed us to set up a powerful tool for *photoselection* of surface and subsurface states. This technique combines the superradiant 2D character of the surface emission, its very good spectral resolution at $T < 80$ K, and its strong sensitivity to gas coating.

b. *Excitation Spectra of the Surface Fluorescence.* In Section II.D we investigated the crystal excited fluorescence, and pointed out the fluorescence origin at 25 093 cm^{-1} (Fig. 2.21), whose lineshape reflects the fast (picosecond) relaxation of the created excitons to the bottom of the exciton band with emission, including the polariton-forbidden 0–0 emission line. Figure 3.4 shows three weak emissions, above the fluorescence origin, coinciding with the three reflectivity structures. Furthermore, these emissions display, upon gas coating, exactly the same behavior as the reflectivity structure, which is the pertinent signature of the surface origin of the observed new fluorescence lines. The excitation of the surface fluorescence and its discrimination with gas coating has yielded key information on the surface exciton states, and on the various channels of relaxation to be discussed in Section III.B.

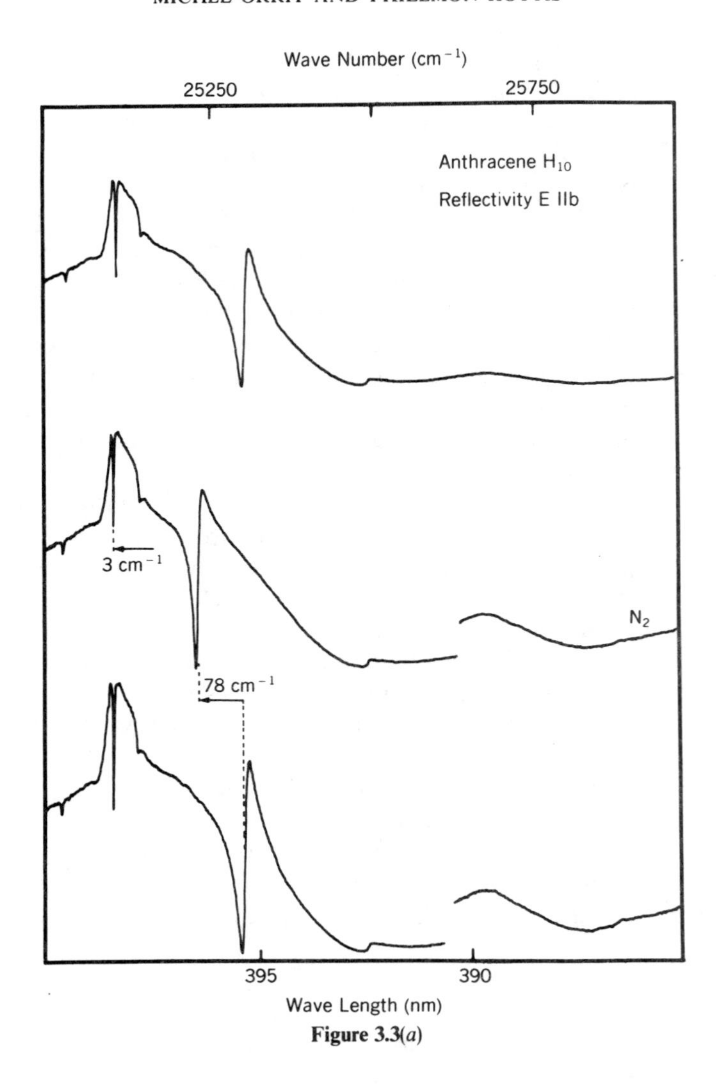

Figure 3.3(*a*)

For a coherent interpretation of the reported experimental data, we need a model of surface excitons, the structures I, II, and III being attributed to excitons confined, respectively, in the first, the second, and the third surface monolayer (see Fig. 3.5). The rapid decay of the van der Waals forces along the c' axis explains the very fast transition, in a few molecular layers, from surface to bulk spectroscopy (the other two faces of the anthracene crystal do not show surface-confined excitons).

As already indicated,[112] the existence of such confined $\mathbf{k} \perp c'$ excitons implies that the exchange interaction energy between excitations of the (a, b)

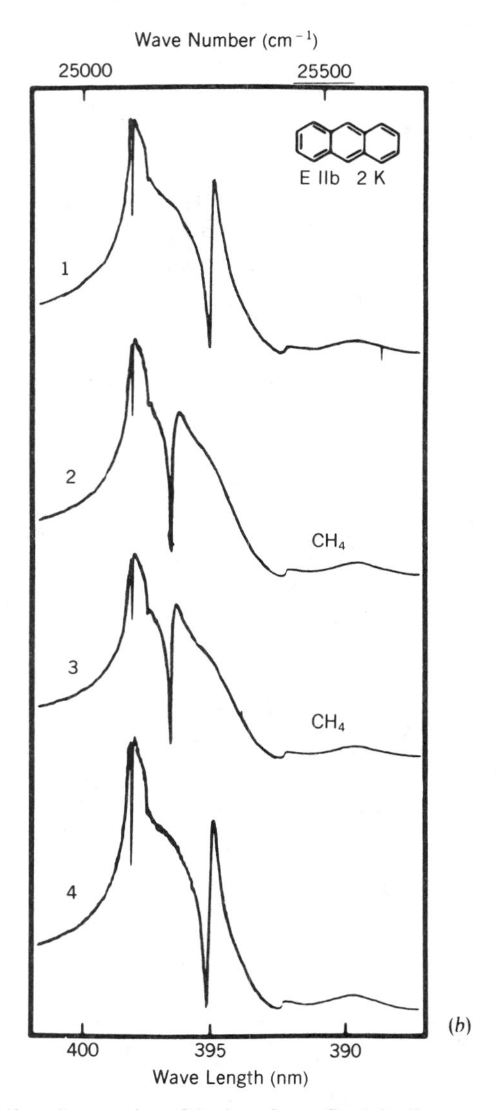

Figure 3.3. Red shift and narrowing of the *b*-surface reflectivity line upon gas condensation on the crystal surface: Left, N_2; right, CH_4.

layers is much weaker than the lack of resonance between zero-order excitations of the concerned layers. If this were not the case, only bulk polaritons could exist, as, for example, for excitons $\mathbf{k} \parallel b$, $\mathbf{k} \parallel a$: excitons cannot be confined in (a, c) and (b, c) crystal monolayers, because, owing to the anisotropy of the excitonic interactions, the corresponding exchange interac-

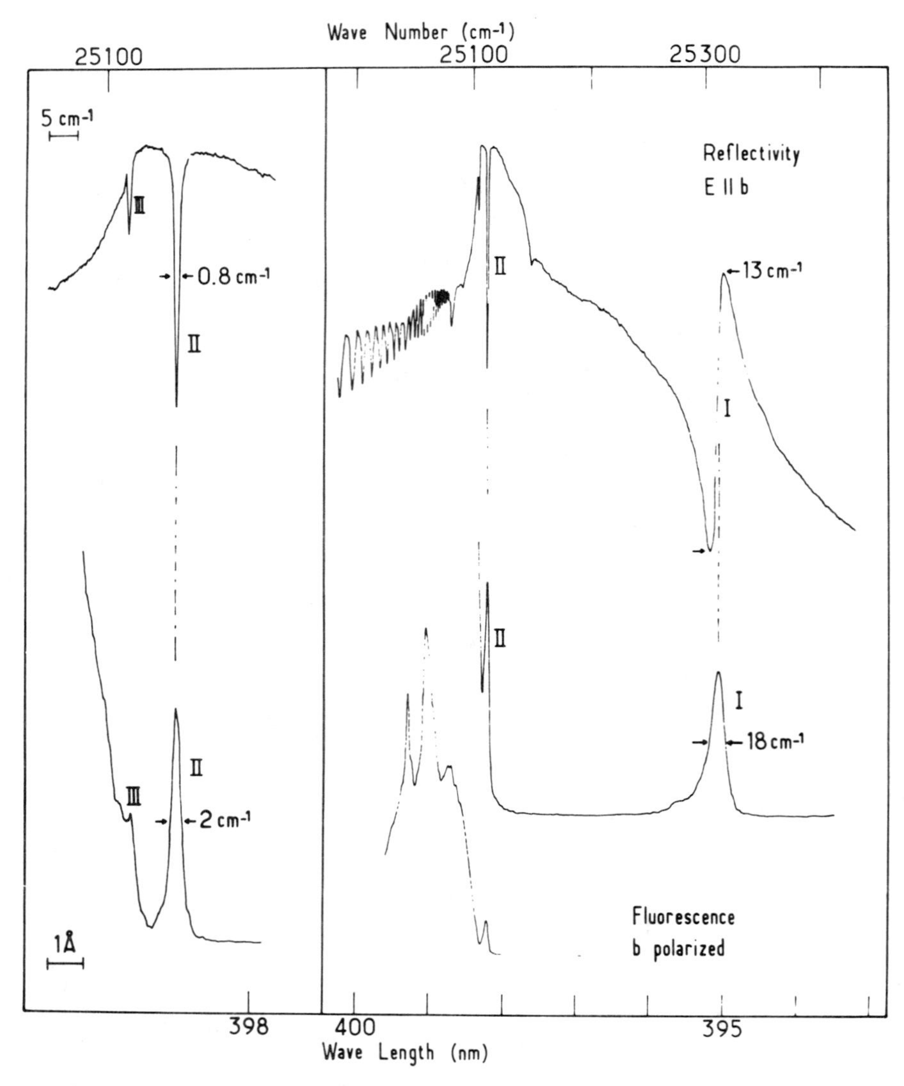

Figure 3.4. Comparison of the reflectivity and the excited fluorescence of one crystal, under the same conditions, at 1.7 K. (The bulk fluorescence is shown to scale out the weakness of the surface fluorescence.) The exact coincidence of the surface fluorescence structures with the reflectivity counterparts reveals the coherent (ultrafast) emission of the surface molecules, subsequent to the intramolecular relaxation of the excited state (cf. Section IV.A).

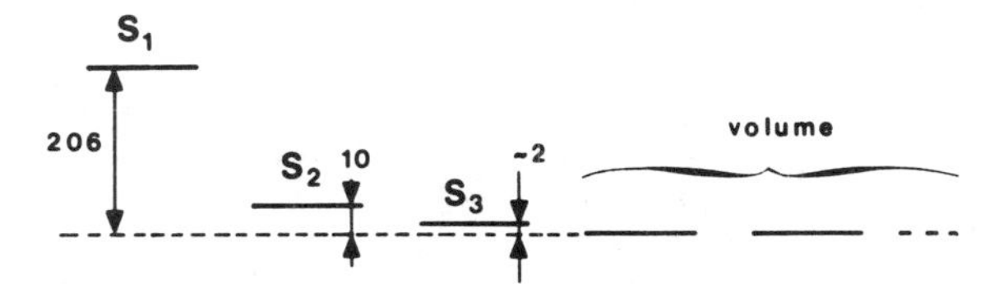

Figure 3.5. Diagram showing the change of the resonance energy of the (001) planes in the transition from the surface plane to a bulk plane. In the model adopted, the resonance energy shifts upwards by 200, 10 and about $2\,\text{cm}^{-1}$ for the first three planes S_1, S_2, and S_3, respectively.

tion energy is very large. The theoretical calculations by Philpott allow one to evaluate at $\sim 2\,\text{cm}^{-1}$ the excitation exchange energy between two nearest-neighbor (a, b) layers, including nondipolar interactions (see Section I.C) and upper-state contributions.[27] These calculations are compatible with the experimental observations that the exciton confined in the third surface layer cannot be separated by more than about $2\,\text{cm}^{-1}$ from one in the fourth layer.

To first order, we consider the molecular structure of the surface layers to be identical to that of the bulk layers. Consequently, all the characteristics corresponding to short-range intralayer interactions (e.g. Davydov splitting, vibrational frequencies, excitonic band structure, vibronic relaxations are similar for bulk and surface layers). In fact, we shall see that even slight changes may be detected. They will be analyzed in Section III.C, devoted to surface reconstruction. Therefore, our crystal model consists of (a, b) monolayers translated in energy relative to the bulk excitation by 206, 10, and $2\,\text{cm}^{-1}$ for the first three layers, as indicated in Fig. 3.5. No other changes are considered in this first-order crystal model.

2. *A Layer of Dipoles Coupled to the Radiation Field*

We consider an ideal surface, an infinite monomolecular layer defining the 2D lattice. First we assume that the lack of resonance with the bulk is large enough for us to assume an isolated layer in the 3D space. In Section III.A.3 we discuss the way this picture changes when the layer lies near a substrate with which it interacts via the radiation field.

We have shown that in the 3D case (Section I.B) the exciton behavior near the middle of the zone ($\mathbf{K} = \mathbf{0}$), and particularly its coupling to the radiation field, may be described using a 3D lattice of point dipoles. The effects of the finite character of the lattice elements and of the nondipolar interactions (analytic at $\mathbf{K} = \mathbf{0}$) will be assumed included in the analytic term $\mathbf{t}(\mathbf{K})$ [cf. (1.63)–(1.64)].

In an analogous way with that in Section I.B, we must calculate the dipole sum, or the radiated field, using the dipole distribution

$$\rho_g(\mathbf{r}) = \sum_{\mathbf{n}} \rho(\mathbf{r} - \mathbf{n}) e^{i\mathbf{K}\cdot\mathbf{n}} \tag{3.1}$$

corresponding to a wave of 2D wave vector **K**, propagating in the plane. Here $\rho(\mathbf{r} - \mathbf{n})$ is the dipole-moment distribution of a dipole located at **n**. The field created by the wave **K** at a point **r** may be calculated using the dipole-sum tensor, which is the solution of

$$\boldsymbol{\phi}_g(\mathbf{r}, \omega) = \frac{-1}{4\pi\varepsilon_0}\left(\frac{\omega^2}{c^2} + \nabla\nabla\right) G_g(\mathbf{r}, \omega) \tag{3.2}$$

and

$$\left(\Delta + \frac{\omega^2}{c^2}\right) G_g(\mathbf{r}, \omega) = -4\pi\rho_g(\mathbf{r}) \tag{3.3}$$

Following a Fourier transformation necessary for the solution of (3.2)–(3.3), we are looking for the tensor $\boldsymbol{\phi}(\mathbf{k}, \omega)$. Then, the summation (3.1) introduces restrictions on the component of the wave vector **k** contained in the 2D layer ($\mathbf{k}_\parallel$): $\mathbf{k} = \mathbf{k}_\parallel + \mathbf{q}$, with $\mathbf{k}_\parallel \cdot \mathbf{q} = 0$, leading to

$$\tilde{\boldsymbol{\phi}}_g(\mathbf{k}, \omega) = \frac{1}{\varepsilon_0} \frac{(\omega^2/c^2 - \mathbf{k}\mathbf{k})}{(\omega^2/c^2 - k^2)} \frac{(2\pi)^2}{S_0} \sum_G \delta(\mathbf{K} + \mathbf{G} - \mathbf{k}\ )\tilde{\rho}(\mathbf{k}) \tag{3.4}$$

where S_0 is the unit-cell surface of the 2D lattice. On transition back to the direct space **r**, the only surviving integration is the one in the direction of **k** normal to the 2D layer, i.e. **q**, which provides the contribution

$$\begin{aligned} \boldsymbol{\phi}_g(r, \omega) = \frac{1}{2\pi\varepsilon_0 S_0} \sum_{\mathbf{G}} \int dq \frac{\omega^2/c^2 - (\mathbf{K} + \mathbf{G} + \mathbf{q})(\mathbf{K} + \mathbf{G} + \mathbf{q})}{\omega^2/c^2 - (\mathbf{K} + \mathbf{G} + \mathbf{q})^2} \\ \times e^{i(\mathbf{K} + \mathbf{G} + \mathbf{q})\cdot\mathbf{r}} \tilde{\rho}(\mathbf{K} + \mathbf{G} + \mathbf{q}) \end{aligned} \tag{3.5}$$

In the summation on the reciprocal-lattice vectors **G**, only the contribution $\mathbf{G} = \mathbf{0}$ is important in the vicinity of $\mathbf{K} = \mathbf{0}$, because of rapid variation with **K**. The other terms, with $\mathbf{G} \neq \mathbf{0}$, may be included in the analytic part of the dipole sum, $\mathbf{t}(\mathbf{K})$. In addition, the variation of the term $\mathbf{G} = \mathbf{0}$ vs **r** being slow, we may safely take $\mathbf{r} = 0$ (even for nondipolar systems). The final result is

$$\boldsymbol{\phi}_g(\mathbf{K}, \mathbf{r} = 0, \omega) = \mathbf{t}(\mathbf{K}) + \frac{1}{2\pi\varepsilon_0 S_0} \int_{-\infty}^{+\infty} dq \frac{\omega^2/c^2 - (\mathbf{K} + \mathbf{q})(\mathbf{K} + \mathbf{q})}{\omega^2/c^2 - (K^2 + q^2)} \tag{3.6}$$

[with $\tilde{\rho}(\mathbf{k}) = 1$ for the point dipoles].

The sum (3.6) contains an integral over a wave vector $\mathbf{q}$ normal to the 2D lattice. Furthermore, since we have shown in Section I.A that this integral is simply that on the photon continuum, one concludes that this continuum reduces (in view of the conservation of the layer wavevector $\mathbf{K}$) to the one-dimensional continuum of photons $\mathbf{K}+\mathbf{q}$, with $\mathbf{q}\perp\mathbf{K}$. Thus, this effective continuum has an energy minimum threshold corresponding to the lowest-energy photon ($\mathbf{q}=\mathbf{0}$) of energy $E_m = \hbar c/K$. The obvious conclusion is that the surface exciton will be stable or unstable according as $\omega \gtrless c|\mathbf{K}|$ (cf Section III.A.2.b below). The integral (3.6) has a divergent part $\hat{\mathbf{u}}\hat{\mathbf{u}}\int_{-\infty}^{\infty} dq$ ($\hat{\mathbf{u}}$ is the unit vector normal to the lattice). This divergence has the same origin as that encountered in equation (1.38): It is connected with the point character of the lattice dipoles. As this divergent part corresponds to a global shift, independent of $\mathbf{K}$ but depending on the fixed direction of the transition dipole, we may safely neglect it. Integration of (3.6) using residue theory (only the poles with positive imaginary part contribute) provides the expression

$$\boldsymbol{\phi}_g(\mathbf{K},\omega) = \frac{-i}{2\varepsilon_0 S_0}\frac{(\omega^2/c^2)(1-\hat{\mathbf{u}}\hat{\mathbf{u}}) - K^2(\hat{\mathbf{K}}\hat{\mathbf{K}}-\hat{\mathbf{u}}\hat{\mathbf{u}})}{(\omega^2/c^2 - K^2)^{1/2}} + \mathbf{t}(\mathbf{K}) \tag{3.7}$$

with $\operatorname{Im}(\omega^2/c^2 - K^2)^{1/2} > 0$, $\hat{\mathbf{K}}$ denoting the unit vector of $\mathbf{K}$.

a. *Coulombic Dispersion of the* 2D *Exciton.* In the vicinity of $\mathbf{K}=\mathbf{0}$, the retarded effects ($\omega/c \to 0$) being neglected, the expression (3.7) becomes real:

$$\boldsymbol{\phi}_g(\mathbf{K},\omega=0) = \frac{1}{2\varepsilon_0 S_0} K(\hat{\mathbf{K}}\hat{\mathbf{K}}-\hat{\mathbf{u}}\hat{\mathbf{u}}) + \mathbf{t}(\mathbf{K}) \tag{3.8}$$

This expression shows also a nonanalytic dispersion at $\mathbf{K}=\mathbf{0}$, but which is continuous: The nonanalyticity of the 2D exciton has regressed by one order compared to the 3D exciton; cf. Section I.B.1. Furthermore, as we are interested in effects around $\mathbf{K}=\mathbf{0}$ for the b component of the anthracene crystal, it is legitimate to replace the whole unit cell by a transition dipole

$$\mathbf{d} = \mathbf{d}_{\parallel} + d_{\perp}\hat{\mathbf{u}} \tag{3.9}$$

with a $\mathbf{K}$ dispersion

$$\mathbf{d}\cdot\boldsymbol{\phi}(\mathbf{K},\omega=0)\cdot\mathbf{d} = \frac{1}{2\varepsilon_0 S_0} K[(\mathbf{d}_{\parallel}\cdot\hat{\mathbf{K}})^2 - d_{\perp}^2] + \mathbf{d}\cdot\mathbf{t}(\mathbf{K})\cdot\mathbf{d} \tag{3.10}$$

For the b surface exciton, with its dipole contained in the 2D lattice layer, the nonanalytic part of the dispersion has the value $(d^2/2\varepsilon_0 S_0)K\cos^2(\mathbf{K},\mathbf{d})$. For the optical wave vectors, this dispersion is less than 1 cm^{-1}, leading to the conclusion that *the nonanalyticity effect is much less pronounced in 2D excitons than in 3D crystal excitations.* This property may be further analyzed in the vicinity of the bottom of the excitonic band by expanding the $\mathbf{t}(\mathbf{K})$ interactions in powers of K; the dispersion becomes

$$E(K) = E_0 + \alpha K \cos^2 \theta + \beta K^2 \tag{3.11}$$

with $\theta = (\mathbf{K}, \mathbf{d})$. The corresponding isoenergetic curves have been drawn in the (K_a, K_b) plane in Fig. 3.6. It is clear that for the lower energies the excitonic wave vectors are oriented preferentially perpendicular to $\mathbf{d}$. Furthermore, the shape of the density of states is determined from (3.11): in the vicinity of $K = 0$ it varies as $(E - E_0)^{1/4}$, showing (for all 2D lattices with strong dipolar transitions) a behavior smoother than that of a 2D exciton with analytic dispersion in K^2 and a density of states undergoing a discontinuity at E_0.

b. *Effect of Retarded Interactions.* Using (3.7) and (3.9), it is straightforward

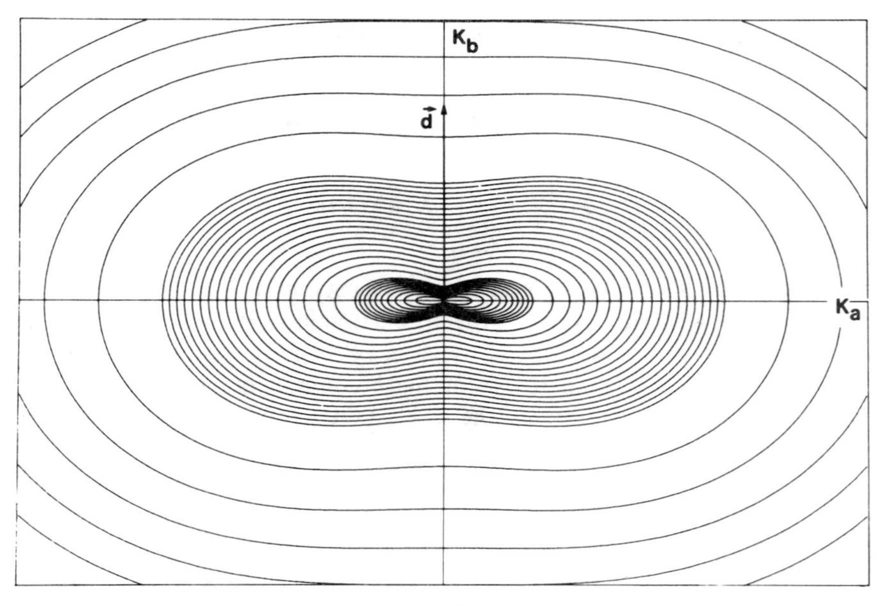

Figure 3.6. Isoenergetic contours in the middle of the Brillouin zone, for a 2D exciton with purely coulombic dispersion. For the large wave vectors, the parabolic dispersion in K^2 prevails and the contours are quasi-circular, whereas in the vicinity of the middle of the zone, the coulombic dispersion in $K \cos^2 \theta$ tends to make $\mathbf{K}$ perpendicular to the transition dipole $\mathbf{d}$. Compare these contours with the 3D exciton, Fig. 2.18.

to obtain the complete expression for the complex renormalization of the excitonic energy $R_{\mathbf{K}}(\omega)$:

$$R_{\mathbf{K}}(\omega) = \frac{-i}{2\varepsilon_0 S_0} \frac{d^2(\omega^2/c^2 - K^2 \cos^2\theta) + K^2 d_\perp^2}{(\omega^2/c^2 - K^2)^{1/2}} + \mathbf{d}\cdot\mathbf{t}(\mathbf{K})\cdot\mathbf{d} \tag{3.12}$$

Thus, $R_{\mathbf{K}}(\omega)$ is imaginary for $K < \omega/c$ and real for $K > \omega/c$; these two cases correspond, respectively, to a radiatively unstable 2D exciton and to a radiatively stable 2D polariton. Solution of equations (1.3) and (3.12) provides the complete description of the 2D polariton dynamics. We analyze below the 2D polariton at different orders of exciton–photon coupling.

To the lowest order of exciton–photon interaction, we are looking for the excitonic solutions slightly perturbed by the photon continuum. For this case, we may put $z = \hbar\omega_0$ in (1.33), so that for each excitonic wave vector K we obtain an equation providing a complex excitonic energy ($\hbar = 1$):

$$z^2 - \omega_0^2 - 2\omega_0 R_{\mathbf{K}}(\omega_0) = 0 \tag{3.13}$$

leading to the approximated solution

$$z = \omega_0 + R_{\mathbf{K}}(\omega_0) \tag{3.14}$$

Following (3.14), we find for $K < \omega/c$ a radiatively unstable exciton emitting photons. This property is easily interpreted on the basis of the conservation, during the emission, of the 2D component of the total wave vector. Furthermore, since to second order the energy of the final photon must be that of the exciton, $\hbar\omega_0$, the simultaneous conservation of momentum and of energy leads to the radiative excitonic states $K < \omega_0/c$. For the remainder of the excitonic branch, $K > \omega_0/c$, the energies are real, and are shifted to lower values, by their coupling to the underlying effective photon continuum. The variation of the renormalization energy $R_{\mathbf{K}}(\omega_0)$ vs K is illustrated in Fig. 3.7 for a transition dipole contained in the 2D layer ($d_\perp = 0$) and for various angles between $\mathbf{K}$ and d .

For the radiative excitons, Im $R_K(\omega_0)$ gives the radiative half width of the 2D excitons. For $K = 0$, the value of this radiative half width is

$$\text{Im}\, R_K(\omega_0) = \tfrac{1}{2}\Gamma_0 = \frac{d^2}{2\varepsilon_0 S_0} \frac{\omega_0}{c} \tag{3.15}$$

with an order of magnitude about $c^2/\omega_0^2 S_0$ times larger than the corresponding value for an isolated dipole, (1.39). This factor, λ_0^2/S_0, represents the

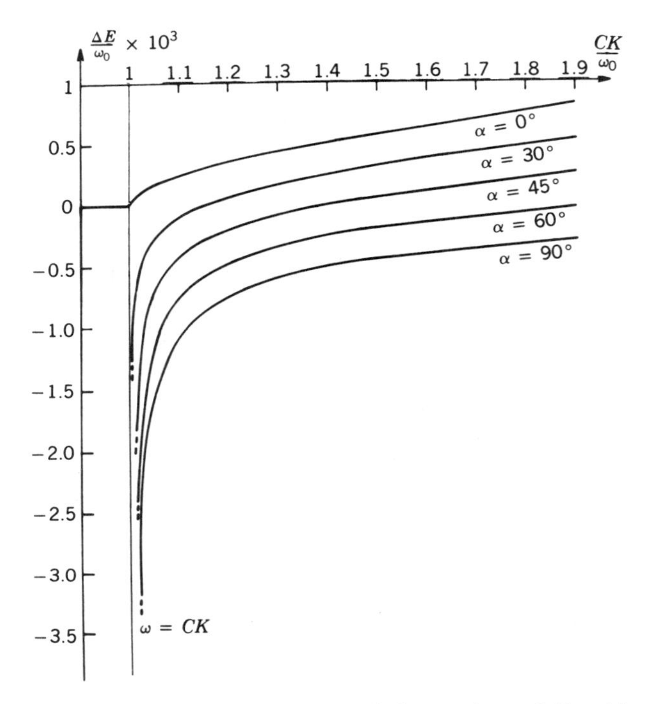

Figure 3.7. Dispersion of the 2D monolayer polariton: real part (left) and imaginary part (right) of the excitonic energy renormalization $R_{\mathbf{K}}(\omega)$, calculated to second order in the exciton–photon coupling, vs the excitonic wave vector K (in units of ω_0/c) for various angles α between **K** and the transition dipole (assumed to lie in the plane). We note the divergence of $\mathrm{Im}\, R_{\mathbf{K}}$ for $K \lesssim \omega_0/c$, and of $\mathrm{R_e}\, R_{\mathbf{K}}$ for $K \gtrsim \omega_0/c$, requiring the inclusion of higher-order terms.[126]

enhancement of the spontaneous-emission power by the excitonic coherence of the **K** wave. The range of this coherence being λ_0^2, the emitted wavelength, (3.15) indicates that about λ_0^2/S_0 surface dipoles cooperate constructively to emit one photon. Using the anthracene parameters, we find that the enhancement factor is of the order of 3×10^5 and leads to a radiative width for the surface excitons of 10 to 20 cm^{-1} (cf. Section III.A.), or to a lifetime of the order of 0.5 to 0.25 ps. We must immediately remark that the excitonic coherence originates from the strong intralayer intermolecular interactions (term $\mathbf{d}\cdot\mathbf{t}(\mathbf{K})\cdot\mathbf{d}$), of the order of 100 cm^{-1}, leading to a dispersion band of 400 to 500 cm^{-1}. Consequently, this strong coherence, and also the surface 2D emission, are relatively insensitive to nonradiative sources of broadening or to the surface disorder. They characterize a special class of crystals: see Section IV.

The order of magnitude of the interaction $R_{\mathbf{K}}(\omega_0)$ of a 2D layer, although very much larger than that of an isolated dipole, is about 50 times weaker than

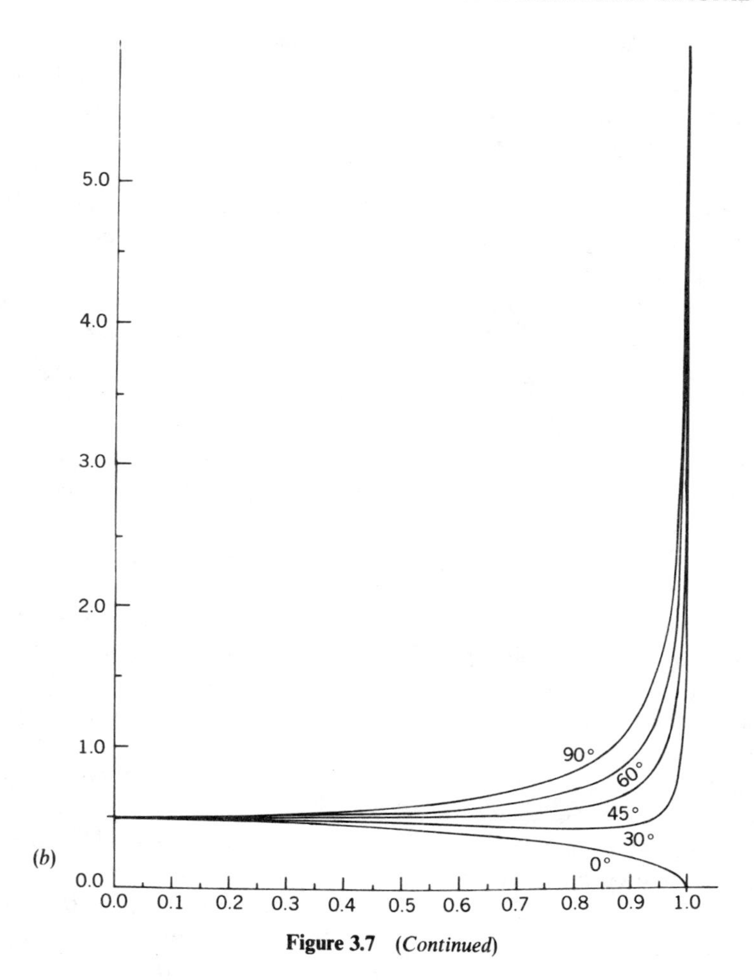

Figure 3.7 *(Continued)*

the corresponding interaction of the 3D lattice, so that the bulk polariton stopping band is quite large ($400\,\text{cm}^{-1}$). For this reason, the second-order exciton–photon-interaction perturbative approach is a very good approximation. Nevertheless, in the vicinity $K = \omega_0/c$, $R_{\mathbf{K}}(\omega_0)$ diverges, leading to the necessity of calculation of the exact complex energies.

c. *The Exact Solution: The 2D Polariton and the Temporal Description.* In the 2D case, the evaluation of the determinant (1.33) is equivalent to finding particular solutions for each wave vector **K**:

$$z^2 - \omega_0^2 - 2\omega_0 R_{\mathbf{K}}(z) = 0 \tag{3.16}$$

The real solutions of (3.16) provide all the eigenenergies of the matter–radiation system. Apart from a discrete state, with $z < cK$, investigated below, (3.16) provides an infinity of real solutions for each $z > cK$. The corresponding eigenstates are those of the effective photon continuum, generated by the photons $\mathbf{K}+\mathbf{q}$, scattered by the excitonic state of wave vector $\mathbf{K}$. Obviously, the eigenenergies alone are not sufficient to characterize the effective photon continuum (perturbed by its coupling to 2D excitons). The most general physical quantities needed for this purpose are the S-matrix elements,[123] providing the probability amplitudes $\langle \mathbf{K}+\mathbf{q}, \varepsilon | S | \mathbf{K}+\mathbf{q}', \varepsilon' \rangle$ for scattering a wavepacket around $\mathbf{K}+\mathbf{q}$ to another wave packet around $\mathbf{K}+\mathbf{q}'$, via interaction with the 2D matter layer. These elements may be given the form

$$\begin{aligned}\langle \mathbf{K}+\mathbf{q}, \boldsymbol{\varepsilon} | S | \mathbf{K}+\mathbf{q}', \boldsymbol{\varepsilon}' \rangle &= \delta_{qq'}\delta_{\varepsilon\varepsilon'} \\ &\quad - \frac{2i\pi}{\hbar} \langle \mathbf{K}+\mathbf{q}, \boldsymbol{\varepsilon} | H_I | \mathbf{K} \rangle \langle \mathbf{K} | H_I | \mathbf{K}+\mathbf{q}', \boldsymbol{\varepsilon}' \rangle \\ &\quad \times G_{\mathbf{K}}(\hbar\omega_{\mathbf{K}+\mathbf{q}} + io)\delta(\omega_{\mathbf{K}+\mathbf{q}} - \omega_{\mathbf{K}+\mathbf{q}'})\end{aligned} \tag{3.17}$$

where H_I is the general exciton–photon interaction [defined in equation (1.26)] and $G_{\mathbf{K}}$ the propagator of the pure excitonic state $|K\rangle$. Thus, the calculation of this propagator is our main goal. Furthermore, it allows us to calculate the emission radiation of an exciton assumed to be prepared at $t = 0$. Also, we remark that in (3.17) the δ function induces the selection rule $\mathbf{q}' = \pm\mathbf{q}$, so that the matrix S describes, respectively, the transmission (+) and the reflection (−) of photons by the 2D layer.

In the above second-order approximation, $G_{\mathbf{K}}(\omega)$ is calculated by introducing an excitonic complex energy, which is an approach adopted by many authors.[124] However, in the vicinity of the region $cK = \omega_0$ we must take account of the rapid variation of $R_{\mathbf{K}}(z)$ in the solution of (3.16) and in the calculation of (3.17).[125] This study has been presented in a detailed manner in Ref. 126; here we reproduce the main results. The propagator $G_{\mathbf{K}}(z)$ is given by

$$G_{\mathbf{K}}(z) = \left(\frac{z^2 - \omega_0^2}{2\omega_0} - R_{\mathbf{K}}(z) \right)^{-1} \tag{3.18}$$

It allows us to calculate successively:

the time-dependent emission probability of the prepared pure exciton,

$$A_{\mathbf{K}}(t) = -\frac{\hbar}{2\pi i} \int_{-\infty}^{+\infty} G_{\mathbf{K}}(\omega + io) e^{-i\omega t} \, d\omega \tag{3.19}$$

the probability of creation of one photon of energy $\hbar\omega$ during the emission (or, more precisely, the probability of a final state of energy $\hbar\omega$),

$$P_{\mathbf{K}}(\omega) = \frac{1}{2\pi} |G_{\mathbf{K}}(\omega)|^2 \operatorname{Im} G_{\mathbf{K}}(\omega) \tag{3.20}$$

the reflection and transmission amplitudes for the 2D layer,[125]

$$r(\omega) = \frac{R_{\mathbf{K}}(\omega)}{(z^2 - \omega_0^2)/2\omega_0 - R_{\mathbf{K}}(\omega)} \tag{3.21}$$

$$t(\omega) = \frac{(z^2 - \omega_0^2)/2\omega_0}{(z^2 - \omega_0^2)/2\omega_0 - R_{\mathbf{K}}(\omega)} \tag{3.22}$$

Equations (3.21)–(3.22) satisfy the conservation law for elastic matter–radiation interactions, $|\mathbf{r}_{\mathbf{K}}(\omega)|^2 + |\mathbf{t}_{\mathbf{K}}(\omega)|^2 = 1$, and the very useful relation $\mathbf{t}_{\mathbf{K}}(\omega) = 1 + \mathbf{r}_{\mathbf{K}}(\omega)$, which accounts for the symmetry of the matter system (when the dipoles lie in the 2D layer) and which is valid for elastic and inelastic interactions. They are obtained if the expression for $\mathbf{R}_{\mathbf{K}}(\omega)$ has only a pure imaginary part, the real part being included in the eigenenergy $\hbar\omega_0$.

Figures 3.8 and 3.9 illustrate the time variation of the emission probability $A_{\mathbf{K}}(t)$ for an exciton $|\mathbf{K}\rangle$, and the spectrum $P_{\mathbf{K}}(\omega)$ of the emitted photons for various values of the ratio cK/ω_0.

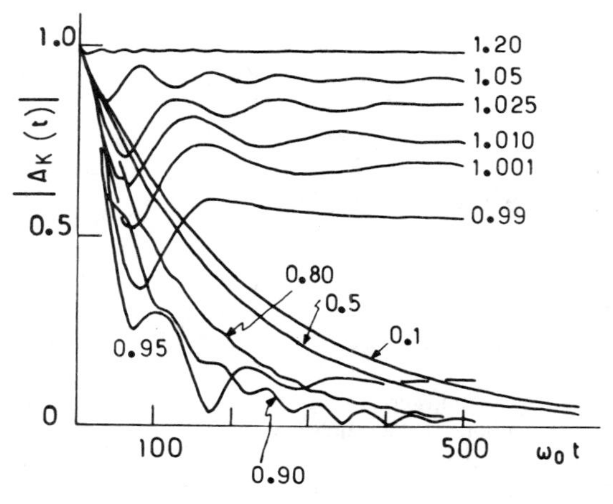

Figure 3.8. The exciton decay in photon and polariton states: The time evolution (in units of $\omega_0 t$) of a 2D exciton **K** created at $t = 0$ ($\mathbf{K} \perp \mathbf{d}$). This decay, illustrated for various wave vectors (in units of ω_0/c), is purely exponential for $K < \omega_0/c$, but exhibits very complex transient oscillatory behavior in the region $K \sim \omega_0/c$. For $K > \omega_0/c$ the 2D exciton is radiatively stable.

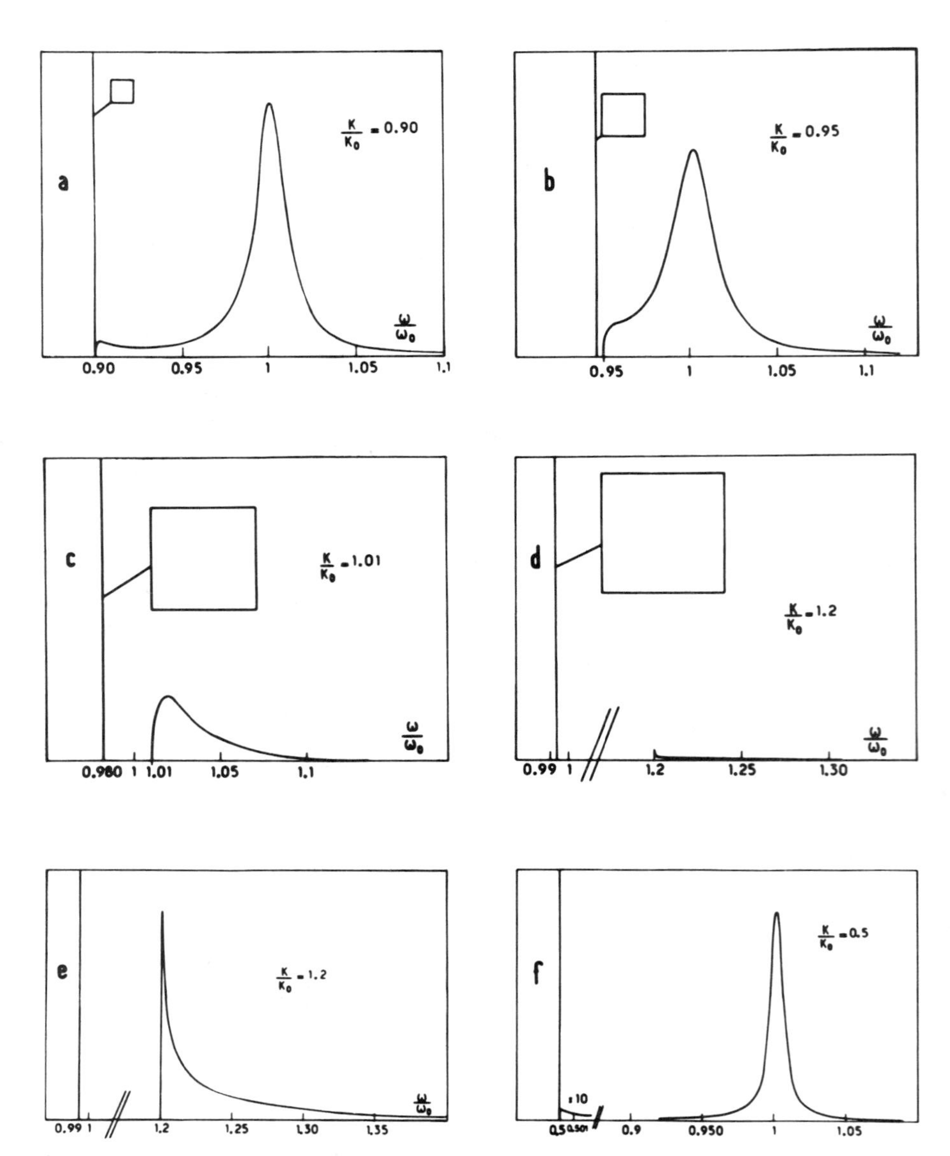

Figure 3.9. The Fourier transform of spectra of Fig. 3.8: Projection $P_{\mathbf{K}}(\omega)$ of the pure excitonic state on the eigenstates of the coupled system of an exciton **K** and the effective photon continuum in a 2D lattice, for various values of the wave vector **K** ($\mathbf{K} \perp \mathbf{d}$). The vertical peak represents a discrete state, whose weight is represented by a rectangle (*a–d*). For an exciton with $K < K_0$, the continuum band (matter-contaminated photons) dominates the spectrum, with a quasi-lorentzian resonance (*a, f*). For $K > K_0$, the discrete state dominates (*d, e*). In the intermediate region (*b, c*) the spectrum reflects the complicated behavior of its Fourier transform; cf. Fig. 3.8.

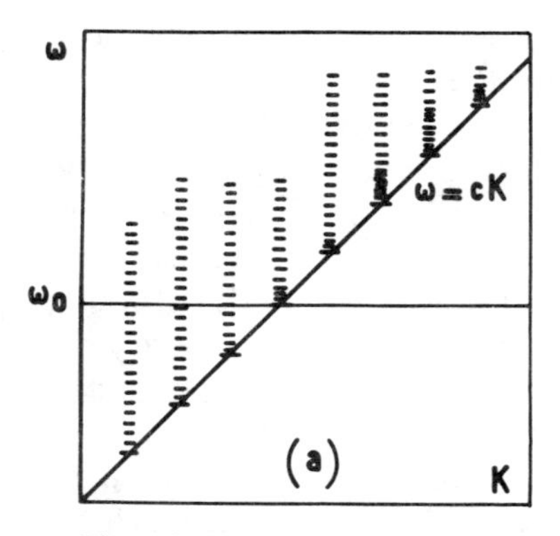

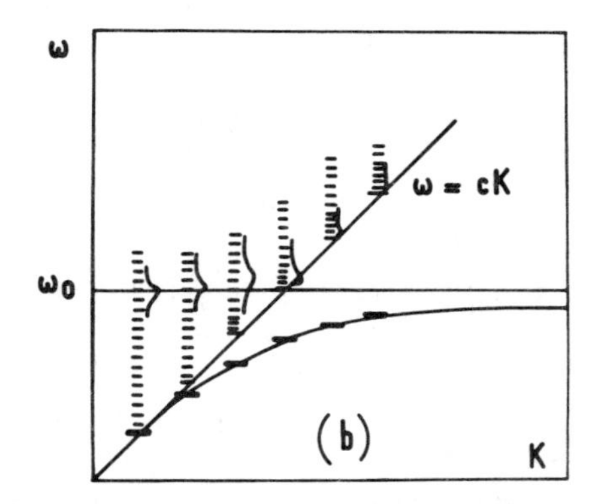

Figure 3.10. Scheme of the 2D polaritons and radiatively very unstable 2D excitons in the coupled system of an exciton **K** and an effective photon continuum: (*a*) The two subsystems are not coupled. (*b*) The coupled system with a discrete state split off below the continuum, called the 2D polariton; excitonic solutions exist only in a small segment of the Brillouin zone, $0 < K < \omega_0/c$.[126]

The above results may be interpreted as follows: Figure 3.10 shows the dispersion diagrams $\omega(K)$ for the uncoupled and the coupled matter–radiation systems. Thus, the coupling induces, for $cK < \omega_0$, a splitting off of the lower state of the effective continuum, repelled to lower energies by its interaction with the matter state $|\mathbf{K}\rangle$.[126]

This split-off discrete state rejoins, for $cK \gg \omega_0$, the exciton energy $\hbar\omega_0$; it behaves qualitatively in the same way as the lower branch of the 3D polariton.[33–35] For this reason we call it the 2D polariton. It is the projection of the exciton $|\mathbf{K}\rangle$ on this 2D polariton (radiatively stable) that constitutes (1) the finite limit value of the curves $A_{\mathbf{K}}(t)$ for $t \to \infty$ (Fig. 3.8), and (2) the weight of the discrete peak in the spectrum $P_{\mathbf{K}}(\omega)$ (Fig. 3.9). The transition, in the 2D polariton branch, between the photon and the pure exciton characters occurs around the value $K_0 = \omega_0/c$ in an area of width $\Delta K = \Gamma_0/c$ (with $\Gamma_0 = 15\,\text{cm}^{-1}$). Thus, the 2D polariton may be considered as a photon mode trapped in the 2D lattice, where it acquires its own dispersion.[115,116,126] Therefore, the 2D polaritons cannot be excited by free photons, but they may be coupled to evanescent waves, by ATR for example.[115,116]

A second type of eigenstates, illustrated in Fig. 3.10, is the band of states formed by the exciton-contaminated photon continuum. Far from the critical area $cK \sim \omega_0$, this band presents a lorentzian resonance (cf. Fig. 3.9), whose temporal instability (cf. Fig. 3.8) is described by an exponential decay. Thus, the exact solution leads back to that of second-order perturbation theory, obtained in Section III.A.2.b above.

On the contrary, in the vicinity $cK = \omega_0$, when the exciton gets near the low threshold of the effective photon continuum (the simple case of an irregular continuum), the excitonic resonance broadens and changes shape while the 2D polariton builds up (cf. Fig. 3.9b, c). The corresponding temporal behavior, with a transient oscillatory regime, shows quantum beats between a damped state (the exciton) and an undamped state (the 2D polariton), as illustrated in Fig. 3.8. This oscillatory response corresponds to grazing-incidence photons for which the nearby directions are strongly resonating; the characteristic angle at which such phenomena may be observed is about $\Gamma_0/\omega_0 \sim 10^{-3}$ rad.

In contrast with the region $cK > \omega$, the region $cK < \omega$ shows no analogy with the 3D case. Instead of a complete upper polariton branch, a new phenomenon appears, which may be represented, equivalently, either as a radiatively unstable exciton (3.19)–(3.20) or as scattering by the exciton-contaminated continuum states (3.21)–(3.22).

A similar analysis may be made for the 1D case.[126] The interaction energies $R_{\mathbf{K}}^{(1)}(\omega)$ then show an enhancement of the order of $\lambda/a \sim 10^3$ relative to the parameters of the isolated point dipole, which is easily overwhelmed by nonradiative processes. Using the simple "golden rule" image, we may say that the transition dipole shows a huge increase from 0D to 3D systems, while the effective photon continuum density varies in an opposite (in fact complementary) way as follows, with radiative-power optimization for 2D systems:

$$g^{(0)}(\omega) = \left\{ \frac{L^3}{2\pi c^3}\omega^2, \right. \qquad g^{(2)}(\omega) = \begin{cases} \dfrac{L}{2\pi c}\dfrac{\omega}{(\omega^2 - c^2K^2)^{1/2}} & \text{for} \quad \omega > cK \\ 0 & \text{for} \quad \omega < cK \end{cases}$$

$$g^{(1)}(\omega) = \begin{cases} \dfrac{L^2}{2\pi c^2} & \text{for} \quad \omega > cK \\ 0 & \text{for} \quad \omega < cK \end{cases} \qquad g^{(3)}(\omega) \propto \delta(\mathbf{K} - \mathbf{k})$$

This study of coupling of collective excitations (excitons) to the radiation field shows the transition, as a function of the dimensionality, from a spontaneous-emission regime of independent point dipoles to the 3D crystal polaritons (Section I.B.3.b), whose emission is phonon- or multiphonon-assisted.[85] In the case of a 2D lattice, which modelizes the monolayer surface, the essential point is the intrinsic strong emission power of the surface excitons, determining an expected picosecond lifetime. This emission power is fairly insensitive to site defect trapping and, more generally, to site energy fluctuations of various origins, when its statistical width lies below a threshold value. This threshold value is about $15\,\text{cm}^{-1}$ for the first singlet state of anthracene, while it may increase to more than $100\,\text{cm}^{-1}$ for the second singlet state S_2 or for dye-molecule monolayers;[118,147] see also Section IV.

3. *The Influence of the Substrate*

One of the most obvious perturbations of the surface excitons by the bulk crystal is the short-range coulombic interactions causing surface-exciton transfer to the bulk. We discard these perturbations on the grounds both of theoretical calculations[27] and of the experimental observations, which show the presence of a surface exciton (the second subsurface exciton S_3; see Fig. 3.2) resolved at about 2 cm^{-1} above the bulk-exciton resonance.

a. *Coupling of Surface to Bulk via Radiation.* A second cause of surface-exciton perturbation is the retarded interactions due to the presence of the bulk: The photon continuum, to which the surface exciton is coupled, is strongly modified by the presence of a semiinfinite refringent space, the volume of the crystal. This interaction continues to be the main perturbation of the emitting surface exciton in the optical region ($0 < K < \omega_0/c$).

Therefore, our model of a surface with substrate is based on interactions, of purely retarded type, between surface and bulk dipoles via the radiation field, or, more formally, on a renormalization by the bulk matter of the photon continuum to which the surface is coupled, with consequent modifications of its radiative properties.

The reflectivity of the surface–bulk system is calculated in the framework of classical electrodynamics, using the notion of multiple-wave interference (a purely quantum-mechanical calculation, reproducing exactly the same results and following the argument in Section I, has also been performed[125]). In what

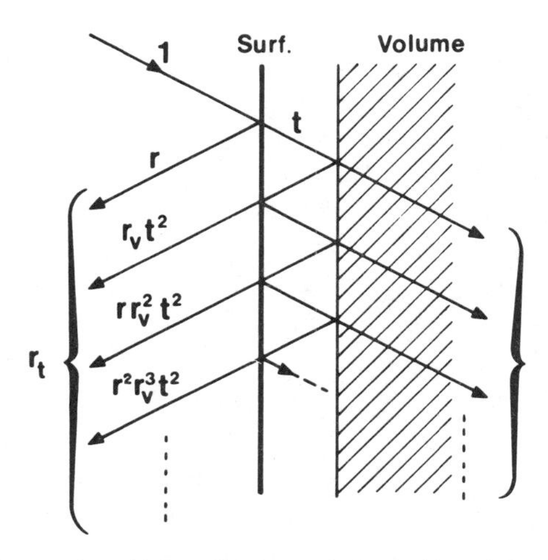

Figure 3.11. Scheme of multiple reflections of the incident beam (amplitude 1) on a surface–bulk system. The amplitude of the surface (bulk) reflection is r (r_v). The transmission amplitude of the surface is t. (Dephasing due to the surface-to-bulk path is negligible.)

follows we use the classical approach to maintain contact with the current literature.

We denote by $r_v(\omega)$ the bulk reflectivity amplitude and by $r(\omega)$ and $t(\omega)$ the reflection and transmission amplitudes of the bulk free surface, calculated in (3.21)–(3.22). The scheme of Fig. 3.11 allows us to obtain the total reflectivity amplitude $r_t(\omega)$ with

$$r_t = r + \theta^2 t^2 \frac{r_v}{1 - \theta^2 r r_v} \tag{3.23}$$

(θ is the phase change of the incident wave due to the surface bulk path, which is to be neglected, so that we take $\theta = 1$.) Using the relation $t(\omega) = 1 + r(\omega)$, related to the symmetry of the 2D lattice [cf. (3.21)–(3.22)], we find a simple form for the total reflectivity:

$$r_t = r + (1 + r)^2 \frac{r_v}{1 - r r_v} = \frac{r + r_v + 2 r r_v}{1 - r r_v} \tag{3.24}$$

Introducing the explicit expression (3.21) for $r(\omega)$, and putting $E = (z^2 - \omega^2)/2\omega_0$, which is roughly the detuning from the isolated-surface excitation, we obtain (with the notation $R_{\mathbf{K}} = iB$)

$$r_t = r_v + (1 + r_v) \frac{iB(1 + r_v)}{E - iB(1 + r_v)} \tag{3.25}$$

Here $r_t(\omega)$ shows the effect of the surface reflectivity, which appears as a lorentzian line, centered at the surface resonance, if we neglect the variation of $r_v(\omega)$ with ω around the surface resonance ($\pm 10\,\mathrm{cm}^{-1}$). The surface excitations are renormalized relative to the bulk-free surface, leading for coupled surface excitons to a frequency shift Δ_s and to a new radiative width Γ_s, both quantities simply related to the complex amplitude of the bulk reflectivity:

$$\Gamma_s = \Gamma_0 (1 + \mathrm{Re}\, r_v)$$
$$\Delta_s = \frac{\Gamma_0}{2} \mathrm{Im}\, r_v \tag{3.26}$$

We notice that $\mathrm{Re}\, r_v(\omega)$ and $\mathrm{Im}\, r_v(\omega)$ are obtained with very satisfactory accuracy by KK analysis of the reflectivity spectrum, as indicated in Section II.

The frequency shift Δ_s, of a few reciprocal centimeters, may be neglected for the surface (S_1) excitons, but it will be evaluated for the two subsurfaces, S_2 and S_3, whose resonances lie near the bulk resonance. The renormalization of the radiative width (3.26) is very important, especially for surfaces with small energy gaps (the case of S_2 and S_3), for which the bulk reflectivity amplitude is strong and negative and drastically reduces the emission power of the surface. The surface emission is quenched when $r_v(\omega) \to -1$, and the surface layer belongs to the 3D polariton with 0–0 forbidden emission.

Therefore, as a general trend, Γ_s decreases when the energy gap between surface and bulk states is made weaker; Figs. 3.1–3 provide a perfect illustration of the expression (3.26) for the bulk effect on the surface emission. A more detailed analysis of the bulk effect will be given below. However, this reduction of the surface radiative width may be interpreted classically as the destructive interference between the emission of the surface and that of its electrostatic image in the bulk.[140] The bulk reflectivity amplitude $r_v(\omega)$ is quasi-metallic near resonance and at low temperatures.

The emission of the surface exciton, coupled to the bulk polaritons, may also be calculated using the above scheme, which, here also, coincides exactly with the quantuum-mechanical results. Instead of the lorentzian emission (3.20), we obtain an emission proportional to

$$\mathscr{E}(E) = \left| \frac{1 + r_v}{E - iB(1 + r_v)} \right|^2 \tag{3.27}$$

which is not lorentzian because of the excitation dependence of $r_v(\omega)$. However, it may be shown that the energy dependence of $r_v(\omega)$ over the surface-emission width is very weak and could by no means account for the observed strong asymmetry of the observed emission of surface I (cf. Fig. 3.4). The origin of this asymmetry in the excitation of the emitting surface state will be analyzed below in Section III.B.3.

The expression (3.25) allows us to predict that at resonance ($E = 0$) the total reflectivity amplitude is -1, amounting to a 100% reflection yield. In reality, a careful examination of the experimental data shows that the maximum reflection yield of structure I is not more than 75% for our best crystals at low temperatures ($T \sim 2$ K). This observation indicates the existence of damping (nonradiative) processes causing the broadening (homogeneous or inhomogeneous) of the surface exciton. In Section IV we shall consider causes of inhomogeneous broadening; for the present, we restrict ourselves to homogeneous broadenings due to processes investigated in Section II.C, i.e. phonons responsible for the fast broadening of the bulk excitons with increasing temperature.

As indicated in Section II.C.1, in calculating the contribution of the exciton motion in the presence of phonons to the exciton self-energy, retarded interactions may be neglected because they operate only in the optical region (middle of the Brillouin zone). This approximation is valid for short times, when relaxation to lower polaritons may be neglected; it would not be valid on the nanosecond time scale of the bulk fluorescence (see Section II.D). Let us assume that the exciton self-energy, due to its coupling to the phonons, is a known function $\Sigma_e(z)$ for a given wave vector **K**. Then, the above theory of the reflectivity amplitudes is generalized by replacing the surface exciton resonance $\hbar\omega_0$ with $\hbar\omega_0 + \Sigma_e(z)$. This simple substitution is valid because the "branching" of the retarded interactions does not modify the self-energy $\Sigma_e(z)$.

Upon replacement of E by $E - \Sigma_e(z)$ in (3.21)–(3.22), the relation $t = 1 + r$ being always valid, the conservation relation $|r(\omega)|^2 + |t(\omega)|^2 = 1$ is transformed to

$$|r|^2 + |t|^2 = 1 - \frac{2\Gamma_e\Gamma_0/4}{(E - \Delta_e)^2 + \frac{1}{4}(\Gamma_e + \Gamma_0)^2} \leqslant 1 \tag{3.28}$$

for surface excitons damped by phonons, with the parameters $R_{\mathbf{K}} = i\Gamma_0/2$, $\Sigma_e = \Delta_e + (i/2)\Gamma_e$. The relation (3.28) indicates, on one hand, the competition between the radiative (Γ_0) and the nonradiative (Γ_e) channels, and on the other hand the dissipation of the radiative energy by internal processes or processes specific to the surface layer.

Using (3.25) with the new surface-damped amplitudes, r and t, we calculated the interference surface structures, I, II, and III, which depend only on two adjustable parameters: the resonance energy E_s of the surface exciton which lies near the center of the observed structures, and the homogeneous nonradiative width Γ_e of the surface exciton, assumed to depend on the energy z. The two remaining parameters in the expression (3.25)—the complex amplitude $r_v(\omega)$ of the bulk reflectivity, and the isolated-surface radiative width—are known and are inserted into the model. The amplitude $r_v(\omega)$ is determined for each temperature by KK analysis (Section II.3), while Γ_0, given by (3.15), may be calculated from the knowledge of d , the b transition dipole, which is deduced from the bulk polariton dispersion. The values (2.122′), determining the polariton dispersion at large wave vectors, provide the set of values $\Gamma_0 \sim 17$ and $340\ \text{cm}^{-1}$ of the polariton stopping bandwidth. A second set of parameters determined near resonance ($\varepsilon = 2.38$ and $A = 951\ \text{cm}^{-1}$) provides the values $\Gamma_0 \sim 14$ and $400\ \text{cm}^{-1}$ respectively. This latter set of values appears in much better agreement with the experimental estimates[67]; hence, we use the value $\Gamma_0 = 14\ \text{cm}^{-1}$ for our simulation with (3.25).

The simulations of structure I, quite isolated from the bulk resonance, and of structures II and III, quite near the bulk resonance, need different approaches, which will be discussed separately.

b. *The First-Surface Reflectivity Signature.* Figure 3.12 illustrates the shape

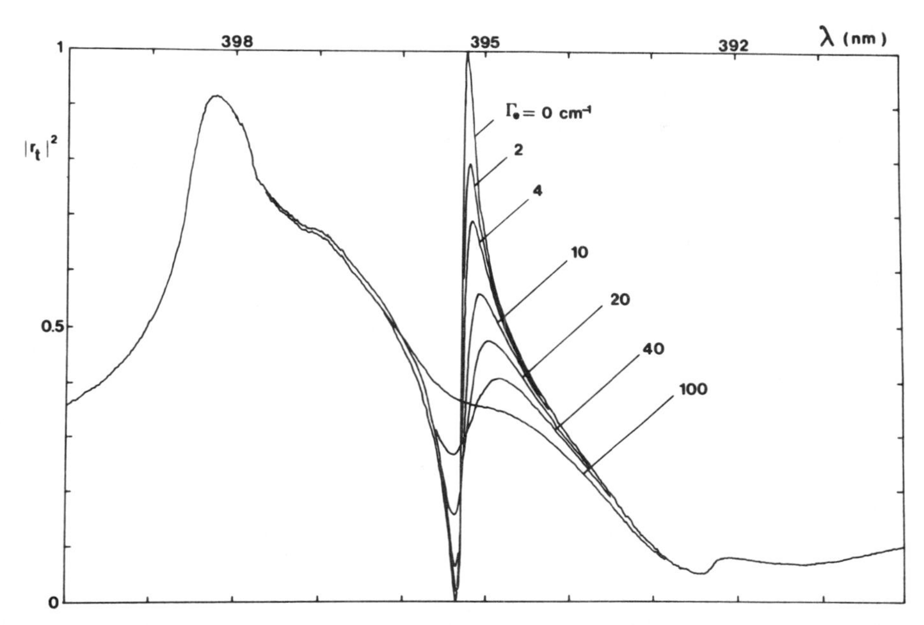

Figure 3.12. Simulation of the *b*-polarized (0–0) reflectivity of the anthracene crystal using the bulk reflectivity amplitude derived from a Kramers–Kronig analysis (Section II.C). The total reflectivity is calculated from the scheme of Fig. 3.11 and (3.24)–(3.25) for various values of the nonradiative broadening parameter Γ_e. Comparison with spectra of our best crystals gives the value $\Gamma_e^0 = 3\,\text{cm}^{-1}$ for $T = 1.7$ K.

variation of the surface signature in (3.25) as a function of the nonradiative damping rate Γ_e^I, varying from 0 to $100\,\text{cm}^{-1}$. The quantity most sensitive to the variation of Γ_e^I is the maximum of the interference structure, although other quantities, such as the interference minimum and the width and shape of the signature, turn out to be very useful for comparison with the experimental data. For a low-temperature spectrum such as that of Fig. 3.12, our best crystals exhibit a structure maximum of 76%, revealing a damping rate of about 3 to $4\,\text{cm}^{-1}$.

When increasing temperature ($T > 2$ K), so introducing thermal disorder, the structure broadens rapidly. For each bulk reflectivity spectrum at a given temperature T, determining the radiative surface width by (3.26), we looked for the value of $\Gamma_e^I(T)$ allowing the best reproduction of the experimental spectrum. The various values obtained for $\Gamma_e(T)$ vs T are plotted in Fig. (3.13) and compared with the width of the bulk exciton obtained by KK analysis (in Section II.C.3).[70,127]

The order of magnitude of $\Gamma_e^I(T)$ and the width of the bulk polariton (which has negligible radiative contribution) appear quite comparable. We may

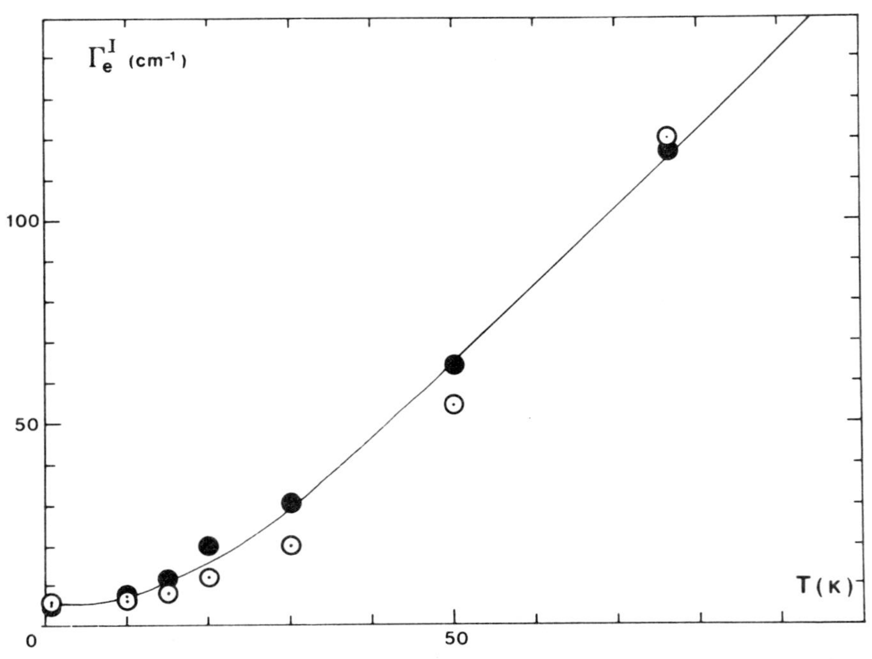

Figure 3.13. Temperature evolution of the parameter $\Gamma_e(T)$ of the first-surface exciton (hollow circles). The full circles indicate the values obtained for the bulk exciton (Fig. 2.14). The quasi-coincidence at $T \sim 0$ is fortuitous, but the surface states appear less broadened than the bulk states in the region 30–50 K, which could correspond to a decay at the surface of the density of interplane phonons coupled to the exciton (see Section III.B).

conclude that for high temperatures ($T > 15$ K) the excitonic broadening is dominated by (a, b) intralayer processes. However, a slight discrepancy, in the region between $T = 20$ K and $T = 50$ K, seems to indicate that the surface exciton is less thermally broadened than the bulk exciton. In the low temperature range ($T < 15$ K), the residual value (3 to 4 cm^{-1}) of the surface-exciton damping rate Γ_e^I is presumably due to relaxation channels specific to this surface. This residual rate, although coinciding with that of the bulk exciton (cf. Fig. 3.13), has not the same origin, as will be pointed out in the investigation of the structures II and III of the subsurfaces. Obviously, for crystals of poor quality, the surface structure may show low-temperature residual broadening of more than 30 cm^{-1}, but this is outside the framework of our investigations of the intrinsic properties, where either residual defects make second-order contributions or the defects are under control for a specific perturbation of the intrinsic properties.

c. *Combined Interface Effects on the Surface Excitons.* Upon gas (N_2, CH_4) coating, the surface structure I shifts to lower energies ($\sim$ 80 and $\sim$ 100 cm^{-1}). This shows strong, nonresonant coupling between the condensed-gas layer and the 2D exciton, itself coupled to the bulk polaritons (coupling of the gas layer to the bulk may safely be neglected, as suggested by the unperturbed bulk structure upon gas coating).

The new surface structure, with double interface, may be simulated with the expression (3.25), accounting for the combined effects of the double interface. The condensed-gas layer additively stabilizes and broadens the surface structure, respectively by δ_G^I and by a contribution to Γ_e^I. Furthermore, by lowering the surface-to-bulk energy gap by δ_G^I (80 or 100 cm^{-1}), the gas layer has the effect of increasing the coupling of the surface excitons to the bulk polariton, resulting in a drastic reduction of the surface radiative channel according to (3.26). The combined effects of the double interface are shown with high accuracy in the experimental variation of the reflectivity upon gas coating: see Fig. 3.3. Furthermore, with an appropriate KK analysis, the above combined effects of the double interface are reproducible using the total-reflectivity expression $r_t(\omega, T)$ of (3.25), allowing one to distinguish coherent narrowing effects and incoherent broadening effects within the observed global narrowing. Indeed, Fig. 3.3 shows that the structure-I shift is accompanied by a strong narrowing of the shape (coherent interference effects are dominant) with a small decrease of the interference-structure maximum, indicating an increase of the nonradiative rate Γ_e^I (incoherent effects) induced by the condensed-gas layer. Our simulations show that at $T \sim 2$ K the rate Γ_e^I increases from 4 to 6 cm^{-1}, accounting for the decay of the maximum of the reflection structure, while the radiative width decreases from 14 to 6 cm^{-1}, consistent with (3.26) and with the observed reduction of the total width upon gas (N_2) coating. A better resolution of the surface-reflectivity structure could provide more data for a better description of this double-interface problem. This is currently being investigated by means of energy exchanges and transitions through three phases.[106]

d. *The Signatures of Subsurfaces* II *and* III. The surface has been considered to be nonresonant with the bulk, developing a well-resolved resonance line. In contrast, the subsurfaces II and III interact strongly with the bulk resonance, with the consequence that their shape is affected as well as their width and their position relative to the bulk resonance. Tentative simulations, with bulk-reflectivity maxima of 95% to 98%, and with widths compatible with our study in Section II, show surface constructive peaks with 100% maxima. Experimentally, surface constructive peaks II and III do not show up even in high resolution and at very low temperature (1.6 K), leading to the conclusion that the observed pattern (Fig. 3.2) of structures II and III is compatible with the

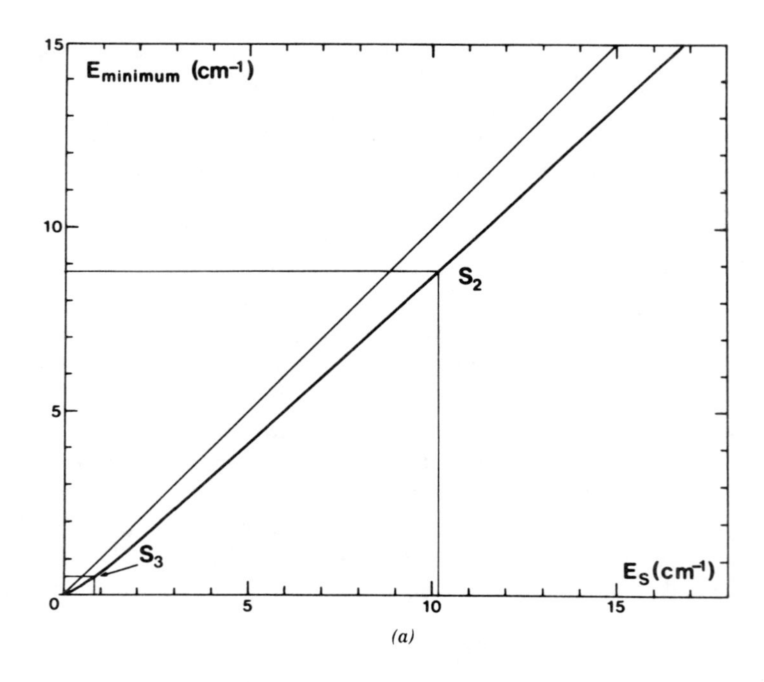
E minimum (cm-1)
S2
S3
ES(cm-1)
0
5
10
15

(a)

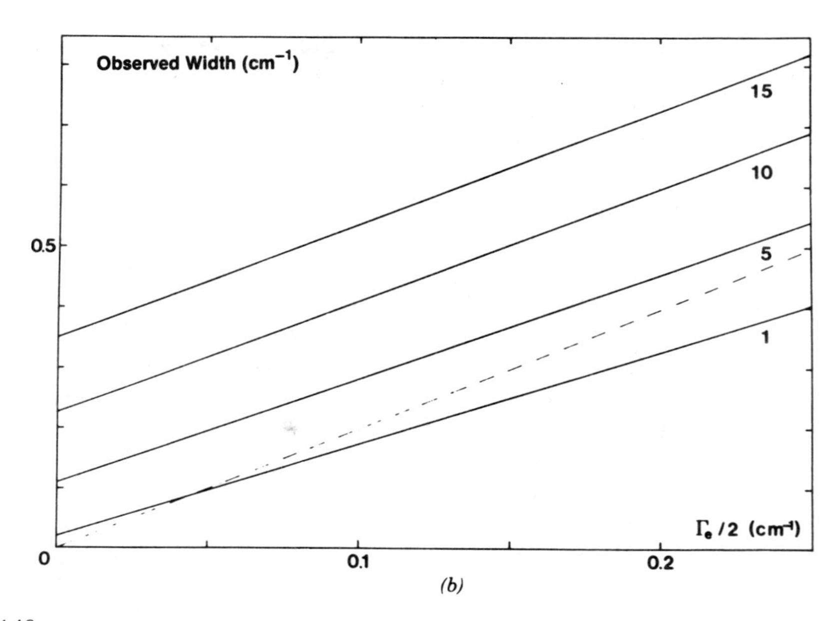
Observed Width (cm-1)
15
10
5
1
0.5
Γe /2 (cm-1)
0
0.1
0.2

(b)

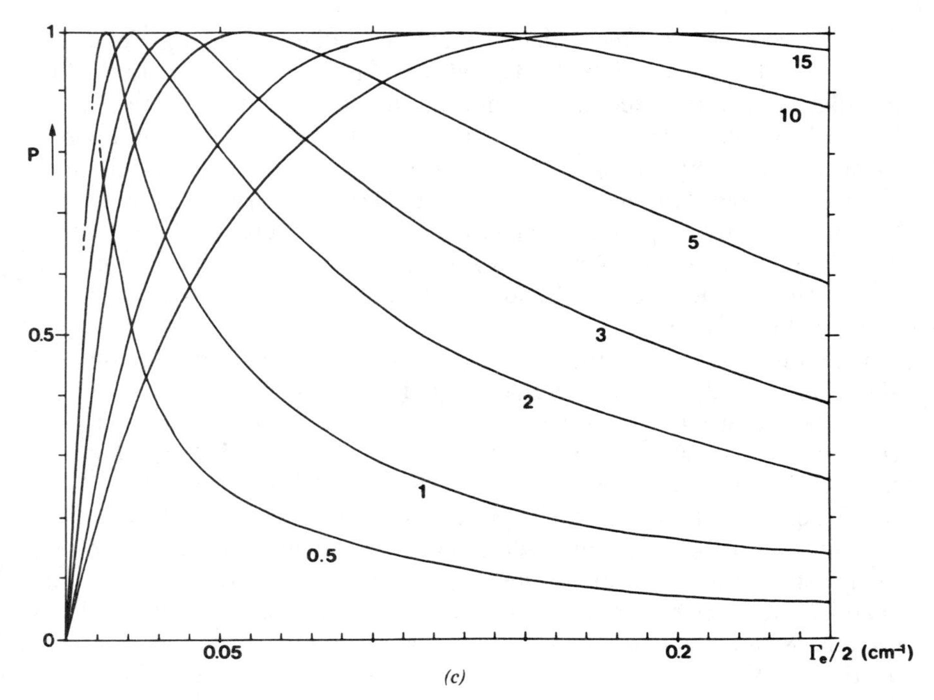

Figure 3.14. (*a*) Variation, vs the resonance energy E_s of the surface, of the position of the minimum of the reflectivity dip of the surface–bulk system without damping Γ_e (bold line). (The light line would be obtained without the energy renormalization of the surface exciton by the bulk; the shift may reach 1 to 2 cm^{-1}.) The positions of the minimum do not depend significantly on the parameter Γ_e; they are drawn for a value $\Gamma_e = 0.05\,\text{cm}^{-1}$. (*b*) Observed width of the reflectivity dip vs Γ_e, for variable detuning (in reciprocal centimeters) of the surface relative to the bulk. The broken line would be obtained for an observed width equal to Γ_e. (*c*) Variation of the magnitude of the observed dip vs Γ_e, for variable detuning (in reciprocal centimeters) of the surface relative to the bulk. No structures are observed without damping ($\Gamma_e = 0$); the structures appear and get deeper with increasing Γ_e, they reach a maximum ($P = 1$), and then they decay as Γ_e increases further.

bulk-reflectivity maximum remaining near 100% over several reciprocal centimeters. More precisely, we conclude that at low temperatures and near resonance, $\gamma(\omega)$ in (2.122) is negligible. Therefore, we have adopted, as a simple model, an undamped bulk-reflectivity amplitude interfering via (3.24) with the surface-reflection amplitude at a position $\delta^{\text{II,III}}$ above the bulk resonance and with a damping rate $\Gamma_e^{\text{II,III}}$. Our simulations provide the expected dip signatures on the bulk reflectivity (100%), practically of lorentzian form, which we characterize by three parameters: the dip position, its width at half magnitude, and the magnitude of the dip. Our simulations suggest very small damping rates $\Gamma_e^{\mathrm{II,III}}$, much smaller than the radiative rate Γ_0. In these conditions, the position of the dip depends only on the resonance energy of the investigated surface exciton and not on the nonradiative width $\Gamma_e^{\mathrm{II,III}}$. Figure 3.14*a* illustrates this dependence: the gap between the surface resonance and the dip may reach a value of about 1 to 2 cm^{-1}. The bulk 0–0 transition energy of the anthracene crystal (cf. Section II.D) is not known with high accuracy. If, to a first approximation, we locate it at 0.5 cm^{-1} below the dip III, then we obtain for the subsurfaces resonances, with the use of Fig. 3.14*a*, $E_s^{\mathrm{III}} = E_v + 0.8\,\mathrm{cm}^{-1}$ and $E_s^{\mathrm{II}} = E_v + 10\,\mathrm{cm}^{-1}$, with an uncertainty of about 1 cm^{-1}.

Once the structures' positions are determined, their widths and magnitudes allow us to evaluate the corresponding damping rates Γ_e. Figure 3.14*b* illustrates the variation of the structures' width with Γ_e for various resonance energies of the surface exciton. The variation is approximately linear, but the slope of the variation ranges between 1 and 0.5 and leads to the important conclusion that in surface excitons in near-resonance with the bulk, the radiative width Γ_0 and the nonradiative width Γ_e do not enter additively (as for structure I) to build up the observed structure. Thus we need a special analysis, rather elaborate, to provide the two rates Γ_0, Γ_e of the surface exciton. With the above values of δ^{II} and δ^{III}, we obtain, from the observed widths $\gamma^{\mathrm{II}} = 0.70\,\mathrm{cm}^{-1}$ and $\gamma^{\mathrm{III}} = 0.38\,\mathrm{cm}^{-1}$, for the nonradiative damping rates the bounds $\Gamma_e^{\mathrm{II}} < 0.50\,\mathrm{cm}^{-1}$ and $\Gamma_e^{\mathrm{III}} < 0.46\,\mathrm{cm}^{-1}$. However, the introduction of the apparatus resolution ($0.3\,\mathrm{cm}^{-1}$) allows us to estimate the intrinsic widths $\gamma^{\mathrm{II}} = 0.46\,\mathrm{cm}^{-1}$ and $\gamma^{\mathrm{III}} = 0.08\,\mathrm{cm}^{-1}$, leading, with the use of Fig. 3.14*b*, to the following upper bound for the damping rates:

$$\Gamma_e^{\mathrm{II}} < 0.25\,\mathrm{cm}^{-1} \quad \text{and} \quad \Gamma_e^{\mathrm{III}} < 0.08\,\mathrm{cm}^{-1}$$

The last parameter for the characterization of the Γ_e surface excitons is the magnitude of the dips. Its variation with Γ_e is illustrated in Fig. 3.14*c*. The dip magnitude, which is zero for $\Gamma_e = 0$, grows to a maximum value 1 for a value of Γ_e depending on the resonance energy of the surface; then it decays slowly on further increase in Γ_e. For the second surface, the observed dip, of magnitude 0.77,[67] is smaller than the intrinsic dip magnitude (apparatus resolution $0.3\,\mathrm{cm}^{-1}$). This provides a lower bound for the damping rate: $\Gamma_e^{\mathrm{II}} > 0.1\,\mathrm{cm}^{-1}$; see Fig. 3.14*c*. For the third surface, the observed dip, of magnitude 0.33, also leads to a larger intrinsic value. Although it is difficult to guess the part of the curve at $\delta = 1$, we arrive at a lower bound for the damping rate, $\Gamma_e^{\mathrm{III}} > 6 \times 10^{-3}\,\mathrm{cm}^{-1}$. This is a weak but nevertheless probable result; its improvement would require high resolution in this part of the spectrum.

To summarize, the main conclusions we are able to draw from this investigation of the surface exciton dynamics, for our best crystals at $T = 2\,\mathrm{K}$, are the following:

First surface: resonance $E_e^{\text{I}} = 25\,305 \pm 5\,\text{cm}^{-1}$, damping rate $\Gamma_e^{\text{I}} \sim 3.5 \pm 0.5\,\text{cm}^{-1}$.

Second surface: resonance $E_e^{\text{II}} = 25\,103 \pm 1\,\text{cm}^{-1}$, damping rate $\Gamma_e^{\text{II}} = 0.25 \pm 0.10\,\text{cm}^{-1}$.

Third surface: resonance $E_e^{\text{III}} = 25\,093 \pm 1\,\text{cm}^{-1}$; damping rate $\Gamma_e^{\text{III}} = 0.1\,\text{cm}^{-1}$.

Gap from the resonance to the bulk:

$$E_S^{\text{II}} - E_V = 10 \pm 1\,\text{cm}^{-1}$$

$$E_S^{\text{III}} - E_V = 0 \text{ to } 2\,\text{cm}^{-1}$$

$$E_S^{\text{II}} - E_S^{\text{III}} = 9.4 \pm 0.4\,\text{cm}^{-1}$$

The difference $E_S^{\text{II}} - E_S^{\text{III}}$ may be estimated with very good accuracy from the observed positions of the dips II and III, distant by $8.3 \pm 0.2\,\text{cm}^{-1}$. With the use of Fig. 14*c*, and taking into account the uncertainty of the position of III, we obtain $E_S^{\text{II}} - E_S^{\text{III}} = 9.4 \pm 0.4\,\text{cm}^{-1}$.

The values of the widths used above are those obtained with our best crystals. Therefore, we have to examine two lines of attack:

1. To prove either that these widths are fundamental limits related to the intrinsic crystal, or that a better quality of crystal could provide finer structures.
2. To understand by which mechanisms, and to what extent, these intrinsic values may be affected.

4. *Origins of Broadening of the Surface Exciton: Intrinsic Broadening*

Broadening of electronic states in molecular crystals originates either from relaxation to other states of lower energy (at very low temperatures) or from the presence of a continuum of various configurations, connected with the presence of disorder. We consider here the broadening in the absence of disorder, the so-called intrinsic or homogeneous broadening. Disorder is examined in Section IV.

In the last subsection, we invoked phonons to explain the nonradiative broadening of the surface structures. However, at very low temperature, the surface state at the bottom of the excitonic band cannot undergo broadening either by phonon absorption or by phonon creation; the phonon bath at 2 K does not suffice to account for the 3- to 4-cm^{-1} nonradiative width of the first surface resonance. Nevertheless, we assume the intrinsic nature of this broadening, since it is observed, constant, for all our best crystals.[67,120]

The only lower states to which the surface exciton can relax are the bulk states. However, we have seen in Section III.A.1 that for the surface exciton vectors $\mathbf{K} \sim \mathbf{0}$, the surface-to-bulk coupling is about $2\,\mathrm{cm}^{-1}$ (very small).[27] It is possible to calculate the order of magnitude of the probability for the surface exciton to directly relax, by transition to the bulk states and relaxation in the bulk:

$$P_{S\to V}^{\mathrm{direct}} = \left| \frac{J_{SV}}{E_S - E_V(k=0)} \right|^2 P_V(E_S) \tag{3.29}$$

where $J_{SV} \sim 2\,\mathrm{cm}^{-1}$ is the surface-to-bulk coulombic interaction, with $E_S(k=0) - E_V(k=0) \sim 200\,\mathrm{cm}^{-1}$, and $P_V(E_s)$ is the probability per unit time of relaxation of a bulk exciton at an energy $\sim 200\,\mathrm{cm}^{-1}$ above $E_V(k=0)$. From experiments, we have $P_V(E_s^1) \sim 100\,\mathrm{cm}^{-1}$, so that with good reliability we find an order of magnitude for direct relaxation, which is $10^{-2}\,\mathrm{cm}^{-1}$. This cannot explain the observed width of 3 to $4\,\mathrm{cm}^{-1}$.

To explain the observed width, it is necessary to look for strong surface-to-bulk interactions, i.e. large magnitudes of surface-exciton wave vectors. Such states, in our experimental conditions, may arise from virtual interactions with the surface polariton branch, which contains the whole branch of $\mathbf{K}$ vectors. We propose the following indirect mechanism for the surface-to-bulk transfer: The surface exciton, $\mathbf{K} = \mathbf{0}$, is scattered, with creation of a virtual surface phonon, to a surface polariton ($\mathbf{K} \neq \mathbf{0}$). For $\mathbf{K} \neq \mathbf{0}$, the dipole sums for the interaction between surface and bulk layers may be very important (a few hundred reciprocal centimeters). Through this interaction the surface exciton penetrates deeply into the bulk, where the energy relaxes by the creation of bulk phonons. The probability of such a process is determined by the diagram

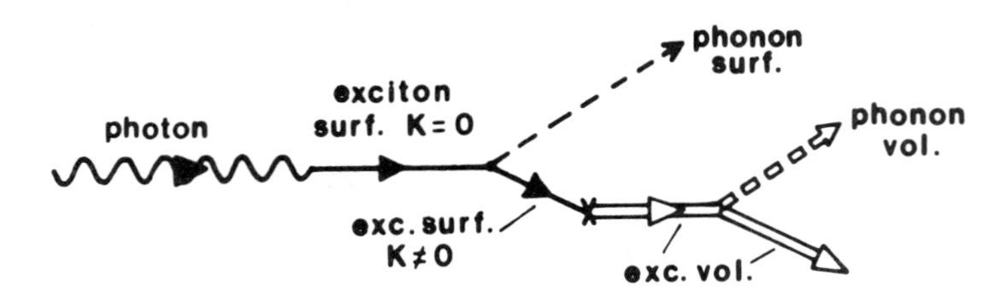

Figure 3.15. Diagram of a nonlocal surface-exciton transfer, corresponding to the optical creation of a surface exciton followed by its relaxation to the bulk. The essential virtual stage is the scattering of a surface phonon ($\mathbf{K} \neq \mathbf{0}$) and the creation of a surface polariton with a large wave vector ($\mathbf{K} \neq \mathbf{0}$), producing large interaction energies with the bulk.[121] Then relaxation in the bulk is ultrafast.

in Fig. 3.15; it may be calculated in the same way as (3.29):

$$P_{S\to V}^{\text{phonons}} = \sum_{\mathbf{K}, V'} \left| \frac{\langle S, K=0 | H_{ep} | S, K; \hbar\Omega_p \rangle \langle S, \mathbf{K} | J_{sv} | V', \mathbf{K} \rangle}{[E_S - E_S(\mathbf{K}) - \hbar\Omega_p][E_S - E_V(\mathbf{K}) - \hbar\Omega_p]} \right|^2 P_V(E_S - \hbar\Omega_p) \tag{3.30}$$

(The sums run over surface wave vectors K and bulk layers V'.)

A rough estimation of the coupling energies involved an (3.30) gives for the first surface exciton a relaxation $P_{S\to V}(E_S^I) \sim 5\,\text{cm}^{-1}$. This order of magnitude is consistent with the observed (3 to 4 cm^{-1}) value for Γ_e^I. The calculation has been performed as follows: The bulk layers being identical, only the sum $\sum_{v'} |\langle S, \mathbf{K} | J_{SV} | V', \mathbf{K} \rangle|^2$ has to be calculated; it is equal to $(1/2N)\sum_{\mathbf{q}} J_{\mathbf{K}+\mathbf{q}}^2$, with N the number of the bulk layers and $J_{\mathbf{K}+\mathbf{q}}$ the dispersion at the exciton vector $\mathbf{K}+\mathbf{q}$. A rough estimation of the parameters in (3.30) gives the following values: 80 cm^{-1} for P_V [the width of the a component in Fig. 2.8 (1a)], 60 cm^{-1} for H_{ep} (evaluated in Section II), 200 for the denominator $E_S - E_S(\mathbf{K})$ (the surface exciton bandwidth being about 400 cm^{-1}), and 100 cm^{-1} for $E_S - E_V(K)$ (the first surface exciton, about the middle of the 400-cm^{-1} bulk-exciton band; the energy of the phonon is neglected). Lastly, the average may be evaluated from the nonanalytic terms (2.139) and (1.57) as the principal source of interaction. This gives an average over $\mathbf{K}$ and over $\mathbf{q}$: $\frac{1}{2}\langle J_{\mathbf{K}+\mathbf{q}}^2 \rangle \sim 7 \times 10^3\ (\text{cm}^{-1})^2$. All this calculation leads to the indicated probability $P_{S\to V}^{\text{phonons}} \sim 5\,\text{cm}^{-1}$.

This interpretation suggests that the surface-exciton relaxation to the bulk, although weak, competes with the radiative process to give the observed reflection yield of 76% at the surface. Furthermore, the surface-exciton transfer is nonlocal and does not favor the first subsurface exclusively, since for $\mathbf{K} \neq \mathbf{0}$ the transfer integrals to a large number of bulk layers could be important. As to the excitation spectra of the subsurface [a priori weak, in view of (3.29)–(3.30)], we must note that the excitation equivalence is broken, since the subsurface emission, 10^2 to 10^3 times faster than the bulk emission, compensates the low density of the emitting subsurface molecules and shows a surprisingly a good quantum yield excited via S_1, whose origin is tentatively attributed here to the nonlocal transfer mechanism (3.30).[141]

As to the relaxation of the excitons in surfaces II and III by the bulk effect, a consistent discussion is as follows: The bulk relaxation probability at the energy level E_{S_2}, $P_V(E_{S_2})$, has been evaluated, at low temperatures, as few reciprocal centimeters and attributed to acoustical phonons (cf. Section II). With $J_{SV} \sim 2\,\text{cm}^{-1}$, the direct probability (3.29) is evaluated as 0.2 cm^{-1}, compatible with the observed value,[141] while the phonon-assisted process (3.30) is drastically depressed and becomes comparable to direct transfer.

As to the excitons of the second subsurface, in near-resonance with the bulk polaritons (nevertheless observed because of the very narrow bulk transition), the direct transfer process is probably dominant. However, it is possible that even at $T = 1.7$ K the thermal broadening (including pure dephasing) dominates the width of structure III.

To summarize, the nonlocal transfer relaxation mechanism to the bulk accounts naturally for the residual nonradiative width (3 to 4 cm^{-1}) of the surface exciton, as being an intrinsic property, specific to the first surface; transfer to the bulk implies the creation of a surface phonon and the activation of the surface polariton branch. This mechanism accounts also for the very small residual damping rates of the two subsurface excitons (respectively, $\Gamma_e^{\text{II}} \sim 0.25$ cm^{-1} and $\Gamma_e^{\text{III}} \sim 0.1$ cm^{-1}). When the temperature increases, the widths of the three surface layers increase as rapidly as that of the bulk (cf. Section II.C). That suffices to explain the successive disappearances of structure III at 5 K, structure II at 35 K, and structure I above 77 K.[1,67,120] We notice that in the process (3.30), when the phonon population is created, or when the exciton processes occur via intrasurface band relaxation (i.e. at large values of the **K** vectors), the reduction factor of virtual processes vanishes and (3.30) may yield surface-to-bulk relaxation rates in the femtosecond range. The thermal analysis of the second- and third-surface excitons will yield information on the relative importance of the broadening mechanisms operating between 0 and 30 K: acoustical-phonon broadening according to our theoretical estimations,[127] and surface disorder, which has the concomitant effect of enhancing the relaxation to the bulk and reducing the radiative channel of the surface[119] (see also Fig. 4.4).

B. Mechanisms of Intrasurface Relaxation

In Section III.A.1 we did not discuss the way the surface emission is excited. The radiative behavior of the surface shows that emission (normal to the surface) is observed as soon as the $\mathbf{K} = \mathbf{0}$ state is prepared. This state may be prepared either by a short (~ 0.2 ps) resonant pulse, or by relaxation from higher, optically prepared excited states. It is obvious that the quantum yield of the surface emission will critically depend on the excitation, owing to intrasurface relaxation accelerated by various types of fission processes (see Fig. 2.8) and in competition with fast irreversible transfer to the bulk (3.30), which is also a surface relaxation, at least at very low temperatures. Thus, the surface excitation spectra provide key information both on the upper, optically accessible surface states and on the relaxation mechanisms to the emitting surface state $\mathbf{K} = \mathbf{0}$.

In the present subsection, we investigate similarities and differences between corresponding surface and bulk states, with the purpose of singling out spectral and dynamical properties specific to the surface states. In what follows, we describe very succinctly the experimental method of discriminating surface states[61,131] and the formalism of the surface-state theory adapted from the general form in the 3D case (cf. Section II.D).

1. *Excitation Spectra of the Surface Emission*

a. *Experimental Procedure.* The excitation source is a dye laser pumped by a N_2 laser. According to the dye used (BBQ, PBD) the excitation energy varies from 25 400 to 27 700 cm^{-1}, as determined by monitoring the excitation of the 0–0 and the 0–1 vibronic spectral regions of the first singlet transition of our sample. The excitation width varies from 3 to 1 cm^{-1}, in the best conditions, with a broad contamination from the laser fluorescence. Therefore, the finest structures of the excitation spectrum will have a width of a few reciprocal centimeters.[131] We selected, on the first surface emission band, a subband (window) of a few reciprocal centimeters near the maximum. The observed structure in the excitation spectrum, for a variable excitation energy, must be connected with the observed energy and width (and not with the maximum of the emission structure I; see Fig. 3.20 below). This procedure is analogous to the experiment of Raman scattering (where the excitation is fixed and the detection varies); it is often called Raman excitation spectroscopy.[77,104] We must underline that, in what follows, at least one of the quantum transition steps, the observation, is resonant, so that our experiments may also be termed spontaneous resonant Raman scattering, along with the other possibilities of spontaneous secondary emission, such as fluorescence, described in Section II.D.

Since at low temperature the surface emission is well resolved from the bulk emission, and transition back to the surface is impossible, the excitation spectrum of the surface emission reveals only surface-state structures. This technique allows one to discriminate the surface states which, in other kinds of experiment, are masked by the bulk states.

b. *Theoretical Description.* We have to study the spontaneous secondary emission of surface excitons vs the excitation energy. Since the exciton–photon interaction, of the order of 10 cm^{-1}, is weak relative to the exciton–phonon coupling (the contrary is true for the bulk excitation: see Section II.D), we apply the result (2.131) for weak exciton–photon coupling. Furthermore, at very low temperatures, we may put for the populations $p(R) = 0$, except for the phonon vacuum.

In these conditions, the transition from the absorbing excitonic state to the emissive excitonic state will involve at least one phonon, if we discard the optical response of the surface investigated in Section III.A. It is convenient to separate the probability $P(\hbar\omega_1 \rightarrow \hbar\omega_2)$ of the secondary emission into a

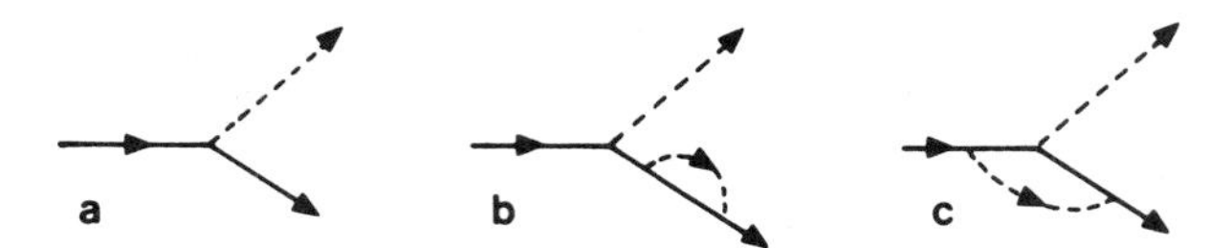

Figure 3.16. A few interactions contributing to the Raman–Stokes process (with creation of one phonon). The propagators are labeled with the notation of the text. The process *a* is of the first order; *b* could be included by renormalization of the created exciton; *c* would require the renormalization of the interaction (or the vertex).

sequence of processes involving one phonon (Raman process, in the absence of pure dephasing, since we work at $T \sim 0$), and involving two, three, or more phonons (hot luminescence, etc.). This approach is valid only for weak exciton–phonon coupling, since multiphonon processes could contribute to the Raman-process amplitude, as, for instance, in the scheme pictured in Fig. 3.16. However, we may partially take account of the higher-order terms of Fig. 3.16 by renormalizing the exciton propagators involved in the terms of the series expansion of H_{ep} in the resolvent $\langle e, R|G|e'R'\rangle$ of (2.131). The probability of observing one photon $\hbar\omega_2$ consecutive to an excitation with a photon $\hbar\omega_1$ may be written

$$P(\omega_1 \to \omega_2) = \frac{|\langle \hbar\omega_1 | H_{e\gamma} | e\rangle\langle e'|H_{e\gamma}|\hbar\omega_2\rangle|^2}{|\hbar\omega_1 - \tilde{E}_e|^2 |\hbar\omega_2 - \tilde{E}_{e'}|^2} \Gamma_{ee'}(\hbar\omega_1 - \hbar\omega_2) \tag{3.31}$$

$$\Gamma_{ee'} = \frac{2\pi}{\hbar} \sum_{\{p_i\}} |\langle e|T|e', \{p_i\}\rangle|^2 \delta(\hbar\omega_1 - \hbar\omega_2 - E_{\{p_i\}}) \tag{3.32}$$

is the total probability of relaxation of exciton e to exciton e' with dissipation to the bath of the energy $\hbar\omega_1 - \hbar\omega_2$, and the $\tilde{E}$ are the exciton complex energies renormalized by the exciton–phonon coupling. T denotes the t matrix describing the $e \to e'$ transition with creation of n phonons $\{p_i\}$, and approximated by an expansion in powers of H_{ep}. [Here, we assume that only one exciton e is prepared and that only one exciton e' is observed. This is true in the case of an excitation and an observation normal to the surface: (3.31) may be extended by summation on e and e'.] Decomposing (3.32) into one-phonon and multiphonon processes, we may write

$$\Gamma_{ee'} = \Gamma^R_{ee'}(\hbar\omega_1 - \hbar\omega_2) + \Gamma^F_{ee'}(\hbar\omega_1 - \hbar\omega_2) \tag{3.33}$$

where:

Γ^R characterizes the amplitude of the Raman process, with Dirac peaks at the various frequencies of the phonons coupled to the exciton e by H_{ep}. This process must satisfy the selection rules on the energy, the wave vector, and the polarization imposed by the incident and the observed photons. Thus Γ^R shows peak variations with the energy $\hbar\omega_1 - \hbar\omega_2$.

Γ^F characterizes luminescence processes. Since at least two phonons are created, no strict selection rules hold; compared to Γ^R, Γ^F will have a smooth energy dependence. Besides, as the observed energy increases, the number of relaxation channels increases with the available energy, leading to a rapidly increasing function of $\hbar\omega_1 - \hbar\omega_2$.

2. *The Vibronic Excited States of the Surface*

This subsection is devoted to the description of the upper excited states appearing in the excitation spectrum of the surface emission. The way this excitation is performed will be examined in Section III.B.3 below, in connection with the theory of Section III.B.1.b. The upper states examined are associated with the first singlet transition, b- and a-polarized, and are of two types: purely electronic and vibronic states in an extended sense, as resulting from the coupling of electronic excitations to vibrations or to lattice phonons.

a. *The Second Davydov Component of the Surface Exciton.*[61] In Section III.A, we studied a lattice with one dipole per unit, b-polarized. The surface exciton presents, as its bulk counterpart, a second transition, a-polarized. It is the second Davydov component of the surface exciton at about $220\,\mathrm{cm}^{-1}$ above the b component, which is visible in reflectivity as well as in fluorescence spectra. The a component does not emit, and its structure in the reflectivity spectrum is very weak: see Fig. 3.7, upper detail. This behavior is understood by analogy with its bulk counterpart, which is phonon-broadened by about $80\,\mathrm{cm}^{-1}$. This width is much larger than the radiative width, estimated at a few reciprocal centimeters according to the ratio of the oscillator strengths of the b and a transitions. Thus, the a surface component relaxes very rapidly to the lower states of the excitonic band, particularly to the b exciton.

Fig. 3.17 shows the b-emission excitation spectrum, with an a-polarized excitation, in the region $E_{sb}^{00} + 220\,\mathrm{cm}^{-1}$ and above. We observe a strong and broad peak, attributed here to the a surface component, following refs. 117, 67. This peak is strongly asymmetric, broadened on the high-energy side. The value of the component indicated on the maximum ($25\,523\,\mathrm{cm}^{-1}$, at $222\,\mathrm{cm}^{-1}$ above the b component) may be falsified by the asymmetry of the band. (This will be examined in Section III.B.3 below.) We find that the surface-exciton Davydov splitting is quite comparable with its bulk counterpart ($222\,\mathrm{cm}^{-1}$ vs $224\,\mathrm{cm}^{-1}$), to the accuracy of our experiments. Furthermore, the two 0–0

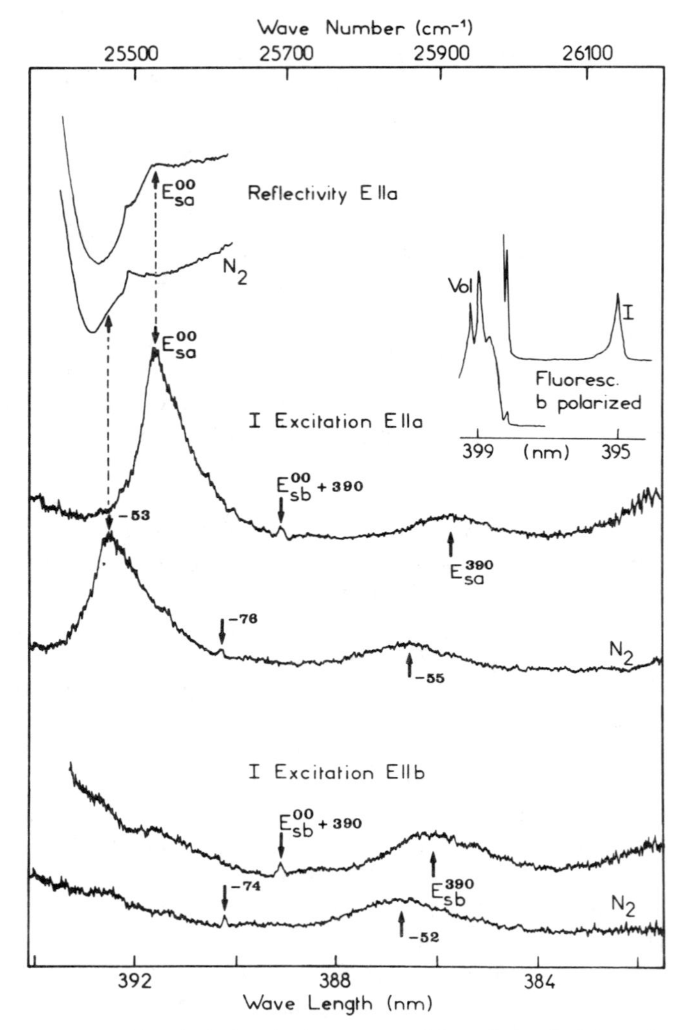

Figure 3.17. *a*- and *b*-polarized excitation spectra of the emission I of the surface of a single anthracene crystal uncoated and coated with condensed gas (N_2) at $T = 1.7$ K. The negative numbers indicate the red shifts of the various structures upon gas coating. (The upper left part gives the detail in the reflectivity of the observed structures.)

surface transitions are found shifted to higher energy by 206 cm^{-1} relative to the bulk 0–0 *b* transition.

Upon gas coating (N_2) of the crystal surface, while the *b* component falls by about 78 cm^{-1}, the *a* component falls by only 57 cm^{-1}: see Fig. 3.17. Thus, the Davydov splitting increases by more than 10% upon gas coating, its value passing from 222 to 243 cm^{-1}. Correlatively, the *a* component broadens, but remains asymmetric. The increase of the Davydov splitting may appear surprising.[1] Indeed, the identity of its values in the bulk and in the surface excitons leads one to think of the (001) layers as rigid entities whose excitation energy is simply shifted, without modification of the intralayer interactions in the surface layer. However, the layer of condensed gas, which is less polarizable than the anthracene layer itself and does not produce a complete shift (78 cm^{-1} instead of 200 cm^{-1}), nevertheless modifies the surface intramolecular interactions. This essential phenomenon of surface nuclear reorganization will be discussed in detail in Section III.C.

b. *The Vibronic 390- and 1400-cm^{-1} Surface States.* As indicated in Section II.B, the characteristic structures visible in the bulk vibronic spectra are the broad vibronic peaks (more or less broadened by their coupling to the two-particle-state continuum), while the low-energy threshold of this continuum is located at one vibration quantum above the bottom of the excitonic band (E_{sb}^{00}). As shown for the bulk states (Section II.B.3), the positions of the vibronic structures on the *a*- and *b*-polarized spectra are consistent, to a very good approximation, with the vibron model, without introducing any collective coupling. Thus, we restrict ourselves to the vibron approximation for the two principal intramolecular vibration modes, 390 and 1400 cm^{-1}.

The excitation spectrum of the vibronic 0–1 region (390 cm^{-1}) is illustrated in Fig. 3.17, in the two polarizations. One finds two broad peaks: the *b*-polarized at $E_{sb}^{390} = 25\,897 \pm 6$ cm^{-1}, and the *a*-polarized at $E_{sa}^{390} = 25\,922 \pm 6$ cm^{-1}, assigned to the two Davydov components of the vibron 390.[61] The observed splitting, 25 ± 12 cm^{-1}, appears larger than the pointing error of these broad structures. At a lower energy, practically the same energy for the two polarizations ($E_{sb}^{00} + 390$ cm^{-1}), one finds a weak, sharp peak corresponding to the bottom of the excitonic band plus 390 cm^{-1}, the vibration energy quantum: This sharp structure is the surface analog of that of the two-particle-state continuum of the bulk. The variation of this structure (sharp peak in surface against step structure in the bulk) will be discussed in Section III.B.3 below.

The excitation spectrum of the vibronic 0–1 1400-cm^{-1} region is shown in Fig. 3.18. One finds, in the two polarization spectra, a broad but sharp peak, at $E_{sb}^{1400} = 26\,817 \pm 6$ cm^{-1} in the *b* polarization, and at $E_{sa}^{1400} = 26\,920 \pm 6$ cm^{-1} in the *a* polarization. The observed splitting of 103 ± 12 cm^{-1} is comparable to that of the bulk, 94 ± 12 cm^{-1}. These peaks are attributed to the surface vibron (1400 cm^{-1}). The low-energy threshold of the two-particle states is missing; it is probably masked by the noise, as also indicated for its bulk counterpart. This threshold structure, relative to the vibronic peak

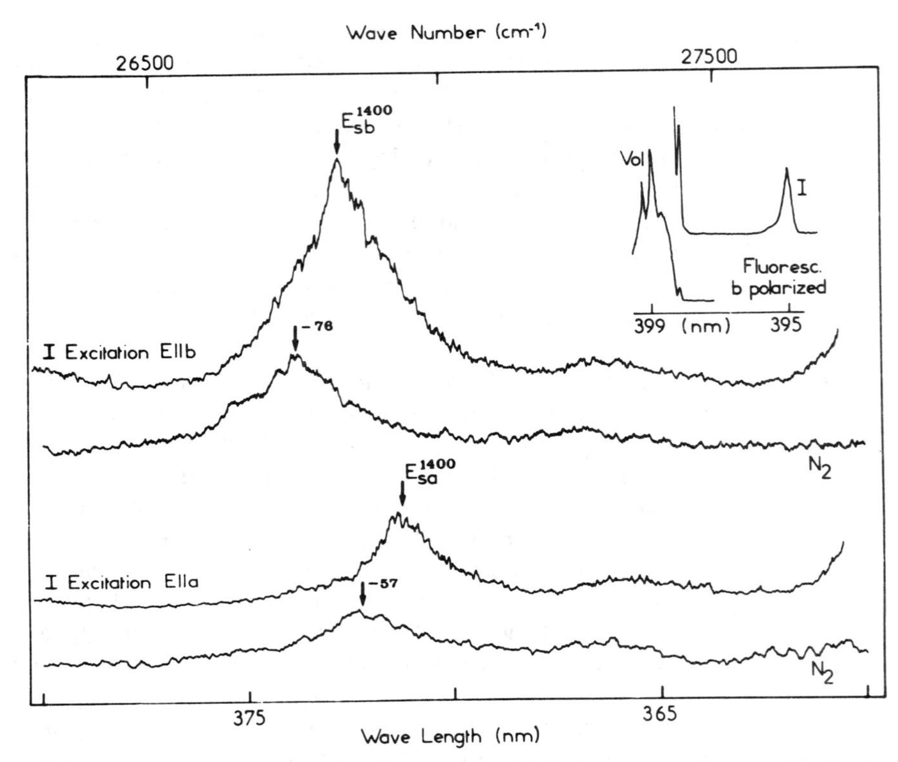

Figure 3.18. *a*- and *b*-polarized excitation spectra of the emission I of the surface in the 0–1 1400-cm^{-1} region of a single anthracene crystal, uncoated and coated with condensed gas (N_2) at $T = 1.7$ K.

magnitude, as much weaker than that observed for the 390-cm^{-1} vibron in the reflectivity spectrum.

The position of the various vibronic structures will allow us to determine, as we did for the bulk (Section II.B.3.b), the center of gravity of the three vibronic bands 0–0, 0–390, 0–1400, using the vibration frequencies at the excited state, 390 and 1397 cm^{-1}. The two determinations (2.85) agree in locating the center of gravity CG_s^{00} of the surface excitonic band in the near vicinity of the *a* component E_{sa}^{00}.[61] This is in contrast with the center of gravity of the bulk excitonic band, CG_b^{00}, which we found located at 40 cm^{-1} below the *a* component (E_{va}^{00}); see Section II.B.3. This difference appears significant, because it is larger than the error limit of our experiments.

This difference between bulk and surface excitonic bands seems to indicate that, inspite of the near-identity of the Davydov splittings, the intermolecular interactions are slightly modified for the surface molecules. Furthermore, to a first approximation, we may think of this modification as occurring in the interactions between equivalent molecules. This discussion is pursued in Section III.C on surface reconstruction.

Figures 3.17–18 illustrate the changes of the vibronic excitation spectra upon gas coating: All the structures investigated in the preceding subsection shift downwards and reveal their surface origin. However, the shift is differential: The narrow peaks, $E_{sb}^{00} + 390\,\text{cm}^{-1}$, on the *a* and *b* polarizations shift by about $-76\,\text{cm}^{-1}$, as does E_{sb}^{00}, in agreement with their assignment as thresholds of surface two-particle-state continua. The vibronic peaks show more complex behavior: The "390" peaks, E_{sa}^{390} and E_{sb}^{390}, shift as the *a* component (or as the center of gravity of the excitonic band). On the contrary, the "1400" peaks, E_{sa}^{1400} and E_{sb}^{1400}, show a differential shift (respectively, -57 and $-76\,\text{cm}^{-1}$) with a subsequent increase of the Davydov splitting, by more than 20%, upon gas coating. (The difference with the 10% increase of the 0–0 splitting is not significant, owing to the error limit.) The new surface excitonic band, upon gas coating, shows a new center of gravity very close to the *a* component. This change would indicate a modification of the surface upon gas coating, with predominant changes in the interaction between nonequivalent molecules. The above spectral analysis provides the sequence of surface and bulk excitonic states shown in Fig. 3.19.

c. *A Monolayer Surface Phonon.*[131] The excitation spectrum below the *a* component is shown in Fig. 3.20 for the *a* and *b* polarizations. The dominant structure is a strong and very narrow peak (much narrower than the *b* emission line) in the *b* polarization at $45.1\,\text{cm}^{-1}$ above the observed emission energy. The complex shape of this structure is well defined in the experimental spectrum and will be analyzed in Section III.C. The structure is interpreted as the low-energy threshold of creation of a surface two-particle state, either phonon–exciton or phonon–photon through a virtual exciton. The phonon involved is the A_g-symmetry libration mode (around the normal axis) observed at $49.39\,\text{cm}^{-1}$ in the bulk reflectivity spectrum (Fig. 2.8, arrow) at the same temperature. We notice the frequency decrease (10%) of the surface phonon with respect to its bulk analog (cf. Section III.C below). According to our discussion in Section II.C, we expect no vibron associated with this phonon mode, the absorption being roughly the product of a lorentzian factor centered at E_{sb}^{00} and the density of excitonic states at the energy $E - 45\,\text{cm}^{-1}$.

Upon gas coating, the structure $E_{sb}^{00} + 45\,\text{cm}^{-1}$ transforms radically into a loose band, with little structure, roughly shifted as E_{sb}^{00}. Thus, upon gas coating, the surface phonon mode is strongly broadened, in contrast with the excitonic structure involving electronic excitation interactions. The nature, homogeneous or inhomogeneous, of the phonon mode broadening could give us key information on the detailed interactions between the film of the

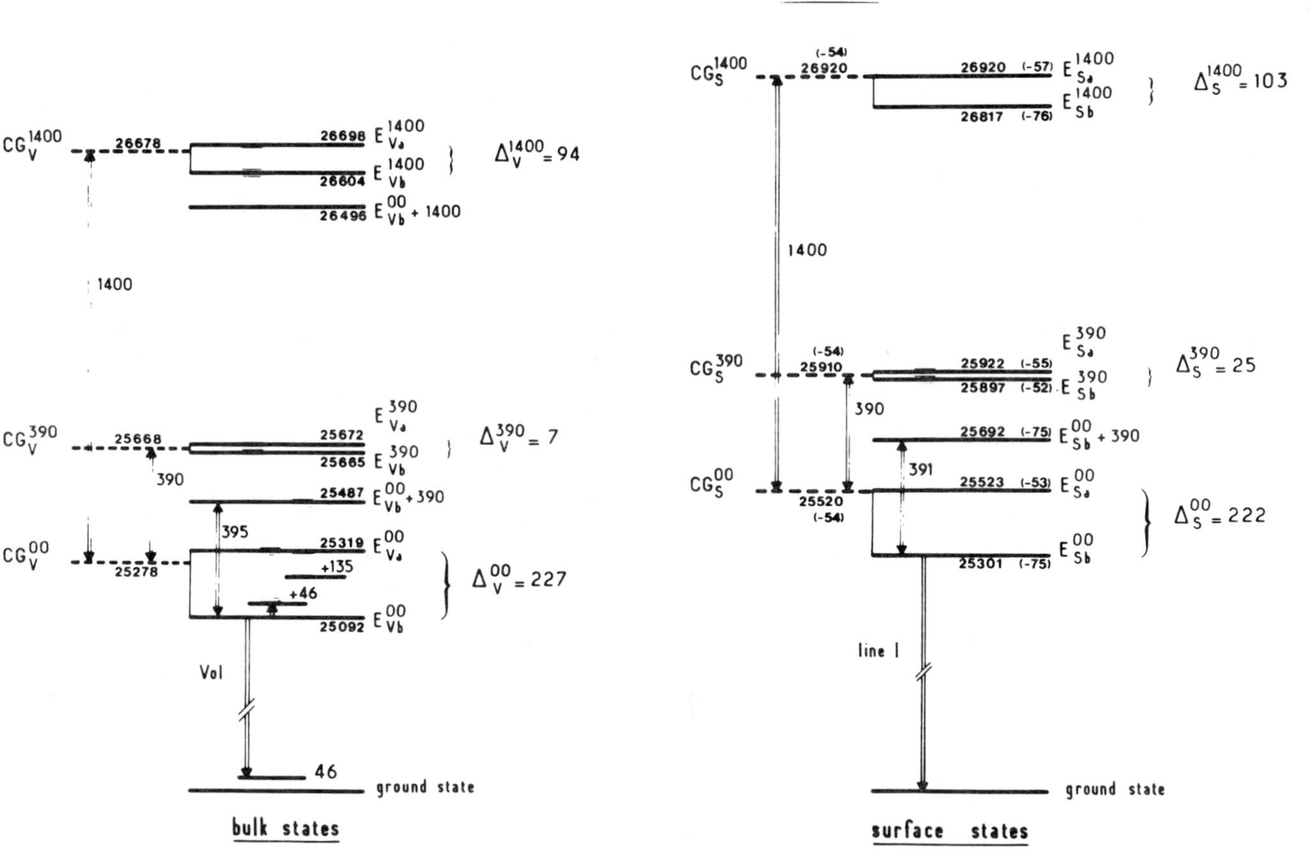

Figure 3.19. Diagram of the optical resonance energies of the bulk (left part) and of the surface (right part). We note the translational quasi-equivalence between bulk and surface energy levels, except for the centers of gravity of the excitonic bands, which are slightly blue-shifted ($40\,cm^{-1}$) for the surface excitonic bands. The negative numbers indicate the red shifts of the surface energy levels upon gas coating at 1.7 K.

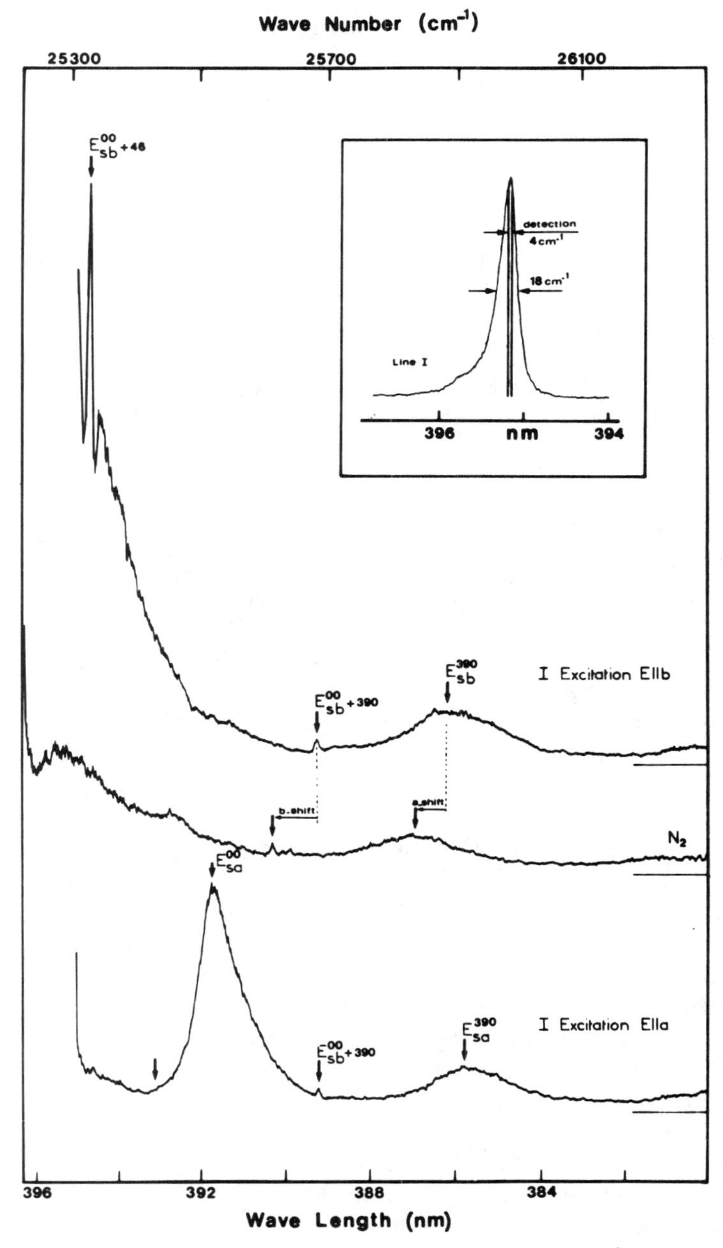

Figure 3.20. *b*- and *a*-polarized excitation spectra, detected by a 4-cm^{-1} window (see inset), of the surface emission at 5 K. In the region at 40 to 50 cm^{-1} above the detected emission; we notice in the *b* polarization a strong Raman peak, at 45.1 cm^{-1} above the detected energy, followed by a characteristic profile. Upon coating with condensed gas (N_2), this spectrum is strongly distorted. The *a*-polarization spectrum shows no structure in this region; the arrow indicates a possible Raman structure corresponding to the B_g lattice mode (140 cm^{-1}) of the surface. The inset indicates the frequency and the width of our detection in the emission line I.

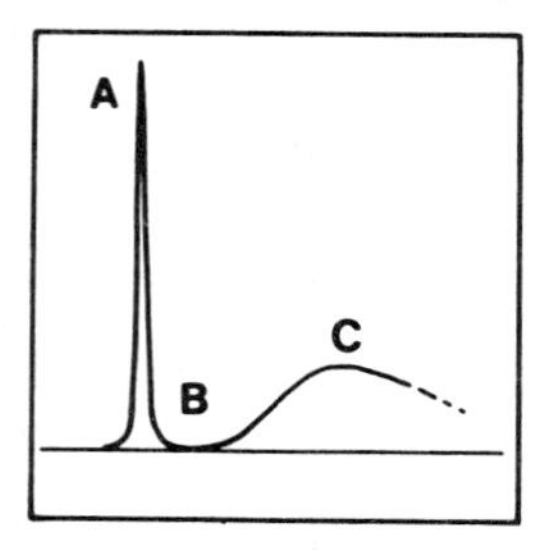

Figure 3.21. Scheme of the expected excitation profile due to the relaxation mechanisms illustrated in Fig. 3.22. We expect (*A*) a Raman peak (quasi-resonant), (*B*) a dip where the intrasurface relaxation Γ_{is} is quenched by the relaxation to the bulk Γ_B, and (*C*) a bump Γ_{is} competing with Γ_B and overhelmed at higher energies Γ_{is}, depending on the exact exciton vibration coupling, and different for the modes at 390 and 45 cm^{-1}.

condensed-gas molecules (which can be chosen specifically), and on the possibility of investigating modes of the condensed film via resonant Raman scattering on the surface 0–0 strong transition coupled to the condensed-gas layer.

Research on other lattice phonons has not yielded results up to now. For instance, the corresponding B_g 140-cm^{-1} surface mode (cf. Section II.D.2) on the *a* component was not observed, in spite of a careful inspection (this research continues under better resolution conditions). The main reasons that single out the behavior of the 45-cm^{-1} phonon from the other modes present in the Raman spectrum (see Fig. 3.21) are of two kinds: (1) this libration mode strongly modifies the transition dipole to the first singlet state; (2) for the *b* polarization the virtual excitation E_{sb}^{00} is necessary to "absorb" one photon, and that corresponds to an important lorentzian factor (wing of the large homogeneous width of the state E_{sb}^{00}) which favors this low-frequency mode. As for the harmonics 2×45 cm^{-1}, or other combinations with this mode, the suppression of the conservation of the wave vectors allows all the phonon-band states to be excited, leading to broad bumps, one of which is visible in the Fig. 3.20. The possible dilution of the phonon states in the bulk will be examined in Section III.B.3 on relaxation mechanisms.

3. *Relaxation Mechanisms*

In this subsection we analyze the way the theory in Section III.B.1.b allows us to interpret the structures of the excitation spectra and of the emission spectra. Since the intrasurface relaxation proceeds first by the creation of phonons and of vibrations, we consider one of the resulting modes, of energy $\hbar\Omega_0$, without specifying it, and we call it "vibration".

The one-vibration processes, termed also Raman scattering [see (3.33)], accounts for the structure $E_{sb}^{00} + 45$ cm^{-1}, $E_{sb}^{00} + 390_b^a$ cm^{-1}, which are "discrete" peaks and whose observed width is fixed by the excitation and the observation conditions (not by the intrinsic width of the excited state), as well as by the intrinsic width of the created mode (smaller than 1 cm.[120,121,131]) The intensity of these peaks depends on the coupling to these vibrations, and on the resonance factors of the exciting and emitted photons relative to the electronic excitons involved in the process (3.31). For example, it is the resonance factor $(\hbar\omega_2 - \tilde{E}^{00}_{sb})^{-2}$ that allows us to understand the way the surface states are photoselected in our technique. In the case of the 45-cm^{-1} phonon the excitation factor $(\hbar\omega_1 - \tilde{E}^{00}_{sb})^{-2}$ decreases rapidly with increasing frequencq of the created phonon, and this explains the small probability of observing high-frequency phonons with our technique. The component a, or B_g, of this mode introduces a very small excitation factor $(\hbar\omega_1 - \tilde{E}^{00}_{sa})^{-2}$, which explains the absence of its structure in the a-polarized excitation spectrum. As for the 390-cm^{-1} vibration, assuming a hypothetical weak exciton–vibration coupling, the factors $(\hbar\omega_1 - \tilde{E}^{00}_{sb})^{-2}$ and $(\hbar\omega_1 - E^{00}_{sa})^{-2}$ are in a ratio of 4 to 1, which is almost counterbalanced by the ratio of the oscillator strengths: The two components are of comparable magnitude, weak, but clearly visible on the excitation spectra. A more refined theory would introduce the vibron–two-partical-state coupling, according to our analysis in Section II.B, which applies to the analysis of the Raman structure associated with the 1400-cm^{-1} mode. A comparison with the bulk peaks, where the Raman peak is evaluated relative to the vibronic peak, provides an idea on the importance of the surface Raman peak and accounts for a possible dilution of the Raman structure in the background continuum created by other channels of relaxation.

Relaxation by creation of two or more phonons will be considered as providing a continuous excitation spectrum as a function of the dissipated energy $\hbar\omega_1 - \hbar\omega_2$ (3.33). Obviously, this approximation must be valid below 30 to 40 cm^{-1}, since only acoustical phonons contribute to the relaxation mechanisms. When $\hbar\omega_1 - \hbar\omega_2$ approaches zero, so does $\Gamma^F_{ee'}(\Delta E)$.[127] In the vicinity of an energy ΔE_0, we may expand $\Gamma^F(\Delta E)$ to the first order in E, neglecting the dependence on the excitonic states e and e', which is justified for a local exciton–phonon coupling (Section II.A).

First, let us examine the Raman structures at 45 and 390 cm^{-1}. If the first step is the creation of one vibration $\hbar\Omega_0$, the created exciton after this step is a **k** state of the band, of energy $\hbar\omega_1 - \hbar\Omega_0$ (which must be larger than the energy E^{00}_{sb} for this transition to be real). For the observation of the emission, a relaxation must occur between the **k** exciton and the observed emission state, the relaxation energy being $\hbar\omega_1 - \hbar\omega_2 - \hbar\Omega_0$. According as this relaxation rate is or is not larger than the relaxation rate (3.30) to the bulk, the secondary surface emission will or will not be observed. Indeed, we observe in the experimental spectrum a dip between the Raman peak and a continuum of surface emission: see Fig. 3.21.

The position and the width of this dip, at about $10\,\mathrm{cm}^{-1}$ above the Raman peak, indicate the energy gap above which the intrasurface relaxation, assisted by acoustical-phonon creation, competes with the surface-to-bulk relaxation. If we figure 3 to $4\,\mathrm{cm}^{-1}$ for the relaxation rate to the bulk (for $\mathbf{K} \sim \mathbf{0}$ wave vectors; cf. Section III.A.3), we conclude that the intrasurface relaxation, at $10\,\mathrm{cm}^{-1}$ above the emitting state, is comparable. This conclusion on the acoustical-phonon relaxation is consistent with the theoretical estimates[121,127] (cf. Section III.A.4) and the experimental values derived by KK analysis of the bulk reflectivity (Section II.C.3b).

Figure 3.22 summarizes the relaxation scheme leading to the observed structures above the surface peaks at 45 and $390\,\mathrm{cm}^{-1}$. When the excitation sweeps through values larger than the observed photon and the main coupled vibration $\hbar\omega_1 > \hbar\omega_2 + \hbar\Omega_0$ (which is the case for excitation of the second Davydov component, which, as shown in Section II.C.3a, relaxes preferentially by the creation of one 140-cm^{-1} phonon), the relaxation probability may be written

$$\Gamma^F_{ee'}(\Delta E) = \Gamma^F(\Delta E_0) + (\Delta E - \Delta E_0) \times b \tag{3.34}$$

where ΔE_0 is the Davydov splitting ($\Delta E_0 = E^{00}_{sa} - E^{00}_{sb}$) and b is a parameter characterizing the variation of Γ^F around ΔE_0. The relaxation expression (3.31) then allows us to plot, at fixed $\hbar\omega_1$, the emission spectrum of the surface, and at fixed $\hbar\omega_2$, the excitation spectrum of the emission. Therefore, we find

$$\text{emission } (\hbar\omega_1 \text{ fixed}) \propto \frac{\Gamma^F(\hbar\omega_1 - \hbar\omega_2)}{|\hbar\omega_2 - \tilde{E}^{00}_{sb}|^2} \tag{3.35}$$

$$\text{excitation } (\hbar\omega_2 \text{ fixed}) \propto \frac{\Gamma^F(\hbar\omega_1 - \hbar\omega_2)}{|\hbar\omega_1 - \tilde{E}^{00}_{sa}|^2} \tag{3.36}$$

Equation (3.35) allows us to explain the strong asymmetry of the first surface emission (cf. Fig. 3.4): The lorentzian line is distorted by the variation of Γ^F, being broadened on the low-energy side where Γ^F increases. Thus, the observed asymmetry is connected with the preparation of the excited state and not with an intrinsic property of the surface state. The counterpart of this asymmetry is the strong broadening of the high-energy side of the *a* component (see Fig. 3.17), which is very broad; the denominator $|\hbar\omega_1 - E^{00}_{sa}|^{-2}$ itself probably behaves asymmetrically. The variation of Γ^F shifts the maximum of the excitation spectra by about $10\,\mathrm{cm}^{-1}$, leading for the Davydov splitting of the surface exciton to a value of about 5% lower than its bulk counterpart.

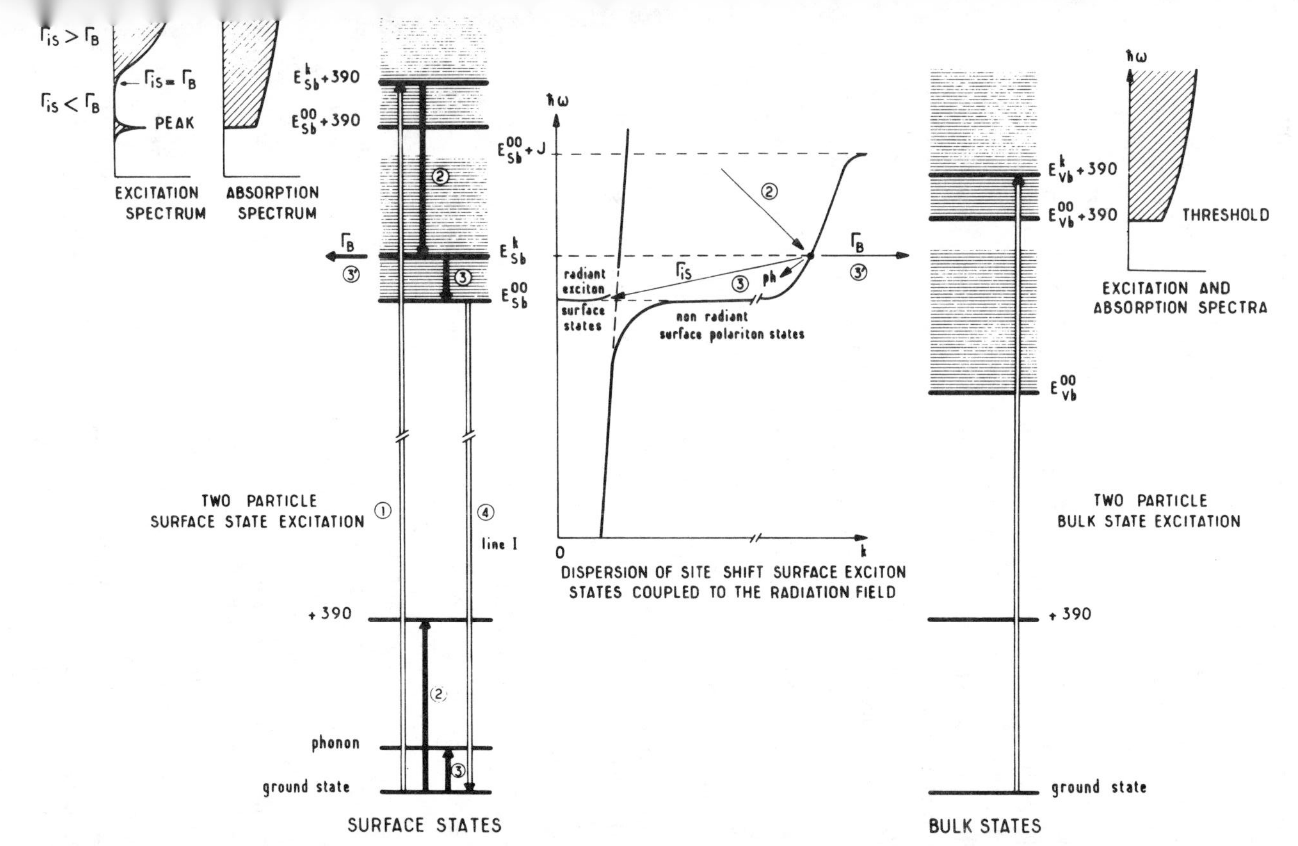

Figure 3.22. Model relaxation of the surface excitons created inside or near the threshold of a two-particle-state continuum (illustrated for the 390-cm^{-1} mode). After excitation (1) at the energy $E^{k}_{sb} + 390\,\mathrm{cm}^{-1}$, a two-particle state is created (2) by fission. Then the exciton may relax along two competing paths: an intrasurface channel (3) leading to emission, and a nonradiative channel (3′) to the bulk (eventually to its fluorescence), with respective probabilities Γ_{is} and Γ_B. Therefore, the surface emission efficiency depends on the ratio Γ_{is}/Γ_B, which determines the observed profile. When the excitation occurs at exactly $390\,\mathrm{cm}^{-1}$ above the detection, we observe the very narrow Raman peak.

The high-energy excitation, $\hbar\omega_1 > \hbar\omega_2 + \hbar\Omega_0$, is also due to the vibronic components at 390 and 1400 cm^{-1}. If we privilege the vibronic relaxation by fission and the creation of one vibration $\hbar\Omega_0$, then the relaxation of the incident photon $\hbar\omega_1$ leads to a state about 100 cm^{-1} above the observed emission. The excitation spectrum due to the vibronic component at 1400 cm^{-1} (see Fig. 3.18) shows that the relaxation by creation of other vibrations contributes also, since no threshold structure is observed around the value $\hbar\omega_2 + 1400\ cm^{-1}$. This conclusion is also consistent with the vibronic analysis of the bulk (Section II.B.3).

To summarize, the mechanisms of the surface relaxation at very low temperature are of two types. The first are Raman processes with creation of one phonon with well-defined wave vector, hence with well-defined energy. These processes generate narrow peaks on the excitation spectra. The second are relaxation processes which involve many phonons and produce a broad band; the main one is usually a vibronic fission in its two-particle-state continuum. These processes determine the femtosecond relaxation rates analyzed for the bulk vibronic relaxation in Section II. The essential feature of these mechanisms is the rapid growth of the relaxation rate with the relaxed energy, which accounts for the asymmetry of the emission line and the quenching of the relaxation below a rather strong gap of about 10 cm^{-1}, where the intrinsic intrasurface relaxation rate is about 5 cm^{-1}. The origin of the gap is not the decaying density of surface phonons, but the presence of the substrate bulk with the strong, **k**-dependent surface-to-bulk relaxation, which allows the Raman peak to show up.

C. Reconstruction of the Crystal Surface

The observations discussed in Sections III.A and B may be interpreted using the model of a perfect surface, without crystal defects. We restrict ourselves to this model in the present section.

We have observed in Section III.B a translational near-equivalence in energy between bulk and surface states, the latter being destabilized by 206 cm^{-1}. This destabilization, which makes surface molecules closer to isolated molecules than bulk ones, may have two origins: the lack of stabilization due to interplane interactions[1] and/or the decay of the intrasurface stabilization terms due to the reconstruction of the surface.

The reconstruction, or relaxation, of the crystal surface is currently being investigated. It has been observed for certain inorganic materials by X-ray diffraction or with an electrontunneling microscope.[110] In these systems the disruption of strong bonds causes reconstruction. In contrast, with organic crystals disruption of the weak van der Waals bonds leads to a discrete surface reconstruction, not yet observed to the best of our knowledge.[107] A fortiori, for the anthracene crystal (001) surface, which is a plane of easy cleavage, the reconstruction is presumably very limited.

Our low-temperature spectroscopic technique may be, as we have already indicated, very sensitive to surface-state modification. However, it provides only an overall and indirect observation of the crystal-surface reconstruction. For this reason, the conclusions we are proposing here need further confirmation by calculations and by complementary experimental techniques.

1. *The Ground-State Surface*

a. *General Considerations.* The simplest way to imagine a crystal surface is as a bulk layer. This image, valid for layers of easy cleavage, turns out to be wrong in general. For instance, certain surfaces contain one atom over two of the subsurface layer.[132] Thus, if the surface is a reticular plane, as a consequence of the translational-symmetry disruption, none of the bulk symmerty elements is necessarily conserved on the surface. Also, the bulk periodicities may be replaced by multiple periodicities or entirely different periodicities, as in the case of a surface with adsorbed atoms, or the 7×7 reconstruction of the Si(111) surface.[110]

In the particular case of the (001) plane of the anthracene crystal, we assume that the surface structure is very little different from that of a bulk layer. Indeed, the creation of two surfaces by cleavage is very easy: The energy cost is very low and needs no molecular displacements of large amplitude. Thus, we adopt, for the (001) layer, the simplest assumption: a 2D layer with periodicity, parallel to the plane of vectors **a** and **b**, preserved on the surface. We assume further that the molecules are rigid and that the symmetry plane (a, c) persists. Under these conditions, the surface layer has the monoclinic structure of a bulk layer, and the only parameters susceptible to modification are:

1. the intermolecular distance along the normal **n** to the c' plane;
2. the monoclinic angle β;
3. the molecular orientation inside the unit cell, defined by the direction cosine of the molecule.

Such a model of surface modification has been used by Filippini, Gramaccioli, and Simonetta[133] to investigate the naphthalene crystal (001) face. Their calculations, based on a Kitaigorodsky-type method,[134] accounted for the reconstruction of the surface and the subsurface layers of the crystal. The principle of the calculation is to minimize the cohesion potential written as a sum of atom–atom potentials.[134] The potentials include an attractive part in R^{-6} (of van der Waals or dispersion forces) which is of the type discussed in Section I.A.2, and a repulsive part which is generally given the form e^{-cR} to account for the impenetrability of the electronic clouds. The potentials are determined for each type of pair C····C, C····H, or H····H, by adjusting parameters so as to reproduce, as well as possible, properties such as the crystallographic structure and the phonon energies. The adjusted parameters vary with the optimized property and the investigated sample. These calculations[133] predict a lengthening of 10^{-2} Å along the c' axis, and a molecular reorientation with an upper limit of 0.5°; the variation of β is not specified. These values are sensitive to the potentials used (for instance, the sign of the molecular reorientation), but their orders of magnitude agree in predicting a moderate reconstruction of the (001) surface. Although these orders of magnitude are transferrable to the case of the anthracene crystal, it is more instructive to evaluate them directly.

b. *A Simple Model of Surface Reconstruction.* To analyze the transition from a bulk layer to a surface layer, we must start by suppressing the interaction with a half space (the missing interactions,[1]) and then observe the subsequent relaxation of the layer to its new equilibrium. The total energy of such a transformation, assumed reversible, measures the surface tension of the crystal along the cleavage plane (this energy may be measured[135]). In this transformation, the suppressed short-range molecular interactions are complex and depend on the positions of a large number of nuclei. On the contrary, beyond the next nearest neighbor, the potentials will vary slowly and therefore can be replaced by a macroscopic potential. The model we adopt here, without further justification, consits in replacing, from the very nearest neighbors, the interactions with the missing half space by the potential of a macroscopic stress. The purpose of this simplistic model will be to provide orders of magnitude for strains and reconstructions undergone by the crystal surface.

Since the range of the cohesive forces is much larger than the range of the repulsive forces, the surface layer will suffer, relative to bulk layers, a dilatation strain. In our model of quasi-independent (001) layers, we are interested only in intralayer quantities, which are thus amenable to the calculation of effects of the dilatation strain, to which the surface layer is assumed to respond linearly, with the elastic constatns of the bulk. Dilatation along the c' axis will provide variations on our three parameters (the c' distance, the monoclinic angle β, and the orientation of the surface molecules in their unit cells) as well as on the frequencies of the associated librations.

Since we are performing a macroscopic calculation, the discrete structure of the surface layer may be neglected. For the calculation of orders of magnitude of the stress, a rough approach is sufficient: Each reticular plane is replaced by a set of two planes connected elastically, on which the various forces are applied; see Fig. 3.23. The two planes, coupled elastically, represent the limits of the π electronic clouds of the anthracene molecules. The forces on them are mainly the attractive van der Waals forces between the reticular planes, and the repulsive forces between the hydrogen atoms.

With this picture of the interactions, we are led to a unidimensional model

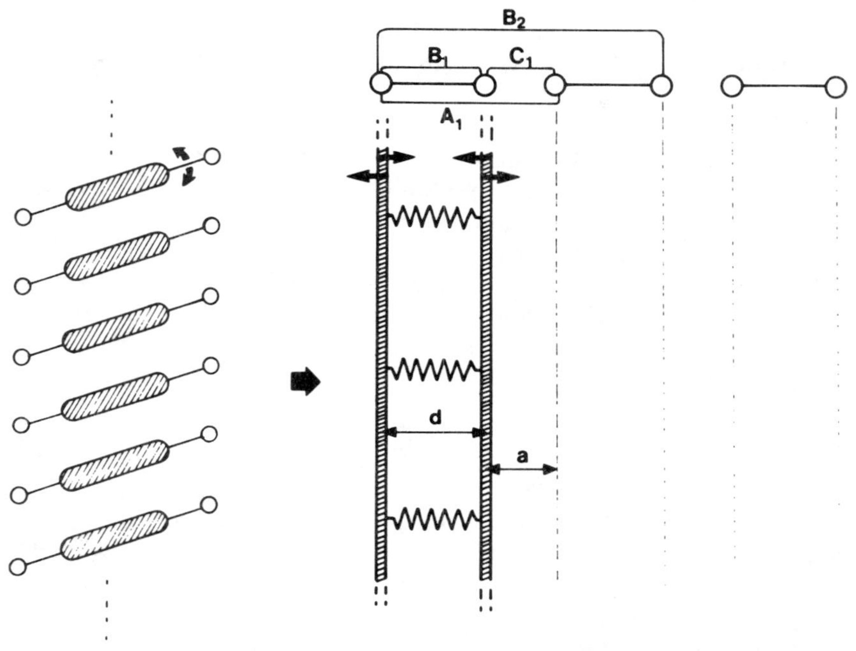

Figure 3.23. Model of surface plane for the evaluation of the surface-molecule reorientation and its effects on the first substrate planes. The changes in the molecular orientations are replaced by a compression or dilation of a set of two planes connected elastically; the missing forces are assumed to act only on the hatched planes. The parameters of the model are the distance d between two hatched planes and the distance a separating two nearest-neighbor hatched planes belonging to two different crystal planes. The interaction forces between planes of different "molecules" are indicated with the notation in the text [cf. (3.38)].

of "diatomic molecules": The dilatation of the surface "molecules" will correspond to reorientations in the surface layers of the anthracene crystal. For instance, as illustrated in Fig. 3.23, the attractive interaction B_2 between two nearest-neighbor layers causes a compression of the surface molecule. (The same phenomenon in the Xe crystal has led[42] to the conclusion that the pair of Xe atoms are compressed in the bulk by the attractive forces caused by the next nearest neighbor and beyond; the order of magnitude of the resulting pressure is 1 kbar.) Thus, using the above model and neglecting repulsive forces and deformation induced by the calculated stresses (as a zero-order approximation), we find for the missing stress on the nth surface layer, where only attractive forces are present,

$$\begin{aligned} \tau_n = A_n + 3A_{n+1} + \cdots + (2p+1)A_{n+p} + \cdots \\ + (B_{n+1} + C_{n+1}) + \cdots + p(B_{n+p} + C_{n+p}) + \cdots \end{aligned} \quad (3.37)$$

with

$$A_n = A(na + nd), \qquad B_n = A[nd + (n-1)a], \qquad C_n = A[na + (n-1)d] \tag{3.38}$$

where d is the thickness of the "molecule", a the distance separating two "molecules", and $A(r)$ the force per unit area derived from the attractive potential between two elementary layers (the n dependence of τ_n will be discussed in Section III.C.2 below).

For the calculation of $A(r)$, the attractive van der Waals force between two layers separated by r, we start with an isotropic dispersion interaction of the form $\mathscr{D}_f = B/r^6$ between point charges. After integration over one layer we find a force per unit area

$$\tau = \frac{2\pi}{S_0^2} B r^{-5} \tag{3.39}$$

where $1/S_0$ indicates the "charge" density. Using the isotropic model, we find that the stabilization of the ground state, $\mathscr{D}_f \sim 100$ kJ/mole, originates mainly from the four neighbors of each molecule: $\mathscr{D}_f = 4Br_0^{-6}$, where r_0^3 is a volume of the order of the molecular volume. Then S_0 represents the surface per molecule in the (001) layer. With $r_0^3 \sim 300$ Å, $r_0 \sim 10$ Å, (3.39) gives $\tau = 4 \times 10^8$ Pa, or 4 kbar. We must indicate that this value is very rough (within one or two orders of magnitude), because of the very rapid variation of the van der Waals forces.

Another estimate of τ may be obtained by transferring the parameters calculated for the naphthalene crystal.[133] They provide as surface cohesion energy a value 85% of that of the bulk:

$$\tau = \frac{1}{S_0}\frac{\partial}{\partial r}(\mathscr{D}_f^v - \mathscr{D}_f^s) = 4\frac{\mathscr{D}_f^v - \mathscr{D}_f^s}{S_0 r} \sim 4\,\text{kbar} \tag{3.39'}$$

This estimate, probably better than (3.39), provides stresses still of the order of a few kilobars.

Now, we may evaluate the surface reconstruction generated by the suppression of a few kilobars of stress along the c' axis, using the deformation coefficients under hydrostatic pressure. The authors of Ref. 136 have determined, in neutron-scattering experiments, the variation of the crystallographic parameters of anthracene:

$$\frac{-1}{a_0}\frac{\partial a_0}{\partial p} = 5.0 \times 10^{-3}/\text{kbar}, \qquad \frac{-1}{b_0}\frac{\partial b_0}{\partial p} = 2.86 \times 10^{-3}/\text{kbar}$$
$$\frac{-1}{c_0}\frac{\partial c_0}{\partial p} = 2.8 \times 10^{-3}/\text{kbar}, \qquad \frac{\partial \beta}{\partial p} = -8.2 \times 10^{-2}\,\text{deg/kbar} \tag{3.40}$$

Unfortunately, the molecules reorientation is not known. For a 4-kbar stress the coefficients measured at room temperature (they are presumably smaller at low temperatures) provide a variation of 0.1 Å for the c' dimension, and a variation of 0.3° for the angle β. These values agree in order of magnitude with those calculated for the naphthalene crystal.

c. *Surface Variation of the 49-cm^{-1} Phonon.* We have seen in Section III.B.2.c that the surface phonon at 45.1 cm^{-1} appears as a sharp peak below its bulk counterpart. This means that its frequency difference with the bulk must be larger than the bandwidth of the bulk phonon dispersion along the c' axis. This condition preserves the surface phonon from dilution in the bulk. The phonon dispersion bandwidth along the c' axis has been determined by neutron scattering on perdeuterated anthracene, and found to be 1.6 cm^{-1},[137] so that we may safely neglect any dynamic effects of the bulk and take the value 45.1 cm^{-1} as being the intrinsic frequency of the surface phonon.

At this stage, we dispose of a well-defined physical quantity, considerably affected by the bulk-to-surface transition, varying from 49.4 cm^{-1} (50.2 cm^{-1} for the center of the band, on taking into account the dispersion along the c' axis) to 45.1 cm^{-1}. This change of frequency originates from the alteration of the libration potential in the surface layer. If we separate, in this potential, the bulk part from the suppressed part, and we expand the two terms in powers of the libration angle θ (around the N axis), we obtain

$$\begin{aligned} V_v(\theta) &= V_0 + b\theta^2 + c\theta^3 + \cdots \\ v(\theta) &= v_0 + \alpha\theta + \beta\theta^2 + \cdots \end{aligned} \tag{3.41}$$

The molecular reorientations, assumed to affect only the angle θ, are caused by the linear term in θ. As to the frequency change, it has two origins: the curvature of the suppressed potential, $\beta\theta^2$, and the anharmonicity $c\theta^3$ of the bulk librations. To lowest order, the reorientation is $\Delta\theta = \alpha/2b$ and the new curvature of the potential is $b + \beta - 3c\alpha/2b$. We may expect that the modification due to β has, as an upper limit, the phonon dispersion along the c' axis. Thus, the frequency variation originates mainly from the intralayer anharmonicity; we neglect the effect of β. (This approximation amounts to supposing that the variation in θ of the intralayer potential is more rapid than that of the interlayer potential v, which seems quite valid.)

The coefficients $\partial\nu/\partial p$ of the phonon frequency variation, in Raman spectroscopy under hydrostatic pressure, have been measured at room temperature and at 30 K by Nicol, Vernon, and Woo.[138] For the 49-cm^{-1} mode at 30 K they found $\partial\nu/\partial p \sim 1$ cm^{-1}/kbar (and at room temperature ~ 2 cm^{-1}/kbar). The observed frequency change, -4.3 cm^{-1}, corresponds here also to a depression of the order of 4 kbar, in agreement with the previous estimates. If we evaluate the angular range of strong variation of $V(\theta)$ out to about 10°, where the anharmonicity becomes very sensitive, we obtain

$$\Delta\theta = -\frac{\alpha}{2b} = \frac{2}{3}\frac{b}{c}\frac{\Delta v}{v} = 7\frac{\Delta v}{v} \text{ degrees} \tag{3.42}$$

which gives $\Delta\theta \sim 0.7°$, in very good agreement with the above estimates. In order to figure more precisely the molecular reorientations in the crystal surface, we must have a knowledge of the variation of the molecular orientation in response to a stress directed along the c' axis.

To summarize, the various estimations tend to show that the reorientation of the surface molecules is of the order of 1° for the anthracene crystal. These reorientations, in conjunction with the strong anharmonicity of the intermolecular modes in the molecular crystals, suffice to explain the observed strong attenuation of the frequency of the surface mode A_g of lowest energy, which passes, at 5 K, from 49.4 to 45.1 cm^{-1}.

In the next subsection, we envisage the effects that reorientation of the surface molecules at the ground state may have on the surface excitons. There are a large number of accurate data on these effects, which are analyzed in the present work.

2. *The Surface Excitons*

We have seen in Section II that the bulk exciton is free (the lattice does not move) owing to the strong J interactions present in the excited state. This conclusion remains valid for the surface exciton, so that we may study the electronic exciton in the reconstructed surface configuration of the preceding subsection. (Owing to the short lifetime of the electronic and vibronic excitons, additional relaxation of the surface during the excitation lifetime is excluded.) Modification of the surface spectroscopic structures relative to the bulk may have two origins: (1) the modification of the nuclear structure; (2) the suppression of the electronic interactions in the excited states due to the missing half space. For obvious reasons, we call the mechanisms corresponding to these two origins *soft surface* and *rigid surface*, respectively.

a. *Possible Origins of Destabilization.* As in Section III.C.1.b, we may estimate orders of magnitude of the above effects:

1. Soft surface: Assuming the surface-layer depression to be known, we derive the resulting surface-exciton shift due to the variation of the term $D_e = \mathscr{D}_e - \mathscr{D}_f$, which is the transition-energy stabilization. The variation $\partial D_e/\partial p$ of the stabilization under hydrostatic pressure is known experimentally at

room temperature; its value is $70\,cm^{-1}$/kbar.[139] Thus, a depression of -4 kbar corresponds to an upward shift of $300\,cm^{-1}$ for the surface exciton. Even if $\partial D_e/\partial p$ is weaker at low temperatures, we find order-of-magnitude agreement with the observed value of $206\,cm^{-1}$.

2. Rigid surface: The potential created by one layer at a distance r, which stabilizes the excited state of a molecule, may be calculated as in Section III.C.1.b: $\Delta D_e = (\pi/2S_0)B_e/r^4$, where B_e is proportional to the difference in the molecular polarizability between the ground state and the excited state and may be calculated from the bulk stabilization $D_e = 2500\,cm^{-1}$. With the values used above, we obtain $\Delta D_e \sim 350\,cm^{-1}$, still in the same range of values with the observed shift ($206\,cm^{-1}$).

Therefore, the effects of the two mechanisms contribute to the surface exciton shift almost equally. However, the sum of these two contributions (whose existence is beyond doubt) clearly overestimates the observed value. A more refined estimation of the rigid-lattice mechanism, taking into account the anisotropy of the molecules, might significantly reduce the contribution of this effect.

b. *Spatial Dependence of the Destabilization.* In each one of the above models, one may calculate the variation of the destabilization in the transition from the first surface layer to the second surface layer, and then to the third surface layer. The quantity which we compare with the experiments will be the ratio $\rho = (D_1 - D_2)/(D_2 - D_3)$ of the energy gaps associated with the transitions $S_1 \to S_2$ and $S_2 \to S_3$. The experimental data in Section III.A.3 yield the experimental ratio $\rho_e = 21.5 \pm 1$.

1. Soft surface: We assume in this model that the surface-layer energy shifts are proportional to the dilatation strain (3.37) caused by the missing interactions.[1] With attractive forces of the type (3.39) in r^{-5}, the ratio ρ may be calculated for various values of the parameters a and d of the model (the spacing and size of the "molecules", Fig. 3.23). It is concluded that ρ is practically independent of these parameters (from $\rho = 25.2$ for $a = d = 0$ it becomes 19.8 for $a = d = 1$). The typical values $a = 6$ and $d = 10$ lead to the ratio $\rho = 20$. On this point, we make the remark that only forces in r^{-5} are capable of yielding values of ρ compatible with the experimental value $\rho_e = 21.5 \pm 1$; forces in r^{-6} give $\rho = 45$, while forces in r^{-4} give $\rho = 9.4$, both values very far off. [Besides, for forces in r^{-5} one could expect an attenuation of the force on the surface of the order of $\frac{1}{32}$ in the transition from surface S_1 to surface S_2; this does not occur because of the complementary terms in the sum (3.37).]

2. Rigid surface: In this model, the values of $D_1 - D_2$ and $D_2 - D_3$ are simply the interactions with the first and with the second surface layer,

respectively, the other interactions being eliminated in the subtraction. For layers without thickness one could obtain $\rho = 16$, but $D_1 - D_2$ depends strongly on the structure of the layers in contact. Using an isotropic model with layers of thickness d and spacing a, the ratio ρ varies from $\rho = 16$ to $\rho = 30$ and beyond (for $a \to 0$, $\rho \to \infty$). For $d = 8$, $a = 6$, the ratio takes the value $\rho = 27.3$, which is larger than the experimental value ρ_e. An anisotropic model would diminish the value of ρ, but also the value ΔD_e. Therefore, it appears that the rigid-surface model is not capable of accounting on its own for the observed sequence of shifts. We must notice here also that only the exponent -4 is able to account for the spatial dependence of the interactions (the value -3 gives $\rho \gtrsim 8$, while -5 gives $\rho \gtrsim 32$).

c. *Surface Excitonic Dispersion.* Our experimental data lead to the conclusion that while the nearest-neighbor layer makes a strong contribution to the stabilization of the surface exciton, its expected contribution to the dispersion in the optical region ($\mathbf{K} \sim \mathbf{0}$) is very weak (see Section III.A.4.a) because of the weak interlayer interactions. In the rigid-surface model we must expect to observe strictly the same Davydov splitting for the bulk and the surface excitons. Indeed, we have observed almost the same Davydov splittings (maybe a slight reduction in the surface exciton: see Section III.B.3), while interactions among equivalent molecules are significantly modified. If this is confirmed, we have certain proof of the existence of a specific surface reconstruction. To evaluate the reconstruction effects, we again make use of the idea of a depression on the crystal surface layer. The pressure variation of the excitonic band structure has been investigated, at room temperature, by Sonnenschein, Syassen, and Otto.[139] Apart from a red shift of about $70\,\text{cm}^{-1}$/kbar, they observed an increase of the Davydov splitting (on sums among nonequivalent molecules) of about $20\,\text{cm}^{-1}$/kbar, while interactions among equivalent molecules appear not to change. Therefore, the depression picture is found to be wrong in these experiments, since it predicts a decrease of $60\,\text{cm}^{-1}$ in the Davydov splitting (corresponding to a global shift of $206\,\text{cm}^{-1}$), and no change in the interactions among equivalent molecules, which in our experiments appear to increase by about $40\,\text{cm}^{-1}$ on the surface. Thus, our interpretation appears uncertain on that point.

Let us now consider the effect of gas coating. The Davydov splitting increases by $20\,\text{cm}^{-1}$, while the interactions among equivalent molecules remain constant; see Section III.B.2.b. The mechanism of the Davydov-splitting increase may be of the rigid-surface type or of the soft-surface type.[61] However, it is difficult to imagine, in the first case, how a condensed-gas layer of molecules with moderate polarizability could produce on the surface-molecule interactions an effect which is hundreds of times larger than that of the nearest-neighbor layer of anthracene. On the contrary, the excess of pressure induced by the condensed-gas layer on the crystal surface provides a consistent explanation of both the global red shift of the bulk and the increase of the Davydov splitting, with the interactions among equivalent molecules remaining constant. The excess of pressure due to the gas layer is of the order of 1 kbar, as estimated either by the variation $\partial D_e/\partial p$ or by the variation $\partial \Delta E/\partial p$ of the Davydov splitting.

3. *Surface Reconstruction or No Surface Reconstruction*

The discussion of Section III.C.2 has shown strong points in favor of a significant reconstruction (in the ground state) of the crystal monolayer surface, in connection with spectroscopic observations on free surfaces and on surfaces clothed with a condensed-gas layer. The most important of these points are the following:

1. The variation in the surface layer (by $-4.3\,\mathrm{cm}^{-1}$) of the A_g phonon frequency is interpreted as a direct proof of surface reconstruction.
2. Theoretical calculations provide orders of magnitude of the observed reconstruction.
3. The modification of the surface excitonic dispersion is inexplicable in the absence of reconstruction.
4. The response of the surface spectroscopic structures to gas coating is also inexplicable in the absence of reconstruction.

If none of these points provides, on its own, a definitive proof of surface reconstruction, together they strongly suggest that surface reconstruction is the most natural explanation of all our observations.

Furthermore, surface reconstruction may be viewed, for the variation of certain properties, simply as a response to "depression", or to a strain of dilatation, in the surface layer. This approach has been applied with success to account for the surface phonon, and for the global destabilization (blue shift) of the surface exciton. In contrast, the modification of the surface excitonic structure is not amenable to such a simple image of the reconstruction.

The orders of magnitude of the surface variation in the crystallographic parameters are also determined by an estimation in agreement with the experiments: 0.01 to 0.1 Å for the variation along the c' axis, and 0.5° to 1° for the reorientation of the surface molecules and for the monoclinic axis angle β. A few points remain for further elucidation:

1. The exact shape of the reconstructed surface (in effect, the nature of the "missing stresses"). The information provided by the analysis of the surface excitonic structure seems to indicate displacements only among equivalent molecules.

2. The importance of purely electronic contributions ("rigid surface") to the destabilization of the surface exciton and to its spatial dependence.
3. The nature of the surface molecules' reorientation upon gas coating and the structure of the shifted excitonic band.

In spite of the good agreement with experiments provided by the soft-surface model, we must not forget the roughness of the two models. The preliminary results we are proposing must be improved by more realistic calculations, and additional experimental techniques must be looked for. For example, the above two models assume a bulk transition (E_{vb}) at 2.8 cm^{-1} below S_3 (the third surface resonance) for the soft surface and at 3.1 cm^{-1} below S_3 for the rigid surface. Therefore, comparison with experiments must take into account the weak bulk exciton dispersion along the c' axis. Also, the delocalization of the surface S_1 phonon (45 cm^{-1}) on surface S_2, observed very recently in our laboratory,[141] could allow us to refine and to test the soft-surface model and to obtain more insight into the spatial dependence of the destabilization and the reconstruction mechanism.

D. Conclusion

This section has been devoted to the study of the surface excitons of the (001) face of the anthracene crystal, which behave as 2D perturbed excitons. They have been analyzed in reflectivity and transmission spectra, as well as in excitation spectra of the first surface fluorescence. The theoretical study in Section III.A of a perfect isolated layer of dipoles explains one of the most important characteristics of the 2D surface excitons: their abnormally strong radiative width of about 15 cm^{-1}, corresponding to an emission power 10^5 to 10^6 times stronger than that of the isolated molecule. Also, the dominant excitonic coherence means that the intrinsic properties of the crystal can be used readily in the analysis of the spectroscopy of high-quality crystals; any nonradiative phenomena of the crystal imperfections are residual or can be treated validly as perturbations. The main phenomena are accounted for by the excitons and phonons of the perfect crystal, their mutual interactions, and their coupling to the internal and external radiation induced by the crystal symmetry. No ad hoc parameters are necessary to account for the observed structures.

The model of an isolated layer was refined by introducing substrate effects: by coupling the surface 2D excitons to the bulk polaritons with coherent effects modulating the surface emission and incoherent **k**-dependent effects damping the surface reflectivity and emission, both effects being treated by a KK analysis of the bulk reflectivity. The excitation spectra of the surface emission allowed a detailed analysis of the intrasurface relaxation dominated by resonant Raman scattering, by vibron fission, and by nonlocal transfer of the surface excitation to the bulk with competing picosecond and femtosecond rates.

The observation of a surface phonon, with strongly altered frequency relative to its bulk anolog, and surface-structure modifications upon gas coating, allowed us to achieve some insight into specific aspects of the surface reconstruction and to propose models of surface reconstruction compatible with the observed surface spectroscopic structures.

IV. SPECIFIC PROPERTIES OF DISORDERED 2D EXCITONS

In the last two sections we analyzed spectral and relaxation properties of 3D and 2D strong dipolar excitons in high-quality crystals at low temperatures in terms of the strong excitonic coherence of band width $\sim 500\,\mathrm{cm}^{-1}$, preserving the properties of the quasi-ideal crystal structure (what we called the intrinsic surface–bulk system) in the presence of weak disorder Δ ($\Delta \ll 500\,\mathrm{cm}^{-1}$). The disorder was either neglected or treated as a specific perturbation of the intrinsic properties (i.e., the intrinsic cyrstal eigenstates form the relaxation basis set). This was done for the reflectivity and the emission spectra, where the residual static disorder was neglected, while the thermal fluctuations, up to $T = 80\,\mathrm{K}$, where taken into account. Also, the gas-coating effect on the surface states was analyzed exactly, although, even for this weak perturbation ($\sim 80\,\mathrm{cm}^{-1}$), the surface reconstruction was not given an explicit description.

The general case of an excitonic coherence, comparable to the disorder width Δ ($\Delta \sim J$), is that, rather than renormalizing the disorder in band states to define a single basis of dressed excitons, we must recognize that it is scattered by the disorder into a random distribution of contracted excitonic domains (with decreased radiative rates γ_n relative to the induced exciton $\mathbf{k} \sim \mathbf{0}$). The consequence is the relaxation of the induced exciton by trapping in "aggregates" and eventually into crystal traps,[102,142] owing to a possible overwhelming of the radiative rates γ_n by nonradiative processes. This situation prevails in a large variety of systems, as (1) singlet excitons in naphthalene crystal with natural disorder observed as shallow traps,[142] (2) all triplet excitons in crystals of poor quality,[67,120] and generally (3) weakly coupled binary systems, the simplest of which is an isotopic mixed crystal where the low-concentration partner determines a perfect diagonal disorder.[38,39,142,143] The disorder (even in its zero-order approximation, diagonal disorder) creates a random topology in the solid over which the exciton–photon interaction averages to provide the macroscopic effects of the elastic or inelastic interaction that we observe in our spectroscopic data. The physical approach, which is that of the renormalization-group scheme, to finding this average is to consider first, on a small length scale the

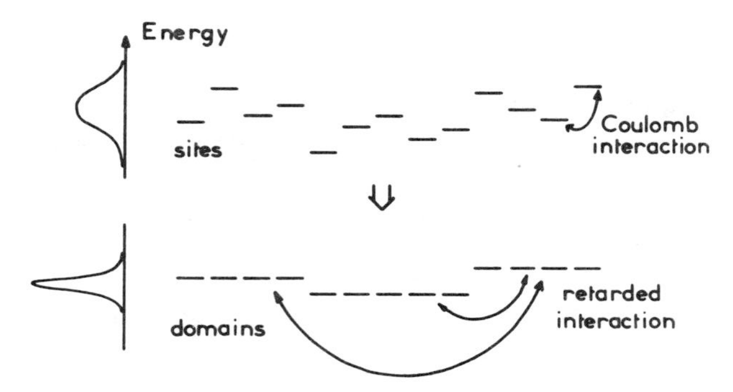

Figure 4.1. Scheme showing the partition of the hamiltonian (4.1) into the energy distribution of sites (microscopic disorder) and that of coherent domains (extended disorder).

microscopic disorder Δ opposed by the coulombic excitonic interactions that favor coherence, and then to include the resulting extended disorder[122] Δ_n in the long-range retarded interactions. This last step is practically specific to strong 2D dipolar excitons radiating long-range fields: the microscopic disorder in the site energy distribution is reduced by the coulombic interactions to a narrower energy distribution of random aggregates (in fact, we consider only their radiative levels), which are coupled only via retarded interactions (or transverse photon exchange in a range of λ) capable of producing coherent collective emission*; see Fig. 4.1. Qualitatively, the hamiltonian of the system is partitioned in two parts: one local part including all the coulombic interactions diagonalized into aggregates in the presence of microscopic disorder, and one nonlocal part including the exchanged photon states. The competition which occurs between the local spread of the energy distribution of the aggregates and the long-range coherence from the light field determines the absorption and the emission spectra of the sample.

Therefore, the problem of the disorder in the intermediate case characterized by the related double condition $\Delta \sim J$ and $\Delta_n(\Delta, J) \sim \Gamma_0$ is the absence of a single natural basis set. Instead, we are faced with the hard task of proposing a model of energy distribution (and topology) of aggregates, itself built up on an average of model configurations of coulombic interactions, and a self-consistent coherence of photons exchanged by these aggregates. This task has been taken on in the last decade, during which the density of states, exciton interactions, and transport and percolation phenomena in various disordered systems have been properly modelized, subject to the approximation, however, of the present level of the experimental

*Often referred to as Dicke's superradiance,[144] but with a very low excitation density.

accuracy.[38,49,145] In this section we propose also to examine a "coherence percolation" which amounts to the conditions that a random assembly of aggregates undergoes a coherent radiative transition.[118]

In what follows, we present in Section IV.A a theory of the effects of weak disorder on the retarded interactions of 2D strong dipolar excitons, and in Section IV.B we analyze the effects of stronger disorders on the coulombic interactions, calculating the density of states and absorption spectra in 2D lattices, in the framework of various approximations of the mean-field theory.

A. A Transition to Coherence in the Spontaneous Emission of 2D Disordered Excitons

1. *Experimental Observations*

The lifetimes of molecular fluorescence emissions are determined by the competition between radiative and nonradiative processes. If the radiative channel is dominant, as in the anthracene molecule, the fluorescence quantum yield is about unity and the lifetime lies in the nanosecond range. In molecular assemblies, however, due to the cooperative emission of interacting molecules, much shorter lifetimes—in the picosecond or even in the femtosecond range—can theoretically be expected; an upper limit has been calculated for 2D excitons [see (3.15) and Fig. 3.7] and for N-multilayer systems with $100 > N > 2$.[78] The nonradiative molecular process is local, so unless fluorescence is in resonance by fission (Section II.C.2), its contribution to the lifetime of the molecular-assembly emission remains constant; it is usually overwhelmed by the radiative process.[118,121] The phenomenon of collective spontaneous emission is often related to Dicke's model of superradiance,[144] with the difference that only a very small density of excitation is involved. Direct measurement of such short radiative lifetimes of collective emissions, in the picosecond range, have recently been reported for two very different 2D systems:

1. The (001)-surface monolayer emission of the anthracene crystal[1,67,117] (at $T = 2\,\mathrm{K}$, with a disorder width $\sim 1\,\mathrm{cm}^{-1}$) has been shown by Aaviksoo and coworkers[146] to be faster than 2 ps;
2. The fluorescence decay of cyanine dye aggregates (showing at room temperature disorder widths of 500 to $1000\,\mathrm{cm}^{-1}$) has been studied by Müller and Dorn[147]; they found it to vary from 10 to 5 ps between $T = 300\,\mathrm{K}$ and $T = 140\,\mathrm{K}$.

These direct data confirm our previous analysis (Section III) of the surface reflectivity and of the zero-phonon surface fluorescence, which predicts a lifetime[1,61,79,148] of the surface exciton of about 0.3 ps, corresponding to a radiative width $\Gamma_r = 14\,\text{cm}^{-1}$ and to a nonradiative width $\Gamma_{nr} \sim 3.5\,\text{cm}^{-1}$.[121] This result is expected from the theory of emission of a perfect 2D exciton,[124,126] where the coherence extends over the whole crystal surface, a number $\nu = \lambda^2\sigma$ of molecules cooperating to emit *one photon* (σ being the density of surface molecules) in excitonic resonance within the limits of the radiative width Γ_r (here $\Gamma_r = 14\,\text{cm}^{-1}$) of the emitting entity. (With a disorder fluctuation of $1\,\text{cm}^{-1}$, all the surface molecules are resonating.)[118] However, it may be suspected that the infinite coherence length used in the derivation of Γ_r is too strong an assumption and that the qualitative behavior sketched here will survive only a limited amount of disorder.

As to the monolayers of *J* aggregates of cyanine dyes,[149a] they are known to show, due to the very strong *J* interactions, some extent of excitonic (local) coherence, even at room temperature.[149b] This is borne out by the spectroscopic observations of a narrow absorption band and of the small Stokes shift between absorption and emission. More recently, energy-transfer experiments with *J* aggregates could be understood by means of a coherence domain for the exciton.[150] There, some 10 to 100 molecules appear to cooperate in emission, yielding a shortening by the same factor of the molecular fluorescence lifetime.[147] The same argument, applied to the *J*-aggregate emission investigated by Müller and Dorn,[147] leads to a lifetime in the picosecond range. However, we see here that only a very limited coherence domain (a coulombic aggregate) is invoked; every one of these domains should emit independently of the others, so that no coherence at the length scale λ (domain of retarded interactions) appears. This is compatible with dominant nonradiative broadening of *J* aggregates, which scatters and averages out long-distance coherence (the absorption peaks exhibit widths from 500 to $1000\,\text{cm}^{-1}$, while the radiative width of a perfect 2D cyanine aggregate is about $100\,\text{cm}^{-1}$).

The purpose of this section is to devise a general theory treating the 2D-exciton absorption and emission for any magnitude of disorder within the approximation of its substitutional form. We have to recover the two experimental observations—coherent surface emission and emission by random coulombic aggregates—as limiting cases for strong radiative and strong nonradiative broadenings, respectively.

2. *Model of 2D Disordered Excitons*

Assemblies of molecules can be seen in a first approximation as assemblies of transition dipoles subject to electromagnetic interactions. If only the coulombic part of the interaction is kept, new eigenstates of the assembly are found with real energies (the excitonic states). To account for the spontaneous emission, we have to include the retardation (to consider the exciton–photon coupling), which amounts to taking the classical problem of a set of radiating antennas; we then find eigenstates with imaginary energies, describing the radiative decay of the excited states. For instance, two identical dipoles, at distances much smaller than the wavelength, show two non-degenerate eigenstates: a symmetric one which has twice the radiative width of a single dipole γ_0, and an antisymmetric subradiant one. More generally, a symmetric arrangement of all dipoles allows a simultaneous diagonalization of the electromagnetic interactions for the lowest excited states of the system. The 2D lattice on which we focus now presents such a translation symmetry.

The plane waves of a perfect 2D lattice diagonalize the electromagnetic interactions, giving rise to the excitonic dispersion through the Brillouin zone, and to the surface-exciton–polariton phenomenon around the zone center.[148,126] The corresponding hamiltonian may be written as

$$H = \sum_{\mathbf{K}} [J_{\mathbf{K}} + R_{\mathbf{K}}(z)] |\mathbf{K}\rangle\langle\mathbf{K}| \tag{4.1}$$

where $J_{\mathbf{K}}$ is the coulombic dispersion for wave vector $\mathbf{K}$, and $R_{\mathbf{K}}(z)$ is the complex energy renormalization arising from the retarded part of the interactions.* Whereas in the 2D case $J_{\mathbf{K}}$ varies smoothly around the zone center, the retarded term $R_{\mathbf{K}}(z)$ is important only in this region for $|\mathbf{K}| \sim |z|/hc$.[151] $R_{\mathbf{K}}$ is purely imaginary for $|\mathbf{K}| \leqslant |z|/hc$, and it is of the order of $i(\lambda/2\pi)^2\sigma\gamma_0$, where σ is the surface density of dipoles and γ_0 the radiative width of a single dipole. For the anthracene surface, $J_{\mathbf{K}} \sim 200\,\text{cm}^{-1}$ and $|R_{\mathbf{K}}| \sim 10\,\text{cm}^{-1}$; for the cyanine monolayers $J_{\mathbf{K}} \sim 2000\,\text{cm}^{-1}$ and $|R_{\mathbf{K}}| \sim 100\,\text{cm}^{-1}$. Now, in a real system, the extent of the coherence is limited by the disorder. We assume the disorder to be purely static (the model we develop hereafter can apply to thermal fluctuations in the adiabatic limit[53] where the nuclear frequencies are neglected). We distinguish hereafter between two possible types of disorder.

1. Microscopic disorder: We consider a lattice the sites of which have disordered resonance energies, with a distribution of width Δ_e, but have the same intersite interactions (same dipole orientation and oscillator strength) as the perfect lattice. This is the so-called substitutional disorder model.[122] We assume the disorder width to be smaller than the excitonic bandwidth ($\Delta_e < B_e$) and examine the bottom of the excitonic band, where the emitting and the absorbing $\mathbf{K} \sim \mathbf{0}$ states lie. In a renormalization-group scheme, we split the lattice into isometric "domains" of n sites, on which the excitation is assumed to be localized, and write the substitutional-disorder hamiltonian in this basis: we thereby obtain a new disorder width $\Delta_n \sim \Delta_e n^{-1/2}$ and a

*The two contributions are here separated for the sake of convenience. They can be obtained as one in the unified treatment of the dipolar gauge in Section I.

new excitonic bandwidth $B_n = B_e r^{-2}$, r being the radius of a domain ($r = n^{1/2}$). The effective size of the coherence domain is determined by the condition $B_n \sim \Delta_n$, i.e.,

$$\begin{aligned} n &\sim (\Delta_e/B_e)^{-2} \\ \Delta_n &\sim B_n \sim \Delta_e^2/B_e. \end{aligned} \tag{4.2}$$

This result is equivalent to the renormalized perturbation theory of ref. 53, which these authors found to be at least qualitatively valid for 1D and 2D lattices. Thus, at the scale of our n-domains, we may forget about interdomain interactions B_n and take a localized hamiltonian:

$$H_e = \sum_A E_A |A\rangle\langle A|, \tag{4.3}$$

where A runs over domains of n sites and E_A is a fluctuating resonance energy with distribution width Δ_n. We have thus renormalized away the short-range coulombic interactions* and are left only with the excitonic diagonal disorder (H_e) and the long-range (retarded) interactions.

2. Extended disorder: We particularize to a simple model where large regions of unperturbed lattice (domains) are separated by impenetrable borders of nonresonant sites. It is convenient for the treatment of exciton–photon interaction to assume that all sites (the border sites as well) have identical dipole orientation and oscillator strength. We assume the domains to be large enough so that the coulombic interaction between different domains is negligible.* Coulombic interaction leads to a fluctuating resonance energy (the lower level) for each domain according to its size, shape, etc., so that a hamiltonian of the type (4.3) is again applicable here.

We combine now the disordered part (4.3) with the retarded part of (4.1), which does not depend on the particular resonance energy of one domain or site, but on the transition dipole operators, and therefore is identical for our model to that of the perfect 2D lattice. Thus, we obtain the following effective hamiltonian for the disordered 2D exciton:

*The suppression of the coulombic interactions is possible only at the botton of the excitonic band. Each domain has excited states (standing waves) which were not included in the hamiltonian (4.3). These states lie at an energy B_n above the band botton, and moreover have negligible oscillator strengths. In addition, exciton interdomain diffusion is negligible because of the low phonon densities. Besides, all the excited domains which do not contribute to the coherent state (fast emission) are captured by the strong surface-to-bulk relaxation[119,141] not considered in the hamiltonian (4.3).

$$H_{\text{eff}} = \sum_{A} E_A |A\rangle\langle A| + \sum_{\mathbf{K}} R_{\mathbf{K}}(z)|\mathbf{K}\rangle\langle\mathbf{K}|. \tag{4.4}$$

We preserve the main features of the disorder if we simplify further and assume that the domains lie on a superlattice with N domains, so that the states $|\mathbf{K}\rangle$ take the form

$$|\mathbf{K}\rangle = N^{-1/2} \sum_{A} e^{i\mathbf{K}\cdot\mathbf{r}_A} |A\rangle. \tag{4.5}$$

However, this assumption is not essential. The hamiltonian (4.4) is now defined by the following parameters:

a. $\Gamma_0 = \omega\sigma p^2/2\varepsilon_0 c$ with $R_{K=0} = i\Gamma_0$, p being the transition dipole moment. This relation is correct when the transition dipole lies in the lattice plane. Otherwise more complicated relations are obtained.[111]
b. Δ_n, the width of the energy distribution of the domains [see (4.2)] and the distribution function itself.
c. The number n of sites in a domain, or the surface density of domains $\sigma_d = \sigma/n$.

The behavior of H_{eff} in (4.4), as a function of the three parameters, may be sketched as follows. For large Γ_0 ($\Gamma_0 \gg \Delta_n$), all domains are brought into coherence by the strong field of their neighbors (within λ^2), and we find the optical response of a perfect 2D lattice with fast surface emission for states $|\mathbf{K}| < \omega/c$ (superradiant states) and no emission for $|\mathbf{K}| > \omega/c$ (subradiant states). On the other hand, if the disorder width Δ_n dominates ($\Delta_n \gg \Gamma_0$), then $R_{\mathbf{K}}$ may be treated as a perturbation of the localized states $|A\rangle$, resulting in a radiative rate for domain A

$$i\Gamma_A = \sum_{\mathbf{K}} R_k |\langle A|K\rangle|^2 = in\gamma_0 \tag{4.6}$$

In (4.6), we have used the trace relation[125,126] on the imaginary part of $R_{\mathbf{K}}$: $\sum_{\mathbf{K}} \operatorname{Im} R_{\mathbf{K}} = Nn\gamma_0$. Thus, domain A has a "superradiant" state with n times the radiation width of a single site, which is however much smaller than $\Gamma_{\mathbf{K}}$ ($R_{\mathbf{K}} = i\Gamma_{\mathbf{K}} \sim i\Gamma_0$), since a single domain is smaller than λ^2. Before discussing the intermediate case where Γ_0 and Δ_n are comparable for the surface, we treat a simplified version where the molecular assembly is smaller than the wavelength λ.

3. *A Simple Model*

Consider an assembly of N domains with resonance energies E_A, with a size smaller than λ. In the coupling to radiation, phase differences between domains may be neglected. The emission diagram is that of a single dipole, corresponding to a radiant state $|R\rangle$ totally symmetric with respect to the domains A:

$$|R\rangle = N^{-1/2}\sum_A |A\rangle. \tag{4.7}$$

A hamiltonian corresponding to (4.4) may be derived from a similar analysis:

$$F = \sum_A E_A |A\rangle\langle A| + i\Gamma |R\rangle\langle R|. \tag{4.8}$$

Using a t-matrix analysis, the resolvent of F may be written as

$$(z - F)^{-1} = G_e + G_e|R\rangle t\langle R|G_e$$

with

$$G_e = \sum_A (z - E_A)^{-1}|A\rangle\langle A| \tag{4.9}$$

and

$$t = [(i\Gamma)^{-1} - \langle R|G_e|R\rangle]^{-1}.$$

The complex eigenvalues of F are also poles of t, and therefore they are solutions of the equations

$$\frac{1}{i\Gamma} = \frac{1}{N}\sum_A (z - E_A)^{-1}. \tag{4.10}$$

In the general case, a numerical solution of (4.10) is required. Here, we shall focus on two cases: $N = 2$, and the continuous limit $N \to \infty$.

1. $N = 2$: We consider two oscillators at $E_A = \pm \Delta_n/2$. The solutions of (4.10) are simply

$$z^{\pm} = \tfrac{1}{2}[i\Gamma \pm (\Delta_n^2 - \Gamma^2)^{1/2}]. \tag{4.11}$$

For $\Delta_n > \Gamma$ we have two states with the "molecular" decay rate $\Gamma/2$. For $\Delta_n < \Gamma$ we have two states with the same real energy ($\mathrm{Re}\, z^{\pm} \sim 0$), but with different decay rates (superradiant $\gamma > \Gamma/2$, subradiant $\gamma < \Gamma/2$). We find a sudden qualitative change in behavior for the system for $\Delta_n = \Gamma$: the time decay passes from biexponential for $\Delta_n > \Gamma$ to a decrease with oscillating beats for $\Delta_n < \Gamma$.[153] This transition is not a special feature of the $N = 2$ case, but even survives in the continuous limit, as we shall see now.

2. $N \to \infty$: The difficulty with this limit is that the molecular radiative rate Γ/N of a single domain [see (4.6)] is vanishingly small. Therefore, we

only may check the possibility of existence of a superradiant state with an emission rate of the order of Γ. In the continuous limit, equation (4.10) gives

$$\frac{1}{i\Gamma} = \int \frac{n(x)}{z - x}\, dx, \tag{4.12}$$

where the distribution function $n(x)$ is a parameter of the model. Separating real and imaginary equations in (4.12) and assuming that $n(x)$ is an even function monotonically decreasing for $x > 0$, we find that $\gamma = \operatorname{Im} z \neq 0$ implies $\operatorname{Re} z = 0$: A superradiant state has to be at the center of the band. Exact solutions of (4.12) may be displayed (Fig. 4.2) for a square distribution function between $-0.5\Delta_n$ and $+0.5\Delta_n$, and for a lorentzian distribution with half width Δ_n. The square distribution for $\Gamma > \Delta_n/\pi$ takes the form

$$\gamma = \tfrac{1}{2}\Delta_n \cot(\Delta_n/2\Gamma) \tag{4.13a}$$

and the lorentzian distribution for $\Gamma > \Delta_n$,

$$\gamma = \Gamma - \Delta_n. \tag{4.13b}$$

The analysis of (4.12) shows that in the general case as well, we have a threshold $\Delta_c = g_c\Gamma$ (with $g_c \sim 1$) above which no coherent emission can build up; only an incoherent emission of isolated domains is then possible. Below the threshold ($\Delta_n < \Delta_c$), the coherent state has γ linear in $\Delta_n - \Delta_c$ [the function

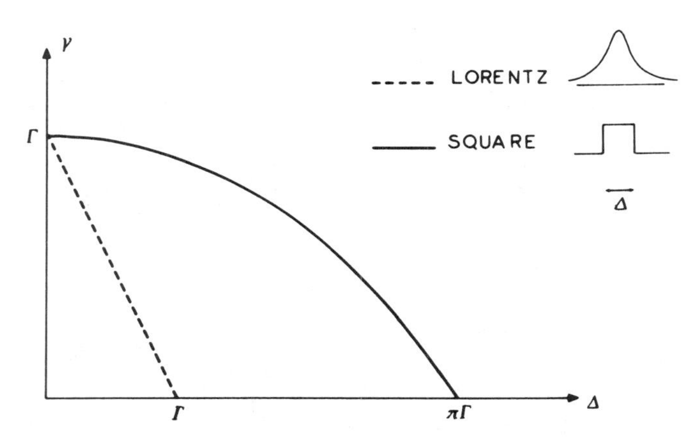

Figure 4.2. Variation of the radiative width γ, or coherent emission rate, of a 2D disordered exciton as a function of the disorder strength Δ. For $\Delta > g_c\Gamma$, the emission becomes incoherent. For all distributions, including the gaussian distribution, there is threshold behavior with a sudden takeoff of the coherent emission.

$\gamma(\Delta_n)$ depends on the actual distribution function $n(x)$]. The projection of the coherent state $|\gamma\rangle$ on the localized states with energy E may be studied using (4.9). In the lorentzian case, we find after normalization

$$|\langle E|\gamma\rangle|^2\,dE = \frac{\gamma\Gamma}{E^2+\gamma^2}\,n(E)\,dE \tag{4.14}$$

This expression indicates that the coherent state mainly forms from the domains with resonance energy in an interval of width 2γ around $E=0$. The oscillator strength of these domains is transferred to the coherent radiant

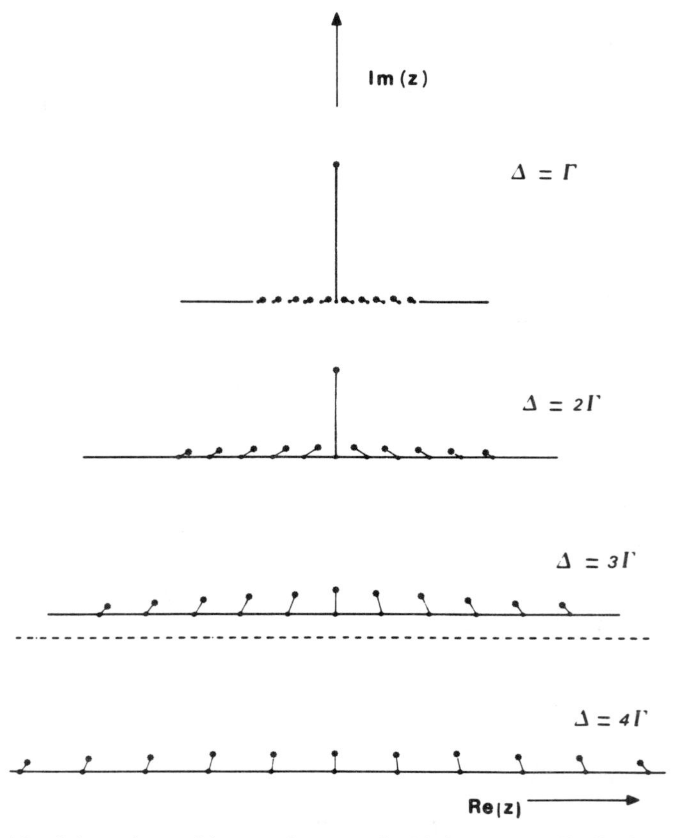

Figure 4.3. Schematic transition to coherence (Fig. 4.2) for a square distribution of domains as the disorder from $\Delta=\Gamma$ to $\Delta<\pi\Gamma$. The extended disorder is opposed by retarded interactions ($|R_e(z)|\rightarrow$): For $\Delta\ll\Gamma$ all the domains have transferred their oscillator strength ($|\mathrm{Im}\,z|\rightarrow_{E\neq 0}0$) to the coherent state at $E=0$, while the strong homogeneous width dominates the emission line.

state, while the other states $|E| > \gamma$ remain localized and keep their emission power of the order of Γ/N $(= n\gamma_0)$; see Fig. 4.3.

4. *Optical Response and Emission from a Disordered 2D Lattice*

We now study the disordered effective hamiltonian (4.4). Since a direct diagonalization of (4.4) is too hard, we shall have to use approximations which are conveniently expressed in the resolvent (or Green's function) formalism. The translation-invariant **K** sum in H_{eff} is restricted to the "optical" wave vectors only (for $|K| \gg \omega/c$, $|R_{\mathbf{K}}| \ll |R_{\mathbf{K}\sim 0}|$). Therefore, it is possible to restrict the problem to this small part of the Brillouin zone using the projector operator

$$P = \sum_{\mathbf{K}\text{opt}} |K\rangle\langle K|. \tag{4.15}$$

The resolvent G of H_{eff} may then be expressed in a t-matrix form analogous to (4.9):

$$G = (z - H_{\text{eff}})^{-1} = G_e + G_e PTPG_e$$

with

$$G_e = (z - H_e^{-1}) = \sum_A (z - E_A)^{-1} |A\rangle\langle A| \tag{4.16}$$

and

$$PTP = PRP(P - PG_e PRP)^{-1}$$

(R being the **K** sum in H_{eff}). The matrix elements of the restriction of G_e are written as

$$\langle \mathbf{K}'|G_e|\mathbf{K}\rangle = \frac{1}{N}\sum_A \frac{e^{i\mathbf{r}_A\cdot(\mathbf{K}-\mathbf{K}')}}{z - E_A} \tag{4.17}$$

These elements are sums of resonating denominators with a random phase depending on the coordinates of particular domains. The averaging of the phase factors is possible whenever there are enough terms resonating in the region of fastest variation of the denominator. This region is determined by $\gamma = \operatorname{Im} z$. We thus look at the sum $(z = E + i\gamma)$

$$S(\mathbf{K} - \mathbf{K}') = \frac{1}{N} \sum_{A,|E_A - E| < \gamma} e^{i\mathbf{r}_A\cdot(\mathbf{K}-\mathbf{K}')} \tag{4.18}$$

which is a sum of $n(E, \gamma)$ terms; $n(E, \gamma) \sim N\gamma/\Delta_n$ is the number of resonating domains for a given z. There are two contributions to S:

1. a δ peak at $\mathbf{K}' = \mathbf{K}$ with $S(0) = n(E, \gamma)/N$ for $|\mathbf{K} - \mathbf{K}'| < 1/L$, where L is the length of the surface sample;
2. a background for $\mathbf{K}' \neq \mathbf{K}$ with $S(\mathbf{K} - \mathbf{K}') \sim [n^{1/2}(E, \gamma)/N]e^{i\phi}$ (ϕ is a random phase). This background is extended to the whole range of the "optical" wave vectors.

To compare these two distributions, we compare the integral of the intensities over their characteristic ranges:

$$\frac{I(\delta \text{ peak})}{I(\text{background})} \sim n(E, \gamma)\left(\frac{\lambda}{L}\right)^2 \sim \frac{\gamma}{\Delta_n}\sigma_d\lambda^2 \tag{4.19}$$

Defining a critical width γ_c by

$$\gamma_c = \Delta_n/\sigma_d\lambda^2 \tag{4.20}$$

we find that the δ peak is dominant for $\gamma > \gamma_c$, while the background dominates for $\gamma < \gamma_c$. The **K**-conserving δ peak, as for the perfect surface, gives rise to specular reflection and to coherent emission, while the background contribution will describe scattering in all directions and incoherent emission. Now, we examine these consequences in more detail for reflection, where γ is imposed by the incident radiation field, and for emission, where γ is determined by the free oscillations (of the domains) of the system itself.

a. *Optical Response.* Up to now we have used a highly idealized model of discrete domains, with perfectly sharp frequencies. However, the transition to each coherence domain is broadened by various damping processes, so that within one wavelength, a large number of domains are resonantly excited by monochromatic incident light. Then, due to the plane lattice geometry, the scattered waves at the incident frequency ω can constructively interfere only in the reflected and transmitted ray directions.

Therefore, it is convenient to introduce a total width $\gamma_n = \gamma_r^n + \gamma_{nr}$ for each domain of energy E_A and to assume that the incident light is perfectly monochromatic: $\text{Im}\, z = 0$.

The role played by γ above (homogeneous width) is now played by γ_n. The reflection (and transmission) amplitude is given by the Green's function G at the energy z of the exciting source.[126] When the transition dipole lies in the lattice plane, we have for the reflection ($r_{\mathbf{K}}$) and transmission ($t_{\mathbf{K}}$) amplitudes

$$r_{\mathbf{K}} = R_{\mathbf{K}}\langle \mathbf{K}|G|\mathbf{K}\rangle, \qquad t_{\mathbf{K}} = 1 + r_{\mathbf{K}} \tag{4.21}$$

(**K** refers to a particular incidence angle, such that the projection of the incident light wave vector in the lattice plane is **K**). From (4.16) we get

$$PGP = PG_eP(P - PRPG_eP)^{-1} \tag{4.22}$$

So for $\gamma_n > \gamma_c$, the amplitude $r_{\mathbf{K}}$ dominates all amplitudes $\langle \mathbf{K}|G|\mathbf{K}'\rangle$. We thus have specular reflection, with negligible scattering because the interference of the large number of excited domains can be constructive only in the direction of the reflected (and transmitted) rays. The reflection amplitude is given by

$$r_{\mathbf{K}} = R_{\mathbf{K}}(g_e^{-1} - R_{\mathbf{K}})^{-1}$$

where

$$g_e = \frac{1}{N}\sum_A (z - E_A)^{-1} \tag{4.23}$$

Defining a self-energy by $\Sigma(z) = z - g_e(z)^{-1}$, the imaginary part of Σ defines a (frequency-dependent) broadening of the 2D transition. This result is equivalent to the mean-polarizability approximation[148,152] applied to the retarded part of the interaction. The approximation is justified here because of the large number of interacting neighbors within a wavelength:

$$n(E, \gamma_{\mathrm{nr}}) = \gamma_{\mathrm{nr}}/\gamma_c \gg 1.$$

Thus we find that, in this limit, the reduction of the exciton mean free path does not affect the radiative broadening $R_{\mathbf{K}}$, as noticed by Agranovitch,[152] since $\mathrm{Im}\,\Sigma$ and $R_{\mathbf{K}}$ enter the reflection amplitude (4.23) additively. This conclusion however does not hold for the emission, as we shall show in Section IV.A.4.b below.

At the opposite limit $\gamma_n < \gamma_c$, sparsely distributed resonant domains scatter the light in all directions of space and the specular reflection becomes negligible. (In the case where $\gamma_r > \gamma_{\mathrm{nr}}$, the condition of weak specular reflection becomes $\Gamma_0 < \Delta_{n'}$.) This behavior of the resonant reflection optical response should be observable for real systems as a function of the temperature, $\gamma_{\mathrm{nr}}(T)$ being[70,127] a rapidly growing function of T; see Fig. 4.4.

b. *Relaxed Emission.* We are interested here in the emission from localized states reached by a fast relaxation of the state prepared by the exciting light: there is no long-range coherence at $t = 0$. This excitation mechanism contrasts with the *resonant* emission, excited by a broad resonant light pulse [one then would observe specularly reflected light whose decay time constant was the reciprocal of the spectral width of the reflection (4.23)].

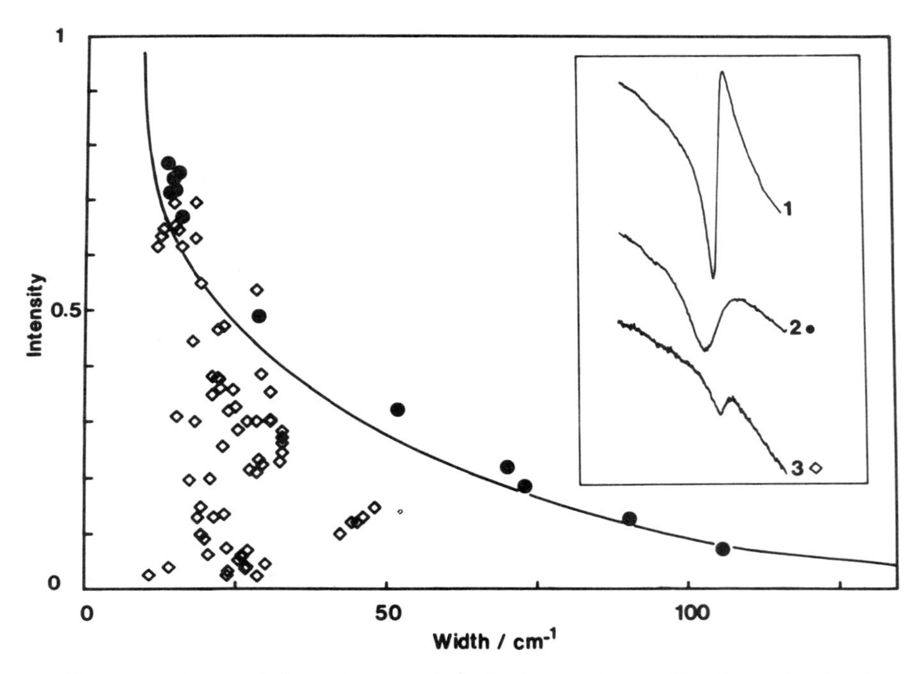

Figure 4.4. Decay of the surface specular reflection vs thermal disorder, static disorder, and surface annihilation caused by photodimerization. The surface reflection intensity of structure I is plotted vs broadening by temperature (full circles) and by photodimerization (hollow diamonds) which causes static disorder and annihilation of surface anthracene molecules. The solid line is deduced from theoretical calculations (2.126) in the adiabatic approximation. The cloud of hollow diamonds suggest that the density σ of unperturbed surface molecules has been reduced below the critical value, with the consequent collapse of the specular reflection; cf. (4.20). The inset shows the perfect surface structure (1), the temperature-broadened surface structure (2), and the structure of a photodimerized surface (3), which allowed us to plot the experimental curves.

The decay function of the emission should be calculated by diagonalizing H_{eff} (4.4), and would present a complicated superposition of exponential decays and frequency beats. Here, we shall be content with discussing the possibility of fast emission, which will follow from the largest imaginary parts of the eigenenergies z_λ of H_{eff}. We proceed along the lines of our model of Section IV.A. Assuming that there exists a "large" γ_λ ($\gamma_\lambda = \text{Im}\, z_\lambda$) with z_λ a pole of T (4.16), $G_e(z_\lambda)$ reduces to its δ-peak contribution (for $\gamma_\lambda > \gamma_c$). As a consequence, $T(z_\lambda)$ diagonalizes on the basis of **K** states:

$$T(z_\lambda) = \sum_{\mathbf{K}} [R_{\mathbf{K}}^{-1} - g_e(z_\lambda)]^{-1} |\mathbf{K}\rangle\langle\mathbf{K}| \tag{4.24}$$

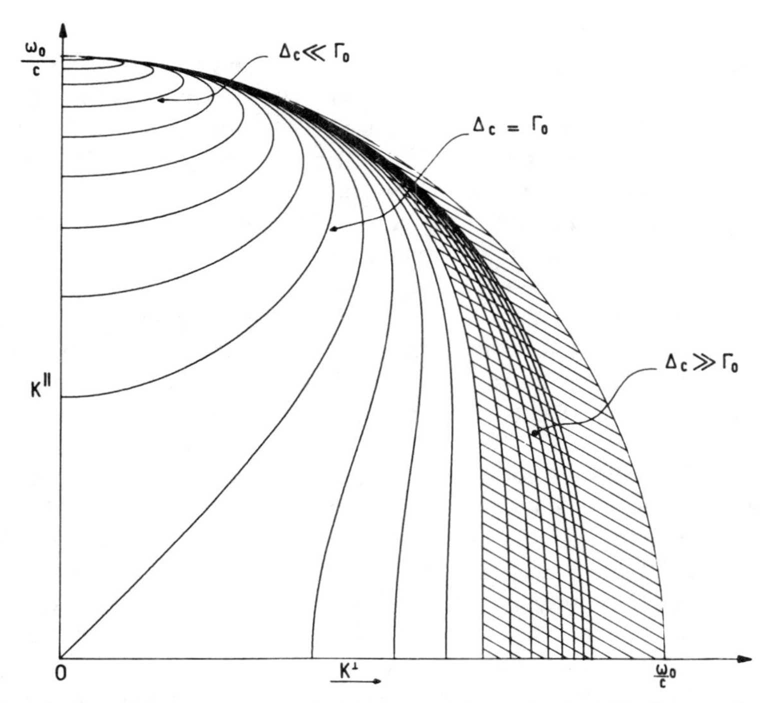

Figure 4.5. Wave vectors around the center of the excitonic Brillouin zone for which coherent emission [solution of equations 4.10 and 4.25] is possible according to the disorder critical value Δ_c. We notice that Γ_0 is the imaginary eigenvalue for $\mathbf{K} = \mathbf{0}$ (emission normal to the lattice plane) and that K'' and $K^{\perp}$ indicate, respectively, components of **K** parallel and perpendicular to the transition dipole moment, assumed here to lie in the 2D lattice. The various curves for constant disorder parameter Δ_c determine areas around the Brillouin-zone center with (1) subradiant states (left of the curve) and (2) superradiant states (right of the curve). We indicate with hatching, for a large disorder ($\Delta_c \gg \Gamma_0$), a region of grazing emission angles and superradiant states for a particular value of Δ.

The search of the poles of T decomposes into independent problems for each **K**, each one similar to (4.10):

$$R_{\mathbf{K}}^{-1} = g_e(z_\lambda) \tag{4.25}$$

According to Section IV.A.3, for each wave vector **K** such that $g_c \Gamma_{\mathbf{K}} > \Delta_n$, there is an imaginary solution of (4.25). Thus, since localized states have a projection on every $|\mathbf{K}\rangle$ state, a very fast photon emission occurs (coherent emission) in the direction determined by **K**. In second-order perturbation theory[124,126] (Fermi's "golden rule"), $\Gamma_{\mathbf{K}}$ takes values between 0 and ∞ for

$|\mathbf{K}| < \omega_0/c$ (Fig. 3.7).* Then, coherent emission is always possible for sufficiently grazing emission angles, in directions normal to the transition dipole (see Fig. 4.5). For a normal or nearly normal emission ($\mathbf{K} = \mathbf{0}$), the transition to the coherent emission regime takes place for $\Delta_n = g_c \Gamma_0$. If our model of static disorder may be extrapolated to dynamical disorder induced by thermal fluctuations, this transition should be observable as a function of temperature. A similar buildup of superradiance below a critical temperature in the condensed phase has been recently reported by Florian et al.[154]

As in the model of Section IV.A.e, eigenstates with small imaginary parts ($\gamma_\lambda < \gamma_c$) exist. These states lead to a much slower regime of incoherent emission, with the rate corresponding to a single domain ($\gamma_\lambda \sim n\gamma_0$). Numerical calculations are probably the best way to study the regime of emission in the intermediate-coupling case ($\Gamma_0 \sim \Delta_n$).[141]

To conclude, we can draw an analogy between our transition and Anderson's transition to localization: the role of extended states is played here by our coherent radiant states. A major difference of our model is that we have long-range interactions (retarded interactions), which make a mean-field theory well suited for the study of coherent radiant states, while for short-range 2D Coulombic interactions mean-field theory has many drawbacks, as will be discussed in Section IV.B. Another point concerns the geometry of our model. The very same analysis applies to 1D systems; however, the radiative width $(\lambda/a)\gamma_0$ of a 1D lattice is too small to be observed in practical experiments. In a 3D lattice no emission can take place, since the photon is always reabsorbed. The 3D polariton picture has then to be used to calculate the dielectric permittivity of the disordered crystal; see Section IV.B.

B. Disordered 2D Coulombic Excitons

In absence of the Bloch theorem,[155] the zero-order disorder effect—termed the minimum, or diagonal, disorder[156]**—enters the coulombic interaction hamiltonian in the following way:

$$H = \sum_{\mathbf{n}} V_{\mathbf{n}} |\mathbf{n}\rangle\langle\mathbf{n}| + \sum_{\mathbf{nm}} W_{\mathbf{nm}} |\mathbf{n}\rangle\langle\mathbf{m}| \tag{4.26}$$

$V'_{\mathbf{n}}$ are the resonance energies of the sites $|\mathbf{n}\rangle$, depending on the exact configuration of the solid; $W'_{\mathbf{nm}}$ are the interactions between sites $|\mathbf{n}\rangle$ and

*For a large value of $\Gamma_{\mathbf{K}}$, the "golden rule" does not apply; a more complete treatment of the exciton–photon interaction has been carried out[126] for the wave vectors $\{(c|\mathbf{K}|/\omega) - 1\} \lesssim (\Gamma_0/\omega_0)^{2/3}$.

**The gap to the perfect random distribution is evaluated, in the case of the isotopic mixed crystal, by comparing the entropy term related to the permutation of the molecules, of the order of $k_f T$, with the elastic energy V_{el} introduced by the substitution of one hydrogenated cell by one deuterated cell. Then at the temperature formation, we find $kT_f \sim 10^2\, V_{el}$, which is consistent with a random distribution.

$|\mathbf{m}\rangle$. Following the distribution function of the random variable $V_{\mathbf{n}}$, the hamiltonian (4.26) leads to various models of diagonalization[122]:

1. the model of Lloyd for a lorentzian random distribution of the $V'_{\mathbf{n}}$;
2. the gaussian model for a gaussian distribution of the $V'_{\mathbf{n}}$;
3. the model of Anderson for an equiprobable distribution between $-\Delta/2$ and $\Delta/2$;
4. the model of the binary alloy, with two species centered at v_A with concentration c_A and at v_B with concentration c_B ($c_A + c_B = 1$).

Hereafter, we shall use mainly model 4, which is defined by two parameters: the concentration c_A of one of the partners, and the ratio W/Δ of the bandwidth in the pure crystal to the "potential" difference $\Delta = v_A - v_B$ between sites A and B. The following limit cases are well specified:

1. $c_A \to 0$ (or $c_A \to 1$) defines the isolated impurity in an almost perfect crystal.

2. $\Delta \ll W$: The sites A and B are practically equivalent; we have a mean crystal whose properties are weighted averages of those of A and B. The singlet excitons of the anthracene isotopically mixed crystal are nice examples of this situation.[129]

3. $\Delta \gg W$: One of the partners plays, relative to the other, the role of a passive host lattice. However, this case is not trivial, because of the difficult problem that arises in the topology of the crystal: the existence, at a given concentration, of a threshold of a topological percolation (for which an infinite coulombic aggregate appears in one of the two species) suffices to show the complexity and the variety of the random binary alloy.

In the intermediate domain of values for the parameters, an exact solution requires the specific inspection of each configuration of the system. It is obvious that such an exact theoretical analysis is impossible, and that it is necessary to dispose of credible procedures for numerical simulation as probes to test the validity of the various inevitable approximations. We summarize, in Section IV.B.1 below, the mean-field theories currently used for random binary alloys, and we establish the formalism for them in order to discuss better approximations to the experimental observations. In Section IV.B.2, we apply these theories to the physical systems of our interest: 2D excitons in layered crystals, with examples of triplet excitons in the well-known binary system of an isotopically mixed crystal of naphthalene, currently denoted as Nd_8–Nh_8. After discussing the drawbacks of treating short-range coulombic excitons in the mean-field scheme at all concentrations (in contrast with the retarded interactions discussed in Section IV.A, which are perfectly adapted to the mean-field treatment), we propose a theory for treating all concentrations, in the scheme of the molecular CPA (MCPA) method using a cell of nine sites, but with an appropriate modulation of the effective medium to relax the artificial strong coupling between the cell and the effective medium, inherent in the MCPA method. Section IV.B.3 is devoted to 3D polaritons in a mixed crystal, as treated in Section II.

1. *The Model of the Binary Alloy*

We summarize below the main theoretical approaches to the binary alloy, i.e. those based on the coherent-potential approximation (CPA).[156]

We consider the hamiltonian (4.26) for a binary system (its extension to an arbitrary number of partners or to continuous disorder creates no difficulty in principle). The Schrödinger solution associated with this hamiltonian describes a coupling of very general occurrence, since it is encountered in conduction and transport phenomena, as well as in kinetics models of disordered systems:

$$i\hbar\dot{u}_{\mathbf{n}} = V_{\mathbf{n}}u_{\mathbf{n}} + \sum_{\mathbf{m}} W_{\mathbf{n}-\mathbf{m}}u_{\mathbf{m}} \tag{4.27}$$

[The symbol $u_{\mathbf{n}}$, which here stands for the component of the wave function on site n, may also represent other quantities, such as a Tyablikov amplitude (Section I.A.3) or a population. The coefficients being adapted to each situation, the form of the system (4.27) is preserved.]

For arbitrary initial conditions, the solution of (4.27) may be derived from the knowledge of the response, at an arbitrary site m, to an excitation localized at $t = 0$ on an arbitrary site $\mathbf{n}$ (this is the Green's function of the system; see Appendix A). As we are interested in spectral data, it is natural to use the Fourier transform of (4.27) corresponding to an initial condition of one excitation localized on site $|n\rangle$:

$$(\hbar\omega - V_{\mathbf{m}})u_{\mathbf{m}}(\hbar\omega) - \sum_{\mathbf{m}'} W_{\mathbf{m}-\mathbf{m}'}u_{\mathbf{m}'}(\hbar\omega) = \delta_{\mathbf{nm}} \tag{4.28}$$

The Green's function $\mathscr{G}$ is written in its matrix form ($z = h\omega + i0$):

$$\mathscr{G}_{\mathbf{nm}}(z) = \langle \mathbf{n}|\frac{1}{z - H}|\mathbf{m}\rangle \tag{4.29}$$

where the matrix $\mathscr{G}(z) = (z - H)^{-1}$ is the resolvent of H, corresponding to a given configuration of the disordered system. In general, knowledge of $\mathscr{G}$ is impossible: a measurement yields only averaged values, since the macroscopic sample we are probing contains a very large number of microscopic configurations. It is important to remark that this approach to the macroscopic system assumes the ergodic hypothesis to be valid: We suppose implicitly, that the real disordered configuration has no influence on the experimental quantities beyond a certain microscopic distance R (R is typically the radius of a sphere inside which the disorder fluctuations—for instance, in the number of impurities in the case of the binary alloy—may be neglected). Thus, on a scale larger than R, the disordered solid appears as homogeneous, and this will lead us naturally to the notion of effective medium. The validity of this hypothesis may be tested experimentally: The physical quantities measured on a scale larger than R must not depend on the sample, or on the specific region of a homogeneous sample.

In a typical experiment on a disordered system, a beam of particles of wave vector $\mathbf{k}$ is scattered to a state of wave vector $\mathbf{k}'$ (in an ordered system, we have $\mathbf{k}' = \mathbf{k} + \mathbf{G}$, $\mathbf{G}$ being a vector of the reciprocal lattice; see Section I). The amplitude of the observed signal is proportional to the $\mathbf{k}'$ emission of the distribution $\sum_{\mathbf{n}} e^{i\mathbf{k}\cdot\mathbf{n}} \mathscr{G}_{\mathbf{nm}}(z)|\mathbf{n}\rangle$ created by $\mathbf{k}$:

$$\mathscr{O}(\mathbf{k}, \mathbf{k}') \propto \frac{1}{N} \sum_{\mathbf{nm}} e^{i(\mathbf{k}\cdot\mathbf{n} - \mathbf{k}'\cdot\mathbf{m})} \mathscr{G}_{\mathbf{nm}}(z) = \langle \mathbf{k} | \mathscr{G}(z) | \mathbf{k}' \rangle \tag{4.30}$$

In particular, the optical absorption amplitude is obtained by observing the $\mathbf{k}'$ vector equal to the very weak vector $\mathbf{k}$ of the excitation. Indicating by $|\mathbf{k} = \mathbf{0}\rangle$ the wave created by the incident photon (and taking into account possible nonanalyticity), the optical absorption is proportional to

$$\mathscr{A}(z) \propto \langle \mathbf{k} = \mathbf{0} | \mathscr{G}(z) | \mathbf{k} = \mathbf{0} \rangle \tag{4.31}$$

The average quantities of the type (4.31), involving a single wave vector, may be calculated [in contrast with the expression (4.30) with $\mathbf{k}' \neq \mathbf{k}$] in an effective mean crystal, of mean resolvent $\langle \mathscr{G}(z) \rangle$ and translationally invariant on the microscopic scale. Thus, the $|\mathbf{k}\rangle$ states are eigenstates of this effective crystal, but with the novelty that the eigenvalues are generally complex, indicating the scattering by the disorder of the wave $\mathbf{k}$ to the other states $|\mathbf{k}'\rangle$. In addition, the eigenenergy will depend, generally, both on the energy z at which we are probing the sample and on the wave vector $\mathbf{k}$:

$$\langle \mathscr{G}(z) \rangle = \sum_{\mathbf{k}} \frac{|\mathbf{k}\rangle\langle\mathbf{k}|}{z - E_{\text{eff}}(\mathbf{k}, z)} \tag{4.32}$$

Separating in E_{eff} the contribution of the intermolecular interactions $W'_{\mathbf{k}}$, the self energy $\Sigma_{\mathbf{k}}$ appears by definition:

$$E_{\text{eff}}(\mathbf{k}, z) = W_{\mathbf{k}} + \Sigma_{\mathbf{k}}(z) \tag{4.33}$$

where

$$W_{\mathbf{k}} = \sum_{\mathbf{m}} W_{0\mathbf{m}} e^{i\mathbf{k}\cdot\mathbf{m}} \tag{4.34}$$

For weak disorders, the correction $\Sigma_{\mathbf{k}}$ tends to zero. Now, introducing the resolvent of the crystal "without disorder" ($V_n = 0$),

$$G_0(z) = \sum_{\mathbf{k}} \frac{|\mathbf{k}\rangle\langle\mathbf{k}|}{z - W_{\mathbf{k}}} \tag{4.35}$$

we may rewrite (4.32)–(4.33) in two equivalent expressions:

$$\langle \mathscr{G}(z) \rangle = G_0 + G_0 \Sigma \langle \mathscr{G} \rangle \tag{4.36}$$

with

$$\begin{gathered} \Sigma = \sum_{\mathbf{k}} \Sigma_{\mathbf{k}} |\mathbf{k}\rangle\langle\mathbf{k}|, \\ \langle \mathscr{G}(z) \rangle = G_0 + G_0 \langle \mathscr{T} \rangle G_0 \end{gathered} \tag{4.37}$$

where the t matrix $\langle \mathscr{T}(z) \rangle$, defined by (4.37), is a configuration average of the real t matrix for a given configuration $\mathscr{T}$, which we define below in Section IV.B.3.b. In order to apply these various quantities, we consider, first, the very simple case of weak disorder, [$\ll W$ (or $V_{\mathbf{n}} \ll W$).]

2. *The Virtual Crystal*

In the case of $\Delta \ll W$, the disorder may be treated as a perturbation of the pure crystal. Starting from the nonperturbed $|\mathbf{k}\rangle$ states, we obtain the energy shift to first order:

$$\langle \mathbf{k} | V | \mathbf{k} \rangle = \frac{1}{N} \sum_{\mathbf{n}} V_{\mathbf{n}} = c_A v_A + c_B v_B \tag{4.38}$$

In the propagator formalism (Green's functions), this result may be obtained by uncoupling the configuration averages:

$$\mathscr{G}(z) = (z - W - V)^{-1} = G_0 + G_0 V G_0 + G_0 V G_0 V G_0 + \cdots \tag{4.39}$$

from which

$$\langle \mathscr{G}(z) \rangle = G_0 + G_0 \langle V \rangle G_0 + G_0 \langle V G_0 V \rangle G_0 + \cdots \tag{4.40}$$

In the approximation in which second-order terms in V are left out, all averages of the type $\langle VV \rangle$ may be replaced by $\langle V \rangle \langle V \rangle$, to give

$$\langle \mathscr{G}(z) \rangle \sim \frac{1}{z - W - \langle V \rangle} \tag{4.41}$$

Hence,

$$\Sigma_k(z) \sim \langle V \rangle = c_A v_A + c_B v_B \tag{4.42}$$

is, in this approximation, independent of both z and $\mathbf{k}$: the effective crystal is obtained simply by a weighted average on the pure crystals A and B. This is the *virtual-crystal* approximation. For a further improvement of this approximation, one could think of continuing the perturbation to a higher order. In fact, the higher-order terms diverge[156] in the vicinity of the van Hove points of the crystal band W. Thus, to go further than the virtual-crystal approximation, we have to resum at least an infinite series of perturbation orders in V.

3. *Dilute Impurities*

We assume in this case that one of the partners exists in very low concentration: $c_A = c \to 0$. We consider the guest impurities in an almost perfect host crystal B, leading to the parameters $v_B = 0$ and $v_A = \Delta$.

a. *The Isolated Impurity.* We assume the impurity localized on site $\mathbf{n} = \mathbf{0}$. The hamiltonian (4.26) simplifies to

$$H = W + \Delta |0\rangle\langle 0| \tag{4.43}$$

Then, $G(z) = (z - H)^{-1}$ is calculated explicitly from $G_0(z) = (z - W)^{-1}$:

$$G = G_0 + G_0 |0\rangle t \langle 0| G_0 \tag{4.44}$$

the scalar t depending on z (see Appendix B):

$$t = \frac{\Delta}{1 - \Delta \langle 0| G_0(z) |0\rangle} \tag{4.45}$$

The poles of $t(z)$ give the eigenenergies of the impurity–crystal coupled system; when Δ is substantially larger than the band width of W, there exists a discrete impurity level, which is solution of

$$\langle 0| G_0(z) |0\rangle \Delta = 1 \tag{4.46}$$

In general, below a critical value Δ_c, depending on the nature of the lattice, the discrete level transforms into a more or less broad resonance in the band W. Nevertheless, for 1D and 2D systems, the discrete level can persist for small values of Δ. Then it lies near the threshold of the band W and cannot be resolved experimentally.

b. *Expansion in t Matrices.* The precedent approximation is, rigorously speaking, valid only for a single impurity. It may be extended to impurities at low concentrations by an uncoupling approximation, which may be described as follows. The hamiltonian of the crystal, containing a large number of impurities on sites **i**, may be given the form

$$H = W - \sum_{\mathbf{i}} \Delta |\mathbf{i}\rangle\langle\mathbf{i}| \tag{4.47}$$

The resolvent of H corresponding to a given configuration may be expressed as a function of the t matrix of this configuration;

$$\mathscr{G}(z) = G_0 + G_0 \mathscr{T} G_0 \tag{4.48}$$

$$\mathscr{T}(z) = V \frac{1}{1 - G_0 V}, \qquad V = \sum_{\mathbf{i}} \Delta |\mathbf{i}\rangle\langle\mathbf{i}| \tag{4.49}$$

Let us separate in G_0 the diagonal propagation, which stays on the initial site, from the nondiagonal propagation, which causes site changes:

$$G_0(z) = g_0(z) + G_0'(z) \quad \text{and} \quad \langle n|g_0(z)|m\rangle = \langle n|G_0|n\rangle \delta_{nm} \tag{4.50}$$

Then

$$\mathscr{T}(z) = V \frac{1}{1 - g_0 V - G_0' V} = \frac{V}{1 - g_0 V}\left(1 - G_0' \frac{V}{1 - g_0 V}\right)^{-1} \tag{4.51}$$

since V commutes with g_0. Then, the matrix $V(1 - g_0 V)^{-1}$ which appears in (4.51) is diagonal on the impurities and embodies all scatterings by individual impurities isolated in the perfect host crystal B:

$$V(1 - g_0 V)^{-1} \equiv T = \sum_{\mathbf{i}} t(z) |\mathbf{i}\rangle\langle\mathbf{i}| \tag{4.52}$$

with $t(z)$ preserving the form derived above in Section IV.B.3.a. Thus, we generate an expansion in powers of t matrices:

$$\mathscr{T} = T + T G_0' T + T G_0' T G_0' T + \cdots \tag{4.53}$$

In this expansion, resummations have been performed relative to (4.39), but we arrive at the same difficulties if we truncate the series, so that we are led to a similar uncoupling approximation:

$$\langle \mathscr{T} \rangle \sim \langle T \rangle + \langle T \rangle G_0' \langle T \rangle + \cdots \tag{4.54}$$

However, this approximation is better than that obtained in Section IV.B.2, in part because (4.54) resums all the successive scatterings on a same site, and in part because G_0' prevents two successive sites from coinciding: hence, the uncoupling is exact to second order. After a resummation of (4.54), we obtain for the self-energy Σ

$$\Sigma = \frac{\langle T \rangle}{1 + g_0 \langle T \rangle} = \sum_{\mathbf{n}} \frac{ct}{1 + \langle 0 | G_0 | 0 \rangle t} |\mathbf{n}\rangle\langle\mathbf{n}| \tag{4.55}$$

The results of this approximation improve with decreasing impurity concentration.

Therefore, we obtain a site-diagonal self-energy, i.e. one that is independent of the wave vector $\mathbf{k}$; but it is complex and z-dependent. This approximation is simple, and it is very useful for the case of small quantities of impurities: it is known as the ATA (average-t-matrix approximation) when the propagator G_0 of the pure crystal B is replaced by the propagator of the virtual crystal of Section IV.B 2. Instead of imposing the propagator of the virtual crystal, one may attempt to include the impurity scattering in this propagator in a self-consistent manner: this is the mean-field theory, whose domain of validity is examined below.

4. *Self-Consistent Mean Field* (CPA)[122,156–162]

When the impurity concentration increases, the isolated-impurity model suffers serious drawbacks. On one hand, at long distances the lattice in which impurities are immersed is not that of a pure B crystal, but that of the mixed crystal; and on the other hand, the presence of other impurities in the near vicinity of one impurity dramatically perturbs, in each given configuration, the level of the isolated impurity investigated above (when it exists). Although physically there is no reason to make a fundamental distinction between these two effects (short- and long-range coulombic interaction), we shall consider propagation at long distance separately. As indicated in Section IV.B.1, beyond a microscopic scale R, the mean propagator $\langle \mathscr{G}(z) \rangle$ describes the physical properties of a mixed crystal. In spite the great complexity of its dependence on $\mathbf{k}$ and on z, $\langle \mathscr{G} \rangle$ is, by definition, the propagator of a perfect crystal, invariant under translation. Our present

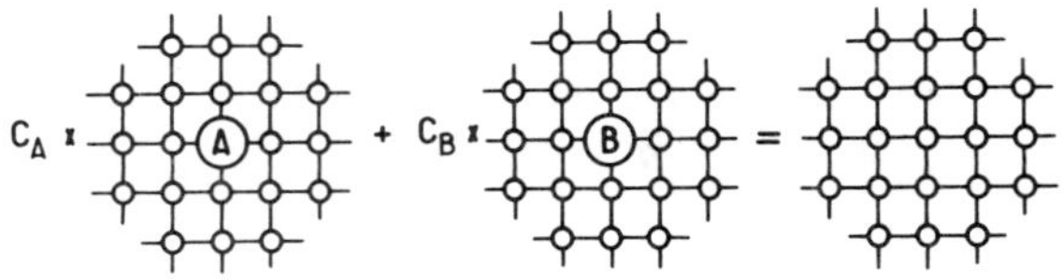

$C_A \times$ (A) $+ C_B \times$ (B) $=$

Figure 4.6. The CPA consists in extending the translational invariance of the mean lattice around a single site, occupied either by an A or by a B molecule. The condition of self-consistency of the mean lattice is obtained from the average of the propagations around this single site, which must be identical to the propagation in the mean lattice.

approximation will consist in extending this translational invariance to the microscopic scale around a given site, whether occupied by an A or a B molecule. In each case, the problem is amenable to treatment as a case of the isolated impurity (Section IV.B.3.a). Then, the approximate mean propagator will be determined by the following self-consistency condition: On averaging over the central site occupation, we must get back this mean propagator. An equivalent condition is to neglect the mean scattering by a "real" single site of a wave propagating in the mean crystal. These two conditions are illustrated in the scheme shown in Fig. 4.6.

Obviously, this procedure is an approximation, since it uses a translationally invariant medium as a model for very complex structures generated by short-distance configurations such as impurity aggregates.

The scheme of Fig. 4.6 corresponds, for a binary alloy, to the model hamiltonian

$$h = H_{\text{eff}} + \mathscr{V}(0) \tag{4.56}$$

where H_{eff} represents the mean effective medium to be determined. As in (4.33), let us write

$$H_{\text{eff}}(z) = W + \Sigma(z) \tag{4.57}$$

$\mathscr{V}(0)$ is the perturbation, localized around the site 0, when that site in the effective medium is replaced by a real site A or B. A priori, $\mathscr{V}(0)$ is not purely local on site 0, because Σ is not, a priori, local:

$$\mathscr{V}(0) = V_0|0\rangle\langle 0| - \sum_{\mathbf{n}} \Sigma_{0\mathbf{n}}(|0\rangle\langle \mathbf{n}| + |\mathbf{n}\rangle\langle 0|) \tag{4.58}$$

Now, if we require that the mean scattering by the real site vanish:

$$\langle t(z)\rangle = 0 \tag{4.59}$$

where

$$t(z) = \mathscr{V}(0)\left(1 - \frac{1}{z - H_{\text{eff}}}\mathscr{V}(0)\right)^{-1} \tag{4.60}$$

then it is possible to show[157] that this condition results in $\Sigma_{0\mathbf{n}} = 0$ for $\mathbf{n} \neq \mathbf{0}$, i.e., the self-energy must be purely local. The solution $\Sigma(z)$ does not depend on the wave vector $\mathbf{k}$, and that is consistent with the local nature of the approximation. Thus, a scalar $\sigma(z) = \Sigma(z)$ suffices to characterize the obtained effective medium: It is the "coherent potential" of the approximation, called the CPA. For the binary alloy, the condition (4.60) determining the expression for $\sigma(z)$ takes the form

$$c_A \frac{v_A - \sigma}{1 - (v_A - \sigma)g(z - \sigma)} + c_B \frac{v_B - \sigma}{1 - (v_B - \sigma)g(z - \sigma)} = 0 \tag{4.61}$$

with

$$g(Z) = \frac{1}{N}\sum_k \frac{1}{Z - W_k} \tag{4.62}$$

Furthermore, we may remark that the condition (4.60) for the vanishing of the mean scattering results from an uncoupling approximation of the type (4.59), where the true condition $\langle \mathscr{T} \rangle = 0$ is replaced by the local condition $\langle T \rangle = 0$. Owing to this, *if we apply* ATA *in the effective medium, the interaction of* ATA *converges on the* CPA.[158]

The CPA method has important properties apart from its relative simplicity. It is analytic in z,[158] and thus respects the elementary physical constraints: causality, Kramers–Kronig relations, sum rules, positive definite spectrum, etc. What is more, it is universal: this method describes the virtual-crystal limit $\Delta \ll W$, the isolated-impurity limit $c_A \to 0$ or $c_A \to 1$, and the isolated-molecule limit $W \to 0$, with the correct contribution of each molecular level.[122] Indeed, the CPA may be derived[159] from this last limit, as well as from that in the locator formalism.[122]

However, in each case where the microscopic environment plays an important role, the CPA treatment becomes inappropriate. This is specifically the case for $\Delta > W$ with moderate concentrations c: the continuous spectrum provided by the CPA does not reproduce the aggregate structure (see Sections IV.C.2, IV.C.3a. below) (except for the separation into two bands A and B); in this case, even for describing the spectra, it is necessary to take the local environment into account.

C. Nonlocal Approximations for the Naphthalene Triplet Exciton

It has been concluded that the CPA embodies in a macroscopic way the microscopic environment of a site. This approximation improves with increasing number ν of sites interacting via W, since the disorder on a large number of interacting neighbor sites will tend to have an average close to its macroscopic value (see Section IV.A). It has been demonstrated that the CPA result is the zero-order approximation of an expansion in powers of $1/\nu$, the next term being in $1/\nu^2$.[160] For long-distance interactions, the mean-field approach will be satisfactory. On the contrary, for short-range interactions, or interactions having a strong component at short distances (such as electron exchange interactions or coulombic dipolar interactions) the near vicinity of one site will be, in general, very different from the effective macroscopic medium. These effects are particularly strong for the case $\Delta \gg W$, i.e. when one of the partners of the alloy is non resonant relative to the other partner. To use a suggestive terminology, for a weak concentration of "guest traps" in the "host" lattice, the observed structures, as satellites around the isolated trap line, are connected to well-defined aggregates, whose order increases with increasing "guest" concentration. Our main interest is to investigate, by a mean-field theory, the transition between the aggregate regime and a pure crystal regime of guest molecules, which occurs at intermediate concentrations.

The discussion of this transition will be exemplified with the triplet excitons of the well-known mixed crystal Nd_8–Nh_8, which has long been investigated in our laboratory.

1. *The Triplet Exciton in an Isotopically Mixed Crystal*[163,164,145]

We summarize here the triplet-exciton theory only for the properties involved in high-resolution (0.1 cm^{-1}) optical absorption spectroscopy; in particular, the triplet spin is left out.

a. *Band Structure of the Triplet Exciton.* Because of the strong anisotropy of the triplet-exciton interactions and their short range, the layered structure of the naphthalene crystal may be replaced by an appropriate pileup of independent (a, b) planes. Therefore, we may consider 2D triplet excitons interacting in an (a, b) plane, in competition with disorder in the 2D lattice. Two interactions have to be considered, both between nearest neighbors: equivalent (along the a axis) and nonequivalent, as shown in Fig. 4.7:

$$\begin{aligned} V_{11} &= +0.5\,\text{cm}^{-1} \\ V_{12} &= -1.2\,\text{cm}^{-1} \end{aligned} \tag{4.63}$$

These values were determined by high-resolution spectroscopy of excitation spectra of mixed crystal at very low concentrations. They are in good agreement with the Davydov splittings in the pure crystals (the slight difference, -5%, in the deuterated naphthalene is attributed to an appreciable

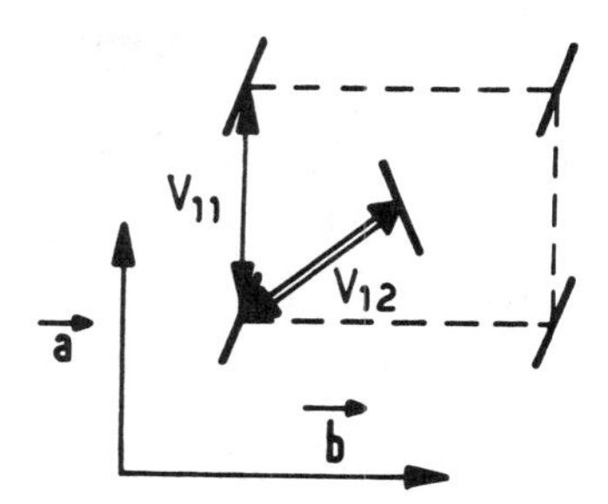

Figure 4.7. Scheme of interactions between neighboring molecules, equivalent along a (V_{11}) and nonequivalent (V_{12}), in the lattice of the naphthalene triplet exciton.

quantity of natural impurity, for instance Nd_7h in Nd_8). The resonance energies of the molecules Nd_8 and Nh_8 in crystalline phase differ by about $100\,\text{cm}^{-1}$, so that, given the values (4.63), we are in a well-defined disorder limit case: a nonresonant binary system with $\Delta \gg W$.

The data may be compared with those generated by the first singlet transition of naphthalene[142]: The structure of the couplings is more complex, and the band denser.[142,165] In particular, the 2D approximation must be abandoned for spectral as well as for dynamical problems.[142] Indeed, the excitonic bandwidth is of the order of $100\,\text{cm}^{-1}$, whereas we have $\Delta \sim 160\,\text{cm}^{-1}$, so that, with $\Delta \geqslant W$, we are dealing with the intermediate case.

The triplet physical observables of our interest are the following:

1. The optical absorption, measurable via the excitation spectra of secondary emissions (delayed fluorescence,[143] phosphorescence,[142] EPR triplet spectra,[166] etc.). The absorption is the response of the crystal to an excitation spatially homogeneous at the microscopic scale:

$$\begin{aligned}\mathscr{A}(E) &= \operatorname{Im}\langle \mathbf{k}=\mathbf{0}|\mathscr{G}(E^+)|\mathbf{k}=\mathbf{0}\rangle \\ &= \operatorname{Im}\langle \mathbf{k}=\mathbf{0}|\langle\mathscr{G}(E^+)\rangle|\mathbf{k}=\mathbf{0}\rangle\end{aligned} \tag{4.64}$$

2. The density of states $n(E)$, or the number of states per unit energy:

$$n(E) = -\frac{1}{N\pi}\operatorname{Im}\operatorname{Tr}\mathscr{G}(E^+) = -\frac{1}{\pi}\operatorname{Im}\langle 0|\langle\mathscr{G}(E^+)\rangle|0\rangle \tag{4.65}$$

The experimental determination of the density of states is not an easy task. Considering vibronic regions, $n(E)$ may be measured indirectly in the 0–1 transitions (two-particle spectra, as discussed in Section II.B) or in 1–0 transitions (hot bands).

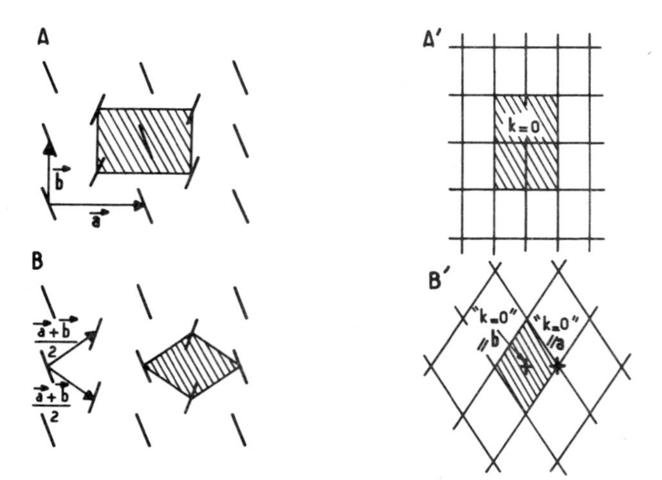

Figure 4.8. Equivalence of the photodynamic properties of the naphthalene lattice with a lattice with one molecule per cell. *A*: The naphthalene lattice with two molecules per cell (**a**, **b**), and its reciprocal lattice *A'*. *B*: The equivalent lattice with one molecule per cell, and its reciprocal lattice *B'*.

b. *The Pure Crystal.* We now discuss the exciton dispersion in the crystal. It is well known that the two molecules of the crystal cell are equivalent in all their dynamical properties and that the real crystal is equivalent, for our study, to a crystal with one molecule per cell as illustrated in Fig. 4.8.

The dispersion of the wave vector **k**, taking into account (4.38), takes the form

$$W_{\mathbf{k}} = 4V_{12}\cos\frac{\mathbf{k}\cdot\mathbf{a}}{2}\cos\frac{\mathbf{k}\cdot\mathbf{b}}{2} + 2V_{11}\cos\mathbf{k}\cdot\mathbf{b} \tag{4.66}$$

and the density of states must be calculated from the function

$$g(E^+) = \frac{1}{N\pi}\sum_{\mathbf{k}}\frac{1}{E - W_{\mathbf{k}} + i0} \tag{4.67}$$

One of the two integrals (4.67) may be evaluated analytically; the second one (elliptic) may be calculated numerically by the trapezoidal method with the introduction of a small imaginary part in (4.67) to avoid spurious oscillations due to the finite number of integration points. The resulting density-of-states function is shown in Fig. 4.9. The band is asymmetric because of the equivalent term V_{11}; it exhibits three van Hove singularity points: two discontinuities at the boundaries, and one logarithmic divergence corres-

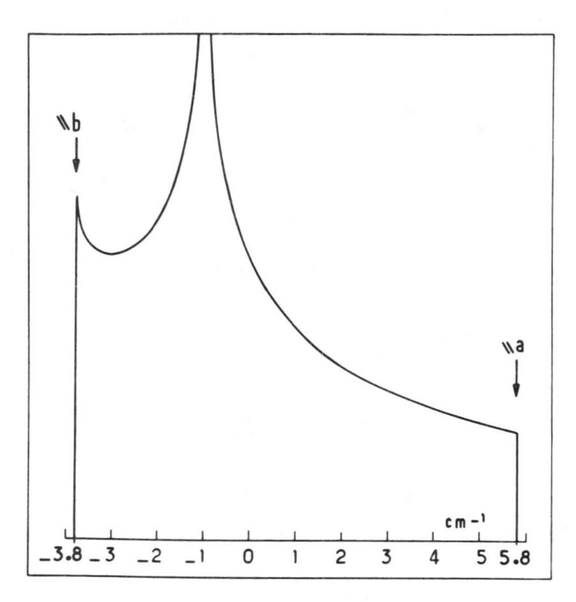

Figure 4.9. Density of states of the naphthalene Nh_8 triplet excitonic band. We note the asymmetric form due to the coexistence of equivalent and nonequivalent interactions, and the three van Hove points connected with the long-range order: two discontinuities at the boundaries (at -3.8 and $5.8\,\mathrm{cm}^{-1}$) ane one logarithmic divergence (at $-1\,\mathrm{cm}^{-1}$), characteristic of 2D dispersion. The optical absorption occurs at the two ends of the band (the **a** and **b** Davydov components).

ponding to a saddle in the surface W_K, at the point $-2V_{11} = -1\,\mathrm{cm}^{-1}$. Using (4.64), we find that the absorption in the pure crystal occurs at the two boundaries of the band, at $4V_{12}+2V_{11}$ for the b excitonic component, and at $-4V_{12}+2V_{11}$ for the a excitonic component.

c. *Expansion in a Basis of Aggregates.* For a low concentration of guests, the general shape of the host band is little affected, except around the singularities: The discontinuities are softened, while the divergence is suppressed owing to the disruption of the long-range order. On the contrary, satellite lines appear around the energy v_A of the guest molecule, associated with the formation of van der Waals aggregates the dominant line is the "monomer" located at the pole of equation (4.46), framed (in the absorption) by lines of weaker intensity associated with aggregates of decreasing concentration and increasing order (dimers, trimers, ..., n-mers).[142,163] It is very easy[167] to treat the problem of one guest aggregate A immersed in a pure crystal B, by generalizing the treatment of the isolated impurity (discussed in

Section IV.A.3.a) with the hamiltonian

$$H = W + \sum_{\mathbf{i}} \Delta_{\mathbf{i}} |\mathbf{i}\rangle\langle\mathbf{i}| \tag{4.68}$$

In our problem, the $\Delta_{\mathbf{i}}'$ are all identical, so that defining the projector $P = \sum_{\mathbf{i}} |\mathbf{i}\rangle\langle\mathbf{i}|$, we have

$$H = W + \Delta P \tag{4.69}$$

which, in matrix notation, gives a t matrix analogous to (4.45):

$$t = \Delta P \frac{1}{1 - PG_0P\Delta} \tag{4.70}$$

PG_0P is also an $n \times n$ matrix (where n is the number of guest molecules forming the aggregate), whose elements between two guest molecules are given by

$$\langle\mathbf{i}|G_0|\mathbf{i}'\rangle = \frac{1}{N} \sum_{\mathbf{k}} \frac{e^{i\mathbf{k}\cdot(\mathbf{i}-\mathbf{i}')}}{z - W_{\mathbf{k}}} \tag{4.71}$$

The diagonalization of the small matrix $PG_0(z)P$, or of $t(z)$, allows us to determine, after iteration, the eigenenergies of the aggregates immersed in the pure crystal B, as well as the corresponding eigenfunctions. (Defining the self-energy $P\Sigma(z)P$ by $PGP = [P(z - \Delta) - P\Sigma P]^{-1}$, one obtains, using $G = G_0 + G_0tG_0$ [cf. equation (4.44)] $P\Sigma(z)P = z - [PG_0(z)P]^{-1}$; the eigenenergies may be calculated $[z_\lambda = \Delta + \sum_\lambda(z_\lambda)]$ by iteration from $z = \Delta$; the eigenvector corresponding to z_λ is the eigenvector of $PG_0(z_\lambda)P$ or of $t(z_\lambda)$.)

For the case $\Delta \gg W$, we obtain a good approximation[145,163] by considering only the nearest neighbors of nonresonant host molecules surrounding the aggregate. Then, the probability of a given aggregate allows us to calculate the spectra of density of states and of absorption for guest concentrations up to 20 or even 30%, when the low-order aggregates (monomers, dimers, trimers) are important for the characterization of the spectra (Fig. 4.10).

Therefore, the expansion of the crystal states in the subspace of guest aggregates turns out to be useful at low guest concentrations, while for intermediate concentrations the approximation breaks down for the following reasons:

1. When the guest concentration increases, enumeration of the aggregates very soon becomes impossible when their size exceeds a few sites; in the vicinity of the percolation concentration (50% for the naphthalene triplet),

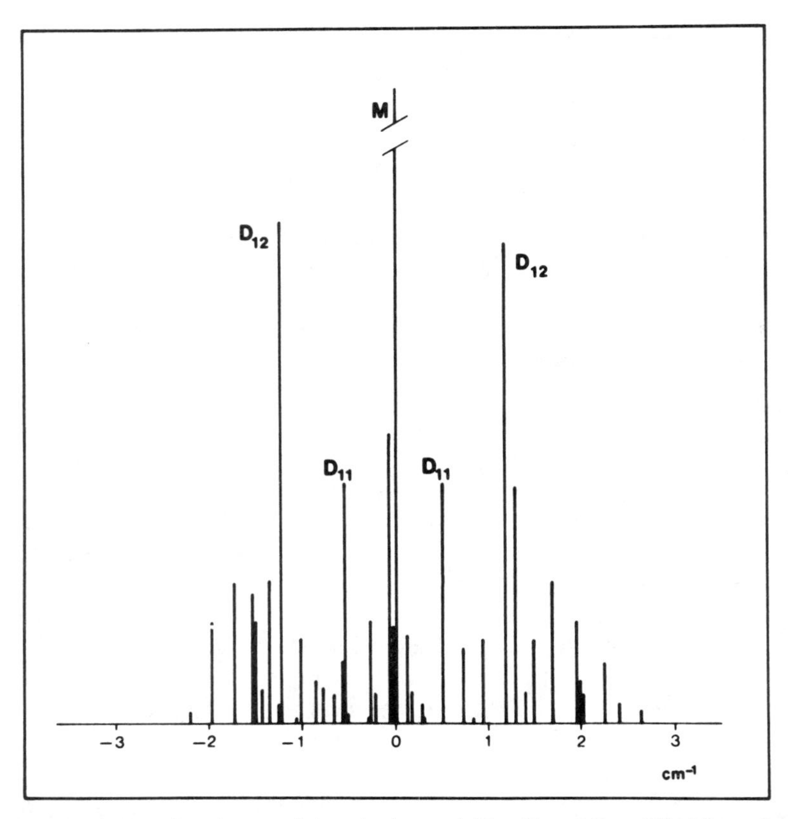

Figure 4.10. Density of states of the mixed crystal Nh_8–Na_8 with $c = 30\%$ Nh_8 molecules. This distribution is obtained by an expansion in aggregates[163–165] around the resonance of Nh_8. (The Nd_8 band is at $100\,\text{cm}^{-1}$ above this distribution.) The expansion was carried up to and including tetramers (four contiguous Nh_8 molecules). (The intensity of certain tetramers is comparable to that of trimers; this casts some doubt on the validity of the expansion in aggregates at $c \sim 30\%$.)

an infinite aggregate appears, with a spatial structure and physical properties which are very complex[122] (in the 2D lattice, only one infinite aggregate may exist above the percolation threshold[122]). To avoid confusion with current literature on the triplet exciton dynamics in random 2D lattices, it is useful to underline that we consider here a topological percolation, between nearest guest neighbors coupled by W, neglecting at low concentrations couplings W via lattice tunneling (termed also superexchange interactions[143]), which involve much more complex hamiltonians[145] (see below). From the point of view of topological percolation, the lattice is equivalent to a triangular lattice, for which the percolation threshold is 50%.[122] At higher concentrations, the infinite aggregate evolves towards the pure crystal A, and it is obvious that the expansion in aggregates loses all meaning and cannot account for this critical behavior.

2. Following the experimental spectral resolution used to investigate the density-of-states distribution, we may, or we may not, content ourselves with the picture of aggregates immersed in a pure crystal B. Coupling of the guest aggregate to neighbor guest molecules (separated by at least one host site) leads to fine structure of the aggregate levels with a width $\sim V^2/\Delta$ [where V is the typical interaction; cf. (4.63)]. This substructure of the aggregate levels reflects coherent transfer by superexchange. For the triplet crystal, $V/\Delta \sim 10^{-2}$, so that this very fine structure, randomized by the intrinsic or the thermal disorder at low temperatures, must be discussed in conjunction with the exciton incoherent transfer.[102] For the naphthalene singlet excitons, with $V^2/\Delta \sim 1\ \text{cm}^{-1}$, tunneling or superexchange interactions must be included in the hamiltonian in order to study the spectral spread, or fine structure, of the aggregates, which accounts for the coherent transfer among zero-order aggregates.[102,142]

The real signature of coherent transfer is quantum beats when the lattice-disorder spread is not prohibitive. More modestly, we define it, just as a conceptual link, as exciton transfer among zero-order aggregates by phonon creation or annihilation or by elastic interaction between phonon and electronic-exciton subsystems (measured by the pure dephasing). With the introduction of super exchange, we are creating various environments for various guest aggregates, which complicates the enumeration of the aggregates, so that a mean-field theory could be more useful in handling the superexchange effects.

The conclusion is that, although attractive, the expansion in aggregates must be abandoned for intermediate and high concentration of one of the partners. Before proposing a theoretical approach leading to a better approximation of these cases, we discuss briefly the numerical simulation procedures, with their advantages and disadvantages, for calculating density-of-states and absorption in disordered systems.

2. *Numerical Simulations of Disordered Triplet Excitons*[169]

Numerical simulations were performed on grids of 2000 to 10000 sites with appropriate boundary conditions[168,169] to assure the equivalence of the sites involved. After a brief summary of the mathematical methods used, the results are discussed in the form of histograms and absorption spectra. Because of the similarity of behavior around the energies v_A and v_B, we investigated only the high-energy species. As we remarked, owing to the asymmetry of the excitonic band, there is no simple relation (translation or symmetry) between the spectra around A and B. However, the differences being due to super exchange effects, they are of the order of $V^2/\Delta \sim 0.1\ \text{cm}^{-1}$, which is comparable to the resolution of our numerical simulations and of our theoretical calculations. Therefore, we consider an approximate equivalence under an energy shift Δ between guest and host spectra for corresponding concentrations. Estimations of the density of states and of the optical absorption were made by a numerical treatment of the hamiltonian (4.26) for sample model crystals. A random-number generator was used to place the A and B sites with average concentrations c_A and c_B respectively in grids of $N = m \times n$ molecules in the equivalent lattice of Fig. 4.8.

a. *Density of States.* The density of states for a given grid was determined by repeated application of the modified negative-eigenvalue algorithm of Dean and Martin (Dean and Bacon[171]), which may be stated as follows. The number $n(E)$ of eigenvalues of H less than a given value E is equal to the number of negative terms of the sequence (e_i), $i = 1, \ldots, N$, determined by the following steps:

1. $L^{(1)} = H - EI$, where I is the identity matrix.
2. Generally, let Z_p be the matrix obtained by striking out the first row and column of $L^{(p)}$. Let Y_p be the first row vector of $L^{(p)}$. Then form

$$M^{(p+1)} = Z_p - \frac{Y_p^t Y_p}{L_{11}^{(p)}}, \qquad p = 1, \ldots, N-1.$$

3. The required sequence is $e_p = L_{11}^{(p)}$, $p = 1, \ldots, n$.

The density of states follows from the difference $n(E) - n(E')$ for closely spaced values of the energy.

Advantage may be taken of the band structure of H to reduce the amount of work involved. Each model grid was joined edge to edge along the longer sides to reduce the boundary effects across the shorter section of the grid. This determined a banded structure of the matrix H with band width $\sim m$, the shorter side. On shifting the seam one unit left or right, the band structure of H becomes exactly regular. As the above algorithm preserves the matrix band width, only a matrix of size $m \times m$ need be kept in the computer core.

The number of multiplications required for one complete cycle of the algorithm is of order

$$Nm^2 = m^3 n$$

A complete solution of the eigenvalue problem would require about $N^3 = m^3n^3$ operations, so, provided the step in the energy E is not too small, appreciable gains in computing time may be achieved. The saving in core space is also important enough to justify the method even when there is no saving of time. A hundred complete cycles with $m = 50$, $n = 100$ required ~ 50 min CPU on an IBM 3350.

Figure 4.11 shows results for crystals of 50×100 sites. Statistical fluctuations of the density of states, with resolution 0.1 cm^{-1}, were then very small from crystal to crystal, so no averaging was attempted.

At low concentrations, we observe a central spike, due to isolated A molecules, or monomers, flanked by satellites due to dimers of equivalent (± 0.5 cm^{-1}) and inequivalent (± 1.2 cm^{-1}) pairs. As the concentration is increased, further satellites appear due to the presence of trimers, tetramers, and so on. At around $c_A \sim 30\%$, the continuous background of energy levels of large aggregates becomes apparent, and it finally dominates at $c_A \geqslant 50\%$.

b. *Absorption Spectra.* In order to calculate the absorption spectrum of a mixed crystal, we must possess its eigenstates and eigenenergies. Neglecting the eigenstate projections on B sites, which are smaller than those on A sites by a factor of order W/Δ ($\ll 1$), the actual crystal hamiltonian (4.26) may be replaced by an effective A-site hamiltonian, containing all nearest-neighbor A–A interactions and more distant ones due to tunneling through nB sites, $n = 1, 2, \cdots$:

$$v \sim W(W/\Delta)^n.$$

Eigenstates derived by diagonalization of this approximate hamiltonian are sufficiently accurate to describe the aggregate absorption, though the small, longrange tails, which are important for problems of energy transport, are rather underestimated, compared with results of exact calculations described elsewhere (Brown et al.[145]). Diagonalization of the A-site hamiltonian is then equivalent to an aggregate expansion of the absorption, without the fuss of identifying all the possible aggregates, a hopeless task at concentrations over $c_A \sim 50\%$. Rather, by diagonalizing the effective hamiltonian of a large model crystal, we have a sample of the eigenstates at the corresponding concentration. Figure 4.11 shows the result of averages over five model crystals with about 450 traps (species A) at each concentration. The main features of the spectra are estimated to be accurate to about 10%. The eigenstates were calculated by Householder reduction to tridiagonal form[185] followed by the QR algorithm, using routines from EISPACK library.[186]

The absorption spectrum was calculated as

$$I(E) = \frac{1}{N\pi} \sum_k \frac{\lambda |\mu_k|^2}{(E - E_k)^2 + \lambda^2}$$

where E_k and μ_k are the energy and the transition moment of the kth eigenstate. λ is the linewidth, set at $0.1\ \mathrm{cm}^{-1}$, an experimental upper bound for the triplet states of traps in mixed naphthalene (Lemaistre et al.[170]). The transition moment follows from the partial sums over the sites of type 1 and 2 in each cell:

$$\mu_k = \left(\sum_{i_1} C_{ki_1} \right) \mu_1 + \left(\sum_{i_2} C_{ki_2} \right) \mu_2$$

where C_{km} is the projection of the kth state on site m.

3. *Nonlocal Extensions of the* CPA

This subsection is devoted to shedding light on the limitations of the one-site CPA of our systems, and, in addition, to describing the inherent difficulties of the self-consistent treatment of the local environment of one site. Then we discuss the sole (to our knowledge) analytical extension: the molecular CPA (MCPA).

a. *Limitations of the* CPA. The solution of equation (4.61) in the case of couplings (4.63) in triplet naphthalene necessitates the calculation of $g(z)$ via (4.62) following Section IV.C.1.b with the use of a complex argument. Then, (4.61) may be resolved by iteration using Newton's method, the derivative being easily calculated. The results were first obtained and discussed in detail by Hoshen and Jortner.[171,172] Our own results, using slightly different parameters, are presented in Fig. 4.11 and compared with the numerical simulations.

The CPA provides smooth lineshapes at all concentrations, without aggregate substructures. The spectra are very satisfactory at high concentrations ($c > 70\%$), as illustrated in Fig. 4.11: the van Hove singularities are smoothed, and in absorption the Davydov component peaks get closer and broader under the influence of the disorder. At the intermediate concentrations ($70\% > c > 30\%$), the CPA reproduces very roughly the envelope of the spectra, but without exhibiting the various fine and well-resolved structures due to aggregates. Finally, it is at low concentrations ($0 < c < 30\%$) that the discrepancies with numerical simulations are the most stiking; instead of a forest of peaks (mainly due to monomers and dimers), the CPA gives a single band, with a width proportional to $\sqrt{c_A}$.[121] The CPA absorption

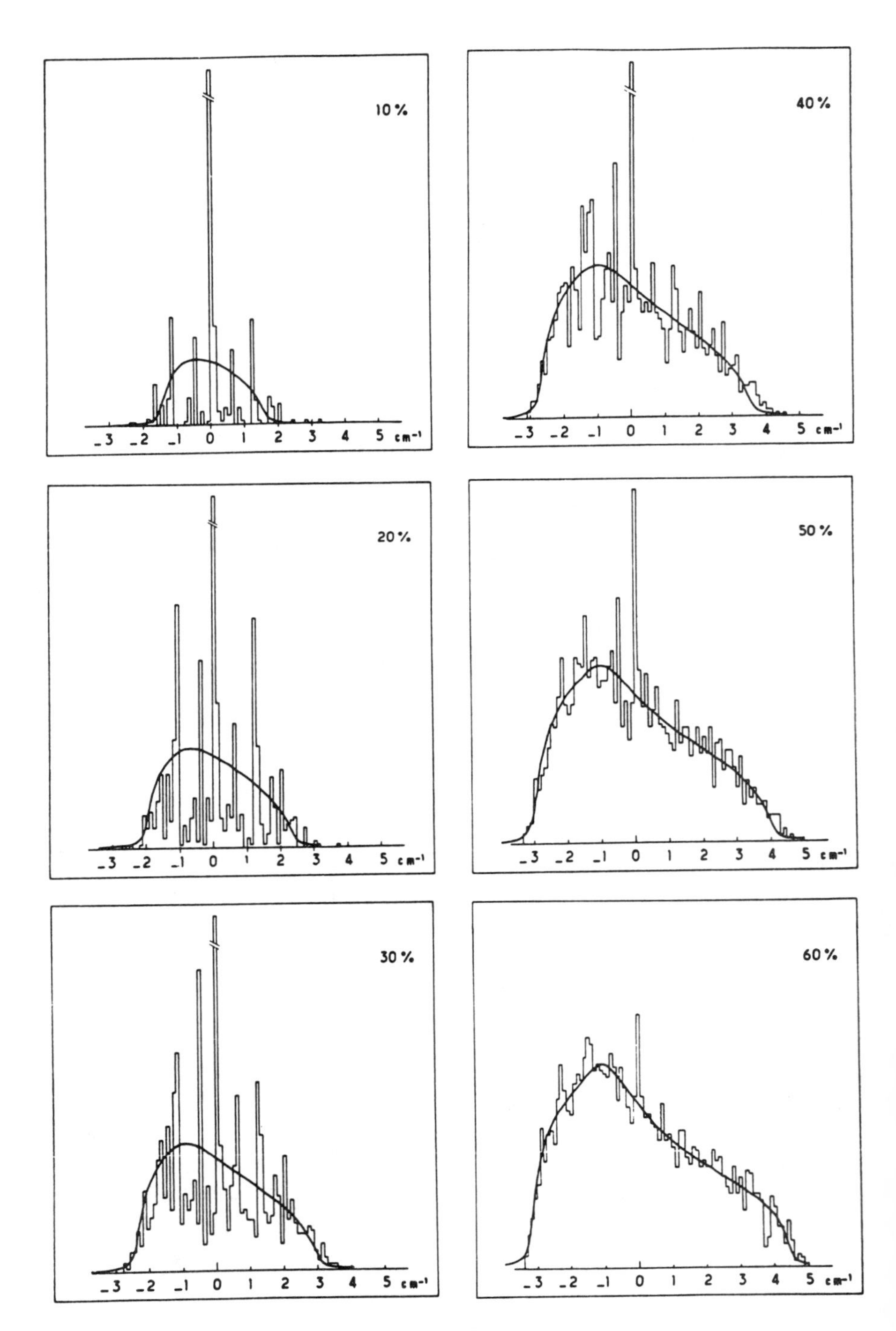
10 %
−3 −2 −1 0 1 2 3 4 5 cm⁻¹
40 %
−3 −2 −1 0 1 2 3 4 5 cm⁻¹
20 %
−3 −2 −1 0 1 2 3 4 5 cm⁻¹
50 %
−3 −2 −1 0 1 2 3 4 5 cm⁻¹
30 %
−3 −2 −1 0 1 2 3 4 5 cm⁻¹
60 %
−3 −2 −1 0 1 2 3 4 5 cm⁻¹

Figure 4.11

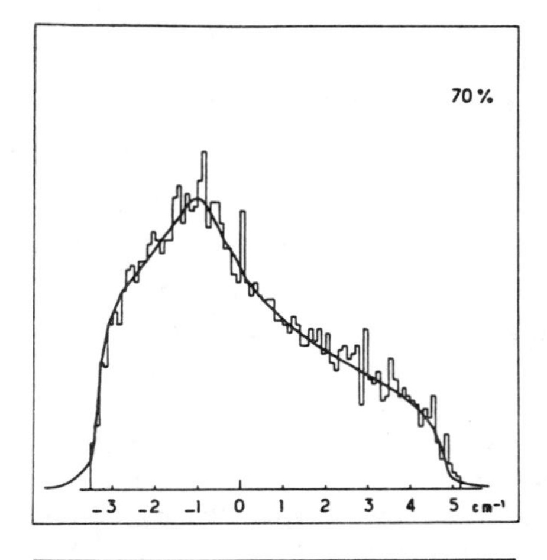

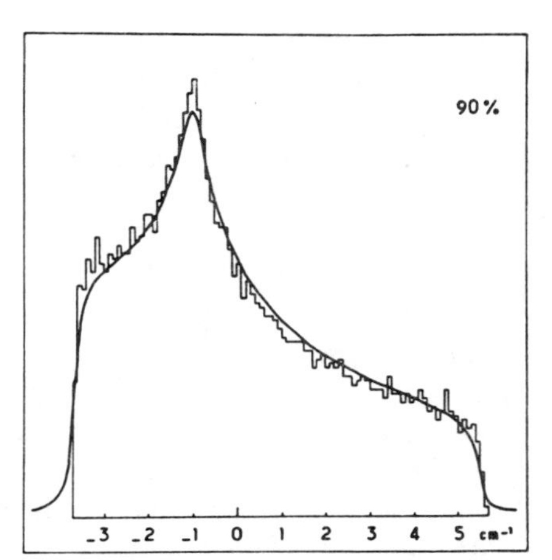

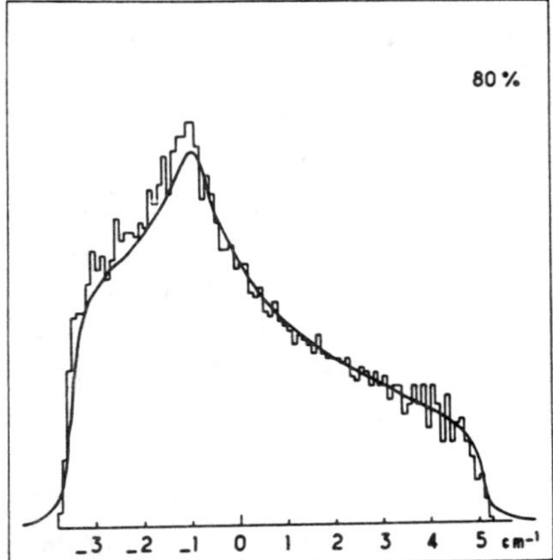

Figure 4.11. The density of triplet states of the trap (Nh_8) band, simulated by numerical calculations on large grids of 5000 sites and 0.1-cm^{-1} resolution, and calculated with the CPA method (smooth curve) for trap concentrations varying form $c = 10\%$ to $c = 90\%$. The CPA curves are satisfactory only at $c > 70\%$.

spectra are equally broad and do not correspond to an expansion in aggregates.

In connection with these discrepancies in the general lineshape of the spectra, the physical description of the disordered system by the CPA is erroneous for the following reasons: Whereas, in the real mixed crystal with low concentration, the guest states are localized (with possible coherent effects due to superexchange), the CPA describes the mixed crystal, on the scale of the site, as a homogeneous medium where the intermolecular interactions are those of the pure crystal, hence too strong for localized states, and where consequently the plane waves, although damped, are eigenstates. For these reasons, the CPA must be abandoned for low and intermediate concentrations, for which we have to include in the theoretical approach the large variety of microscopic environments of the sites. In contrast, for high concentrations ($c > 70\%$), the CPA is quite well adapted to the study of the perturbation of the crystal band by traps in low concentrations,[173] as, for example, natural isotopic defects, which may amount to 10–20% in the samples currently utilized.

b. *New Difficulties Connected with the Microscopic Environment of the Site.* The microscopic environment of one site, particularly with resonating impurities present in the vicinity of a given guest site, greatly modifies the response of the site in various ways, depending on the topology of the inspected configuration. Thus, the use of any perturbation expansion around the CPA appears excluded, because each configuration, at least for its immediate environment, must be inspected separately and calculated exactly. Then, an essential question is the distance scale beyond which the fluctuations are amenable to a macroscopic average description. In other words, we have to know the radius of the local effective environment. This radius, rigorously speaking, is infinite for the study of physical properties involving long-range interactions (conduction, percolation, quantum delocalizations, etc.), but it can be taken as finite when describing, with poor resolution, more local quantities, such as the density of states or optical absorption. Indeed, at low concentration the immediate vicinity suffices for an expansion in aggregates, whereas at high concentrations the one-site CPA proves very satisfactory.

Let us assume, in the spirit of the CPA, that we have an effective medium into which a given aggregate is immersed—a pair of guest molecules, for example. This problem is that of the isolated impurity treated in Section IV.C.1.c. If, furthermore, we wish to make an uncoupling approximation, neglecting correlations between neighbor aggregates, a frontier problem arises. Indeed, to uncouple, for instance, scattering by two neighboring pairs, we must isolate one pair from all the others, which is physically impossible, since common sites are shared by many pairs (see Fig. 4.12). This difficulty in treating aggregates of many sites is not present in the isolated-site case

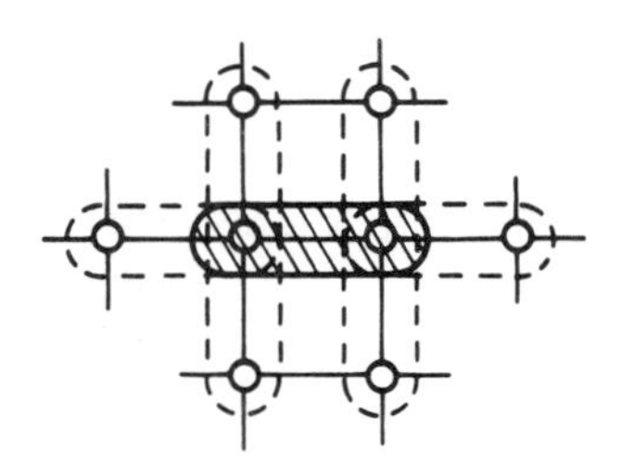

Figure 4.12. Scheme illustrating the impossibility of isolating one pair in a lattice of pairs.

(isolating one site from all the others does not exclude any neighbors. This is the essential reason for the weakness of self-consistent theories proposed for the pairs: These theories provide no analytic solutions in the cases of strong disorder ($\Delta > W$), which lead to discontinuities, to multiple solutions (multivaluations), etc.[156]

The simplest way to overcome these difficulties is to accept losing a certain number of pairs, or aggregates, and to partition the real crystal into a superlattice, paved with supermolecules (cells), which will be treated in a self-consistent way, as the one-site CPA does. This approximation is known[174] as the molecular CPA, or MCPA.

c. *The Limits of the Molecular* CPA (MCPA). We partition the real lattice into distinct cells, paving the real crystal; then we isolate one particular cell and treat it as immersed in a medium of effective cells (see Fig. 4.13). It is important to realize that this procedure leads, because of boundary effects, to an underestimation, not only of aggregates larger than the cell (which are simply ignored), but also of smaller aggregates, like pairs, among which only those entirely included in the cell are considered to contribute.

This method is absolutely analogous to one-site CPA on replacing the scalar quantities, such the self-energy, with $n \times n$ matrices. Therefore, one may show[158] that the MCPA is also analytic and satisfies the necessary elementary conditions: positivity of the spectrum, uniqueness of the solution, sum rules, etc. The self-consistency relation is obtained in a manner analogous to that of the one-site CPA (Section IV.A.4); it will be detailed in the calculation of the $n \times n$ matrix elements.

The cells are indexed by $\mathbf{i}$; they contain N_c sites, indexed by α. The

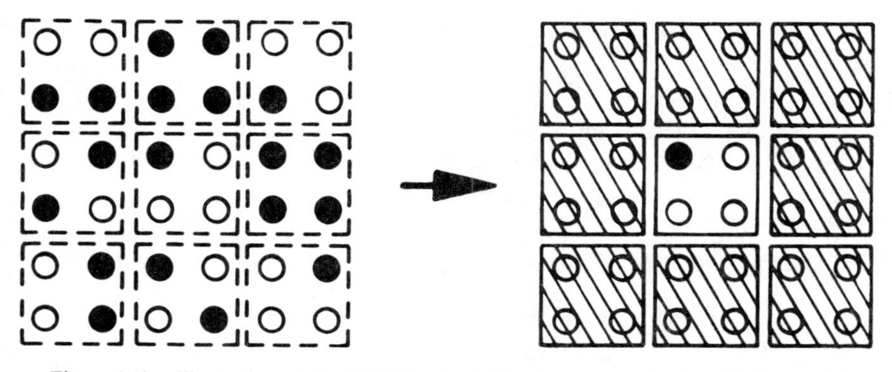

Figure 4.13. Illustration of the MCPA. The lattice is uncoupled into cells (here of four sites) forming a pavement. The real configuration around a single cell is replaced by a lattice of effective (identical) cells, determined by the vanishing of the mean scattering on the central cell.

projector on cell i is written

$$P_{\mathbf{i}} = \sum_{\alpha} |\mathbf{i}\alpha\rangle\langle\mathbf{i}\alpha| \tag{4.72}$$

One demonstrates[158] as in Section IV.A.4 that the coherent potential (or self-energy) is a local matrix, that is to say, without matrix elements between two different cells:

$$\Sigma = \sigma \sum_{\mathbf{i}} P_{\mathbf{i}} \tag{4.73}$$

where σ is an $N_c \times N_c$ matrix, internal to each cell.

The intermolecular interactions W must be written on the basis of cells. These interactions are independent of the nature of the interacting cells. For that reason, we shall calculate them on the basis of delocalized states defined by the Fourier transform*:

$$|\mathbf{K}\alpha\rangle = \frac{1}{\sqrt{N_c}} \sum_{\mathbf{i}} e^{i\mathbf{K}\cdot\mathbf{i}} |\mathbf{i}\alpha\rangle \tag{4.74}$$

On this basis, the matrix elements of W are given by

$$\langle \mathbf{K}\alpha | W | \mathbf{K}'\beta\rangle = \delta_{\mathbf{KK}'} \sum_{\mathbf{d}} e^{i\mathbf{K}\cdot\mathbf{d}} \langle 0\alpha | W | \mathbf{d}\beta\rangle \tag{4.75}$$

where the summation on $\mathbf{d}$ is over the cells which are neighbors of the cell $\mathbf{i} = \mathbf{0}$, in particular the cell $\mathbf{i} = \mathbf{0}$ itself. In the same basis, the elements of Σ are independent of $\mathbf{K}$:

$$\langle \mathbf{K}\alpha | \Sigma | \mathbf{K}'\beta\rangle = \delta_{\mathbf{KK}'} \langle \alpha | \sigma | \beta\rangle \tag{4.76}$$

which corresponds to the local character of Σ.

Then, we have to treat the case of a real cell immersed in a medium of effective cells, described by the operator Σ. The corresponding t matrix on the cell $\mathbf{i} = \mathbf{0}$ is written [with $g_0 = P_0(z - W)^{-1} P_0$]

$$t_0(\beta) = P_0(V_\beta - \sigma) \frac{1}{i - g_0(V_\beta - \sigma)} \tag{4.77}$$

where V_β is the real distribution of the resonance energies, assigning the configuration β probability $p(\beta)$. By uncoupling of t matrices, the analog of

*These states are defined in the theory of excitons with many molecules per cell; see refs. 155, 156.

(4.54), we evaluate the mean scattering, or mean t matrix:

$$\langle t \rangle = \sum_{\beta} p(\beta) t_0(\beta) \qquad \left(\sum_{\beta} p(\beta) = 1 \right) \tag{4.78}$$

Hence, the correction to the self-energy is

$$\Delta\sigma = \langle t \rangle \frac{1}{1 + g_0 \langle t \rangle} \tag{4.79}$$

For a self-consistent potential, we must have $\Delta\sigma = 0$. However, as the solution of $\langle t \rangle = 0$ by Newton's method is too laborious, we obtain σ by ATA iteration, i.e. by the recurrence formula

$$\sigma^{(n+1)} = \sigma^{(n)} + \langle t^{(n)} \rangle \frac{1}{1 + g^{(n)} \langle t^{(n)} \rangle} \tag{4.80}$$

whose convergence to the MCPA solution has been demonstrated.[158]

Starting from a medium of effective cells, the calculation of the physical quantities is immediate. For the density-of-states distribution, we take the trace on all the wave vectors $\mathbf{K}$ of the superlattice and on all the sites of the cell ($\mathcal{N}$ is the number of cells of the superlattice):

$$n(E) = -\frac{1}{\mathcal{N} N_c \pi} \operatorname{Im} \sum_{\mathbf{K}\alpha} \langle \mathbf{K}\alpha | \langle G \rangle | \mathbf{K}\alpha \rangle \tag{4.81}$$

For the absorption spectrum, since the translational invariance of the initial lattice is lost, the wave vector is defined modulo a vector of the reciprocal superlattice; thus, we must bring the absorption wave vector $\mathbf{k}$ ($\mathbf{k} = \mathbf{0}$ and $\mathbf{k} = (2\pi/b)\hat{\mathbf{b}}$) to a vector of the first zone of the superlattice:

$$\langle \mathbf{k} | \langle G \rangle | \mathbf{k} \rangle = \frac{1}{\mathcal{N}} \sum_{\alpha\beta} e^{i\mathbf{k}\cdot(\alpha-\beta)} G_{\alpha\beta}(\mathbf{k}_0) \tag{4.82}$$

where $\mathbf{k}_0$ is the representation of $\mathbf{k}$ in the first zone of the superlattice.

This MCPA calculation has been carried out for the triplet naphthalene mixed crystal, with a square cell of nine sites (3 × 3). The density-of-states distribution for $c_A = 30\%$ is shown in Fig. 4.14. One first notices that the smoothed shape disagrees generally with the numerical-simulation spectrum. The finest structure, at the maximum of the spectrum, corresponds to a configuration with a guest molecule A completely surrounded by host

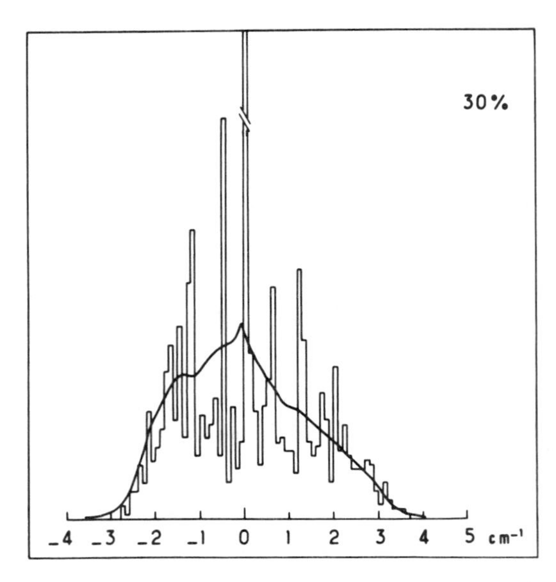

Figure 4.14. Comparison of the MCPA density-of-states distribution with numerical results. There is improvement over the CPA calculations.

molecules; the two broad bumps in the wings correspond to the so-called *AB* (or "12") dimers. The MCPA response, with a cell of nine sites, is much more structured than that of the one-site CPA, but it turns out rather disappointing in view of the lengthy and elaborate calculations required. There is no doubt that a larger lattice would provide a better approximation, but then we would have to give up averaging over all configurations; besides, one may legitimately ask whether the slow convergence rates of such large matrices do not effectively transform the MCPA procedure into a numerical-simulation approach.

For 1D lattices, MCPA with a cell of seven sites provides excellent results, in very good agreement with numerical simulations.[175] Therefore, we must explain, and learn from, the failure of MCPA to treat the relatively simple 2D disordered systems equally well. The inside interactions of the cell (3×3) being calculated exactly, the origin of the failure must reside in the treatment of its interaction with the outside effective medium, made of identical effective cells. In particular, we see that the sole configuration which produces a fine structure similar to the numerical simulation is the monomer on the central site, isolated from the boundary layer of nonresonant host molecules. On the contrary, the aggregates which are in contact, by at least one site, with the effective medium (which is quasi-resonant in the approximation of the scattering width, or of the imaginary part of σ) produce broad structures. A preliminary conclusion is that the MCPA becomes useful only when the boundary (edge) effects become negligible. This condition is fulfilled for the 1D system using cells of seven sites,[175] whereas in our 2D system, the boundary affects eight out of nine sites. Therefore, in order to continue to use small cells, we must modify the interaction with the effective medium, since the latter always contains components resonating with the energy of any aggregate of the cell, and since the interaction between the aggregate and the effective medium is the interaction W of the pure crystal (hence too strong). In what follows, we take advantage of this analysis, especially the notions of resonance and of effective interaction with the outside medium. To do so, we recall the model where these notions have been utilized: the model of percolation.

4. *Percolation and* HCPA

In the problem of the binary alloy with nonresonant components, the major role, for low trap concentrations, is played by practically isolated aggregates, i.e., those interacting very weakly with their environment. Thus, the mean-field theory must reproduce this limit situation of isolated aggregates and recover the fine-structure spectra of the numerical simulation. The effective crystal being always resonant at the energy of the traps, or at that of the aggregates, the only way to attenuate the interaction is to modulate the intermolecular interactions in the effective medium. Actually, the general lineshpae, in CPA as well as in numerical calculations, shows a narrowing of the band around the energy of the guest A at low concentrations. The problem of finding an effective interaction, or "conductance", in a random lattice is closely connected with that of the effective conductivity in a random lattice of resistances, and more generally, with the problem of percolation on the lattice.

a. *Percolation of Bonds and Sites.* Let us consider the following problem of electrokinetics.[122] In a perfect lattice of resistances, we take away randomly a fraction $1 - c_L$ of resistances: In what way will the macroscopic conductivity of the lattice vary with c_L? The problem is that of percolation of bonds. For small values of $c_L (c_L < c_L^p)$, the equivalent conductivity of the lattice is zero; only isolated aggregates of resistances are found. For a critical concentration of percolation, $c_L = c_L^p$, a first infinite aggregate connects the borders of the macroscopic sample. For $c_L > c_L^p$, the effective conductivity increases progressively up to the value for a perfect lattice.

A mean-field theory (Kirkpatrick[176]) manages to account for this percolation phenomenon. In the framework of the CPA, a real resistance is considered immersed in a perfect effective medium, with the requisite that this substitution will not induce, on average, an additional potential difference. The effective conductivity obtained in this way is very satisfactory: It shows a percolation threshold, while the results in the vicinity of percolation are not different from those of numerical simulations. (The behavior around the percolation threshold has been analyzed by means of renormalization-group theory[177]).

An analogous problem is that of percolation of sites, which is created when nodes, or A sites, are randomly suppressed in the lattice, along with the resistances connecting them to their neighbors. Qualitatively, the behavior is similar to that of the bond percolation, but with a higher percolation threshold, $c_s^p > c_L^p$.[122] (For a square lattice, $c_s^p = 0.59$ and $c_L^p = 0.50$.) The best mean-field theory for treating the problem of equivalent conductivity consists, not in considering one site in an effective medium, but simply in inserting a single bond into a lattice of effective bonds.[156,178] The probability of encountering a "passing" bond must then be related to the concentration of the remaining A sites:

$$c_L = c_A^2 \tag{4.83}$$

since a given passing bond must connect two A sites.

The transition to a model of random bonds neglects the correlations imposed by construction between passing bonds around an A site, as is shown distinctly in Fig. 4.15, where two lattices satisfying the equivalence (4.83) are compared: one lattice with random bond distribution, and one lattice with random site suppressions. Aggregation of bonds is easily discernable in the case of site percolation. However, as a matter of fact, these correlations have no importance in the case of conductivity, so that we may obtain a good approximation when leaving them out.[178] It will be shown below that this approximation is questionable for more local properties where the microscopic arrangements of bonds may be crucial.

We consider now the way the above classical considerations on percolation may be transferred to the quantum treatment of a random 2D lattice.

b. *Pure Nondiagonal Disorder and* HCPA. The quantum analog of the percolation model in electrokinetics will be a lattice with equivalent sites and random interactions between neighbor sites. More specifically, the interactions will take the value w with probability c_L and the value 0 with probability $1 - c_L$. In this case, the sample contains isolated aggregates (localized excitations) at concentrations below the percolation threshold

Figure 4.15. Random square lattice of sites (A) and bonds (B), corresponding to the same concentration of passing bonds (solid lines). The concentration of sites is about 71% in lattice A, and that of bonds 50% in both lattices. Thus, the B lattice is at a percolation concentration, whereas the A lattice is above the percolation threshold.[176–177] Note the aggregation of bonds in lattice A relative to lattice B.

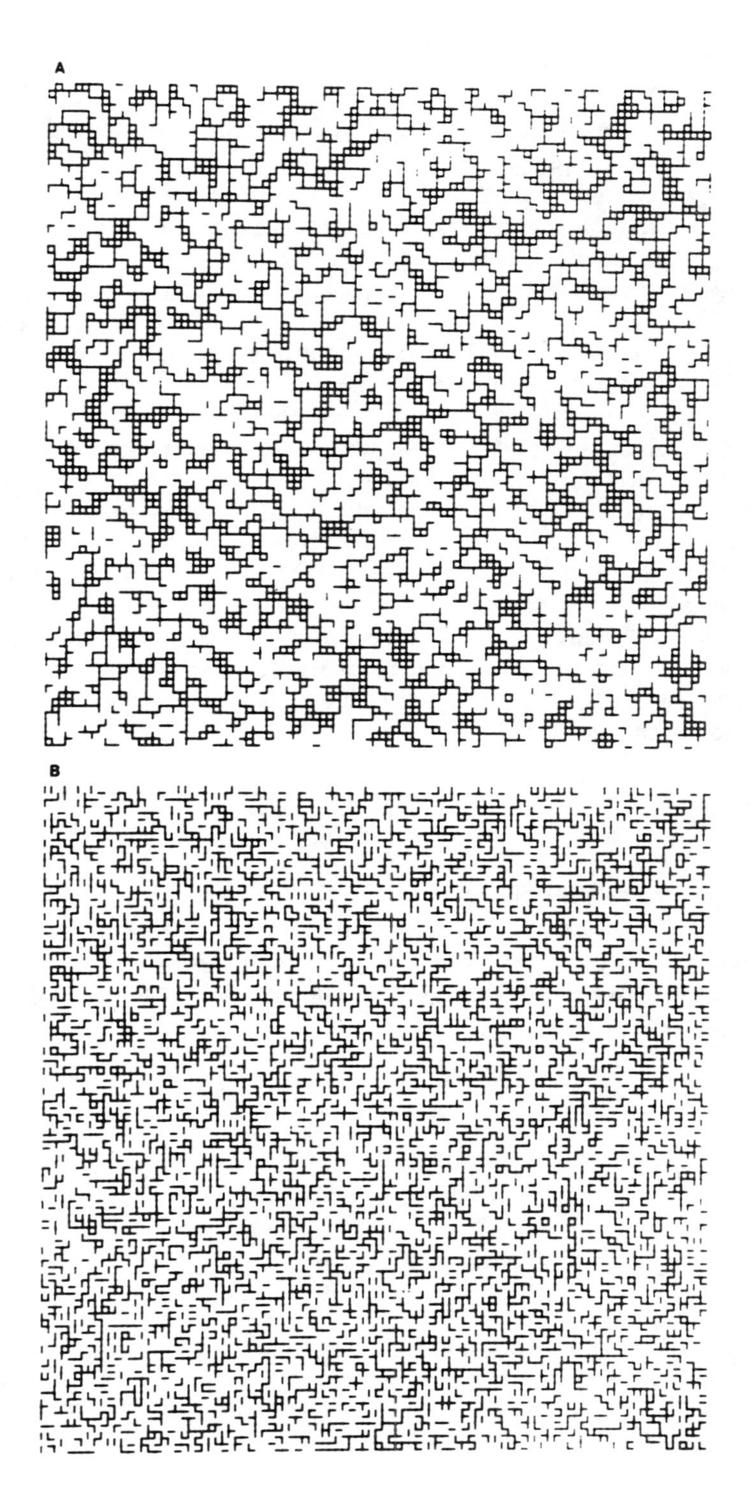
A
B

c_L^p, and one or more aggregates, with the possibility of extended states, at concentrations above c_L^p.[179]

As opposed to the binary alloy studied in the above sections, these models involve a pure nondiagonal disorder where the V_n's (4.26) are all equal and the W_{nm}'s take random values.

The corresponding mean-field approximation consists in canceling, in the lattice of effective bonds, the average scattering on a real (passing or stopping) bond. This approximation has been recently proposed by Yonezawa and Odagaki[180] under the name *homomorphic CPA* (HCPA), for bonds or more general aggregates. In comparison with the calculation of an effective resistance, an important distinction appears: in the present case, we have to include the energy z at which the lattice response is calculated. In other words, a system of the type (4.28) must be solved, and the resistivity is its response at $z = h\omega \to 0$.[176]

Let us consider the lattice of bonds illustrated in Fig. 4.16. The real bond introduces the interaction w_L (with values w or 0), but no diagonal term. In contrast, the effective bond introduces an interaction $w_{\text{eff}}(z)$ and a "bond potential" $s(z)$ at the two ends. In the case of a lattice with all bonds equivalent, the effective medium is represented by a single type of bond; the Green's function of this medium takes the form

$$G_{\text{eff}}(z) = \frac{1}{z - \nu s(z) - W_{\text{eff}}(z)} \tag{4.84}$$

where ν is the number of bonds per site. $G_{\text{eff}}(z)$ is the perfect analog of (4.32) with a locally diagonal coherent potential (self-energy) $\sigma(z) = \nu s(z)$, but now the operator $W_{\text{eff}}(z)$ is different from W of the perfect crystal, with an interaction between sites $w_{\text{eff}}(z)$ depending on z and generally complex. The

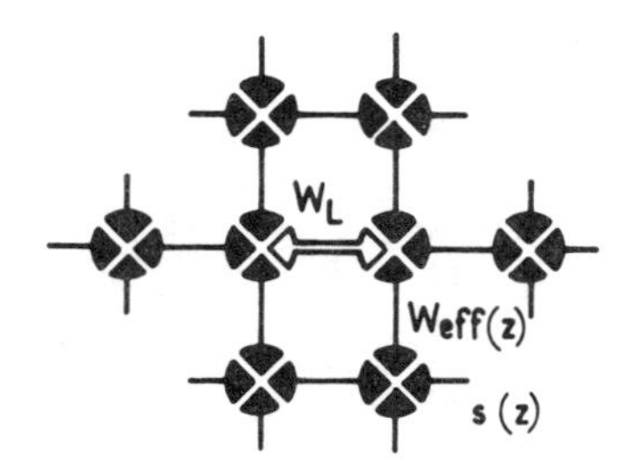

Figure 4.16. Scheme of lattice with effective bonds used in HCPA. A bond is characterized by one interaction $W_{\text{eff}}(z)$ and one bond potential $s(z)$. Self-consistency is obtained by requiring a zero value of the mean scattering on a real bond (interaction w, without potential) immersed in the effective medium.

two unknown functions $w_{\text{eff}}(z)$ and $s(z)$ are then determined by the condition of zero mean scattering around a single bond.

The scattering-potential difference between real and effective bonds is a 2×2 matrix

$$\begin{pmatrix} -s(z) & w_L - w_{\text{eff}} \\ w_L - w_{\text{eff}} & -s(z) \end{pmatrix} \tag{4.85}$$

The corresponding t matrix is easily calculated by the use of the symmetry of (4.85). The cancellation conditions are

$$\langle t^L_{\pm}(z) \rangle = \langle \{ [-s \pm (w_L - w_{\text{eff}})]^{-1} - (g_0 \pm h_0) \}^{-1} \rangle = 0 \tag{4.86}$$

(The average is over w_L: w with probability c_L, and 0 with probability $1 - c_L$). The functions g_0 and h_0 are matrix elements of $G_{\text{eff}}(z)$:

$$\begin{aligned} g_0 &= \langle 0 | G_{\text{eff}} | 0 \rangle \\ h_0 &= \langle 0 | G_{\text{eff}} | 1 \rangle \end{aligned} \tag{4.87}$$

Sites 0 and 1 are nearest neighbors.

The numerical solution of the system (4.86), by a procedure of the Newton–Raphson type with two variables, requires the calculation of the derivatives $\partial t_{\pm}/\partial s$ and $\partial t_{\pm}/\partial W_{\text{eff}}$. The results we obtained for a square lattice are similar to those by Yonezawa and Odagaki[180] for a cubic lattice. The most striking feature is the existence, at low concentration, of a gap in the density of states,[179] which isolates the zero energy on which a δ peak builds up. Thus the HCPA produces a forbidden region of energy for the transport; the gap and the δ peak disappear at a critical concentration, analogous to the percolation threshold of the mean-field of resistances.

We have extended this model to the naphthalene triplet 2D lattice, where each site has four nonequivalent bonds and two equivalent bonds. Therefore, we have considered the two types of bond separately, with two bond potentials s_e (s_i) and two interactions w_e (w_i) for the effective equivalent (nonequivalent) bonds. These four functions are determined by the vanishing of the mean scattering on each bond separately: see Fig. 4.17. The calculation is quite analogous to the preceding one. The system to be solved involves four equations:

$$\langle t^L_{j\varepsilon} \rangle = 0 \tag{4.88}$$

with $j =$ equivalent or nonequivalent, $\varepsilon = \pm$, and $w_j = 1$ or 0 for passing or

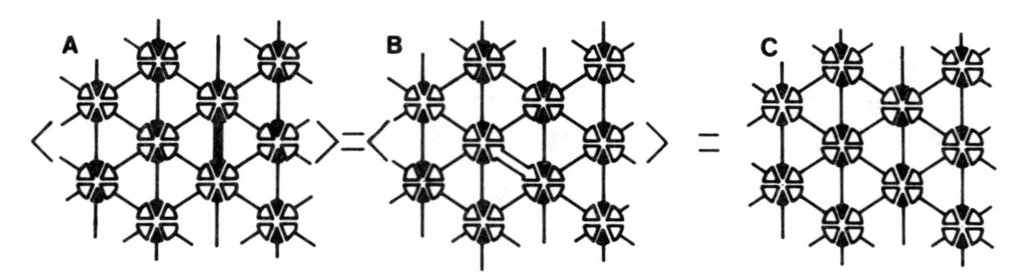

Figure 4.17. Scheme in HCPA in a lattice with two types of bond, as in Nh_8–Nd_8. The mean scattering around each type of bond (equivalent: *A*; nonequivalent: *B*) is set equal to zero in order to obtain the required perfect effective lattice *C*.

stopping bonds:

$$t_{j\pm}^{L} = \{[-s_j \pm (w_{jL} - w_j^{\text{eff}})]^{-1} - (g_0 \pm h_j)\}^{-1} \tag{4.89}$$

where h_j is the element of G_{eff} between nearest equivalent or nearest nonequivalent neighbors. The effective Green's function is determined by the coherent potential

$$\sigma(z) = 4s_i(z) + 2s_e(z) \tag{4.90}$$

and by the effective interactions $w_e^{\text{eff}}(z)$ and $w_i^{\text{eff}}(z)$.

The solution of (4.88), using Newton's method with four variables, is illustrated in Fig. 4.18 for various concentrations c_L of passing bonds. At low concentrations, we find back the abovementioned gap in the density of states, with a more complex structure due to the presence of two types of bond; at high concentrations, the spectrum becomes smoothed and approaches that of a perfect crystal.

c. *Pure Diagonal Disorder.* We consider now the initial diagonal disorder as encountered in the real isotopic mixed crystal. The uncoupling in the HCPA leads to a difficulty: In the square lattice, for example, three bonds are possible (AA, AB, BB), whose superposition in the real lattice gives the real potentials v_A or v_B on each site. The HCPA bond potentials will be roughly $v_A/4$ and $v_B/4$, according to the nature of the bond border.

Up to now, the uncoupling, although arbitrary, remains legitimate. However, if we neglect the correlations between bonds (Fig. 4.15) in calculating an effective bond, we shall arrive at spurious resonances around $v_A/4$ or $v_B/4$. These spurious resonances make the HCPA inapplicable to the case of an arbitrary diagonal disorder.[181] However, apart from the case of a pure nondiagonal disorder ($v_A = v_B = 0$), there is a case where the spurious

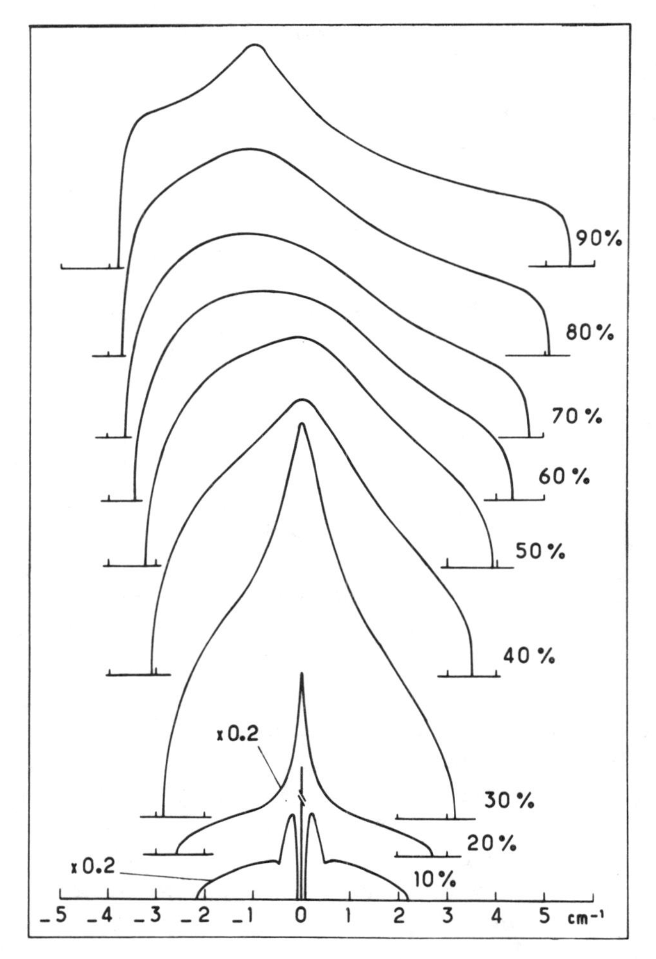

Figure 4.18. Density of states of HCPA with two types of bond (for the Nh_8–Nd_8 lattice) at various concentrations of passing bonds. At low concentration ($c \lesssim 10\%$) one sees a gap around the origin with a δ peak at the origin. At high concentrations the distribution tends to that of a perfect crystal.

resonances are not disturbing: this is the limit case $v_A = 0$ with $v_B \to \infty$. Then, the spurious resonance is repelled to infinity. This case applies to the naphthalene triplet lattice where $\Delta \gg W$, in the approximation of neglecting the shifts in W^2/Δ.

If we consider the species B to be nonresonant, we may obtain this limit case by letting $v_B \to \infty$. An almost equivalent passage to the limit is obtained by letting W (the interaction between B sites and other sites) go to 0, as may be checked, for example, on the expansion of the locator[122] of $\langle n|\mathscr{G}(z)|m\rangle$ for a given configuration:

$$\langle n|\mathscr{G}(z)|m\rangle = \frac{1}{z-v_n}\delta_{nm} + \frac{1}{z-v_n}W_{nm}\frac{1}{z-v_m} + \sum_p \frac{1}{z-v_n}W_{np}\frac{1}{z-v_p}W_{pm}\frac{1}{z-v_m} + \cdots \tag{4.91}$$

Except for the first term of the expansion, the abovementioned transitions to the limit are equivalent. Thus, we obtain in average:

$$\lim_{v_B\to\infty} \langle \mathscr{G}\rangle = \lim_{\substack{W(B,\text{-})\to 0\\ v_B=v_A}} \langle \mathscr{G}\rangle - \frac{c_B}{z-v_A} \tag{4.92}$$

We recover the limit of nonresonant species B: a lattice with pure nondiagonal disorder ($v_A = v_B = 0$) where the bonds of the randomly chosen B sites are suppressed $[W(B,\text{-})\to 0]$ on the condition of zero response at $z = v_A$ of the introduced fictitious B sites. Up to this stage, the equivalence (4.92) remains exact.

From now on, we wish, in the spirit of the site percolation in electrokinetics (Section IV.C.4.a), to neglect the bond correlations. Thus, we consider an effective medium around the energy v_A where the excitation will propagate; it is clear that this medium correctly describes the propagation, but that it will not correctly describe, for example, the density-of-states distribution, since it contains also fictitious B sites at the energy v_A. Therefore, by means of this restriction, the HCPA method is then directly transferable to the naphthalene triplet lattice, with probability $c_L = c_A^2$ of having a passing bond (4.83). The curves of Fig. 4.18 are likewise transferable, but, because of the fictitious B sites, the density of states around v_A is not normalized at the real concentration of the A sites (as was possible for the CPA cases: cf. Fig. 4.11).

5. *One-Cell Scattering in an Effective Medium*

We come back to the scheme (Section IV.C.3.b) of a small cell in a real crystal, assumed to account for the couplings in the configuration at short distances, as immersed in an effective medium representing the mean macroscopic crystal which modifies the response of the cell.

a. *Mean Scattering.* Figure 4.19*A* shows a cell of four sites, immersed in an effective macroscopic medium, now translationally invariant (in contrast with the MCPA effective medium).

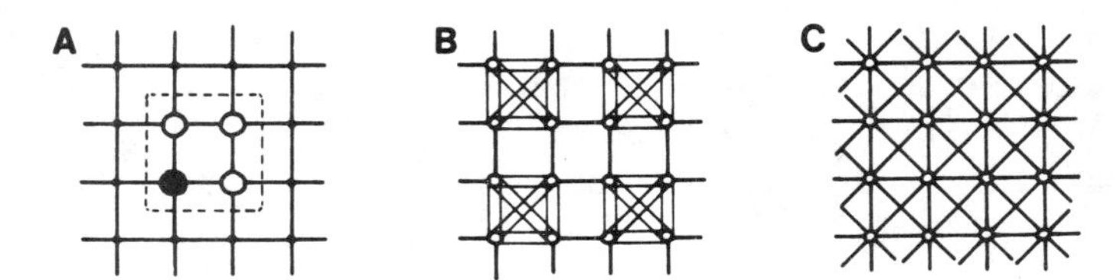

Figure 4.19. Various stages of the calculation of an effective medium from a single cell. (*A*) The cell is immersed in a non-self-consistent medium. (*B*) The mean scattering on the cell *A* allows one to obtain a superlattice of mean cells. (*C*) The translational invariance, broken in stage B, is restored by averaging, at the cost of introducing new interactions between neigbors.

We may perform an uncoupling approximation of the type (4.54), in order to calculate the mean scattering of the cell. This approximation assumes the partition of the medium into uncoupled cells, in the manner of the MCPA. After calculating the mean scattering $\langle t \rangle$, one may calculate, with the use of (4.69), a self-energy $\Delta\sigma$ corresponding to a crystal of effective cells, which is the analog of the MCPA; see Fig. 4.19*B*. At this stage, the translational symmetry is lost, the bonding between cells being achieved by the initial effective medium. In order to restore the translation symmetry of the initial lattice, we have eliminated the spurious periodicities induced by the partition into cells, by averaging the t-matrix elements of the effective crystal. If $\langle t^c \rangle$ is the local t matrix on a cell, with the elements $\langle t^c \rangle_{\alpha\beta}$ between sites α and β, the t-matrix element of the final effective crystal (translationally invariant: Fig. 4.19*C*) between two sites distant by $\boldsymbol{\delta}$ is

$$\langle \bar{t}(\boldsymbol{\delta}) \rangle = \frac{1}{N_c} \sum_{\boldsymbol{\alpha}} \langle t^c \rangle_{\boldsymbol{\alpha}, \boldsymbol{\alpha} + \boldsymbol{\delta}} \tag{4.93}$$

The sum runs over the sites $\boldsymbol{\alpha}$ in the cell, since $\langle t^c \rangle_{\boldsymbol{\alpha}, \boldsymbol{\alpha}+\boldsymbol{\delta}} = 0$ if $\boldsymbol{\alpha} + \boldsymbol{\delta}$ is a site out of the cell (N_c is the number of sites). The elements of $\langle \bar{t} \rangle$ correspond

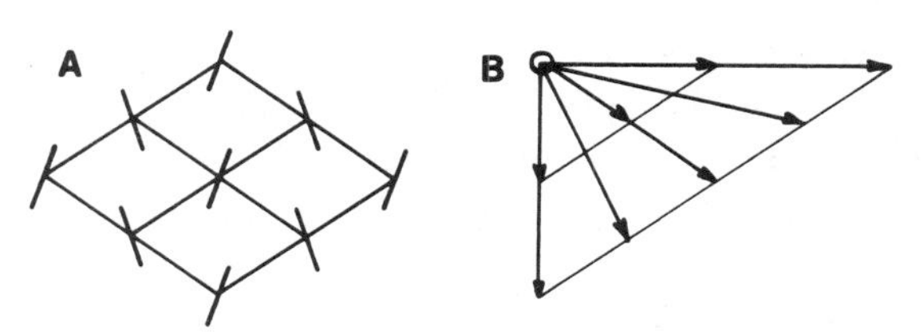

Figure 4.20. (*A*) Cell of nine (3 × 3) sites of the naphthalene triplet lattice, utilized in our numerical simulations. (*B*) Various possible vectors of translation between sites of the cell (modulo the symmetry operations).

to all possible translation vectors $\boldsymbol{\delta}$ in the cell. As an example, Fig. 4.20 shows these vectors for the cell of nine sites used in the present work.

With the matrix $\langle \bar{t} \rangle$ known, we derive the physical properties of the system (density of states, absorption, etc.). The self-energy, in our approximation, is obtained via

$$\Sigma = \langle \bar{t} \rangle (1 + g \langle \bar{t} \rangle)^{-1} \tag{4.93'}$$

where g is the matrix of the Green's function between sites of the cell in the HCPA effective medium.

The approximation (4.93) is not self-consistent; it resembles the (cellular) ATA in its crystal cells, but has an initial effective medium whose choice is free in order to satisfy additional conditions.

b. *The Choice of the Effective Medium.* The self-consistency of the above theory would require that the initial effective medium be identical to the one present at the end of the operations. Unfortunately, we have not found an appropriate condition of self-consistency preserving the required properties of analyticity, so we content ourselves with using a non-self-consistent theory for which the choice of an initial medium is essential.

The choice of the one-site CPA effective medium will eventually provide spectra very similar to those of the CPA itself: the aggregates, correctly described in the cell, are washed out by the strong imaginary part of the CPA.

Another choice, which we actually use in this work, is the HCPA effective medium of Section IV.C.4.c, with results illustrated in Fig. 4.21. The reason is that at low concentrations the effective interactions in the HCPA effective medium are very weak. This leads to practically isolated aggregates of the cell, and hence to fine structure.

c. *Discussion of the Results.* The calculated spectra show a gradual transition from the spiked aggregate spectrum at low concentrations to the broad band spectrum at high concentration, in agreement with the numerical simulations. This is very satisfactory performance of our modified HCPA theory. In the absorption spectrum, the Davydov components, broadened by disorder, build up smoothly at medium concentrations from shaggy absorption bands due to aggregates. Regarding the details, we must note several shortcomings:

1. Some aggregates, such as equivalent pairs, are seriously perturbed by the effective medium, and at low concentrations the Davydov splitting of these pairs is less than the expected value, $1\,\text{cm}^{-1}$.[121]
2. In general, the aggregate peaks are underestimated, due to the small size of the cell. This failure is not entirely compensated by the effective medium.

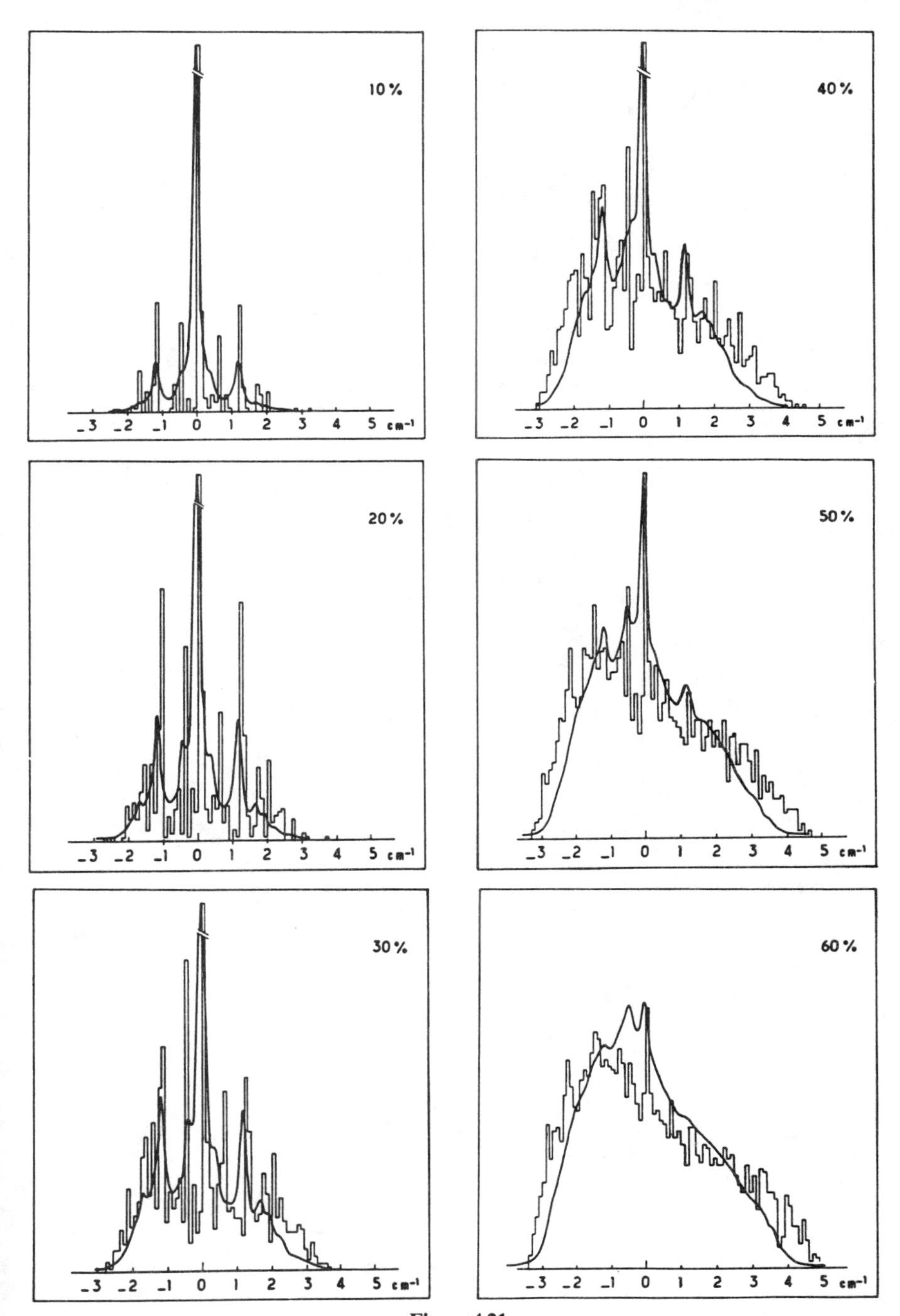

Figure 4.21

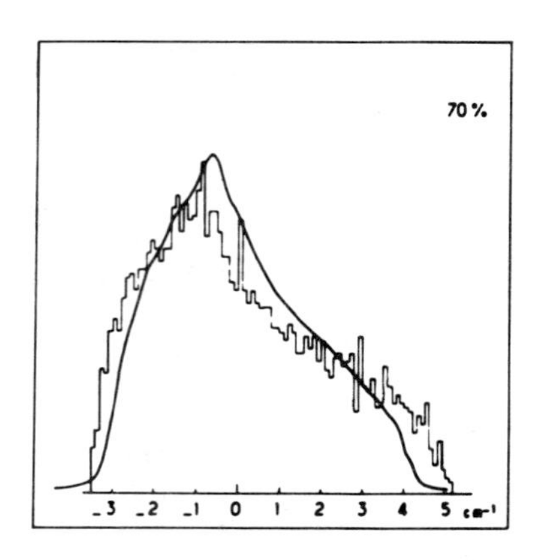

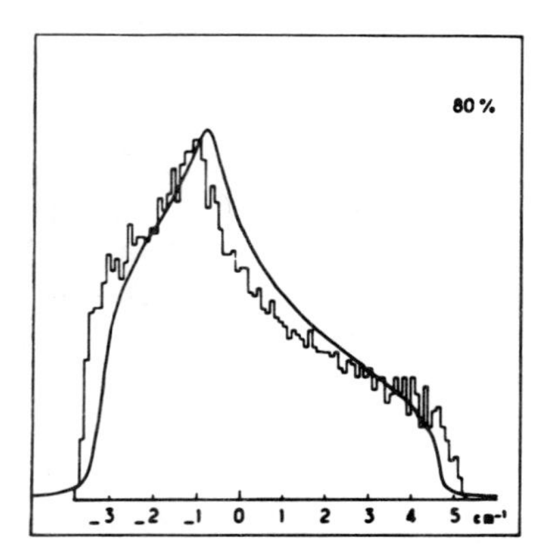

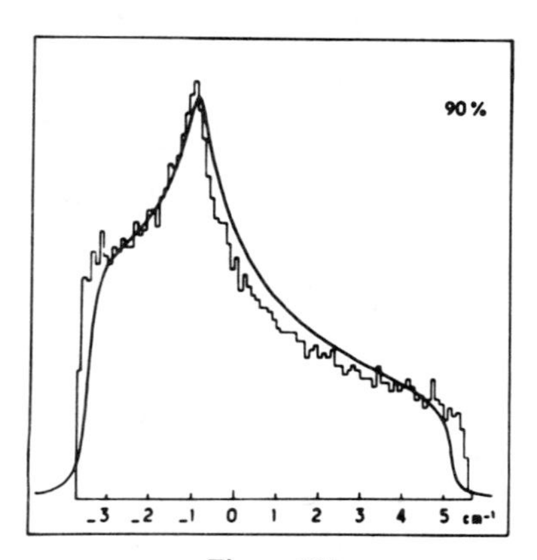

Figure 4.21

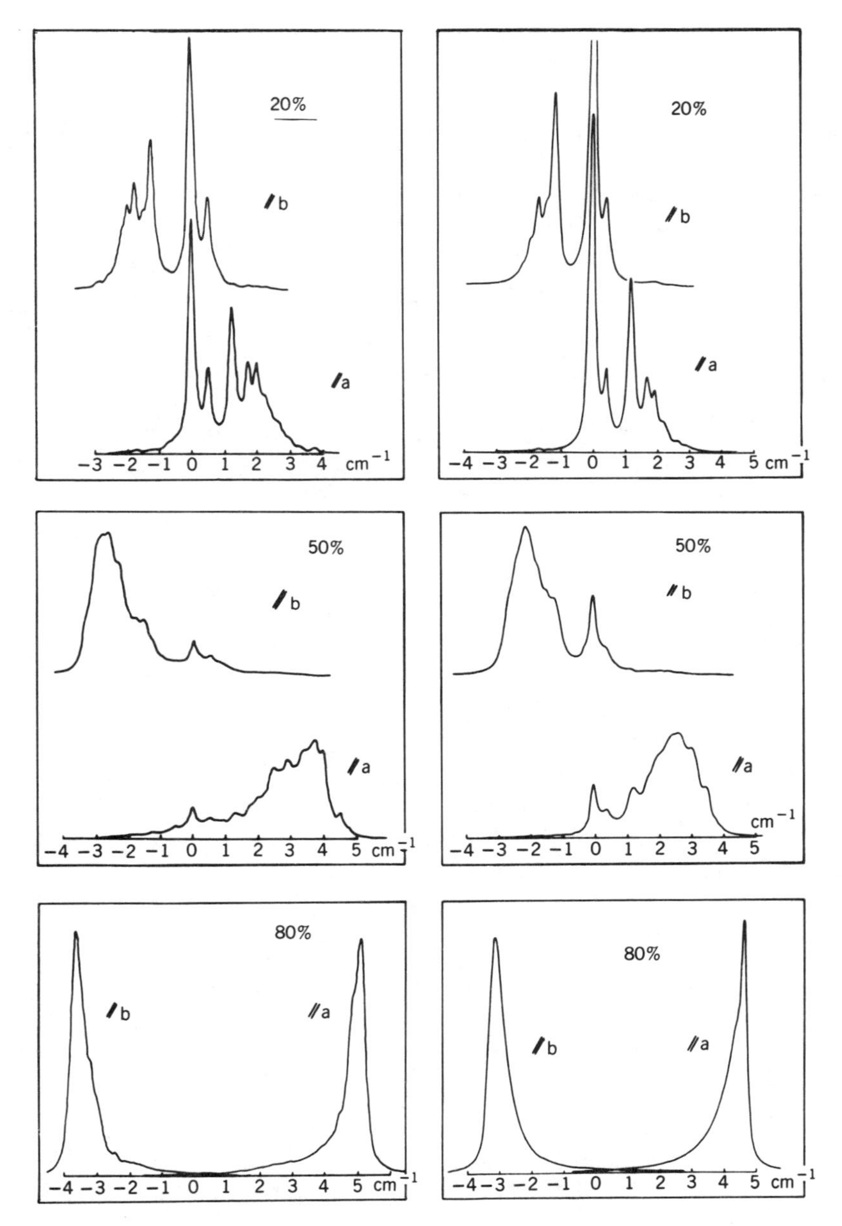

Figure 4.21. Density-of-states distribution and absorption spectra obtained at all concentrations by numerical simulations and our modified HCPA (non-self-consistent), with both the effective site energy and the effective interactions of the mean field being adjustable functions. Comparison with numerical simulations confirms the observed transition from a spectrum of discrete fine resonances to a quasi-continuous spectrum at high concentrations. The comparison points out the norrowing of the band in our method, visible at all concentrations. (Aggregates not contained in the 3 × 3 cell are underestimated.)

3. There is a general shrinking of the band, compared to the numerical simulations. This is clearest at higher concentrations and stems from a similar defect of the HCPA effective medium.

4. We note that the density of states and absorption spectra vary continuously with concentration both in the numerical simulation and in the modified MCPA theory at our resolution of $0.1\,\text{cm}^{-1}$.

The third defect, at least, could probably be attenuated by the choice of a medium better suited than the HCPA, restoring the self-consistency. We were unable to find a good self-consistent condition to derive the required effective medium from the average t matrix. Nonetheless, to be optimistic, we feel that our calculations show the possibility of utilizing the advantages of a mean-field approach on small cells, including small aggregates, which could account for all coarse features of the density of states and of absorption distributions of even strongly disordered crystals.

d. *Concluding Remark.* Notwithstanding the above criticisms, we may draw an optimistic overall conclusion from our calculations: The present advanced capabilities for numerical simulation (computers, intelligent algorithms, etc.) allow us for the first time to set up and to test mean-field approximations. They have the advantage of allowing analytical derivation of complex physical properties of disordered systems. They embody phenomenological physics and connect with the physical phenomenon in an intrinsically more "homogeneous" way than direct numerical simulations can. In the present case, a mean-field approach describing both the broad structures and the finer resonances due to aggregates seems possible if one accepts that both the effective site energy and the effective interactions of the mean field are adjustable functions in the model. We think that self-consistency (not obtained here), which is a good probe on the nature of the disorder, would improve the theory both quantitatively and conceptually.

D. Polaritons in 3D Mixed Crystals

In this subsection we consider mixed crystals in the specific case of 3D $W_{\mathbf{K}}$ interactions: the coulombic dipolar interactions of Section I. The main approximation we made in Section IV.A, on the crystal substitutional disorder, will be illustrated with semiclassical pictures: dipoles and fields.

1. *General Aspects*

The most investigated crystals from the polariton point of view are ionic crystals, studied by IR spectroscopy.[156] (Other applications of polariton theory are time spectroscopy and nonlinear optics in semiconductors[182] and semiconductor-doped glasses.[183]) The mixed crystals are of the type K(Cl, Br) or (K, Na)Cl, as well as semiconductors of the type Ga(As, P). According to the observed spectra with one or two strong reflectivity bands, the mixed crystal is classified as *amalgamation* or *persistence* respectively. This classification is independent on the concentration, as discussed by Onodera and Toyozawa.[159]

In the case of electronic spectroscopy in molecular crystals, the first singlet state of the isotopic mixed crystal of anthracene has been investigated in reflectivity and shown to be of amalgamation type.[129] Actually, while the gap of resonances ($\Delta = v_A - v_B$) is about $70\,\mathrm{cm}^{-1}$, the excitonic bandwidth is estimated at $500\,\mathrm{cm}^{-1}$. Very few mixed crystals have been investigated in the whole range of concentrations,[120] whereas examples of low concentrations (of impurities, for instance) are very numerous: The anthracene crystal contains β-methylanthracene naturally and shows an impurity level below the excitonic band.[120] (This does not suffice, however, to predict that a mixed anthracene–β-methylanthracene crystal will be of the persistence type.)

Thus, we study the binary alloy (cf. Section IV.B) with the intermolecular interactions W of Section I. We are in the case of weak disorder $\Delta \ll W$, e.g. of a virtual crystal with one polariton band, and for $\Delta \gg W$ we have separation into two bands, around v_A and v_B, each one embodying a "polariton" broadened by the disorder.

2. *Disorder Effects on* 3D *Retarded Interactions*

The problem is to discuss the generalized polarizability $\boldsymbol{\alpha}_e$(1.49) with the matrix $\boldsymbol{\alpha}^{-1}$ not commuting with that of the dipolar interactions, $\boldsymbol{\phi}$. To show that the pure retarded interactions may be discarded in the dynamics of mixed crystals, we assume here that the coulombic interactions are suppressed in $\boldsymbol{\phi}$. The interaction tensor is then reduced to its retarded term $\boldsymbol{\psi}$ (1.74). Then the dispersion is given by (1.35):

$$R(\mathbf{k}, z) = \frac{\cos^2\theta}{1 - c^2k^2/z^2} R \tag{4.94}$$

where θ is the angle $(\mathbf{k}, \mathbf{d})$ and R the width of the stopping band, with $R = d^2/\varepsilon_0 V_0$.

The dynamics of the mixed crystal, with purely retarded interactions, is governed by the sums of the type (4.62):

$$g(z) = \frac{1}{N}\sum_{\mathbf{k}} \frac{1}{z - R(\mathbf{k}, z)} \tag{4.95}$$

Calculation of (4.95) shows that $g(z)$ differs from z^{-1} by terms in $(a/\lambda z)^2 R$, i.e., the energy scale at which purely retarded effects show up in the mixed crystal dynamics is extremely reduced, to about $10^{-3}\,\mathrm{cm}^{-1}$ for a stopping band width $R = 1000\,\mathrm{cm}^{-1}$. The physical reason for this disparity resides in the extreme weakness of the purely retarded interactions. Actually, a stopping band of $1000\,\mathrm{cm}^{-1}$ width results from interactions in a volume λ^3, containing about 10^9 molecules, thus appearing in a very restricted domain of wave vectors. In contrast, the disorder effects are essentially local (for the properties investigated here) and cause averaging on all the wave vectors of the band; the purely retarded interactions disappear in this averaging.

Therefore, our problem amounts to treating

$$\boldsymbol{\alpha}_e = [\boldsymbol{\alpha}^{-1} - \boldsymbol{\phi}(\mathbf{k}, \sigma)]^{-1} \tag{4.96}$$

containing only coulombic interactions. Thus, with the replacement of $\boldsymbol{\alpha}_e$ by the polarizability of an effective crystal, translationally invariant (Section I.A.1), the optical response of the mixed crystal will be determined by the transverse dielectric tensor (1.79) of the effective crystal.

3. *A Few Simple Approximations*

We use a semiclassical formalism, close to that of the exact Green's function $\mathscr{G}(z)$ (4.29), but tensorial, since $\boldsymbol{\alpha}_e$ applies to electric dipoles. For example, the response on site $\mathbf{n}$ of an external excitation on site $\mathbf{m}$ takes the form

$$\mathbf{d}_{\mathbf{n}} = \langle \mathbf{n} | \boldsymbol{\alpha}_e | \mathbf{m} \rangle \mathbf{E}_{\mathrm{ext}}(\mathbf{m}) \tag{4.97}$$

Then it is straightforward to apply the discussion of Section IV.B according to the values of the disorder parameters.

a. *The Virtual Crystal.* When the dispersion of the dipole sums $\boldsymbol{\phi}$ is large compared to the difference of the resonance energies, $\omega_A - \omega_B$, the local field

$$\mathbf{E}_{\mathrm{loc}} = \sum_{\mathbf{m} \neq \mathbf{n}} \boldsymbol{\phi}(\mathbf{m} - \mathbf{n}) \mathbf{d}_{\mathbf{m}} \tag{4.98}$$

imposes the coherence of the plane wave on the different sites. The generalized polarizability of the effective crystal is obtained by averaging the small fluctuating term $\boldsymbol{\alpha}^{-1}$:

$$\langle \boldsymbol{\alpha}_e \rangle \sim (\langle \boldsymbol{\alpha}^{-1} \rangle - \boldsymbol{\phi})^{-1} \tag{4.99}$$

In particular, we have only one polariton mode, with an excitonic resonance energy at about $c_A \omega_{eA} + c_B \omega_{eB}$, which is an average of the energies of the pure excitons A and B.

b. *Local Impurity.* This is the most frequently encountered experimental case. We consider it from the general point of view, with the sole restriction that the lattice is not distorted around the impurity. The polarizability tensor of the impurity may differ from that of the host molecules in its resonance frequency, but also in the direction and the magnitude of the transition dipole.

Therefore, consider a crystal with two molecules per cell, equivalent under translational symmetry operations, as in the anthracene crystal. The molecular polarizability tensors of the two host molecules, in positions 1 and 2, for a molecular transition are given the form

$$\boldsymbol{\alpha}_1 = 2\omega_A \frac{\mathbf{d}_1\mathbf{d}_1}{\omega_A^2 - \omega^2}, \qquad \boldsymbol{\alpha}_2 = 2\omega_A \frac{\mathbf{d}_2\mathbf{d}_2}{\omega_A^2 - \omega^2} \tag{4.100}$$

whereas for the B impurity, we have

$$\boldsymbol{\alpha}_0 = 2\omega_B \frac{\mathbf{d}_0\mathbf{d}_0}{\omega_B^2 - \omega^2} \tag{4.101}$$

Assuming a substitutional impurity at position 1 in the cell 0, the generalized tensor (4.69) of the total system becomes

$$\boldsymbol{\alpha}'_e = (\boldsymbol{\alpha}'^{-1} - \boldsymbol{\phi})^{-1} \tag{4.102}$$

with

$$\boldsymbol{\alpha}'^{-1} = \boldsymbol{\alpha}^{-1} - \boldsymbol{\Delta}|01\rangle\langle 01| \tag{4.103}$$

and

$$\boldsymbol{\Delta} = \boldsymbol{\alpha}_1^{-1} - \boldsymbol{\alpha}_0^{-1} \tag{4.104}$$

$\boldsymbol{\alpha}'_e$ is derived straightforwardly from $\boldsymbol{\alpha}_e$ of the pure host crystal following the decomposition (4.44):

$$\boldsymbol{\alpha}'_e = \boldsymbol{\alpha}_e + \boldsymbol{\alpha}_e \mathbf{t} \boldsymbol{\alpha}_e \tag{4.105}$$

where

$$\mathbf{t} = |01\rangle \frac{1}{\boldsymbol{\Delta}^{-1} - \langle 01|\boldsymbol{\alpha}_e|01\rangle} \langle 01| \tag{4.106}$$

In the basis of coulombic excitons which diagonalize $\boldsymbol{\alpha}_e$ (Section I.B.3), we calculate $\langle 01|\boldsymbol{\alpha}_e|01\rangle$ (for molecule 1 in cell 0):

$$\langle 01|\boldsymbol{\alpha}_e|01\rangle = \frac{1}{N}\sum_{e\mathbf{k}} |\langle 1|e\mathbf{k}\rangle|^2 \frac{2\omega_{e\mathbf{k}}\mathbf{d}_{e\mathbf{k}}\mathbf{d}_{e\mathbf{k}}}{\omega_{e\mathbf{k}}^2 - \omega^2} \tag{4.107}$$

$|\langle 1|e\mathbf{k}\rangle|^2$ represents the weight of the exciton $|e\mathbf{k}\rangle$ on molecule 1. According to (4.106), we restrict the matrix elements of the tensor (4.107) to the basis which spans $\mathbf{\Delta}:(\hat{\mathbf{d}}_0, \hat{\mathbf{d}}_1)$; then

$$
\begin{aligned}
\hat{\mathbf{d}}_0\cdot\langle 01|\boldsymbol{\alpha}_e|01\rangle\cdot\hat{\mathbf{d}}_0 &= g_0 \\
\hat{\mathbf{d}}_1\cdot\langle 01|\boldsymbol{\alpha}_e|01\rangle\cdot\hat{\mathbf{d}}_1 &= g_1 \\
\hat{\mathbf{d}}_1\cdot\langle 01|\boldsymbol{\alpha}_e|01\rangle\cdot\hat{\mathbf{d}}_0 = \hat{\mathbf{d}}_0\cdot\langle 01|\boldsymbol{\alpha}_e|01\rangle\cdot\hat{\mathbf{d}}_1 &= g_{01}
\end{aligned}
\tag{4.108}
$$

allow us to calculate $\mathbf{t}$. The g's are complex when ω is inside the exciton band; they are real otherwise. If $\mathbf{t}$ has real poles, we find impurity levels outside the excitonic band (because of relaxation phenomena, only the impurity levels below the excitonic band are eventually observed). The impurity absorption is calculated (see Appendix A) for a direction $\mathbf{E}_i$ of the incident monochromatic field, of wave vector $\mathbf{k}$:

$$I(\omega) \propto \mathrm{Im}\,(\mathbf{E}_i\cdot\langle\mathbf{k}|\boldsymbol{\alpha}'_e|\mathbf{k}\rangle\cdot\mathbf{E}_i) \tag{4.109}$$

or for the absorption related to impurities, by (4.105),

$$I(\omega) \propto \mathrm{Im}\sum_{ee'}\frac{2\omega_{\mathbf{k}e}(\mathbf{E}_i\cdot\mathbf{d}_{\mathbf{k}e})2\omega_{\mathbf{k}e'}(\mathbf{E}_i\cdot\mathbf{d}_{\mathbf{k}e'})}{(\omega_{\mathbf{k}e}^2-\omega^2)(\omega_{\mathbf{k}e'}^2-\omega^2)}\langle\mathbf{k}e|01\rangle\langle 01|\mathbf{k}e'\rangle\mathbf{d}_{\mathbf{k}e}\cdot\mathbf{t}\cdot\mathbf{d}_{\mathbf{k}e'} \tag{4.110}$$

The absorption is observed at the poles of $\mathbf{t}$, at the energy ω_I shifted relative to ω_B (4.101) by the local field. If ω_I is close to an excitonic transition of the host crystal, the absorption resonates strongly for $\omega_I \sim \omega_{\mathbf{k}e}$; in fact, the impurity transition is polarized according to the host excitonic transition in its vicinity.[16] (In other words, the impurity borrows oscillator strength, and more or less spatial extension, from the host band.[184]

c. ATA *and* CPA. After the solution of the isolated-impurity problem, it is easy to extend the result to low concentrations of impurities; as we have done in Section IV.B. Then we obtain the average-t-matrix approximation (ATA) expressed in terms of polarizations: *each B site is treated independently of the others, as an impurity in the pure A crystal.*

In the same way as in Section IV.B.4, the treatment may be given self-consistency by seeking an effective polarizability such that the average t matrix of an A or B site immersed in the effective medium vanishes; one recovers the CPA method. This treatment will not be given in detail, because it is strictly identical to that of Section IV.B.

d. *Approximation of the Average Polarizability*[16]. The CPA is the best one-site approximation. However, it is hard to handle and requires knowledge of the density-of-states distribution of the excitonic band. Simpler approximations may be useful, provided one knows well their defects and limitations. This is true of the average-polarizability approximation we analyze below.

This approximation consists in replacing the real disordered lattice by a crystal, translationally invariant, with molecules of average polarizability $\langle \boldsymbol{\alpha} \rangle$:

$$\langle \boldsymbol{\alpha} \rangle = c_A \boldsymbol{\alpha}_A + c_B \boldsymbol{\alpha}_B \tag{4.111}$$

This amounts to neglecting the modification of the local field by the presence of the impurity A or B:

$$\langle \mathbf{d}_n \rangle = c_A \boldsymbol{\alpha}_A \cdot \mathbf{E}_{\text{loc}}(A) + c_B \boldsymbol{\alpha}_B \cdot \mathbf{E}_{\text{loc}}(B) \sim \langle \boldsymbol{\alpha} \rangle \cdot \mathbf{E}_{\text{loc}} \tag{4.112}$$

In the formalism of Section IV.B, this approximation is applied to the expansion of the locator (4.91) and may be called the average-locator approximation: If we replace in (4.91) all the denominators $(z - V_p -)^{-1}$ by their average, we obtain the expansion for a pure crystal, with resolvent

$$\mathscr{G}_{\text{eff}}(z) = \frac{1}{\langle g \rangle^{-1} - W} \tag{4.113}$$

with

$$\langle g \rangle = \left\langle \frac{1}{z - V_n} \right\rangle = \frac{c_A}{z - V_A} + \frac{c_B}{z - V_B} \tag{4.114}$$

The corresponding self-energy, for the binary alloy ($v_A = 0, v_B = \Delta$), is

$$\Sigma_{\text{loc}\cdot\text{av.}} = z - \langle g \rangle^{-1} = \frac{c_B \Delta}{1 - (1 - c_B)(\Delta / z)} \tag{4.115}$$

to be compared with (4.55):

$$\Sigma_{\text{ATA}} = \frac{c_B \Delta}{1 - (1 - c_B)\Delta \langle 0 | G_0(z) | 0 \rangle} \tag{4.116}$$

The differences with the ATA are the following:

1. The average locator applies to the domain of all concentrations, contrary to the form (4.116) of the ATA, where $\langle 0 | G_0(z) | 0 \rangle$ refers to propagation in the pure A.

2. The average locator leads to a real self-energy, contrary to ATA. For the calculation of the mean scattering, this approximation neglects the bandwidths of A and B around the molecular levels (indeed, for $W \to 0$, $\langle 0 | G_0(z) | 0 \rangle \sim 1/z$). Thus, the average-locator approximation is well adapted to an investigation off resonance (far from the excitonic band) of the properties of a mixed crystal (optical properties, for instance). If the difference in the resonance energies of the two components is large compared to the interactions ($\Delta \gg W$), the average-locator approximation will be valid outside the domain of the pure A and B bands.

The mean-polarizability approximation, discussed in detail by Agranovitch,[16] presents the same advantages (simplicity, arbitrary concentrations, etc.), and the same limitations as the average-locator approximation; in particular, this theory provides two bands of persistence behavior for all values of the parameters. This may be checked on the example of a cubic crystal, where the local field has a very simple form: The modes of the mixed crystal are given by

$$\langle \alpha(z) \rangle = \frac{1}{A} \tag{4.117}$$

where A represents the force of the local field $\mathbf{P}/3\varepsilon_0$. A graphical discussion of this equation is given in Fig. 4.22, and always provides two solutions, even in a strong local field: the one at the energy of the isolated impurity, ω_B [not renormalized by the lattice like the pole of (4.106)] for low B concentrations; the other at the limit of the virtual crystal in very strong fields.

4. *Concluding Remarks on the* 3D *Disordered Lattice*

We may summarize the behavior of dipolar excitons in a mixed crystal, as provided by the CPA method.[156] According to the value of Δ, the mixed crystal will exhibit one or two optical transitions (amalgamation for $\Delta < W$, persistence for $\Delta > W$). In the intermediate case $\Delta \sim W$, the behavior varies with the concentration. For example, one impurity, with its level slightly detached from the excitonic host band, will give, at 50%, a mixed crystal of the amalgamation type. The optical transitions, building up polaritons in the pure crystal, will persist in the presence of mixed-crystal disorder broadening (inhomogeneous broadening). For a single excitonic transition, the transverse dielectric tensor (1.79) transforms in the CPA as follows:

$$\boldsymbol{\varepsilon}^{\perp}(K, \omega) = 1 - \frac{1}{\varepsilon_0 V_0} \frac{\mathbf{d}_e^{\perp} \mathbf{d}_e^{\perp}}{(\omega - \omega_e - \sigma(\omega))} \tag{4.118}$$

(where the resonance approximation is made). The imaginary part of $\sigma(\omega)$

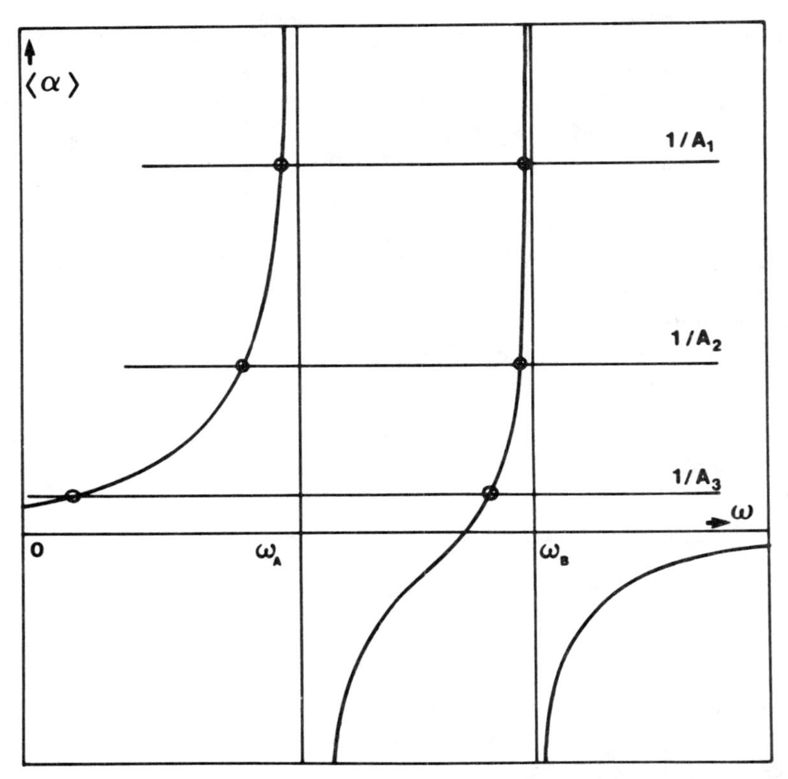

Figure 4.22. Polariton solutions for a 3D mixed crystal in the mean-polarizability approximation (4.117). In strong local field (A_3), one obtains a resonance of the virtual crystal $c_A\omega_A + c_B\omega_B$; another solution, strongly shifted, exists at low frequencies. On the contrary, in weak local fields (A_1), the frequencies of the pure A and B crystals, slightly shifted, are solutions. We note that for $c_B \to 0$, one of the solutions tends, for any strength of the local field A, to ω_B, which is the frequency of B unshifted by the interaction with the lattice A.

reflects the inhomogeneous broadening of the excitonic transition in the vicinity of one solution of the equation:

$$\omega = \omega_e + \mathrm{Re}\,\sigma(\omega) \tag{4.119}$$

We must remark that the broadening is in general nonlorentzian and asymmetric, particularly when the optical transition occurs at the boundary of the excitonic band (as in the case of the anthranene crystal). Lastly, the analyticity of the CPA method assures that (4.118) satisfies the Kramers–Kronig relations, and that the total oscillator strength of the transition, redistributed on the two bands, is conserved.

The final question is whether the CPA approach can be improved. At low or especially intermediate concentrations, an expansion in aggregates could be envisaged (Section IV.B.1.c). However, the density of states must embody a huge number of aggregates,[142] complicated (relative to the triplet lattice) by the long-range dipolar interactions. For these reasons attempts to extend the CPA on small grids are unlikely to succeed. The second aspect of the long-range interactions relates to the optical response at $\mathbf{k} \sim \mathbf{0}$: Interactions between distant aggregates become important, allowing averaging over a large number of configurations. It is possible that the CPA will yield a better approximation for the optical response than for the density of states, because of the long-range dipole–dipole interactions.

These last conclusions are speculative and have to be confronted with:

1. experimental data (such as the optical response of the mixed crystal anthracene–β-methylanthracene);
2. numerical simulations on grids of sufficiently large size to allow the buildup of the optical response of the polariton $\mathbf{k} \sim \mathbf{0}$.

APPENDIXES

A. The Linear Response

Let us consider an isolated system described by the hamiltonian H. The system is assumed in thermodynamical equilibrium at temperature T. Thus, its density operator has the form

$$\rho_0 = \frac{1}{Z} e^{-\beta H} \qquad \left(\beta = \frac{1}{kT}\right) \tag{A.1}$$

We assume further that this system is subjected to a weak external perturbation. This perturbation may be as weak as one wishes; it will be treated semiclassically by a time-dependent hamiltonian (an incident electromagnetic field, for example).

1. *The General Formalism*

Let the perturbation have the form $V(t) = A\psi(t)$, with $\psi(t)$ a scalar time-dependent function and A an operator of the investigated system. The most general observed property is

$$\langle B(t) \rangle = \mathrm{Tr}\,[\rho(t)B] \tag{A.2}$$

in Schrödinger representation, $\rho(t)$ being the density operator of the system at instant t, obeying the evolution equation

$$-i\hbar\dot{\rho}(t) = [\rho(t), H + V(t)] \tag{A.3}$$

In interaction representation (evolution under H alone: $B^I(t) = e^{iHt/\hbar}Be^{-iHt/\hbar}$), restricting to the term linear in ψ, we obtain

$$\langle B(t)\rangle = \frac{1}{i\hbar}\int_{-\infty}^{t} \mathrm{Tr}\, \rho_0[B^I(t), A^I(t')]\psi(t')\, dt' \tag{A.4}$$

which may be written as a convolution integral:

$$\langle B(t)\rangle = \int_{-\infty}^{+\infty} \chi^r_{BA}(t-t')\psi(t')\, dt' \tag{A.5}$$

with the introduction of the generalized susceptibility

$$\chi^r_{BA}(t-t') = \Theta(t-t')\frac{1}{i\hbar}\mathrm{Tr}\,\{\rho_0[B^I(t-t'), A]\} \tag{A.6}$$

Upon Fourier transformation, the two expressions become

$$\langle \tilde{B}(\omega)\rangle = \tilde{\chi}_{BA}(\omega)\tilde{\psi}(\omega) \tag{A.7}$$

$$\tilde{\chi}_{BA}(\omega) = \frac{1}{i\hbar}\int_0^{\infty} e^{i\omega\tau}\,\mathrm{Tr}\,\{\rho_0[B(\tau), A]\}\, d\tau \tag{A.8}$$

$\tilde{\chi}_{BA}(\omega)$ is a complex scalar characterizing the magnitude and the phase of the response of B to the A excitation, at each temperature T and frequency ω. The phase of $\tilde{\chi}_{BA}$ (or its imaginary part) will determine the energy absorption by the system.

In an expansion in the eigenstates of H, $\{|n\rangle\}$, assumed to be known,

$$H|n\rangle = E_n|n\rangle = \hbar\omega_n|n\rangle \tag{A.9}$$

the expression (A.8) transforms to

$$\tilde{\chi}_{BA}(\omega) = \sum_{nm} p_n\left(\frac{B_{nm}A_{mn}}{\hbar\omega - E_{mn} + i0} - \frac{A_{nm}B_{mn}}{\hbar\omega + E_{mn} + i0}\right) \tag{A.10}$$

with $E_{mn} = E_m - E_n$. Furthermore, introducing the resolvent of the operator H,

$$G(z) = (z - H)^{-1} \quad \text{and} \quad z = \hbar\omega + i0 \tag{A.11}$$

we have

$$\tilde{\chi}_{BA}(\omega) = \sum_n p_n[\langle n|BG(E_n + z)A|n\rangle + \langle n|AG(E_n - z)B|n\rangle] \tag{A.12}$$

Equations (A.11) and (A.12) bear the following interpretation: Starting from each state $|n\rangle$ with probability p_n, the system passes, under the perturbation A, to the other states, whose evolution is described by the resolvent G, at the "available" energy $E_n \pm z$ (when the perturbation contains the two frequencies $\pm\omega$); subsequently the system returns to state n by the observation B. An analogous term arises from first observing propagation at $E_n \mp z$, and returning to n by the A perturbation. It is thus clear that A and B play conjugate roles in these processes, and this justifies the notation χ_{BA}. (Only the terms in $E_n + z$ are resonant at the optical frequencies and the usual temperatures.)

The form (A.12) shows that any linear response of the system may be calculated from the resolvent $G(z)$ of the unperturbed hamiltonian H—the energy levels (thermally populated) being assumed known. In the case with only electronic excitations involved ($E_n - E_0 \gg kT$), only the ground state need be considered in the summation on n, and $G(z)$ will provide all the responses of the system (cf. Section IV).

2. *The Optical Absorption*

The optical absorption is defined as follows[135]: it is the decay rate, per unit time, of the density of the pure electromagnetic energy of a plane wave inside the matter medium, and it is equal to $\omega\varepsilon''(\omega)$ [we assume the field directed along an eigendirection of $\varepsilon(\mathbf{K}, \omega)$, so that ε'' will be considered as a scalar quantity]. Actually, this definition is that of the conductivity; hence the name *optical conductivity* given to $\omega\varepsilon(\omega)$. Therefore, we may calculate $\varepsilon(\mathbf{K}, \omega)$, and hence the optical conductivity, from the relations (1.68) in Section I.

Other equivalent forms may be given for the absorption, derived from generalized susceptibilities. Let us consider the response of the medium to the external perturbation $-\mathscr{d}\cdot\mathscr{E}_{\text{ext}}$ (Section I.A.4), $\mathscr{E}$ being of the form $\mathbf{E}_0 e^{iQr}\cos\omega t$. Then the absorption energy, given by the response of $-\mathbf{d}\cdot\mathbf{E}$ to $-\mathbf{d}\cdot\mathbf{E}$, is the following:

$$\frac{\delta W}{dt} = \frac{\omega}{2}\,\mathrm{Im}\,\chi_{-\mathbf{d}\cdot\mathbf{E},\,-\mathbf{d}\cdot\mathbf{E}}(\omega) \tag{A.13}$$

This relation shows that the optical absorption at frequency ω is proportional to the Fourier transform of the autocorrelation function of the dipole $\mathbf{d}$:

$$\mathscr{A}(\omega) \propto \frac{1}{i\hbar} \int_{-\infty}^{+\infty} e^{i\omega\tau} \langle [d(\tau), d(0)] \rangle \, d\tau \tag{A.14}$$

[we have used (A.8), $\langle \cdots \rangle$ representing the thermal average and $\mathbf{d}$ the "useful" component of the electric dipole operator].

In the specific case of optical transitions in a molecular crystal, the transition dipole takes the form $\mathbf{d}_{nf}(B_{nf} + B_{nf}^{\dagger})$ (1.10). Starting from thermally populated states, only the terms in $B_{nf}^{\dagger}$ will resonate during the excitation (the terms in B_{nf} will be active during the transition back to the initial state). Thus, the optical absorption takes the well-known form

$$\mathscr{A}(\omega) \propto \mathrm{Re} \int_{0}^{\infty} d\tau \, e^{i\omega\tau} \langle B_{\mathbf{K}}(\tau) B_{\mathbf{K}}^{\dagger}(0) \rangle \tag{A.15}$$

which embodies the excitonic Green's function (creation of an exciton at $t = 0$, propagation until $t = \tau$, and annihilation).

B. The Resolvent

Under certain conditions, the calculation of the optical response amounts to that of matrix elements of the resolvent G; see (A.12). We illustrate this in two examples where $G(z)$ simplifies and allows us to derive analytical expressions.

1. *Elimination of Certain Degrees of Freedom*

We assume that the space on which H operates is a direct sum of two subspaces characterized by orthogonal projectors P and Q ($P + Q = 1$, $PQ = QP = 0$). If we are interested in propagations in the subspace P alone, it is possible to eliminate quite formally propagations in the subspace Q, for each energy z at which the system is interrogated. Putting

$$G(z) = (z - H)^{-1} \quad \text{and} \quad G_0(z) = (z - PHP - QHQ)^{-1}$$

we obtain the relation

$$G = G_0 + G_0(PHQ + QHP)G \tag{A.16}$$

By projecting (A.16) into the subspace P, and using relations between P and Q, we eliminate the unknown operators, QGQ, PGQ, and QGP, to arrive at

$$PG(z)P = \frac{P}{z - PHP - PHQ[1/(z - QHQ)]QHP} \tag{A.17}$$

The formal solution of (A.17) requires, in each case, the inversion of $(Qz - QHQ)^{-1}$, very often as complicated as that of $z - H$. However, when QHQ inverts easily, (A.17) provides the desired solution directly. Another important advantage of (A.17) is the possibility of a perturbative expansion when PHQ is small relative to the eigenterms of each subspace PHP and QHQ: Then the third term of (A.17) is the self-energy associated with the propagation in P.

2. *A Perturbation Acting Inside One Subspace*

When H has the form $H = H_0 + V$, where H_0 is known and V operates inside the subspace of the projector P, the resolvent $G(z)$ is also easily calculated. Indeed, we have

$$PVP = V \tag{A.18}$$

$$G_0(z) = (z - H_0)^{-1} \tag{A.19}$$

The expansion of G in series of powers of V, using (A.18), provides

$$G = G_0 + G_0 PTPG_0 \tag{A.20}$$

where

$$PT(z)P = PVP + PVPG_0PVP + \cdots = PVP\frac{1}{1 - PG_0PPVP} \tag{A.21}$$

is the t matrix associated with the perturbation V.

The decomposition (A.20) is a particular case of the relations obtained in the problem of particle scattering by a potential (of the Lippmann–Schwinger type). It turns out to be particularly useful for the calculation of the states of impurities, or of impurity aggregates, in a perfect lattice whose propagator G_0 is known.

C. Unitary Transformation

A unitary transformation modifies the eigenvectors of a hamiltonian without changing its spectrum of eigenvalues. Thus, it is possible to eliminate certain terms in H by changing the basis set, and sometimes even diagonalize H completely. To apply a unitary transformation, it is possible either to transform the entire hamiltonian, or to express the initial hamiltonian as a function of state vectors or of transformed operators according to the relations

$$|\lambda\rangle \rightarrow |\tilde{\lambda}\rangle = E^S|\lambda\rangle$$

$$V \rightarrow \tilde{V} = E^S V E^{-S} \tag{A.22}$$

The general expression for the transform $\tilde{V}$ of an operator V, as a function of V and S, is given by an identity relation we shall demonstrate. Let us consider the superoperators $\mathscr{L}_S$ acting on the operators of the Hilbert space, and defined as follows:

$$\mathscr{L}_S V = [V, S] \tag{A.23}$$

or equivalently

$$(S + \mathscr{L}_S)V = VS \tag{A.24}$$

As S (understood as a super operator) commutes with $\mathscr{L}_S$ (because $[SV, S] = S[V, S]$), we have (with an arbitrary function f)

$$Vf(S) = f(S + \mathscr{L}_S)V \tag{A.25}$$

from which we derive in a straightforward manner

$$\tilde{V} = e^S V e^{-S} = e^{-\mathscr{L}_S} V = V + [V, -S] + \frac{1}{2!}[[V, -S], -S] + \cdots \tag{A.26}$$

When the relation (A.26) is truncated, the method of unitary transformation reduces to that of the perturbations. It is sometimes possible to resum (A.26) exactly: This is the case when we apply the transformation (2.35) to the hamiltonian (2.15) containing the linear and quadratic excitation–vibration couplings. The use of (A.26) allows us to obtain the relations (2.38)–(2.39).

The decomposition (2.42) of one exponential into a product of exponentials is a generalization of the formula due to Weyl and Glauber.[187] One finds the expressions for f, g, h, ψ by differentiating the two members of (2.42) with respect to τ and using (A.26) to eliminate the exponentials. One finds a system of differential equations in the variable τ, whose solution satisfying the required limit conditions is (2.43).

Acknowledgements

We are very much indebted to Professor J. M. Turlet and Dr. J. Bernard for helpful discussions and for providing us with material included in this work. Many stimulating private communications with Dr. M. R. Philpott are acknowledged. Dr. R. Brown has performed very extensive numerical simulations to help explore various extensions of the mean-field theory. One of us (Ph. K.) acknowledges stimulating discussion and correspondence with Professors S. A. Rice and V. M. Agranovitch on the distinction between our monolayer elementary excitations and the usual surface polariton modes investigated by those authors.

References

1. J. M. Turlet, P. Kottis, and M. R. Philpott *Adv. Chem. Phys.*, **54**, 303–468 (1983), and references therein on surface-state techniques and surface states.

2. J. Frenkel, *Phys. Rev.* **37**, 17 (1931).
3. R. Peierls, *Ann. Phys.* **13**, 905 (1932).
4. G. H. Wannier, *Phys. Rev.* **52**, 191 (1937).
5. A. S. Davydov, *Theory of Molecular Excitons*, McGraw-Hill, New York, 1962.
6. D. S. McClure, *Solid State Phys.* **8**, 1 (1958).
7. H. C. Wolf, *Solid State Phys.* **9**, 1 (1959).
8. R. M. Hochstrasser, *Rev. Modern Phys.* **34**, 531 (1962).
9. R. M. Hochstrasser Ed., *Molecular Aspects of Symmetry*, Benjamin, New York, 1966.
10. S. A. Rice, in *The Triplet State*, A. B. Zahlan, Ed., Cambridge U. P. 1967, pp. 265–310.
11. S. A. Rice and J. Jortner, in *Physics and Chemistry of the Organic Solid State*, vol. III, 1967, p. 199.
12. A. S. Davydov, *Theory of Molecular Excitons*, Plenum, New York, 1971.
13. A. S. Davydov, *Théorie du Solide*, Editions MIR, Moscow, 1983.
14. J. C. Slater, *Quantum Theory of Molecules and Solids*, Vol. 3, McGraw-Hill, New York, 1967.
15. M. R. Philpott, *J. Chem. Phys.* **52**, 5842 (1970) and references therein.
16. V. M. Agranovitch, *Sov. Phys. Usp.* **17**, No. (1), 103 (1974).
17. D. P. Craig and S. H. Walmsley, *Excitons in Molecular Crystals*, Benjamin, New York, 1968.
18. A. S. Davydov and E. F. Sheka, *Phys. Stat. Sol.* **11**, 877 (1965).
19. G. D. Mahan, *J. Chem. Phys.* **43**, 1569 (1965).
20. M. R. Philpott, *J. Chem. Phys.* **50**, 5117 (1969).
21. S. V. Tyablikov, *Methods in the Quantum Theory of Magnetism*, Plenum, New York, 1967.
22. V. M. Agranovitch, *Sov. Phys. JETP* **37**, 307 (1960).
23. See for example ref. 14, p. 97 ff.
24. C. Cohen-Tannoudji, B. Diu, and F. Laloe, *Mécanique Quantique*, Hermann, Paris, 1973.
25. C. Cohen-Tannoudji, Cours au Collège de France, 1974–1975 (unpublished).
26. C. Cohen-Tannoudji, Cours au Collège de France, 1978–1979 (unpublished).
27. D. W. Schlosser and M. R. Philpott, *J. Chem. Phys.* **77**, 1969 (1982).
28. M. R. Philpott and J. W. Lee, *J. Chem. Phys.* **58**, 595 (1973).
29. M. R. Philpott, *J. Chem. Phys.* **58**, 588 (1973).
30. See the historical part of ref. 14.
31. See, for example, J. D. Jackson, *Classical Electrodynamics*, Wiley, New York, 1975, p. 223.
32. K. Syassen and M. R. Philpott, *J. Chem. Phys.* **68**, 4870 (1978).
33. J. J. Hopfield, *Phys. Rev.* **112**, 1555 (1958). The notion of polariton was first developed in classical electrodynamics by M. Born and K. Huang, *Dynamical Theory of Crystal Lattices*, Oxford U.P., Oxford, 1956.
34. S. I. Pekar, *Sov. Phys. JETP* **6**, 785 (1958).
35. V. M. Agranovitch, in *Optical Properties of Solids*, F. Abeles, Ed., North Holland, 1972, Chapter 6; see also ref. 13, Chapter XI.
36. D. W. Schlosser and M. R. Philpott, *Chem. Phys.* **49**, 181 (1980).
37. A. Honma, *J. Phys. Soc. Jap.* **42**, 1129 (1977).
38. P. W. Anderson, *Phys. Rev.* **109**, 1492 (1958).
39. J. Klafter and J. Jortner, *J. Chem. Phys.* **71**, 1961 (1979).

40. D. Ricard, "Optique non linéaire et surfaces", *Ann. Phys. Fr.* **8**, 273–310 (Masson, Paris, 1983) and all references therein; Y. R. Shen, *The Principles of Nonlinear Optics*, Wiley, New York, 1984, Chapter 25; J. M. Hicks, K. Kemnitz, K. B. Eisenthal, and T. F. Heinz, *J. Phys. Chem.* **90**, 560 (1986); K. Kemnitz, K. Bhattacharyya, J. M. Hicks, G. R. Pinto, K. B. Eisenghal, and T. F. Heinz, *Chem. Phys. Lett.* **131**, 258 (1986) and references therein; F. Hache, D. Ricard, and C. Flytzanis, *J. Opt. Soc. Am. B* **3**, 1647 (1986) and references therein.
41. J. Bernard, M. Hadad, and Ph. Kottis, *Chem. Phys. Lett.* **133**, 73 (1987); *Chem. Phys.*, 118 (1987).
42. M. Ghelfenstein, Thèse d'Etat, Université Paris Sud-Orsay, 1973; D. Ceccaldi, M. Ghelfenstein, and H. Schwarc, *J. Phys. C*, **7** L155 1974; **8** 417 (1975).
43. F. Z. Khelladi, *Chem. Phys. Lett.* **34**, No. 3, 490 (1975).
44. R. M. MacNab and K. Sauer, *J. Chem. Phys.* **53**, 7 2805 (1970).
45. S. L. Chaplot, N. Lehner, and G. S. Pawley, *Acta Cryst. B* **38**, 483 (1982).
46. S. L. Chaplot, G. S. Pawley, B. Dorner, V. K. Jindal, J. Kalus, and I. Natkaniec, *Phys. Stat. Sol.* (*b*) **110**, 445 (1982).
47. M. Suzuki, T. Yokoyama, and M. Ito, *Spectrochim. Acta* **24A**, 1091 (1968).
48. See ref. 13, p. 372 ff.
49. D. P. Craig and L. A. Dissado, *Chem. Phys. Lett.* **44**, No. 3, 419 (1976).
50. H. Sumi, *J. Phys. Soc. Jap.* **36**, 770 (1974); **36**, 825 (1974).
51. A. Matsui, Xth Molecular Crystal Symposium, Québec, Canada, 1982.
52. H. Sumi and Y. Toyozawa, *J. Phys. Soc. Jap.* **31**, 342 (1971).
53. M. Schreiber and Y. Toyozawa, *J. Phys. Soc. Jap.* **51**, 1528, 1537, 1544 (1982).
54. E. I. Rashba, *Sov. Phys. JETP*, **23**, 708 (1966); **27**, 292 (1968).
55. M. R. Philpott, *J. Chem. Phys.* **55**, 2039 (1971).
56. B. Dagorret, M. Orrit, and Ph. Kottis, *J. Phys. Fr.* **46**, 365 (1985); B. Dagorret, Thèse de 3e Cycle, Université de Bordeaux I, 1984.
57. M. R. Philpott, *J. Chem. Phys.* **47**, 2534, 4437 (1967).
58. E. F. Sheka, *Sov. Phys. Usp.* **14**, 484 (1972).
59. D. W. Schlosser and M. R. Philpott, *J. Chem. Phys.* **77**, 1969 (1982) and references to previous work.
60. M. R. Philpott and J. M. Turlet, *J. Chem. Phys.* **64**, 3852 (1976).
61. M. Orrit, J. Bernard, J. M. Turlet, and Ph. Kottis, *J. Chem. Phys.* **78**, 2847 (1983); J. Bernard, M. Orrit, J. M. Turlet, and Ph. Kottis, *J. Chem. Phys.* **78**, 2857 (1983).
62. M. G. Sceats, Ph.D. Thesis, University of Queensland, Australia, 1973; G. C. Morris and M. G. Sceats, *Chem. Phys.* **3**, 164 (1974).
63. M. R. Philpott, *J. Chem. Phys.* **50**, 5117 (1969).
64. C. Cohen-Tannoudji, Cours au Collège de France 1977–1978 (unpublished).
65. Y. Toyozawa, *Prog. Theor. Phys.* **27**, 89 (1962).
66. K. Cho and Y. Toyozawa, *J. Phys. Soc. Jap.* **30**, 1555 (1971).
67. J. M. Turlet, Thèse d'Etat, Université de Bordeaux I, 1979.
68. G. C. Morris, S. A. Rice, and A. E. Martin, *J. Chem. Phys.* **52**, 5149 (1970).
69. L. Landau and E. Lifschitz, *Physique Statistique*, Editions Mir, Moscow, 1966 § 122.
70. J. M. Sajer, Thèse de 3e Cycle, Université de Bordeaux I, 1984.

71. J. Bernard, Thèse d'Etat, Université de Bordeaux I, 1984.
72. A. Matsui, *J. Phys. Soc. Jap.* **21**, 2212 (1966).
73. H. C. Wolf, *Z. Naturforsch.* **13a**, 414 (1959).
74. J. Ferguson, *Z. Phys. Chem. Neue Folge* **101**, 45 (1976).
75. G. S. Pawley, *Phys. Stat. Sol.* **20**, 347 (1967).
76. L. A. Dissado, *Chem. Phys.* **8**, 289 (1975).
77. E. Miyazaki and E. Hanamura, in *Relaxation of Elementary Excitations*, R. Kubo and E. Hanamura, Eds., Solid State Sciences, 18, Springer, 1980.
78. M. Orrit, Thèse de 3e Cycle, Université de Bordeaux I, 1978.
79. M. Orrit, J. Bernard, J. M. Turlet, and Ph. Kottis, *Chem. Phys. Lett.* **95**, 315 (1983).
80. A. Kastler and A. Rousset, *J. Phys. Radium* **2**, 49 (1941).
81. M. T. Shpak and N. I. Sheremet, *Opt. and Spectrosc.* **17**, 374 (1964).
82. E. Glockner and H. C. Wolf, *Z. Naturforsch.* **24a**, 943 (1969).
83. L. E. Lyons and L. J. Warren, *Aust. J. Chem.* **25**, 1411 (1972).
84. M. S. Brodin, M. A. Dudinskii, S. V. Marisova, and E. N. Myasnikov, *Phys. Stat. Sol.* (*b*) **74**, 453 (1976).
85. M. D. Galanin, S. D. Khan-Magometova, and E. N. Myasnikov, *Mol. Cryst. Liq. Cryst.* **57**, 119 (1980).
86. J. Aaviksoo, G. Liidja, and P. Saari, *Phys. Stat. Sol.* (*b*) **110**, 69 (1982).
87. J. Aaviksoo, P. Saari, and T. Tamm, in *Proc. of the 2nd Int. Symp. Ultrafast Phenomena in Spectroscopy*, Reinhardsbrunn, G. D. R., 1980, Vol. 2, p. 479; J. Aaviksoo, A. Freiberg, T. Reinot, and S. Savikin, *J. Lumin.* **35**, 267 (1980).
88. O. S. Avanesjan et al., *Mol. Cryst. Liq. Cryst.* **29**, 165 (1974).
89. N. Karl, *J. Lumin.* **12/13**, 851 (1976).
90. H. Sumi, *J. Phys. Soc. Jap.* **41**, 526 (1976).
91. J. M. Turlet and M. R. Philpott, private communication; see also ref. 1.
92. M. Orrit, Thesis, Université de Bordeaux I, 1984.
93. E. A. Silinsh, in *Solid State Sciences, Vol. 16, Organic Molecular Crystals*, M. Carbone, P. Fulde, and H. J. Queisser, Eds., Springer, Berlin, 1980 Chapter 3.
94. H. C. Wolf, *Z. Naturforsch.* **13a**, 414 (1959); J. Ferguson, *Z. Phys. Chem. N.F.* **101**, 45 (1976).
95. S. V. Marisova, E. N. Myasnikov, and A. N. Lipovchenko, *Phys. Stat. Sol.* **115b**, 649 (1983).
96. G. C. Morris and M. G. Sceats, *Chem. Phys.* **3**, 342 (1974).
97. R. M. Hochstrasser and P. N. Prasad, *J. Chem. Phys.* **56**, 2814 (1972).
98. Y. Toyozawa, in *Solid State Sciences, Vol. 18: Relaxation of Elementary Excitations*, R. Kubo and E. Hanomura, Eds., Springer, Berlin, 1980, p. 3.
99. D. Yarkony and R. Silbey, *J. Chem. Phys.* **63**, 1042 (1976).
100. A. S. Davydov and E. N. Myasnikov, *Phys. Stat. Sol.* **20**, 153 (1967).
101. D. W. Schlosser and M. R. Philpott, *J. Chem. Phys.* **77**, 1969 (1982).
102. C. Aslangul and Ph. Kottis, in *Advances in Chemical Physics*, I. Prigogine and S. A. Rice, Eds., Vol. XLI, 1980, pp. 321–476.
103. M. Schreiber and Y. Toyozawa, *J. Phys. Soc. Jap.* **51**, 1528 (1982).
104. D. Lee and A. C. Albrecht, in *Advances in Infrared and Raman spectroscopy* Vol. 12, R. J. H. Clark and R. E. Hester, Eds., Wiley-Heyden, New York, 1985, Chapter 4; J. S. Melinger and A. C. Albrecht, *J. Chem. Phys.* **84**, 1247 (1986) and all references therein.

105. J. Hager, Ch. Flytzanis, and H. Walther, in *Proc. 8th International Conference in Laser Spectroscopy*, R. Samberg, Ed., Springer 1987; H. Vach, J. Hager, C. Flytzanis, and H. Walther, *ibid.*

106. *Topics in Surface Chemistry*, Plenum, New York, 1978; *Surface Excitations*, V. M. Agranovitch and R. Loudon, Eds., Elsevier, New York, 1984; *Organic Phototransformations in Nonhomogeneous Media*, M. A. Fox, Ed., ACS Symp. Ser., No. 278, *Am. Chem. Soc.*, Washington, 1985; H. Vach, J. Hager, B. Simon, C. Flytzanis, and H. Walther, "Dynamics of Elementary Processes at Surfaces", in *Mat. Res. Soc. Symp.*, Vol. 51, H. Kurz, G. L. Olson, and J. M. Paate, Eds., 1988.

107. L. E. Firment and G. A. Somorjai, *J. Chem. Phys.* **63**, 1037 (1975).

108. International Conference on Ellipsometry, Paris, June 1983.

109. L. A. Nafie and D. W. Vidrine, in *Fourier Transform Infrared Spectroscopy*, Vol. 3, J. R. Ferraro and L. J. Basile, Eds., Academic, New York, 1982.

110. G. Binnig and H. Rohrer, *Helv. Phys. Acta* **55**, 726 (1982); G. Binnig, H. Rohrer, Ch. Gerber, and E. Weibel, *Phys. Rev. Lett.* **50**, 120 (1983).

111. M. Orrit, D. Möbius, U. Lehmann, and H. Meyer, *J. Chem. Phys.* **85**, 4966 (1986).

112. V. I. Sugakov, *Ukr. Fiz. Zh.* **14**, 1428 (1970); V. I. Sugakov and V. N. Tovstenko, *Ukr. Fiz. Zh.* **18**, 1495 (1973).

113. M. S. Brodin, M. A. Dudinski, and S. V. Marisova, *Opt. Spectrosc.* **31**, 401 (1971).

114. J. M. Turlet and M. R. Philpott, *J. Chem. Phys.* **62**, 4260 (1975); M. R. Philpott and J. M. Turlet, *J. Chem. Phys.* **64**, 3852 (1976) and references therein.

115. M. G. Sceats, K. Tomioka, and S. A. Rice, *J. Chem. Phys.* **66**, 4487 (1977).

116. K. Tomioka, M. G. Sceats, and S. A. Rice, *J. Chem. Phys.* **66**, 2984 (1977).

117. J. M. Turlet, J. Bernard, and Ph. Kottis, *Chem. Phys. Lett.* **59**, 506 (1978).

118. M. Orrit and Ph. Kottis, *Phys. Rev. B* **34**, 680 (1986).

119. M. Orrit, J. Bernard, J. M. Turlet, Ph. Kottis, and D. Möbius, in *Dynamics of Molecular Crystals*, J. Lascombe, Ed., Elsevier, 1987, p. 379.

120. J. Bernard, Thesis, University of Bordeaux I, 1984.

121. M. Orrit, Thesis, University of Bordeaux I, 1984.

122. J. M. Ziman, *Models of Disorder* Cambridge U. P., Cambridge, 1979, Chapters 8, 9; R. J. Elliot, J. A. Krumhansl, and P. L. Leath, *Rev. Mod. Phys.* **46**, 465 (1964).

123. C. Cohen-Tannoudji, Lectures, Collège de France, 1975–76 (unpublished).

124. V. M. Agranovitch and D. A. Dubovski, *JETP Lett.* **3**, 323 (1966); M. R. Philpott and P. G. Sherman, *Phys. Rev. B* **12**, 5181 (1975).

125. M. Orrit, Thèse de 3e Cycle, Universitè de Bordeaux I, 1978.

126. M. Orrit, C. Aslangul, and Ph. Kottis, *Phys. Rev. B* **25**, 7263 (1982).

127. J. M. Sajer, M. Orrit and Ph. Kottis, *Chem. Phys.* **94**, 415 (1985).

128. See discussion on the results of M. G. Sceats, in this chapter, Section II.

129. Y. Tokura, T. Koda, and I. Nakada, *J. Phys. Soc. Jap.* **47**, 1936 (1979).

130. M. Hadad, Thèse de 3e Cycle, Université de Bordeaux I, 1985.

131. M. Orrit, J. Bernard, J. M. Turlet and Ph. Kottis, *Chem. Phys. Lett.* **95**, 315 (1983).

132. *La Recherche*, December 1983, p. 1563.

133. G. Filippini, C. M. Gramaccioli, and M. Simonetta, *J. Chem. Phys.* **71**, 89 (1979).

134. E. A. Silinsh, in *Solid State Sciences, Vol. 16, Organic Molecular Crystals*, M. Cardona, P. Fulde, and H.-J. Queissen, Eds., Springer, 1980.

135. L. Landau and E. Lifschitz, *Théorie de l'élasticité*, Editions Mir, Moscow, 1966 pp. 67–68; Moscow; *Electrodynamique des Milieux Continus*, Editions Mir, Moscow, 1966.

136. S. A. Elnahwy, M. El Hamamsy, and A. C. Damask, *Phys. Rev. B* **19**, 1108 (1979).

137. S. L. Chaplot, G. S. Pawley, B. Dorner, V. K. Jindal, J. Kalus, and I. Natkaniec, *Phys. Stat. Sol.* (*b*) **110**, 445 (1982).

138. M. Nicol, M. Vernon and J. T. Woo, *J. Chem. Phys.* **63**, 1992 (1975).

139. R. Sonnenschein, K. Syassen, and A. Otto, *J. Chem. Phys.* **74**, 4315 (1981); J. Kalinowski and R. Jankowiac, Xth Molecular Crystal Symposium, Québec, 1982.

140. M. Orrit, J. Bernard, J. Gernet, J. M. Turlet, and Ph. Kottis, in *Recent Developments in Condensed Matter Physics*, Vol. 2, J. T. De Vreese, L. F. Lemmens, V. E. Van Doren, and J. Van Royen Eds., Plenum, New York 1981.

141. L. David, Thesis, University of Bordeaux I, 1988; L. David and M. Orrit, in preparation.

142. K. E. Mauser, H. Port, and H. C. Wolf, *Chem. Phys.* **1**, 74 (1973); H. Port, D. Vogel, and H. C. Wolf, *Chem. Phys. Lett.* **34**, 23 (1975); F. Dupuy, Ph. Pee, R. Lalanne, J. P. Lemaistre, C. Vaucamps, H. Port, and Ph. Kottis, *Mol. Phys.* **35**, 595 (1978); Ph. Pee, J. P. Lemaistre, F. Dupuy, R. Brown, J. Megel, and Ph. Kottis, *Chem. Phys.* **64**, 389 (1982); F. Dupuy, Y. Rebiere, J. L. Garitey, Ph. Pee, R. Brown, and Ph. Kottis, *Chem. Phys.* **110**, 195 (1986); R. Brown, J. L. Garitey, F. Dupuy, and Ph. Pee, *J. Phys. C* (in press).

143. D. N. Hanson and G. W. Robinson, *J. Chem. Phys.* **43**, 474 (1965); J. Hoshen and J. Jortner, *J. Chem. Phys.* **53**, 933; **56**, 5550 (1972); R. Kopelman, E. M. Monberg, and F. W. Ochs, *J. Chem. Phys.* **19**, 143 (1977); R. Brown, J. P. Lemaistre, J. Megel, Ph. Pee, F. Dupuy, and Ph. Kottis, *J. Chem. Phys.* **76**, 5719 (1982); Ph. Pee, Y. Rebiere, F. Dupuy, R. Brown, Ph. Kottis, and J. P. Lemaistre, *J. Phys. Chem.* **88**, 956 (1984).

144. R. H. Dicke, *Phys. Rev.* **93**, 99 (1954).

145. See, for instance, M. Lax, in *Multiple Scattering and Waves in Random Media*, P. L. Chow, W. E. Kohler, and G. C. Papanicolaou, Eds., North Holland, 1981, and references therein; J. Klafter and M. S. Schlesinger, *Proc. Nat. Acad. Sci. U.S.A.* **83**, 848 (February 1986), and references therein; R. Brown, Thesis, University of Bordeaux I, 1987; R. Brown et al. *J. Phys. C* **20**, L649 (1987); **21** (1988) in press. (This last work thoroughly discusses the applicability of fractal theory to isotopically mixed crystals as disordered system. Serious criticism is presented both of the analysis of the experimental data and of the fundementals of their description in the present theory of fractals. For this reason we omit all works treated there.

146. J. Aviksoo, Private communication.

147. H. P. Dorn, Dissertation, University of Stuttgart, 1985; H. P. Dorn and A. Müller, *Chem. Phys. Lett.* **130**, 426 (1986).

148. V. M. Agranovitch, in *Surface Excitations*, V. M. Agranovitch and R. Loudon, Eds., Elsevier, New York, 1984, p. 551.

149. (a) G. Scheibe, Angew. *Chem.* **50**, 212 (1937); E. E. Jelley, *Nature* **10**, 631 (1937); (b) H. Kuhn, D. Möbius, and H. Bücher, in *Physical Methods of Chemistry*, Weissberger and B. Rossiter, Eds., Wiley, New York, 1972, Vol. 1, Part 3B, p. 589.

150. D. Möbius, and H. Kuhn, *Isr. J. Chem.* **18**, 375 (1979).

151. See Ref. 126 Part I, Figs. 6–8.

152. V. M. Agranovitch, *Usp. Fiz. Nank.* **112**, 143 (1974) [*Sov. Phys.—Usp.* **17**, 103 (1975)].

153. M. T. Raiford, *Phys. Rev. A* **9**, 1257 (1974); K.C. Liu and T. F. George, *Phys. Rev. B* **32**, 3622 (1985).

154. R. Florian, L. O. Schwan, and D. Schmid, *Phys. Rev. A* **29**, 2707 (1984).

155. C. Kittel, *Introduction to Solid State Physics* (3rd ed.), Wiley, 1966, Chapter 9.
156. R. J. Elliot, J. A. Krumhansl, and P. L. Leath, *Rev. Mod. Phys.* **46**, 465 (1974).
157. F. Ducastelle, *J. Phys. F: Metal Phys.* **2**, 468 (1972).
158. F. Ducastelle, *J. Phys. C: Solid State Phys.* **7**, 1795 (1974).
159. Y. Onodera and Y. Toyozawa, *J. Phys. Soc. Jap.* **24**, 341 (1968).
160. L. Schwartz and E. Siggia, *Phys. Rev. B* **5**, 383 (1972).
161. F. Ducastelle, *J. Phys. C: Solid State Phys.* **4**, 175 (1971).
162. S. Kirpatrick, B. Velicky, and H. Ehrenreich, *Phys. Rev. B* **1**, 3250 (1970).
163. R. Brown, Thèse de 3e Cycle, Université de Bordeaux I, 1982.
164. Ph. Pee, Thesis, University of Bordeaux I, 1982.
165. M. Beguery, Thèse de 3e Cycle, Université de Bordeaux I, 1983.
166. J. P. Lemaistre and Ph. Kottis, *J. Chem. Phys.* **68**, 2730 (1978); **76**, 872 (1982).
167. J. Hoshen and J. Jortner, *J. Chem. Phys.* **56**, 4138 (1972).
168. P. Dean and M. D. Bacon, *Proc. Roy. Soc. London* **A283**, 64 (1964).
169. M. Orrit and R. Brown, *J. Phys. C* **18**, 5585 (1985).
170. J. P. Lemaistre, Ph. Pee, F. Dupuy, R. Brown, and Ph. Kottis, *Chem. Phys. Lett.* **89**, 207 (1982).
171. J. Hoshen and J. Jortner, *J. Chem. Phys.* **56**, 933 (1972).
172. J. Hoshen and J. Jortner, *J. Chem. Phys.* **56**, 5550 (1972).
173. Ph. Pee, J. P. Lemaistre, F. Dupuy, R. Brown, and Ph. Kottis, *Chem. Phys.* **64**, 389 (1982).
174. M. Tsukada, *J. Phys. Soc. Jap.* **32**, 1475 (1972).
175. N. H. Butler, *Phys. Rev. B* **8**, 4499 (1973).
176. S. Kirkpatrick, *Rev. Mod. Phys.* **45**, 574 (1973).
177. D. Stauffer, *Phys. Rep.* **54**, 1 (1979).
178. B. P. Watson and P. L. Leath, *Phys. Rev. B* **9**, 4893 (1974).
179. S. Kirkpatrick and T. P. Eggarter, *Phys. Rev. B* **6**, 3598 (1972).
180. F. Yonezawa and T. Odagaki, *Solid. State Comm.* **27**, 1199, 1203, 1207 (1978).
181. J. van der Rest, P. Lambin, and F. Brouers, in *Recent Developments in Condensed Matter Physics*, Vol. 2, J. T. Devreese, L. F. Lommens, V. F. van Doren, and J. van Royen, Eds., Plenum, New York, 1981, p. 41.
182. G. M. Gale, F. Vallee, and C. Flytzanis, *Phys. Rev. Lett.* **57**, No. 15, 1867 (1987).
183. P. Roussignol, D. Ricard, J. Lukasik, and C. Flytzanis, *J. Opt. Soc. Am. B* **4**, 5 (1987).
184. E. I. Rashba and G. E. Gurgenshvili, *Sov. Phys. Sol. State* **4**, 759 (1962).
185. B. T. Smith, Ed., *Lecture Notes in Computer Physics*, Vol. 6, Springer, Borlin 1976.
186. A. S. Davydov, *Quantum Mechanics*, Pergamon, Oxford, 1965.
187. See ref. 13 (p. 408), 24 (pp. 174–175), and 18 p.

INTERMOLECULAR AND INTRAMOLECULAR POTENTIALS

Topographical Aspects, Calculation, and Functional Representation via a Double Many-Body Expansion Method

A. J. C. VARANDAS

Departamento de Quimica, Universidade de Coimbra, Coimbra, Portugal

CONTENTS

I. INTRODUCTION

The concept of potential-energy surface (or just potentials) is of major importance in spectroscopy and the theoretical study of molecular collisions. It is also essential for the understanding of the macroscopic properties of matter (e.g., thermophysical properties and kinetic rate constants) in terms of structural and dynamical parameters (e.g., molecular geometries and collision cross sections). Its role in the interpretation of recent work in plasmas, lasers, and air pollution, directly or otherwise related to the energy crisis, makes it of even greater value.

Figure 1 shows the interrelationship between the molecular potential and the various theoretical and experimental areas of chemical physics. For example, the theoretical study of an elementary chemical reaction consists of (1) calculating the potential related to the supermolecule (or just molecule) which results from the species participating in the molecular collision process,

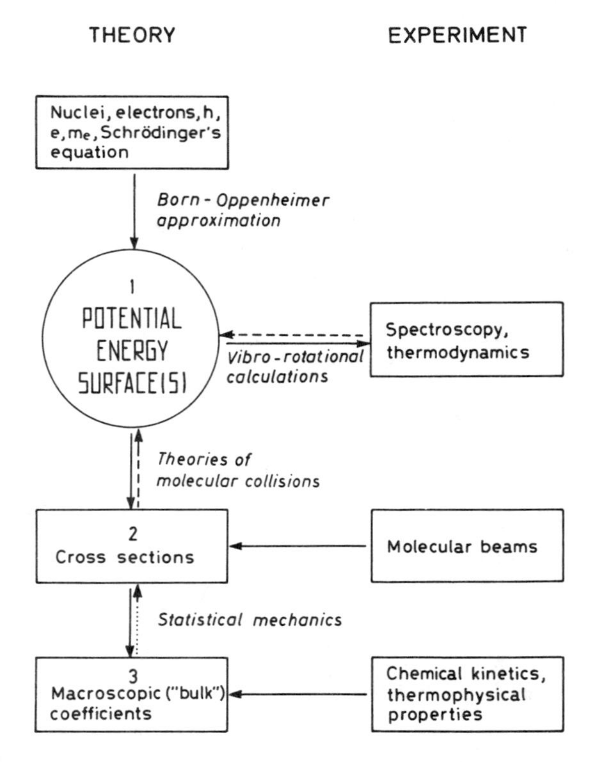

Figure 1. The molecular potential in its interrelationship with theory and experiment (adapted from ref. 4).

followed by (2) calculating of the microscopic kinetic parameters (cross sections) as a function of the initial state of the reactants and the final state of the products, and (3) calculating the macroscopic kinetic parameters (bulk properties) by integrating the cross section with respect to distributions related to the initial state of the reactants. Thus, only in favorable cases and for diatomic molecules is it possible to determine the potential-energy curve by directly inverting the levels of its vibrational–rotational energy spectrum[1] or the data from measurements of inelastic collision cross sections[2] and thermophysical properties.[3] However, for polyatomic systems, the determination of the molecular potential energy from experimental data is performed almost entirely in an indirect way using trial and error: adoption of a potential model, calculation of the dynamical properties, comparison with experiment, modification of the model, etc., until good agreement is achieved between the theoretical prediction and the experimental results. Figure 1 shows these stages in establishing the relation between theory and the various experimental fields.

This chapter refers only to stage 1 of Figure 1, that is, the molecular potential. Even so, the breadth of the subject still prevents a balanced treatment of all recent progress in this field. Therefore, emphasis is given to the work in the author's group during the past few years, particularly on the calculation and functional representation of potential-energy surfaces. Also, for further simplicity, special attention is given to triatomic systems. In fact, although the shape and functional dependence of the potential-energy curve which governs the interaction of two atoms is well established,[5] the same does not apply to the topographical features of the potential-energy surface of a triatomic molecule, even for the simplest case. This approach is also justified by the experimental importance of triatomic systems and the modest complexity of its theoretical treatment.

Although this chapter is written in the form of a review, it does include some new results not reported elsewhere. Since we shall be primarily concerned with our own work, we direct the reader to the vast literature both in the field of nonbonding interactions (intermolecular potentials[6–11]) and in the field of bonding or valence-type interactions, the latter being more frequently encountered in spectroscopy and studies of chemical reactivity (intramolecular potentials[12–19]).

The structure of this chapter is as follows. In Section II, after the concept of potential-energy surface and the coordinate systems in which the potential can be represented have been introduced, we describe the most important topographical characteristics of the molecular potential function. The general aspects, which refer to the calculation of the potential energy by ab initio methods, are analyzed in Section III. The need to develop efficient methods for the calculation of the potential function and the corresponding gradient in dynamical studies is then discussed. Although brief, mention is made in Section IV of some of the most popular methods for the functional representation of the potential energy, particularly the London–Eyring–Polanyi–Sato (LEPS) method[20] and its variants based on the approximate solution of Schrödinger's equation. However, emphasis will be placed on the many-body expansion (MBE) method of Murrell et al.[16] and on an extension of the latter recently proposed by the present author, which has been called the double many-body expansion (DMBE).[17,21] Examples of the application of the DMBE theory to van der Waals molecules and other more stable molecules are given in Section V. The main conclusions are gathered in Section VI. A general strategy for the calculation and global representation of the potential energy of small polyatomic systems, irrespectively of the type of interaction, is there suggested for the first time.

II. GENERAL PRINCIPLES

A. The Concept of Molecular Potential

Although the concept of potential has a classical meaning, it is in the Born–Oppenheimer approximation that it finds significance in the context of the present work. Born and Oppenheimer[22] recognized that for the great majority of molecular collisions with chemical interest, the nuclei move much more slowly than the electrons, and hence their motions can be treated as separable. The concept of potential-energy surface stems from this separation.

In principle, the solution for the molecular problem of N nuclei and n electrons can be found in the time-independent Schrödinger equation,

$$[\mathscr{H}(\mathbf{r},\mathbf{R}) - E]\Omega(\mathbf{r},\mathbf{R}) = 0, \tag{1}$$

where $\mathbf{r} = \{r_i\} = \{x_i, y_i, z_i\}$ and $\mathbf{R} = \{R_I\} = \{X_I, Y_I, Z_I\}$ are collective variables of the electronic and nuclear coordinates, respectively. Here $\mathscr{H}$ is the nonrelativistic hamiltonian operator, which we represent as

$$\mathscr{H} = \mathscr{H}^0 + \mathscr{H}^1, \tag{2}$$

where

$$\mathscr{H}^0 = -\frac{1}{2}\sum_{i=1}^{n} \nabla_i^2 + \sum_{i=1}^{n}\sum_{j>i}^{n} |\mathbf{r}_i - \mathbf{r}_j|^{-1} - \sum_{I=1}^{N}\sum_{i=1}^{n} Z_I|\mathbf{R}_I - \mathbf{r}_i|^{-1} + \sum_{I=1}^{N}\sum_{J>I}^{N} Z_I Z_J |\mathbf{R}_I - \mathbf{R}_J|^{-1}, \tag{3}$$

$$\mathscr{H}^1 = -\sum_{I=1}^{N} \frac{\nabla_I^2}{2M_I} \tag{4}$$

and M_I and Z_I represent the mass and charge of nucleus I, respectively; the other symbols have their usual meaning.* Since $M_I \gtrsim 2000$ and the nuclear kinetic-energy operator ($\mathscr{H}^1$) has an inverse dependence on M_I, this suggests a perturbation treatment of equation (1) based on the partition of equation (2).[24] Thus, by using Rayleigh–Schrödinger perturbation theory, one obtains at zeroth order†‡

$$[\mathscr{H}^0 - E_i^0(\mathbf{R})]|(\Psi_i^0(\mathbf{r}; \mathbf{R})\rangle = 0, \tag{5}$$

and at first order the total wave function $\Omega(\mathbf{r}, \mathbf{R})$. By expanding now $\Omega(\mathbf{r}, \mathbf{R})$ in terms of the complete basis of zeroth-order functions, and introducing the approximations (Born–Oppenheimer)

$$\Omega(\mathbf{r}; \mathbf{R}) = \sum_i \chi_i(\mathbf{R})|\Psi_i^0(\mathbf{r}; \mathbf{R})\rangle \sim \chi_i(\mathbf{R})|\Psi_i^0(\mathbf{r}; \mathbf{R})\rangle, \tag{6a}$$

$$\langle \Psi_i^0(\mathbf{r}; \mathbf{R})|\nabla_I|\Psi_j^0(\mathbf{r}; \mathbf{R})\rangle \sim 0 \tag{6b}$$

one finally gets

$$\left(-\frac{1}{2}\sum_{I=1}^{N}\frac{\nabla_I^2}{M_I} + E_i^0(\mathbf{R})\right)\chi_i(\mathbf{R}) = E_i\chi_i(\mathbf{R}). \tag{7}$$

Equation (5) is called the electronic Schrödinger equation because it describes the motion of the electrons for the nuclei fixed in the geometry defined as $\mathbf{R}$. For the same reason, the semicolon is used to express the parametric dependence in $\mathbf{R}$ of the eigenvectors of that equation. Equation (7) describes the motion of the nuclei under the influence of the potential function $E_i^0(\mathbf{R})$, which is commonly represented by $V_i(\mathbf{R})$, or simply $V(\mathbf{R})$ if there is no ambiguity with regard to the electronic state i.

B. Coordinates and Potential Representation

Although the most fundamental coordinate system is formed by the cartesian coordinates of each nucleus, the study of $V(\mathbf{R})$ in the absence of external fields may be accomplished only in coordinates (called internal coordinates) which relate to the perimeter or shape of the molecule, disregarding its position or orientation in the space as a whole. Therefore, the discussion and represent-

*Unless stated otherwise, all quantities are expressed in this work in atomic units: a.u. of bond length $a_0 = 0.529177$ Å $= 0.0529177$ nm; a.u. of mass $m_e = 9.109390 \times 10^{-27}$ kg; a.u. of charge $e = 1.602177 \times 10^{-19}$ C; a.u. of angular momentum $\hbar = 1.054573 \times 10^{-34}$ $J\,s$; a.u. of energy $E_h = 27.211652$ eV $= 627.509$ kcal/mol $= 219475$ cm^{-1} $= 4.359748$ aJ.[23]

†For convenience, we use the Dirac notation $|\ \rangle$ to represent the zeroth-order wave functions.

‡For simplicity, the electronic hamiltonian $\mathscr{H}^0$ will be referred to later as H.

ation of $V(\mathbf{R})$ for $N > 2$ requires only $3N - 6$ internal coordinates, whether the molecule is linear or not. One should notice that the number of independent vibrational normal modes for a linear molecule is $3N - 5$, but two of these (those associated with the bending of the molecule in orthogonal planes) are degenerate and therefore indistinguishable as far as the potential energy is concerned. In general, $V(\mathbf{R})$ is obtained as a potential-energy hypersurface in a $3N - 5$-dimensional space. Hence, in the case of a triatomic molecule one has a total of four coordinates: three configurational (e.g., three internuclear distances or two distances* and an angle; see Fig. 2) and the potential energy. It should be noted that for $N > 4$ the number of internal coordinates, $N(N - 1)/2$, exceeds $3N - 6$, which leads to some ambiguity in the selection of a set of internuclear coordinates.

Consider now the problem of graphical representation of $V(\mathbf{R})$ for the specific case of a triatomic molecule. Since the potential-energy hypersurface is four-dimensional (4D), only sections of this hypersurface are representable in the 3D real space. Among the possible representations (examples of which are given in Section V), we pay special attention to diagrams of isoenergetic contours for fixed values of the molecular perimeter using a triangular coordinate system.[25] This system of coordinates, illustrated in Fig. 3, is based

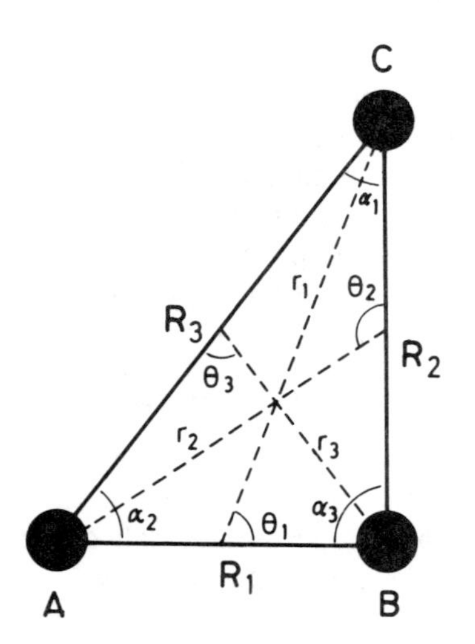

Figure 2. Some relevant coordinate systems for triatomic molecules: (R_1, R_2, R_3), (r_i, R_i, θ_i), and (R_i, R_j, α_k).

*Where convenient, we introduce the mixed notation R_i ($i = 1$–3) or R_{AB}, R_{BC}, and R_{AC} to represent the internuclear distances AB, BC, and AC, respectively. One should not mistake the collective variable $\mathbf{R} = (R_1, R_2, R_3)$ for the one given in the previous section.

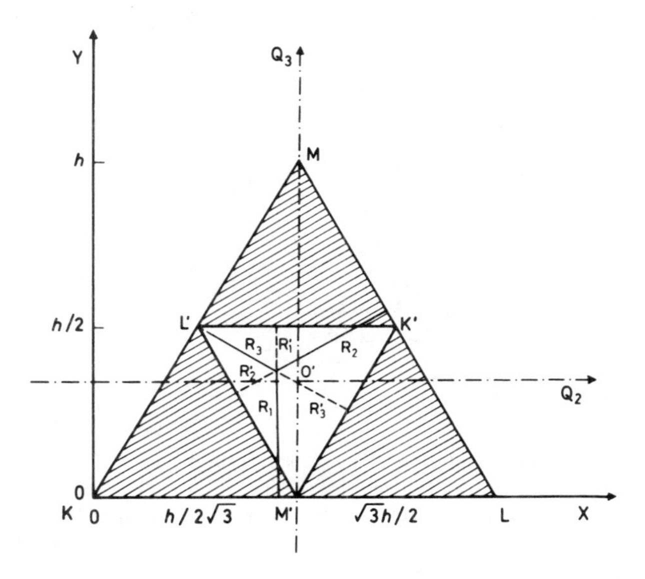

Figure 3. Triangular diagram to represent a triatomic potential-energy surface. For any point of the triangle KLM the molecular perimeter is constant, $R_1 + R_2 + R_3 = \text{constant} = h$. The unshaded area of the $K'L'M'$ triangle corresponds to the space physically accessible to the nuclei for that perimeter. Inside the $K'L'M'$ triangle, one has $R'_1 + R'_2 + R'_3 = h/2$, where R'_i $(i = 1, 2, 3)$ are the Pekeris coordinates.

on the well-known result that the sum of the three distances (taken perpendicularly) from a given point of an equilateral triangle (KLM) to its sides is equal to the height h of that triangle. Thus, if $h = P$, where P represents the molecular perimeter, those distances fit the values of the internal coordinates (R_1, R_2, R_3) for that perimeter. One should notice that none of these internal coordinates R_i is allowed to exceed the sum of the other two,

$$R_i + R_j \geqslant R_k, \qquad i, j, k = 1, 2, 3 \tag{8}$$

which limits the available configuration space of the molecule (with perimeter P) to the triangle $K'L'M'$, with height $h/2$, in KLM. Overall, the volume of $\mathbf{R}$ space which is physically accessible to the nuclei defines a pyramid with the vertex at the origin (which represents the united atom for $R_1 = R_2 = R_3 = 0$) and an infinite height, as illustrated in Fig. 4. This pyramid coincides with the octant of $\mathbf{R}$ space where $R_i \geqslant 0\,(i = 1, 2, 3)$. It is important to notice the relation between the coordinate system $\mathbf{R}'$ (that of Pekeris[26]) and the system $\mathbf{R}$, which

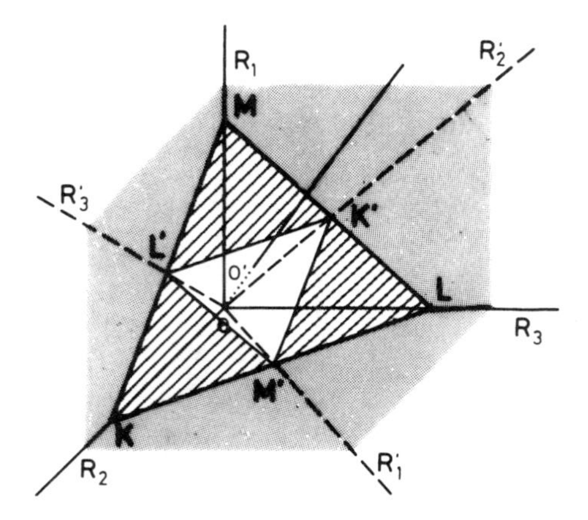

Figure 4. Interatomic coordinates (**R**) and Pekeris coordinates (**R**′). The unshaded pyramid with the vertex at the origin and infinite height delimits the total volume of configuration space physically accessible to the nuclei, which is identified with the octant $R_i' \geqslant 0$ $(i = 1, 2, 3)$ of the Pekeris coordinate system. In terms of the Q coordinates, Q_1 has a constant value and Q_2 and Q_3 are orthogonal coordinates, with the former being parallel to the sides KL and $K'L'$ of the triangles KLM and $K'L'M'$, respectively.

may be easily derived from Fig. 3. The result is

$$R_i' = \frac{h}{2} - R_i = \frac{(R_j + R_k - R_i)}{2}, \qquad i, j, k = 1, 2, 3, \tag{9}$$

or in matrix notation

$$\mathbf{R}' = \mathbf{UR}, \qquad \mathbf{R} = \mathbf{U}^{-1}\mathbf{R}', \tag{10}$$

where

$$\mathbf{U} = \begin{bmatrix} -\frac{1}{2} & \frac{1}{2} & \frac{1}{2} \\ \frac{1}{2} & -\frac{1}{2} & \frac{1}{2} \\ \frac{1}{2} & \frac{1}{2} & -\frac{1}{2} \end{bmatrix}, \qquad \mathbf{U}^{-1} = \begin{bmatrix} 0 & 1 & 1 \\ 1 & 0 & 1 \\ 1 & 1 & 0 \end{bmatrix}. \tag{11}$$

It is often convenient to use the symmetry coordinates that form the irreducible basis of the molecular symmetry group. This is because the potential-energy surface, being a consequence of the Born–Oppenheimer approximation and as such independent of the atomic masses, must be invariant with respect to the interchange of equivalent atoms inside the molecule. For example, the application of the projection operators for the irreducible representations of the symmetry point group D_{3h} (whose subgroup

C_{3v} is isomorphic with the permutation group S_3) to the internal coordinate **R**, leads to the **Q** symmetry coordinates

$$\mathbf{Q} = \mathbf{W}\mathbf{R} = \mathbf{W}'\mathbf{R}', \tag{12}$$

where

$$\mathbf{W} = \begin{bmatrix} 1/\sqrt{3} & 1/\sqrt{3} & 1/\sqrt{3} \\ 0 & 1/\sqrt{2} & -1/\sqrt{2} \\ \sqrt{2}/\sqrt{3} & -1/\sqrt{6} & -1/\sqrt{6} \end{bmatrix},$$

$$\mathbf{W}' = \begin{bmatrix} 2/\sqrt{3} & 2/\sqrt{3} & 2/\sqrt{3} \\ 0 & -1/\sqrt{2} & 1/\sqrt{2} \\ -\sqrt{2}/\sqrt{3} & 1/\sqrt{6} & 1/\sqrt{6} \end{bmatrix}. \tag{13}$$

From these symmetry coordinates, Q_1 is associated with the totally symmetric representation A_1', while the pair (Q_2, Q_3) is associated with the doubly degenerate representation E'. The geometric meaning of these coordinates is illustrated in Fig. 5. From the coordinates (Q_2, Q_3) we may define the polar coordinates

$$\rho = (Q_2^2 + Q_3^2)^{1/2}, \qquad \Xi = \arctan(Q_3/Q_2). \tag{14}$$

Figure 6 shows the relation between the several coordinate systems and the molecular conformations for a constant perimeter of the triatomic molecule. Also shown in this figure is the locus of the possible symmetry point groups (we consider the symmetry associated with the group operations applicable to identical nuclei); unless stated otherwise, the symmetry group will be C_s.

The symmetry coordinates show themselves to be particularly useful for the functional representation of the molecular potential. For example, the potential function of a X_3-type molecule must be invariant with respect to the interchange of any internal coordinate R_i ($i = 1, 2, 3$); hence it must be totally symmetric in relation to those coordinates. Thus, in terms of the coordinates Q_i ($i = 1, 2, 3$), such a function can only be written in terms of Q_1 or totally symmetric combinations of Q_2 and Q_3. Such combinations may in fact be obtained by using the projection-operator technique.[16,27] In fact, one can demonstrate[16,27] that any totally symmetric function of three variables is representable in terms of the integrity basis,[28]

$$\Gamma_1 = Q_1, \qquad \Gamma_2 = \rho^2 = Q_2^2 + Q_3^2, \qquad \Gamma_3 = Q_3^3 - 3Q_2^2Q_3. \tag{15}$$

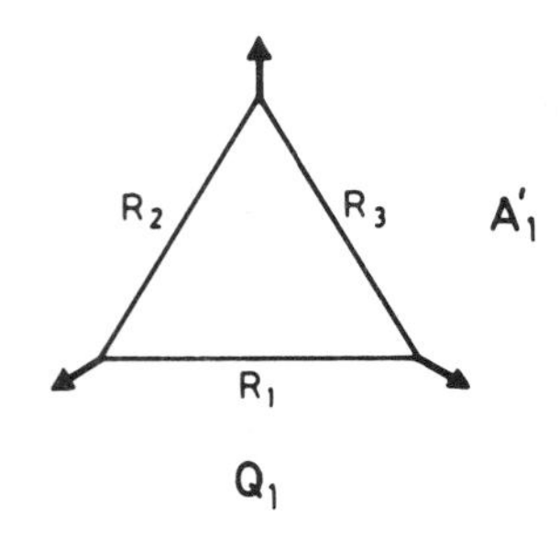

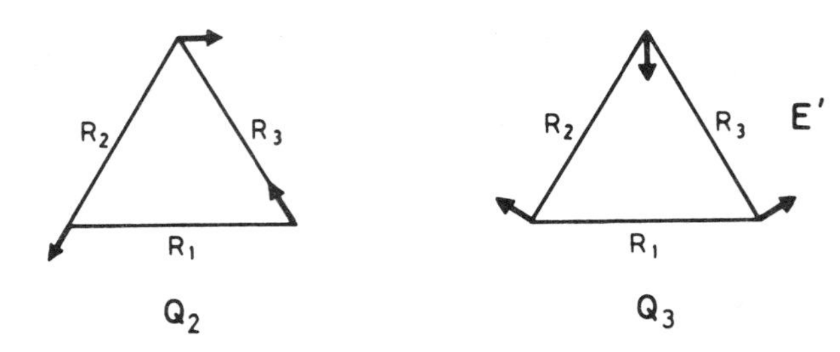

Figure 5. Displacements associated with the Q coordinates which transform as irreducible representations of the group S_3 (which is isomorphic with the symmetry point group C_{3v}, itself a subgroup of D_{3h}). Q_1 represents the totally symmetric breathing mode of the equilateral triangle, and Q_2 represents the antisymmetric distortion coordinate, which is degenerate with the Q_3 bending mode for D_{3h} geometries.

In the case of potentials having a lower permutation symmetry, the coordinates Q_i ($i = 1, 2, 3$) may still be used, but other terms besides those of equation (15) are obviously allowed for the representation. For example, for AB_2-type systems, there are two linear terms, two quadratic terms, and one cubic term with C_{2v} symmetry which may be used for the representation of the potential-energy surface, namely[29]

$$Q_1, \qquad Q_3, \qquad Q_2^2 + Q_3^2, \qquad Q_2^2 - Q_3^2, \qquad Q_3^3 - 3Q_2^2 Q_3. \tag{16}$$

Finally we should mention the hyperspherical coordinate system,[30,31]

$$\Lambda = \mathbf{T}\mathbf{R}^2 = \begin{bmatrix} 1 & 1 & 1 \\ 0 & \sqrt{3} & -\sqrt{3} \\ 2 & -1 & -1 \end{bmatrix} \begin{bmatrix} R_1^2 \\ R_2^2 \\ R_3^2 \end{bmatrix}, \tag{17}$$

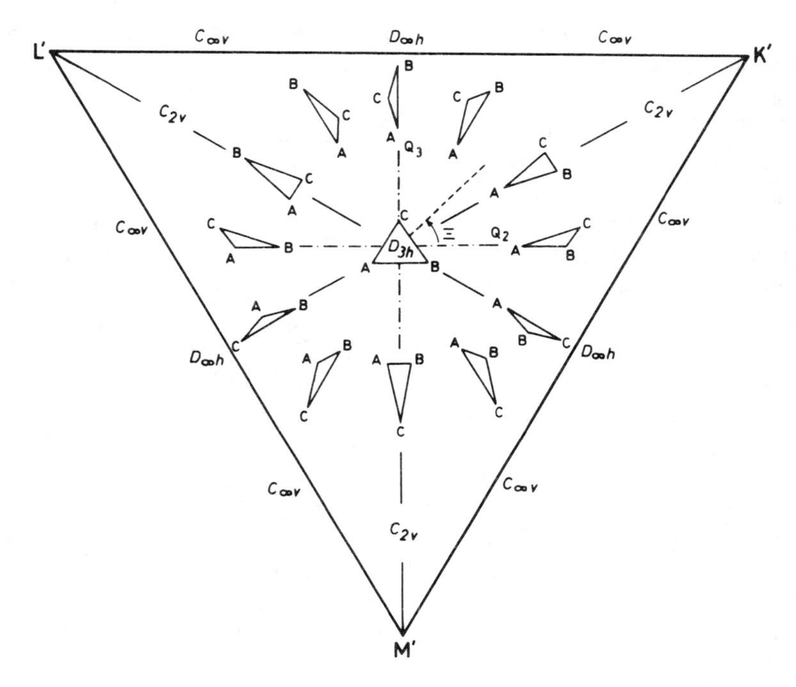

Figure 6. Relationship between coordinates and molecular conformation. Italicized symbols label the loci of the possible symmetry point groups, assuming the nuclei to be identical. Where unassigned, the symmetry point group will be C_s.

from which the auxiliary coordinates

$$
\begin{aligned}
q &= (\Lambda_1/3)^{1/2}, && 0 \leqslant q < \infty, \\
s &= (\Lambda_2^2 + \Lambda_3^2)^{1/2}/\Lambda_1, && 0 \leqslant s \leqslant 1, \\
\varphi &= (\operatorname{sgn} \Lambda_2)[\pi\text{-arccos}(\Lambda_3/s\Lambda_1)], && 0 \leqslant \varphi \leqslant 2\pi,
\end{aligned}
\tag{18}
$$

may be defined. Notice that, for the system of coordinates of equation (17), the space accessible to the nuclei consists of a cone with infinite height and with vertex at the origin. Therefore, instead of the triangular section for the coordinate system of equation (12), we obtain a circular section for a fixed value of the molecular size defined by the sum of the squares of the internuclear distances.[30]

C. Topographical Features

The importance of the topography of the potential-energy surface in spectroscopy and molecular dynamics is well established. For example, the relations between the force constants ($F_i = d^i V/dR^i$) at the minimum of the potential-energy curve of a diatomic molecule and the spectroscopic parameters ω_e, $\omega_e x_e$, B_e, α_e, etc., which characterize the vibrational–rotational spectrum are well known.[16,32] The geometry R_m and depth ε of the potential well also characterize the type of interaction: a deep minimum at small values of R is observed in valence interactions with the formation of a chemical bond; a shallow minimum at intermediate values of R occurs in van der Waals interactions. In the most important case, the potential-energy minimum separates the repulsive region, typically at small values of the internuclear coordinate R, from the attractive region, typically at large values of R. It should be noted that for $R = 0$ the potential becomes infinite due to the Coulomb interaction between the nuclei, and that for $R \to \infty$ it reduces to the van der Waals interaction. Obviously, the number of Coulomb singularities increases with the number N of nuclei in a polyatomic system as $N(N-1)/2$, i.e., one singularity for each pair of nuclei.

Moreover, the number and type of stationary points is even larger in the case of a polyatomic molecule, as can be shown from the analysis of the structure of the hessian matrix of the force constants ($F_{ij} = \partial^2 V/\partial R_i \partial R_j$). Take $\mathbf{R}^0$ as one such stationary point, and expand $V(\mathbf{R})$ near $\mathbf{R}^0$ in a Taylor series (note that grad $V = 0$ for $\mathbf{s} = 0$). Up to quadratic terms, one gets

$$V(\mathbf{s}) = V(\mathbf{0}) + \tfrac{1}{2}\mathbf{s}^\dagger \mathbf{F}\mathbf{s}, \tag{19}$$

and, by rotation of the system of axis so as to diagonalize $\mathbf{F}$,

$$V(\mathbf{s}) = V(\mathbf{0}) + \tfrac{1}{2}\mathbf{Y}^\dagger \mathscr{F}\mathbf{Y} = V(\mathbf{0}) + \frac{1}{2}\sum_{i=1}^{3N-6} Y_i \mathscr{F}_{ii} Y_i, \tag{20}$$

where

$$\mathbf{s} = \mathbf{R} - \mathbf{R}^0, \qquad \mathbf{Y} = \mathbf{G}\mathbf{s}, \qquad \mathbf{G}^\dagger\mathbf{G} = \mathbf{1}, \qquad \mathscr{F} = \mathbf{G}^\dagger\mathbf{F}\mathbf{G}. \tag{21}$$

Thus, the potential-energy variation for infinitesimal distances of the stationary point will depend on the signs of the eigenvalues $\mathscr{F}_{ii}$. For example, it will be a minimum if all $\mathscr{F}_{ii}$ are positive and V increases in any principal-curvature direction (direction of the Y eigenvalues) along which the displacement takes place. In the opposite case, where all $\mathscr{F}_{ii}$ are negative, it will be a maximum. If p such eigenvalues are negative and the remaining $3N$-6-p are positive, then $\mathbf{R}^0$ will be a saddle point of order p; such a stationary point behaves as a maximum for p directions and as a minimum for the others. It should be noted that first-order saddle points are extremely important in chemical kinetics, since only a stationary point of that type (and, if more than one exist, the one with lowest energy) can represent the transition state that links the reactant valley to the products valley in an elementary chemical reaction.[33] Finally, the reactants themselves and the products correspond to stationary points of $V(\mathbf{R})$ having all $\mathscr{F}_{ii}$ positive but one (associated with the reaction coordinate), which is zero.

Another important topographical feature often arises from the fact that polyatomic potential-energy surfaces may intersect in regions of the configuration space where the corresponding wave functions belong to the same (spatial and spin) symmetry.[34-39] The importance of this result in the context of the present work justifies a brief analysis of its essential aspects.

Let Ψ_1 and Ψ_2 represent two (adiabatic) electronic states which become degenerate for a given geometry. By expanding Ψ_1 and Ψ_2 in the usual manner in terms of two (diabatic) basis functions ψ_1 and ψ_2, which are supposed to be orthogonal among themselves and with the other electronic states, the adiabatic state energies are obtained from the secular determinant[35]

$$\begin{vmatrix} H_{11}(\mathbf{R}) - V & H_{12}(\mathbf{R}) \\ H_{12}(\mathbf{R}) & H_{22}(\mathbf{R}) - V \end{vmatrix} = 0, \tag{22}$$

where $H_{ij} = \langle \psi_i | H | \psi_j \rangle$. Thus, one gets

$$V^{\pm} = \tfrac{1}{2}([H_{11}(\mathbf{R}) + H_{22}(\mathbf{R})] \pm \{[H_{11}(\mathbf{R}) - H_{22}(\mathbf{R})]^2 + 4H_{12}^2(\mathbf{R})\}^{1/2}), \tag{23}$$

which shows that two conditions must be fulfilled if intersection is to take place[37]*:

$$H_{11}(\mathbf{R}) - H_{22}(\mathbf{R}) = 0, \tag{24}$$

$$H_{12}(\mathbf{R}) = 0. \tag{25}$$

For a diatomic molecule there is only one configurational parameter (R), and thus it is impossible to satisfy the above two conditions. Hence, the intersection of diatomic curves with the same symmetry can only occur by accident. This result is usually referred as the nonintersecting-state rule or equivalently noncrossing rule.[34] In the general case of a polyatomic molecule, the intersection may occur along a degeneracy manifold of dimension $3N - 8$. For a triatomic molecule, such a degeneracy manifold will be a line in the 3D configuration space. It also follows that, when it exists, such an intersection locus will reduce to a point in the triangular diagram of Fig. 3, since the molecular perimeter stays fixed. But what is the topographical aspect of such sections of the intersection locus?

*Although the sufficiency and independence of the conditions (24), (25) have been questioned,[40-42] the criticisms made have since been refuted.[43]

Let us consider a triatomic molecule with C_{2v} symmetry in such a geometry that two of its electronic states with the same spin but different symmetries, A_1 and B_2, intersect. Consider also a molecular distortion where the molecular perimeter is kept fixed, and let this distortion be represented in terms of the coordinates Q_2 and Q_3 (note that only Q_2, which belongs to the representation B_2, implies displacements that destroy the C_{2v} symmetry of the intersection geometry). With distortion, both states A_1 and B_2 transform according to the same representation A' of the symmetry point group C_s (i.e., only by accident would the matrix element H_{12} be zero), thus inducing the lifting of the degeneracy. On the other hand, the intersection in the original C_{2v} geometry implies that H_{12} vanishes for $Q_2 = 0$. Therefore, if only the linear terms are kept in the analysis and the intersection point is taken as the origin, one gets

$$\begin{vmatrix} V^0 + bQ_3 - V & lQ_2 \\ lQ_2 & V^0 + dQ_3 - V \end{vmatrix} = 0. \tag{26}$$

After substitution of $m = (b + d)/2$ and $k = (b - d)/2$ into equation (26), one obtains

$$V^{\pm} = V^0 + mQ_3 \pm (l^2 Q_2^2 + k^2 Q_3^2)^{1/2}, \tag{27}$$

which represents a double cone with the vertex at the origin.[36] Because this result is typical of the potential-energy surface when represented in a plane of two suitably selected coordinates, such topological features are known as conical intersections (see illustration in Fig. 7). One should note that although the matrix elements are analytic (H_{11} and H_{22} are often referred as diabatic potentials), the adiabatic potentials V^+ and V^- and the corresponding electronic wave functions are not. In fact, one can prove[36,43] that the electronic wave function changes sign when taken along a closed loop which encloses the conical intersection. This result has been used as a criterion to identify a conical intersection for the LiNaK system.[44] For dynamical studies, the total wave function must be continuous and unique, and it has been suggested that this change of sign in the electronic wave function should be compensated by another one in the nuclear wave function.[30]

However, the most-studied conical intersection is the one that occurs in the so-called Jahn–Teller systems.[36] In this case, the intersection corresponds to a high local symmetry of the molecule and it is imposed by that symmetry. The case of a molecule with D_{3h} symmetry is no doubt the most thoroughly studied. In such a system, any arbitrary distortion (Q_2, Q_3) leads to C_s states with symmetry A'. Thus, the intersection coincides with the D_{3h} symmetry axis, where the matrix element that involves the two A' states is zero. In this case the

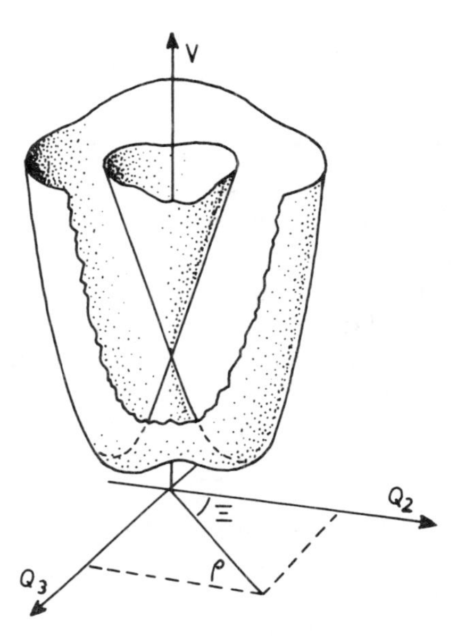

Figure 7. Schematic representation of a conical intersection. In the present case it represents a Jahn–Teller potential for a molecule with a C_3 axis perpendicular to the plane of vibration of the atoms (see text).

generalized Wigner–Eckart theorem establishes[45] $b = -d = l$, which, from equations (14), (27), leads to

$$V^{\pm} = V^0 \pm |l|(Q_2^2 + Q_3^2)^{1/2} = V^0 \pm |l|\rho. \tag{28}$$

Finally, the so-called Renner–Teller systems[46] (see also ref. 47) are characterized by a conical intersection due to two degenerate states which intersect themselves in linear geometries. Note that, for example, on bending the molecule, the Π and Δ states give rise to two other states, one with symmetry A' and the other with symmetry A''. Therefore, states A' can only intersect in regions of the configuration space where the matrix element H_{12} is zero, i.e., in linear geometries. As in the Jahn–Teller systems, the intersection here is also forced by symmetry. Hence, the intersection locus is infinite in extent and coincides with the symmetry element itself, which in this case is the axis $C_{\infty v}$. However, for intersections which are not forced by symmetry, the intersection locus can be either finite or infinite in extent.[38,39] As we shall see in Section IV.A, the type of intersection locus plays an important role in the functional representation of the molecular potential.

III. CALCULATION OF THE POTENTIAL ENERGY BY AB INITIO METHODS: A SURVEY OF THE THEORY

A. Judicious Synthesis of Theory and Experiment

Driven by the impossibility of solving the electronic Schrödinger equation accurately except for simple models and hydrogenic species, chemists turn to approximate methods to calculate the potential energy: ab initio, semi-theoretical, semiempirical, and empirical.

Based exclusively on first principles, the ab initio methods introduce only the approximations which are inherent in the theoretical model and the level of representation adopted for the wave function. For a given model, the best choice of approximation will depend on the value of the interaction energy, hence on the region of configuration space. Thus, in the region of weak interactions, the use of perturbation theory to calculate directly the interaction energy, considered as a perturbation to the energy of the interacting system, is justified. On the other hand, in the region of strong interactions, the obvious approach will be the calculation of the interaction energy by subtracting the energy of the infinitely separated subsystems from the energy of the supermolecule, all energies being calculated using the variational method.* Irrespective of the theoretical model, the ab initio calculations encounter severe difficulties when based on good-quality wave functions. Thus, such calculations are only realistic for systems with a small number of electrons. (It is worth pointing out that in other circumstances the designation "ab initio" may not be synonymous with a good potential energy.[9,13])

In order to carry out a priori theoretical calculations of the potential energy for systems with a large number of electrons, the semitheoretical methods use effective potentials which simulate the core electrons.[19,48] Note here that the inclusion of relativistic effects may be important in the description of the effective potentials in heavy atoms.[49]

The semiempirical and empirical methods have been by far the major source of the potentials used for molecular-dynamics studies. This fact stems from computational problems, which are examined next. First, the number of bielectronic molecular integrals in an ab initio calculation increases as the fourth power of the dimension of the function basis set (d^4) or approximately as n^4, where n represents the number of electrons—the n^4 explosion.[50,51] Second, the number of ab initio points necessary to determine (as a large table) the potential-energy surface, i.e., the number of nuclear geometries for which

*It should be noted that in the region of very strong interactions, for $R \rightarrow 0$, the perturbation solution is still possible (e.g., ref. 5); in this case, the unperturbed system will be the united atom.

the electronic Schrödinger equation has to be solved, varies approximately as X^{3N-6}, where X represents the number of points necessary to characterize one of the $3N-6$ configurational degrees of freedom of the molecule—the X^{3N-6} explosion. Further, although it is often sufficient in spectroscopy to know the molecular potential function in the region close to the equilibrium geometry of the atoms, in molecular dynamics it is essential to know the potential from the reactant valley to the product valley, including the intermediate regions which are physically accessible to the nuclei under the given working experimental conditions. Combining these two difficulties, it is easy to realize that even for a relatively modest value of $X = 10$ the ab initio or semitheoretical calculation of a grid in 12 dimensions (e.g., in the study of the reaction $Cs + CH_3I \rightarrow CsI + CH_3$) is impossible; if one spent 0.1 μs per molecular integral, one million years would be required to perform such a calculation. Finally, there are accuracy requirements which need to be satisfied by the potential-energy surface, especially close to the topographical features that may influence the dynamical process. For example, an error of 1.5 kcal/mol (6 kJ/mol) in the potential-energy barrier height (approximated for convenience by the activation energy) may lead to a kinetic rate constant which differs by a factor of 10 from the true result. Nevertheless, if we bear in mind that the fact that the absolute value of the electronic energy of a small polyatomic system, e.g., HO_2, is of the order of $150E_h$ (about 10^5 kcal/mol), this means that the calculation of the energy must be carried out with an error smaller than 0.001%.

The semiempirical and empirical methods have therefore two important advantages in application to dynamical studies. First, they use available experimental data to calibrate the potential-energy surface, thus allowing reproduction of topographical details which are known to play an essential role in the dynamical process. Second, they give the potential energy in explicit form (an exception is the DIM method[52] in its most general form[53]), thus fulfilling another important requirement of those studies, namely, speed in calculating the potential and its gradient at an arbitrary geometry of the molecule configuration space.[54] Consequently, even in situations where ab initio methods are feasible, it is of major importance to have a functional form which enables the analytical continuation of the potential energy to regions of configuration space not covered by the ab initio calculations.

In conclusion, we share the philosophy H. F. Schaefer III expressed in 1979 (which we believe is still valid in 1986, and very likely to be for the foreseeable future): "We have been convinced for about five years that ab initio electronic structure calculations should not even attempt (except for the very simplest systems) to predict the entire potential energy surface". Since the success of a semiempirical method stems from the judicious combination of theory and experiment, we present a brief survey of the main theoretical methods in the remainder of this section.

B. The Perturbational Method

If the distance between the interacting systems A and B is sufficiently large to enable the overlap of the respective electronic clouds to be disregarded, then the interaction energy may be calculated from the Rayleigh–Schrödinger perturbation theory, using as a basis of functions to describe the wave function of AB

$$\psi_{rs}(i,j) = \Psi_r^{\mathrm{A}}(i)\Psi_s^{\mathrm{B}}(j), \tag{29}$$

where $\Psi_r^{\mathrm{A}}(i)$ represents the wave function of the rth state of system A (and similarly for system B), and $\{i\}$ and $\{j\}$ represent the electrons associated with systems A and B, respectively.

By defining the perturbation operator as

$$U(a,b,i,j) = H_{\mathrm{AB}} - H^0 = H_{\mathrm{AB}} - (H_{\mathrm{A}} + H_{\mathrm{B}}) \tag{30}$$

and taking into account only the terms of order less than two in the perturbational expansion, one gets

$$V_{\mathrm{AB}}(R) = V_{\mathrm{AB,ele}}(\mathbf{R}) + V_{ab,\mathrm{ind}}(\mathbf{R}) + V_{\mathrm{AB,dis}}(\mathbf{R}), \tag{31}$$

where

$$V_{\mathrm{AB,ele}} = \langle \psi_{00} | U | \psi_{00} \rangle, \tag{32}$$

$$V_{\mathrm{AB,ind.}} = -\mathop{\mathrlap{\int}\sum}\nolimits_r' \frac{|\langle \psi_{r0} | U | \psi_{00} \rangle|^2}{E_{r0} - E_{00}} - \mathop{\mathrlap{\int}\sum}\nolimits_s' \frac{|\langle \psi_{0s} | U | \psi_{00} \rangle|^2}{E_{0s} - E_{00}}, \tag{33}$$

$$V_{\mathrm{AB,dis}} = -\mathop{\mathrlap{\int}\sum}\nolimits_r' \mathop{\mathrlap{\int}\sum}\nolimits_s' \frac{|\langle \psi_{rs} | U | \psi_{00} \rangle|^2}{E_{rs} - E_{00}}. \tag{34}$$

Equation (32) represents the first-order electrostatic energy, whereas equations (33), (34)* represent the second-order contributions for the induction and dispersion energies, respectively; $E_{rs} = \langle \psi_{rs} | H^0 | \psi_{rs} \rangle$. If A and B are closed-shell neutral atoms, then the electrostatic and induction energies vanish for large values of r, and the dispersion energy is the first nonzero term of the interaction energy.

For large values of the distance r between the A and B centers the energy contributions (32)–(34) may be expanded as an inverse power series in r (note that in perturbation theory the energies are generally represented in terms of

*As usual, the prime shows that the index does not take the value 0, and $\mathrlap{\int}\sum$ represents a sum over the discrete states and integration over the continuum of the unperturbed system.

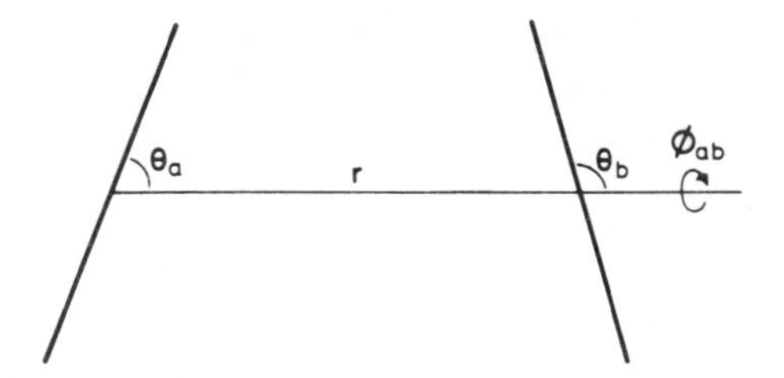

Figure 8. System of spherical polar coordinates for the interaction of two polyatomic molecules.

the polar spherical coordinate system of Fig. 8). For example, if A and B are rigid linear molecules (the z axis is taken as the line joining their centers), the electrostatic energy in equation (32) takes the form[8,9]

$$V_{\mathrm{AB,ele}} = \mu_{\mathrm{A}}\mu_{\mathrm{B}} f_{11}(\omega) r^{-3} + [\mu_{\mathrm{A}}\Theta_{\mathrm{B}} f_{12}(\omega) - \Theta_{\mathrm{A}}\mu_{\mathrm{B}} f_{21}(\omega)] r^{-4} + \cdots, \tag{35}$$

where

$$f_{11}(\omega) = -2\cos\theta_a \cos\theta_b + \sin\theta_a \sin\theta_b \cos\varphi_{ab}, \tag{36}$$

$$f_{12}(\omega) = \tfrac{3}{2}[\cos\theta_a(3\cos^2\theta_b - 1) - 2\sin\theta_a \sin\theta_b \cos\theta_b \cos\varphi_{ab}]. \tag{37}$$

In equation (35), μ_{A} (μ_{B}) and Θ_{A} (Θ_{B}) are the dipole and quadrupole moments of A (B), respectively, and f_{21} may be written as an expression analogous to f_{12} but with indexes a and b permuted.

Similarly, for spherically symmetric interactions, the asymptotic forms of the induction energy and the dispersion energy are

$$V_{\mathrm{AB,ind}} = -\sum_{l_a=1}^{\infty} C_n(l_a, 0) r^{-2(l_a+1)}, \tag{38}$$

$$V_{\mathrm{AB,dis}} = -\sum_{l_a=1}^{\infty}\sum_{l_b=1}^{\infty} C_n(l_a, l_b) r^{-2(l_a+l_b+1)}, \tag{39}$$

where $n = 2(l_a + l_b + 1)$, and only the interaction of the charge of B with the induced electronic density of A is considered; there is obviously an expression analogous to equation (38) for the induction in B. In equations (37), (38), the induction coefficient is related to the static polarizability of A through[8,10]

$$C_n(l_a, 0) = \alpha_{l_a}(0)/2, \tag{40}$$

with the dispersion coefficient $C_n(l_a, l_b)$, associated with the interaction between the 2^{l_a}-order induced multipole of atom A and the 2^{l_b}-order induced multipole of atom B, being given by (e.g., ref. 8)

$$C_n(l_a, l_b) = \frac{1}{2\pi}\binom{2l_a + 2l_b}{2l_a}\int_0^\infty \alpha_{l_a}(i\omega)\alpha_{l_b}(i\omega)\, d\omega. \tag{41}$$

As for the dynamic polarizability, which depends on the imaginary frequency $i\omega$, it can be well approximated by the one-term Padé approximant,[55]

$$\alpha_{l_a}(i\omega) = \frac{\alpha_{l_a}(0)}{1 + (\omega/\Delta\varepsilon_{l_a})^2}, \tag{42}$$

where $\Delta\varepsilon_{l_a}$ is a parameter with the dimensions of an energy. By substituting equation (42) into equation (41) one gets the generalized London formula

$$C_n(l_a, l_b) = \frac{1}{4}\binom{2l_a + 2l_b}{2l_a}\frac{\Delta\varepsilon_{l_a}\Delta\varepsilon_{l_b}}{\Delta\varepsilon_{l_a} + \Delta\varepsilon_{l_b}}\alpha_{l_a}(0)\alpha_{l_b}(0). \tag{43}$$

For practical purposes, the expansions (33), (34) may be truncated after the first few terms, e.g., from $C_{10} = C_{10}(1, 3) + C_{10}(2, 2) + C_{10}(3, 1)$ onwards in the case of the dispersion energy. Values of such dispersion coefficients, which have been calculated using dynamic polarizabilities represented by Padé approximants,[56–60] have been reported in the literature for many systems of practical interest (see also Section V.A).

Let us now turn to the case of intermediate values of r for which, although small, the orbital overlap and electronic exchange effects cannot be ignored. In such a case, a good basis for the representation of the wave function of AB is

$$\Psi_{rs}(i, j) = \mathscr{A}\,\Psi_r^{\mathrm{A}}(i)\,\Psi_s^{\mathrm{B}}(j) = \mathscr{A}\,\psi_{rs}(i, j), \tag{44}$$

where $\mathscr{A} = N^{1/2}(1 + \mathscr{P})$ is the antisymmetrizing operator of the electrons $\{i\}$ and $\{j\}$ (e.g., ref. 6). However, Ψ_{rs} is not an eigenfunction of the zeroth-order hamiltonian operator H^0, which leads to formal difficulties in the perturbation expansion.[6,9,11] In any case, all symmetry-adapted perturbation theories are unanimous in giving the first-order energy as

$$V_{\mathrm{AB}} = \frac{\langle \mathscr{A}\psi_{00}|U|\psi_{00}\rangle}{\langle \mathscr{A}\psi_{00}|\psi_{00}\rangle}. \tag{45}$$

In regions of small orbital overlap equation (45) may be written (after expanding the denominator and keeping terms of order not greater than one in

$\mathscr{P}$) as the sum of the electrostatic energy (32) and the first-order exchange energy $V_{\mathrm{AB,exc}}$,

$$V_{\mathrm{AB,exc}} = \langle \mathscr{P}\psi_{00} | U | \psi_{00} \rangle - \langle \mathscr{P}\psi_{00} | \psi_{00} \rangle \langle \psi_{00} | U | \psi_{00} \rangle. \tag{46}$$

Since Ψ_0^{A} and Ψ_0^{B} are not, in general, eigenfunctions of the respective hamiltonians, an additional term appears in the perturbational treatment, known as the zeroth-order exchange energy or complementary exchange term,[61] $\Delta(r)$. In regions of small orbital overlap, Δ may be written as[62]

$$\Delta = \langle \mathscr{P}\psi_{00} | H_{\mathrm{A}} + H_{\mathrm{B}} | \psi_{00} \rangle - \langle \mathscr{P}\psi_{00} | \psi_{00} \rangle \langle \psi_{00} | H_{\mathrm{A}} + H_{\mathrm{B}} | \psi_{00} \rangle. \tag{47}$$

The fact that this term, at the one-exchange level, should vanish in the Hartree–Fock limit of the zeroth-order wave functions[62] (and, obviously in the exact limit) led us and others to suggest the minimization of Δ^2 as a criterion for optimizing the basis functions in first-order energy perturbation calculations.[63–65] The results have proved to be rather encouraging when using minimal basis sets optimized with this criterion for the interactions He$\cdots$He[63,64] and Ne$\cdots$Ne.[64] However, though a necessary condition, the requirement of a vanishing Δ may not be a sufficient condition to guarantee Hartree–Fock quality of the wave functions.[66]

The antisymmetrizing operator also affects the form of the electrostatic, induction, and dispersion energies. For example, in the case of neutral, spherically symmetric systems, the electrostatic and induction energies (which vanish in the limit of negligible orbital overlap) display a dependence on r similar to the first-order exchange energy,[67]

$$V_{\mathrm{AB,exc}} = K S_{\mathrm{AB}}^2 / r, \tag{48}$$

where S_{AB} represents the overlap integral involving the most external orbitals from centers A and B.

Somewhat different is the case of the induction and dispersion energies. For these the expansion in inverse powers of r is only valid in the limit of vanishing orbital overlap, and in this case the expansions of equations (38), (39) are shown to overestimate the true value of the energy when such orbital overlap is taken into account. Indeed, studies carried out for small systems[68–77] show that the values of the induction and dispersion coefficients decrease with decreasing r. Formally, it is possible to account for this effect by introducing the so-called damping functions as follows:

$$C_n^*(l_a, l_b; r) = \chi_n(l_a, l_b; r) C_n(l_a, l_b) \tag{49}$$

where the damping functions $\chi_n(l_a, l_b; r)$ have a similar r dependence for the various values of n, although in reality the intensity of damping of the dispersion energy may vary with n and the order of the induced moments of A and B. In Section IV.D we shall examine the problem of the functional representation of these damping functions and their application to the semiempirical calculation of the interaction energy.

The application of perturbation theory to many-body interactions leads to pairwise-additive and non-pairwise-additive contributions. For example, in the case of neutral, spherically symmetric systems which are separated by distances such that the orbital overlap can be neglected, the first non-pairwise-additive term appears at third order of the Rayleigh–Schrödinger perturbation treatment and corresponds to the dispersion energy which results from the induced-dipole–induced-dipole–induced-dipole[78]* interaction

$$V_{\mathrm{ABC,ddd}}(\mathbf{R}) = C_9 \frac{1 + 3\cos\alpha_1 \cos\alpha_2 \cos\alpha_3}{(R_1 R_2 R_3)^3} \tag{50}$$

where α_1, α_2, and α_3 are the internal angles of the triangle formed by the three atoms. Also, in the case of C_9, there are available in the literature values of this coefficient for a considerable number of nonbonding interactions,[57,60,79,80] as well as values of some of the dispersion coefficients involving higher-order induced moments for some systems. It should be noted that in many cases of practical interest,[81] the triple-dipole (ddd) term (50) has been shown to be the only one necessary on a quantitative basis, possibly due to fortuitous cancellations of the terms of order higher than C_9 amongst themselves and with the non-pairwise-additive exchange terms.[82,83] In this case, as with the second-order dispersion energies, the orbital overlap and exchange effects tend to reduce the value of the asymptotic expression.[84–86] The problem of the explicit representation of such effects will also be deferred to Section IV.D.

C. The Variational Method

In the variational method (also known as supermolecular method), the interaction energy A $\cdots$ B is calculated from

$$V_{\mathrm{AB}}(R) = E_{\mathrm{AB}} - E_{\mathrm{A}} - E_{\mathrm{B}}, \tag{51}$$

where E_{AB}, E_A, and E_B are calculated by optimization of the functional

$$E = \frac{\langle \Psi | H | \Psi \rangle}{\langle \Psi | \Psi \rangle} \geqslant E_{\mathrm{exact}} \tag{52}$$

*In spite of a reference to an article by Y. Muto[78] in Axilrod's 1951 article[78] (see footnote), and the fact that the ddd energy is often referred to as the Axilrod–Teller–Muto term, the author has not been able to consult Muto's article, because the volume and year given appear inconsistent.

with $\Psi = \Psi_{AB}$, Ψ_A, and Ψ_B, and $H = H_{AB}$, H_A, and H_B respectively, for the calculation of E_{AB}, E_A, and E_B. It should be noted that for large distances $\Psi_{AB} = \mathscr{A}\Psi_A\Psi_B$, which makes equation (52) equivalent to the first-order energy (45). For still larger distances where the orbital overlap may be regarded as negligible, the antisymmetrizing operator $\mathscr{A}$ becomes irrelevant, and equation (51) thus reduces to the first-order electrostatic energy (32).

In its simplest formulation at the level of the Hartree–Fock theory, the wave function has the form of a Slater determinant

$$\Psi = \| \psi_1(1)\psi_2(2), \ldots, \psi_n(n) \|, \tag{53}$$

where the spin-orbitals $\psi_i = \varphi_i \sigma_i$ (the spin functions σ_i may be either α or β), which are obtained by the self-consistent-field (SCF) method, are expanded in terms of a given basis set of functions

$$\varphi_i = \sum_j^d c_{ji} f_j. \tag{54}$$

To complete the description of the Hartree–Fock theoretical model one needs to define the functions f_j in equation (54). In the method of the linear combination of atomic orbitals (LCAO), these functions are approximated with the atomic orbitals. Although exponential functions of the Slater type [Slater-type orbitals (STO), e.g., $f_{1s} = \exp(-\zeta_1 r)$, $f_{2s} = r\exp(-\zeta_2 r)$, etc.] may behave approximately like optimized atomic orbitals, the leading trend in the recent years, particularly for polyatomic calculations, has been towards the use of gaussian-type functions (GTO), e.g., $f_{1s} = \exp(-\zeta_1' r^2)$, $f_{2s} = r\exp(-\zeta_2' r^2)$, etc. Indeed, in the case of GTO functions,[87] the bielectronic molecular integrals may be calculated in a quick and efficient manner in terms of known mathematical functions,[50,51] which contrasts with the slowness of the numerical calculation when using STO functions. As a compromise the functions f_j are usually represented as a linear combination of GTOs, which are known as contracted gaussian functions, e.g., through transformations of the type

$$e^{-r} \rightarrow \sum_k^K d_{1k} \exp(-\zeta_{1k}' r^2). \tag{55}$$

By choosing the contracting scheme, it is possible to determine the contracted gaussian functions which best approximate Slater functions,[88] Hartree–Fock atomic orbitals,[89,90] etc. It should be noted that the size and balance (revealed, for example, by the optimum ratio of the number of s-type and p-type functions) of the basis are definite factors which control the quality of the final result (e.g., ref. 90).

The computational cost of a Hartree–Fock calculation is similar to that of a first-order perturbation calculation, which perhaps explains the tendency to use the supermolecular methodology for calculating weak interactions. In this case, the main handicap is the accuracy required for the variational calculation of the energies of the supermolecule and of the interacting fragments when infinitely separated from each other. As Kutzelnigg, inspired by a famous sentence of Coulson's pointed out,[91] the calculation of the interaction energy in the region of van der Waals as a difference between two large numbers is analogous to the calculation of the mass of a sailor's hat by subtracting the mass of a ship with the bare-headed sailor from the mass of the ship with the hatted sailor. An important step in lessening this difficulty consists in reducing to a minimum the so-called basis-set superposition errors.[92] The simplest way to achieve this goal consists in calculating $V(R)$ from

$$V(R) = E_{\mathrm{AB}} - E_{\mathrm{A}b} - E_{a\mathrm{B}}, \tag{56}$$

where $E_{\mathrm{A}b}$ ($E_{a\mathrm{B}}$) represents the energy of A (B) calculated from the supermolecular wave function Ψ_{AB} once the nuclear charge at position B (A) has been canceled (see ref. 93 and references therein). Finally, the partition of the Hartree–Fock energy into physically meaningful constituents may be achieved by using the scheme originally suggested by Morokuma and coworkers[94] and developed by other authors.[95] One has

$$V_{\mathrm{AB,HF}}(R) = V_{\mathrm{AB,ele}} + V_{\mathrm{AB,exc}} + V_{\mathrm{AB,ind}}, \tag{57}$$

where $V_{\mathrm{AB,HF}}$ is obtained from equation (56) and the remaining terms have the meaning given in Section III.B.

Nevertheless, the single-configuration Hartree–Fock method [which is based on just one Slater determinant, equation (53)] presents some problems, of which the most important is due to the correlation energy not being taken into account. Thus, it is not suitable for calculating the dispersion energy which arises from the correlation between the electrons of the interacting systems. It also leads to a faulty description of the dissociation of covalent bonds where the initial and final states involved in the dissociation process differ in the number of pairs of electrons. A typical example is that of the dissociation of the hydrogen molecule in its ground state ($\tilde{X}\,^1\Sigma_g^+$). In this case, the single-configuration wave function

$$\Psi_{\mathrm{SCF}} = \|\sigma_g \sigma_g\| \tag{58}$$

predicts an equal dissociation probability for $H^+ + H^-$ and for $H\cdot + H\cdot$, whereas the correct result is 100% for $H\cdot + H\cdot$.

In order to make up for those imperfections one needs to turn to post-Hartree–Fock methods. Two variational techniques are worth discussing due to their popularity: the configuration-interaction (SCF CI) method and the multiconfiguration self-consistent-field (MC SCF) method.

In the most popular version of the SCF CI method, in a first stage the basis functions are optimized from a single-configuration Hartree–Fock calculation and then used for the construction of the various configurations (combinations of Slater determinants adapted to the spin symmetry) which make up the correlated wave function

$$\Psi_{\mathrm{CI}} = c_0 \Psi_{\mathrm{HF}} + \sum_i c_{Si} \Phi_i + \sum_j c_{Dj} \Phi_j + \sum_k c_{Tk} \Phi_k + \sum_l c_{Ql} \Phi_l + \cdots, \tag{59}$$

where the indexes S, D, T, and Q label the type of configuration in relation to the number of orbitals in which it differs from the reference Hartree–Fock configuration, namely singly excited configurations (S), doubly excited configurations (D), etc. Hence, for example, c_{Si} represents the coefficient of the ith configuration of type S, and these coefficients are calculated by solving the eigenvalue equation

$$\mathbf{HC} = \mathbf{EC}, \tag{60}$$

where $H_{ij} = \langle \Phi_i | H | \Phi_j \rangle$. Although the setting up of the SCF CI method is simple and general, the fact that the self-consistent field of the basis orbitals is different from that of the Φ orbitals generally requires consideration of a great number of terms in the CI expansion (59). It should be emphasized that the matrix elements H_{ij} between the Hartree–Fock configuration and configurations which involve triple excitations or excitations of higher order are zero, since they differ in more than two spin-orbitals.[51] In perturbational terms, such configurations do not contribute to the first-order wave function as long as the Hartree–Fock configuration represents the zeroth-order wave function. From similar reasoning, the matrix elements which involve single and quadruple excitations should vanish. Furthermore, Brillouin's theorem[96] forbids direct interaction between singly excited configurations and the fundamental Hartree–Fock configuration. Nevertheless, such an interaction may be indirectly allowed via coupling with doubly excited configurations, which may interact with the Hartree–Fock configuration.

CI *SD* calculations are currently almost routine (ref. 13 and references therein), and in many cases (particularly systems with few electrons) they have been used to obtain a large fraction of the correlation energy. However, in very accurate calculations it is necessary to include triple and quadruple excitations, in which case the most common ab initio technique to tackle this problem consists in building up single and double excitations from a multiple reference space. This approach is usually referred as the MR CI (multireference CI) method.[97a]

The so-called problem of size consistency should also be mentioned. The calculated energy for the system A···B should approach, in the limit of an infinite separation between A and B, the sum of the energies of the isolated subsystems calculated separately. For the case of interactions between closed-shell systems, the single-configuration Hartree–Fock method satisfies that requirement. The same goes for a complete CI calculation, since it gives the exact answer. Nevertheless, in general the truncation which has to be carried out for practical reasons in the CI expansion (59) does not. Nonvariational theories which satisfy that principle—e.g., coupled-cluster theories (CCA, CEPA), and the many-body perturbation theory of Moller–Plesset (where the correlation energy is treated as a perturbation, with the zeroth-order energy being that associated to the Hartree–Fock configuration)—have also been used for the calculation of the potential energy; the reader may consult refs. 9, and 51 and references therein for further details.

The MC SCF method usually takes into account a minimum number of configurations capable of assuring some fundamental requirements, this step being followed by the optimization of the basis functions using the self-consistent-field method. For example, in order to describe accurately the dissociation of ground-state H_2 it is only necessary to consider the two-configuration wave function

$$\Psi_{\mathrm{MC\,SCF}} = A_{\mathrm{g}} \| \sigma'_{\mathrm{g}} \sigma'_{\mathrm{g}} \| + A_{\mathrm{u}} \| \sigma'_{\mathrm{u}} \sigma'_{\mathrm{u}} \| , \tag{61}$$

a calculation usually known as extended Hartree–Fock[98]; here σ'_{g} and σ'_{u} are the molecular orbitals for the ground state ($\tilde{X}\,^1\Sigma_{\mathrm{g}}^+$) and the first excited state ($b\,^3\Sigma_{\mathrm{u}}^+$) of the H_2 molecule.

Concerning the set of nonoccupied molecular orbitals used in the construction of the excited configurations, the literature[99] usually distinguishes three subsets: (M1), those necessary to warrant the accurate description of the dissociation into Hartree–Fock fragments; (M2), those which may contribute to the reduction of the molecular energy without modifying the asymptotic limit of the dissociation into Hartree–Fock fragments; and (M3), the ones capable of lowering the asymptotic energy and thus accounting for the intraatomic correlation energy. Nevertheless, one may think that such a partition lacks physical support, since some of the (M2) molecular orbitals may be valence orbitals such as those from the (M1) set. Accordingly, one may consider a different partition[100] in which the orbitals are divided into valence (v) and nonvalence orbitals (nv). The relationship between these two classification schemes, which is schematically illustrated in Fig. 9 through the corresponding MC SCF potentials [for example, MC SCF1 for the MC SCF

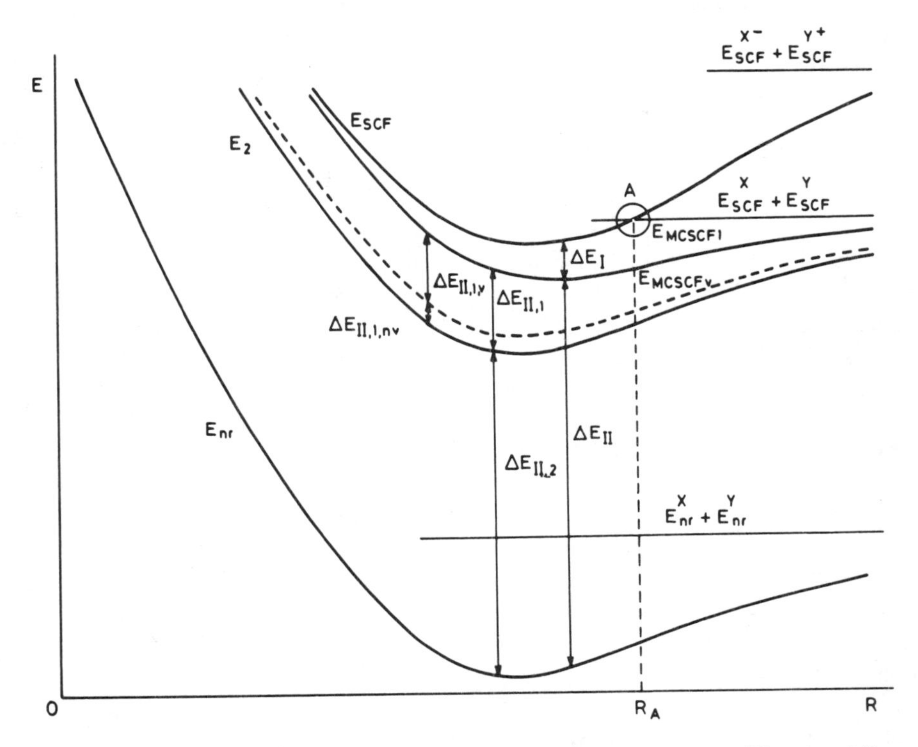

Figure 9. Energies of the various theoretical models for the diatom XY.[101] A, instability point of the single-configuration SCF model; E_1, MC SCF energy obtained from the (M1) orbitals; $E_{\mathrm{MC\,SCFv}}$, energy for the multiconfiguration model of optimized valence orbitals (v); E_2, minimum energy without inclusion of the intraatomic correlation; E_{nr}, nonrelativistic limit of the energy; E_{nr}^X (E_{nr}^Y), nonrelativistic energy of the X (Y) atom; $\Delta E_I = E_{\mathrm{MC\,SCF\,1}} - E_{\mathrm{SCF}}$, etc. (see text).

calculation based on the (M1) set, etc.], has recently been discussed in the context of the correlation force ($F = dV/dR$) in diatomic interactions.[101]

The designation CAS SCF (complete-active-space self-consistent field) has been used for a MC SCF calculation which uses all possible reference configurations obtained from the nonoccupied valence molecular orbitals. The correlation energy so obtained from this CAS SCF model is known as internal or nondynamical correlation, thus differing from the external or dynamical correlation energy, which involves the dispersion energy as well.[97,100] This partition of the correlation energy will be further discussed in Section IV.C. It should be noted that the best variational method used to calculate the external correlation energy is perhaps the MR CI method, which in this case is known as CAS SCF CI. Finally, it should be stated that the CAS SCF method when applied to the ground state of the H_2 molecule is equivalent to an extended Hartree–Fock calculation based on the wave function (61). We use the latter approach in Section IV.C to imply the former.

IV. SEMIEMPIRICAL CALCULATION AND FUNCTIONAL REPRESENTATION OF THE POTENTIAL ENERGY

A. Local versus Global Representations

We have already stressed the importance for dynamical studies of having a quick and efficient method to calculate the potential energy and its gradient at any arbitrary point of the molecular configuration space. Two main techniques have been suggested: local adjustment and global representation.

Using seminumerical techniques of polynomial interpolation, of which the most thoroughly examined has been the one based on cubic spline functions,[102] the local methods consist in splitting the total configuration space into subspaces of such dimension that the potential can be represented by a cubic polynomial function inside them. The calculation of the polynomial coefficients is carried out so as to ensure that the total potential and the first two derivatives are continuous at the boundaries (nodes). It should be noted that splines may either be adjusted to fit the points exactly or have no such restriction, in which case it is possible to give a weight to the adjusted points. The spacing between the nodes may also vary, the optimum distribution being one in which the density is approximately proportional to the curvature of the adjusted surface.[103] In spite of their extreme flexibility in one and two dimensions, the use of cubic splines for three or more dimensions has been rather limited. For example, the number of points required for a $15 \times 15 \times 15$ spline is 3375, which is an order of magnitude higher than the number of points usually calculated by ab initio methods. On the other hand, a cubic-spline fit to a model potential for the system HClH has shown a root-mean-square deviation of about 0.027 eV.[104] Such accuracy has proved insufficient to warrant agreement, point by point, between the classical trajectories run on the model surface and on the spline function which has been adjusted.[104] Furthermore, the use of 3D cubic splines defined in terms of the interparticle coordinates **R** may require ab initio points, and hence the solution of Schrödinger's equation, in unphysical regions of the molecule configuration space where the triangulation rules (8) cannot be satisfied.[105] The use of the Pekeris coordinate system presented in Section I.B may, in this context, be worth investigating. To our knowledge, no such study has been reported in the literature. The actual practice is to combine the use of 1D and 2D cubic splines with other methods. For example, in the 3D version of the rotated Morse-cubic spline method,[106] the potential assumes the form of a

Morse curve with the coefficients depending on two angular variables: one (α) defines the deviation of the molecule from linearity, while the other (ν) defines the angle of rotation of the Morse potential around a fixed point on the atom–atom–atom dissociation plateau (R_M is the distance between this point and the minimum of the Morse curve):

$$V(R_M, \alpha, \nu) = D_M(\alpha, \nu)(1 - \exp\{-\beta_M(\alpha, \nu)[R_M^0(\alpha, \nu) - R_M]\})^2. \qquad (62)$$

In spite of the flexibility introduced into equation (62) by using 2D splines to describe $D_M(\alpha, \nu)$, a comparison of classical trajectories run on a model potential with those run on a rotated Morse-cubic spline fitted to that model does not show good point-by-point agreement (see ref. 54 and references therein).

Since the excessive number of ab initio points which have to be calculated presents a major handicap to the use of local methods, it is essential to have a global method for the representation of the potential-energy surface for the majority of systems of practical interest. In fact, even if we were to have the necessary computation time, only for a very few systems (H_3,[107–112] HeH_2,[113,114] and a few others) would the ab initio calculation reach sufficient accuracy to be of a practical interest for dynamical studies. For example, it proves extremely difficult to calculate potential-energy barrier heights with chemical accuracy ($\leqslant 1$ kcal/mol). It seems more feasible, though, to know the potential topography (i.e., the geometry and force constants) accurately at the saddle point.[108]

With either a semiempirical or an empirical basis, the global methods have in general fewer adjustable parameters than the local methods. Although this may imply less flexibility for the global methods, it does also have the advantage of giving to such methods greater reliability for the extrapolation of the potential to regions of the molecular configuration space where no ab initio or experimental data exist. Amongst these, the LEPS,[12,20] the DIM,[52,53]* and the MBE[16] methods have acquired the greatest popularity. Based on familiar concepts from the study of intermolecular forces, the MBE method will be thoroughly analyzed in the next subsection, and a generalization of it (DMBE) will be presented in Section IV.C. Of the LEPS and DIM methods, which are based on valence-bond theory, only the LEPS method will be commented on in Section IV.B.1, set in the context of the MBE for potential surfaces with conical intersections.

*The LEPS and DIM methods become identical for s^3 systems provided that orbital-overlap effects are disregarded.[115]

B. The MBE Method

The importance of non-pairwise-additive terms in interactions involving clusters of atoms or molecules is well known, and has already been referred to in Section III.B in connection with the dispersion energy. In the present subsection we shall show that the MBE method[16] offers a general strategy for the global representation of $V(\mathbf{R})$.

In principle, the potential energy for a polyatomic system may be written as a many-body expansion[16]

$$V_{\mathrm{ABC}\cdots\mathrm{N}}(\mathbf{R}) = \sum V_{\mathrm{A}}^{(1)} + \sum V_{\mathrm{AB}}^{(2)}(R_1) + \sum V_{\mathrm{ABC}}^{(3)}(R_1, R_2, R_3) + \cdots, \tag{63}$$

where the summations are over all terms of a given type, and where $V_{\mathrm{A}}^{(1)}$ is the energy of atom A in the electronic state resulting from the adiabatic withdrawal of this atom from the AB$\cdots$N aggregate; $V_{\mathrm{AB}}^{(2)}$ is the two-body energy, which is a function of the distance between atoms A and B (which vanishes asymptotically when $R_1 \to \infty$); $V_{\mathrm{ABC}}^{(3)}$ is the three-body energy, which is a function of the sides of the triangle shaped by atoms A, B, and C (which must go to zero when any one of the three atoms is removed to infinity); etc.

In equation (63) we must insist, however, that the potential have only one set of atomic and molecular fragments as the dissociation limit. Hence, the electronic states of those fragments (which are established by the spin–spatial Wigner–Witmer correlation rules[116]) are supposed to be independent of the way the dissociation occurs. This is a typical situation in van der Waals molecules, e.g., $\mathrm{RgX_2}$ (Rg = rare gas, X = H, Li,$\ldots$),

$$\mathrm{RgX_2}(\tilde{X}\,^1A') \to \begin{cases} \mathrm{X_2}(\tilde{X}\,^1\Sigma_g^+) + \mathrm{Rg}(^1S) \\ \mathrm{RgX}(\tilde{X}\,^2\Sigma^+) + \mathrm{X}(^2S) \\ \mathrm{Rg}(^1S) + 2\mathrm{X}(^2S) \end{cases} \tag{64}$$

but can also be realized in the case of some chemically stable molecules, e.g., $\mathrm{HO_2}$,

$$\mathrm{HO_2}(\tilde{X}\,^2A'') \to \begin{cases} \mathrm{O_2}(\tilde{X}\,^3\Sigma_g^-) + \mathrm{H}(^2S) \\ \mathrm{OH}(\tilde{X}\,^2\Pi) + \mathrm{O}(^3P) \\ \mathrm{H}(^2S) + 2\mathrm{O}(^3P) \end{cases} \tag{65}$$

However, there are potentials in which the electronic states of the fragments depend on the way the dissociation limit is reached. One of the most studied examples[117] where this occurs is the ground electronic state of $\mathrm{H_2O}$,

$$
H_2O(\tilde{X}\,{}^1A') \rightarrow \begin{cases} H_2(\tilde{X}\,{}^1\Sigma_g^+) + O({}^1D) & \text{(a)} \\ H_2(b\,{}^3\Sigma_u^+) + O({}^3P) & \text{(b)} \\ OH(A\,{}^2\Sigma^+) + H({}^2S) & \text{(a)} \\ OH(\tilde{X}\,{}^2\Pi) + H({}^2S) & \text{(b)} \\ O({}^1D) + 2H({}^2S) & \text{(a)} \\ O({}^3P) + 2H({}^2S) & \text{(b)} \end{cases} \tag{66}
$$

In this case, fragments (b) are obtained when an atom is removed from the equilibrium molecule along a dissociation path with C_s symmetry, whereas fragments (a) are obtained if the dissociation occurs along a path with a $C_{\infty v}$ symmetry. Thus, there is an intersection of Σ and Π states along the $C_{\infty v}$ projection line, which changes into an avoided intersection for nonlinear geometries.

Another example[118] is $O_3(\tilde{X}\,{}^1A_1)$, whose potential presents three equivalent C_{2v} minima (imposed by symmetry) associated with the equilibrium structure, and a metastable D_{3h} minimum expected from ab initio calculations[119] (for a recent study see, e.g., ref. 120). The O_3 equilibrium geometry correlates with the ground state (a) of the dissociation products,

$$
O_3(\tilde{X}\,{}^1A_1) \rightarrow \begin{cases} O_2(\tilde{X}\,{}^3\Sigma_g^-) + O({}^3P) & \text{(a)} \\ O_2(a\,{}^1\Delta_g) + O({}^1D) & \text{(b)} \\ 3O({}^3P) & \text{(a)} \\ 2O({}^3P) + O({}^1D) & \text{(b)} \end{cases} \tag{67}
$$

which has led to the conclusion[121] that equation (63) would be suitable for the whole configuration space. However, theoretical evidence exists showing that this may not be the case. In fact, ab initio calculations[122] show that the electronic structure of ground-state O_3 for a $D_{\infty h}$ geometry with 1.37-Å nearest-neighbor distance has a ${}^1\Delta_g$ symmetry. In this case, the Wigner–Witmer rules lead to the dissociation (b) with O_2 in the ${}^1\Delta_g$ excited electronic state. To correlate with the ground-state products, such a structure should have a ${}^1\Sigma_g^+$ symmetry. It is therefore likely that a conical intersection of ${}^1\Delta_g$ and ${}^1\Sigma_g^+$ states exists at $C_{\infty v}$. We must note that a C_{2v} distortion changes ${}^1\Delta_g$ and ${}^1\Sigma_g^+$ into $A_1 + B_1$ and A_1, respectively, which justifies the conical intersection of the A_1 states.

The functional representation of the potential-energy surface for the ground state of molecules like H_2O and O_3, which includes the nonanalytic behavior due to the conical intersection, may be represented as the lowest-energy eigenvalue of a 2×2 matrix[16]

$$\begin{bmatrix} V_{11}(\mathbf{R}) & V_{12}(\mathbf{R}) \\ V_{12}(\mathbf{R}) & V_{22}(\mathbf{R}) \end{bmatrix}, \tag{68}$$

where the diagonal elements V_{11} and V_{22} represent the diabatic potentials associated with the dissociation channels (a) and (b), respectively, and V_{12} is the coupling term, which for the above systems should vanish at linear geometries.[117,118] For example, in the case of O_3 one has

$$V_{11} = \sum_{i=1}^{3} V_{\mathrm{OO}}^{(2)}(\tilde{X}\,{}^3\Sigma_g^-; R_i) + V_{11}^{(3)}(R_1, R_2, R_3), \tag{69a}$$

$$V_{22} = V_{\mathrm{O}}^{(1)}({}^1D) + \sum_{i=1}^{3} V_{\mathrm{OO}}^{(2)}(a\,{}^1\Delta_g; R_i) + V_{22}^{(3)}(R_1, R_2, R_3). \tag{69b}$$

In principle, the most general representation of the matrix elements of equation (68) includes a many-body expansion with one-, two-, and three-body terms. However, only two conditions [(a) and (b) in equations (66), (67)] may be used in the determination of the one- and two-body energy terms. If more conditions were to exist, the dimension of the matrix necessary to represent the adiabatic potential would be larger.[16] In what follows we shall argue that the definition of V_{12} must depend on the type of conical intersection, namely on whether its locus is finite or infinite in extent (Section II.C). For the cases of H_2O and O_3 [equations (66), (67)] the conical intersection occurs along the $C_{\infty v}$ projection line, but only for finite values of the molecular perimeter (for H_2O, there is also a simple intersection at the $O + H_2$ asymptotic channel, which is avoided for finite values of the OH distances[117]). For example, one gets for the linear dissociation of O_3

$$V_{11} = V_{\mathrm{OO}}^{(2)}(\tilde{X}\,{}^3\Sigma_g^-; R_1), \tag{70a}$$

$$V_{22} = V_{\mathrm{O}}^{(1)}({}^1D) + V_{\mathrm{OO}}^{(2)}(a\,{}^1\Delta_g; R_1), \tag{70b}$$

which shows that $V_{11} < V_{22}$ at any arbitrary value of R_1. Thus, the intersection is finite in extent. This result can be obtained, for large perimeters, from the 2×2 matrix (68) only if V_{12} vanishes faster than the diagonal terms.[123] In terms of a many-body expansion, one may have an unambiguous solution[16] by imposing the condition that V_{12} is reduced to the three-body term. We are therefore left with the problem of defining V_{12} for potentials where the locus of intersection is infinite in extent. A much-studied potential of this type is the ground state of H_3[27,124–126]; a similar situation may be found for the homonuclear trimers of the alkali metals.[31,45,127–129] We are indeed

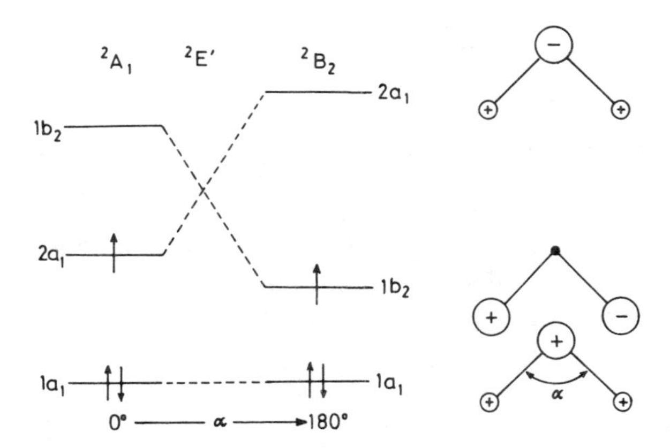

Figure 10. Diagram of molecular orbitals for the H_3 molecule.

allowed by the theory of molecular orbitals to write the electronic configuration of the ground state as $(a')^2(e')^1$, which gives a $^2E'$ state for not very small separations of three ^2S_g atoms on a geometry with D_{3h} symmetry. On the other hand, the Walsh-type diagram of Fig. 10 shows that a small C_{2v} distortion from the degenerate $^2E'$ state leads to a $^2A_1$ state if $\alpha < 60°$, but a $^2B_2$ state if $\alpha > 60°$.* Note that for large distortions the states of the corresponding linear geometries are $^2\Sigma_g^+$ and $^2\Sigma_u^+$. Note also that in the case of H_3, for very small D_{3h} geometries, the $^2E'$ state should evolve for the ^2S_g state of the united atom, $Li(1s^22s^1;{}^2S_g)$. Thus, along the D_{3h} axis, there should occur an intersection of the $^2E'$ state with the $^2A_1'$ state, whose symmetry ($^2A'$) is equal to the symmetry of the components of the $^2E'$ state. In this case, a C_{2v} distortion from D_{3h} geometry always leads to a $^2A_1$ state, even when[26] $\alpha > 60°$. Therefore, for the case of H_3, the dissociation limits are

$$H_3(^2E') \rightarrow \begin{cases} H_2(\tilde{X}\,{}^1\Sigma_g^+) + H(^2S) & (^2A_1), \\ H_2(b\,{}^3\Sigma_u^+) + H(^2S) & (^2B_2). \end{cases} \tag{71}$$

No doubt, if one writes

$$V_{11} = \sum_{i=1}^{3} V_{HH}^{(2)}(\tilde{X}\,{}^1\Sigma_g^+; R_i) + V_{11}^{(3)}(R_1, R_2, R_3), \tag{72a}$$

*For the MM_2' (M, M′ = Li, ..., Cs) trimers, semiempirical calculations[130] based on valence-bond theory suggest that the most stable state is $^2B_2$ ($^2A_1$) with $\alpha > 60°$ ($\alpha < 60°$), provided that the trimer is of type LHH (HLL), where L represents a light atom and H a heavy one. (For example, $LiNa_2$ is of type LHH.)

$$V_{22} = \sum_{i=1}^{3} V_{\mathrm{HH}}^{(2)}(b\,^3\Sigma_u^+; R_i) + V_{22}^{(3)}(R_1, R_2, R_3), \tag{72b}$$

it will not be possible to obtain a conical intersection of the two doublet states for large perimeters if V_{12} is to be represented by a three-body term alone. Nevertheless, the valence-bond treatment of H_3 shows that the intersection extends to infinitely large perimeters. Since the wave function based on valence theory is correct in first order, a choice of two-body terms based on this theory is entirely justified. This approach will be carried out in the next subsection.

1. *The* MBE *Method for* s^3 *Systems*

As is well known (see, e.g., ref. 12), the wave functions for the doublet states of an s^3 system obtained from the basis

$$\Psi_1 = \| s_a\alpha \; s_b\alpha \; s_c\beta \|, \quad \Psi_2 = \| s_a\alpha \; s_b\beta \; s_c\alpha \|, \quad \Psi_3 = \| s_a\beta \; s_b\alpha \; s_c\alpha \| \tag{73}$$

may be chosen as

$$\Phi_1 = \frac{\Psi_2 - \Psi_3}{\sqrt{2}}, \qquad \Phi_2 = \frac{2\Psi_1 - \Psi_2 - \Psi_3}{\sqrt{6}}. \tag{74}$$

On neglecting orbital overlap, the hamiltonian matrix has a structure analogous to equation (26), namely

$$\begin{bmatrix} \mathcal{Q} + \frac{1}{2}(2\mathcal{J}_1 - \mathcal{J}_2 - \mathcal{J}_3) & \frac{1}{2}\sqrt{3}(\mathcal{J}_2 - \mathcal{J}_3) \\ \frac{1}{2}\sqrt{3}(\mathcal{J}_2 - \mathcal{J}_3) & \mathcal{Q} - \frac{1}{2}(2\mathcal{J}_1 - \mathcal{J}_2 - \mathcal{J}_3) \end{bmatrix}, \tag{75}$$

with the eigenvalues being given by [see equation (28)]

$$V_{\pm} = X_{\mathrm{LEPS}} \pm \tfrac{1}{2} Y_{\mathrm{LEPS}}^{1/2}, \tag{76a}$$

where

$$X_{\mathrm{LEPS}} = \mathcal{Q} = \sum_{i=1}^{3} \mathcal{Q}_i, \tag{76b}$$

$$Y_{\mathrm{LEPS}} = \sum_{i,j=1}^{3} (\mathcal{J}_i - \mathcal{J}_j)^2, \tag{76c}$$

and the $\mathcal{Q}$'s and $\mathcal{J}$'s represent diatomic Coulomb and exchange integrals, respectively. For example, for the AB fragment, one has $\mathcal{Q}_1 = \langle s_a(1)s_b(2)|H'|s_a(1)s_b(2)\rangle$, $\mathcal{J}_1 = \langle s_a(1)s_b(2)|H'|s_a(2)s_b(1)\rangle$, where H' is the interaction hamiltonian, $H' = -r_{\mathrm{B1}}^{-1} - r_{\mathrm{A2}}^{-1} + r_{12}^{-1} + R_{\mathrm{AB}}^{-1}$.

In 1929 Eyring and Polanyi[20] introduced the semiempirical treatment of the London equation (76), the method being named LEPS due to the important contribution of Sato[20] in 1955. In one of its most popular versions,[131] the method consists of calculating the $\mathscr{Q}$ and $\mathscr{J}$ integrals from the potentials for the ground singlet and lowest triplet states of the diatomic fragments calculated within the same approximation level as equation (76a). Thus, for H_3, they are written as

$$V_{\mathrm{HH}}^{(2)}(\tilde{X}\,^1\Sigma_g^+; R_1) = \mathscr{Q}_1 + \mathscr{J}_1, \tag{77a}$$

$$V_{\mathrm{HH}}^{(2)}(b\,^3\Sigma_u^+; R_1) = \mathscr{Q}_1 - \mathscr{J}_1. \tag{77b}$$

By using equations (77), the hamiltonian matrix (75) can be cast in a form similar to that of equation (68), (72) by writing

$$V_{11} = \sum_{i=1}^{3} V_{\mathrm{HH}}(\tilde{X}\,^1\Sigma^+; R_i) - \tfrac{3}{2}(\mathscr{J}_2 + \mathscr{J}_3), \tag{78a}$$

$$V_{22} = \sum_{i=1}^{3} V_{\mathrm{HH}}^{(2)}(b\,^3\Sigma^+; R_i) + \tfrac{3}{2}(\mathscr{J}_2 + \mathscr{J}_3), \tag{78b}$$

$$V_{12} = \tfrac{1}{2}\sqrt{3}(\mathscr{J}_2 - \mathscr{J}_3). \tag{78c}$$

Note, however, that all terms in equation (78) are of two-body type, since they decay to zero with a single internuclear distance.

Although the LEPS method can be used to calculate on a priori grounds (i.e., without adjustable parameters) the potential energy of a polyatomic molecule, it is as a global functional form to fit available ab initio or empirical data that the method has deserved greater attention. Basically, this is because the diatomic triplet-state curve (77b) does not play any effective part in dynamical studies of the triatomic system which involve only the ground-state potential-energy surface (i.e., it does not correspond to any atom–molecule asymptotic dissociation limit). Thus, it can be taken as an arbitrary functional form to be determined from a fit to existing ab initio and experimental data on the triatomic (polyatomic, in general) molecule. For example, in the case of H_3, Truhlar and Horowitz[132] used the effective triplet state form

$$V_{\mathrm{HH}}^{\mathrm{eff}}(b\,^3\Sigma_u^+) = \left(\sum_{i=0}^{2} \beta_i R^i\right) e^{-\gamma R}, \tag{79}$$

where β_i ($i = 0, 1, 2$) and γ are adjustable parameters. Of these parameters, two were determined so that, using the "exact" ab initio curve[133] for the ground singlet state of H_2 [equation (77a)], the H_3 potential surface reproduced the location and height of the linear symmetric ($D_{\infty h}$) saddle point for the hydrogen-atom exchange reaction as calculated by Liu and Siegbahn.[107,108] The remaining two parameters were determined from a least-squares fit to the H_3 ab initio points[107,108] for $D_{\infty h}$ geometries. For the same system, we have shown[134] how to determine a functional form for $V_{HH}^{eff}(b\,^3\Sigma_u^+)$ so as to reproduce the full quadratic force field at the H_3 saddle point. The LEPS method, as given by equations (76), (77), has also been used for the calculation of the ground-state potential-energy surface of the alkali-metal trimers.[129,130] In this case, the potential $V_{HH}^{eff}(b\,^3\Sigma^+)$ has been represented by an EHFACE model (see Section IV.D) suitably parametrized to reproduce the experimental atomization energy[135] of Li_3 and generalized, in the light of previous work[136] on the alkali dimers, for the remaining homonuclear and heteronuclear alkali trimers.

The most general many-body expansion of the potential for the ground doublet state of an s^3 system may, in principle, be obtained by adding a three-body term to the diagonal and nondiagonal elements of the hamiltonian matrix (75), say $V_{11}^{(3)}$, $V_{22}^{(3)}$, and $V_{12}^{(3)}$:

$$V_{11} = \sum_{i=1}^{3} V_i^{(2)}(\tilde{X}\,^1\Sigma_g^+; R_i) - \tfrac{3}{2}(\mathcal{J}_2 + \mathcal{J}_3) + V_{11}^{(3)}(R_1, R_2, R_3), \qquad \text{(80a)}$$

$$V_{22} = \sum_{i=1}^{3} V_i^{(2)}(b\,^3\Sigma_u^+; R_i) + \tfrac{3}{2}(\mathcal{J}_2 + \mathcal{J}_3) + V_{22}^{(3)}(R_1, R_2, R_3), \qquad \text{(80b)}$$

$$V_{12} = \tfrac{1}{2}\sqrt{3}(\mathcal{J}_2 - \mathcal{J}_3) + V_{12}^{(3)}(R_1, R_2, R_3). \qquad \text{(80c)}$$

After diagonalization of the resulting 2×2 potential matrix one gets

$$V_{\pm} = X \pm \tfrac{1}{2} Y^{1/2}, \qquad \text{(81a)}$$

where

$$X = X_{\text{LEPS}} + X^{(3)}, \qquad \text{(81b)}$$

$$Y = Y_{\text{LEPS}} + Y^{(3)}, \qquad \text{(81c)}$$

and the notation

$$X^{(3)} = \tfrac{1}{2}(V_{11}^{(3)} + V_{22}^{(3)}), \qquad \text{(81d)}$$

$$Y^{(3)} = (V_{11}^{(3)} - V_{22}^{(3)})[(V_{11}^{(3)} - V_{22}^{(3)}) + 2(2\mathcal{J}_1 - \mathcal{J}_2 - \mathcal{J}_3)] + 4V_{12}^{(3)}[V_{12}^{(3)} + \sqrt{3}(\mathcal{J}_2 - \mathcal{J}_3)] \qquad \text{(81e)}$$

has been used for simplicity.

However, for systems with twofold or higher permutational symmetry, the $X^{(3)}$ and $Y^{(3)}$ terms fail to have the appropriate symmetry, due to the

$\mathscr{J}$-containing factors. Such symmetry restrictions are, of course, inherent in the simple London equation, but are lost when three-body terms are arbitrarily added to the hamiltonian matrix elements. To preserve the symmetry of the London equation, which is accurate to first order, one has to impose $V_{11}^{(3)} = V_{22}^{(3)}$ and $V_{12}^{(3)} = 0$. However, this may reduce the flexibility of the final form for fitting purposes as $Y^{(3)}$ becomes zero. An alternative approach is to formally treat $X^{(3)}$ and $Y^{(3)}$, rather than the individual three-body energy terms in the matrix elements, as the fitting functions. In this case, it is on $X^{(3)}$ and $Y^{(3)}$ that the symmetry requirements are to be imposed. For example, for a system formed from three equal ^2S atoms, $X^{(3)}$ and $Y^{(3)}$ must be totally symmetric on permutation of any atoms. Moreover, the two Riemann sheets for the doublet states must intersect conically along D_{3h} geometries, thus implying that $Y^{(3)}$ must be zero in D_{3h}. A convenient functional form which satisfies the above requirements can in principle be represented in terms of the integrity basis (15) as the product of a polynomial and an exponentially decreasing range-determining factor, namely

$$X^{(3)} = \mathrm{Pol}(\Gamma_1, \Gamma_2, \Gamma_3)\,\mathrm{Range}(\Gamma_1), \tag{82}$$

$$Y^{(3)} = \mathrm{Pol}'(\Gamma_1, \Gamma_2, \Gamma_3)\,\mathrm{Range}'(\Gamma_1). \tag{83}$$

Note, however, that $Y^{(3)}$ cannot contain any terms which depend on Γ_1 alone or are constant, because they would not vanish at D_{3h} geometries. In summary, one has

$$V_{\pm} = X_{\mathrm{LEPS}} + X^{(3)} \pm \tfrac{1}{2}(Y_{\mathrm{LEPS}} + Y^{(3)})^{1/2}, \tag{84}$$

a form similar to that recently suggested by Murrell and the present author.[123] It should be pointed out that in practical applications the square-root argument must remain positive over the complete molecular configuration space so as to have physical meaning.

Consider now the case of AB_2-type systems formed from ^2S atoms. In this case, the only possibility of degeneracy arises for C_{2v} geometries, as anticipated from the London equation. Yet, since the symmetry is not forced, the intersection locus may be infinite or finite in extent. In fact, though the intersection condition for C_{2v} $(R_2 = R_3)$ geometries requires that

$$\mathscr{J}_{\mathrm{BB}}(R_1) = \mathscr{J}_{\mathrm{AB}}(R_2), \tag{85}$$

such intersection will occur in the space physically accessible to the nuclei only if the triangle inequality (8) is verified, which allows one to write

$$2R_2 \geqslant R_1. \tag{86}$$

On the other hand, valence orbitals are more diffuse for larger atoms. Thus, with, e.g., Li and Na, one may write for large R

$$|\mathscr{J}_{\mathrm{NaNa}}(R)| > |\mathscr{J}_{\mathrm{LiNa}}(R)| > |\mathscr{J}_{\mathrm{LiLi}}(R)|, \tag{87}$$

which shows that the values of R_1 and R_2 that satisfy equation (85) also satisfies equation (86) for the case of $NaLi_2$ but not for $LiNa_2$. The intersection locus is therefore infinite in extent for $NaLi_2$,[123,130] but need not be for $LiNa_2$.

We conclude this subsection by noting that a similar approach has been suggested[137] to obtain a two-valued function for the lowest singlet states of an s^4 system. In this case, X and Y [equation (81)] include three-body and four-body energy terms, which can be suitably chosen so as to warrant the topography (which would be correct, at least for large internuclear separations where the valence wave function is a good approximation of the exact wave function) of the LEPS potential as far as the structure of the degeneracy manifold[138] is concerned.

C. The DMBE Method

The conceptual and practical advantages of partitioning the interaction energy into (single-configuration) Hartree–Fock and dynamical correlation contributions are well known from the study of nonbonding interactions.[11,17,82,139] Nevertheless, for bonding interactions the single-configuration Hartree–Fock method is known to be unstable, and it often leads to an incorrect description of the dissociation limits. Thus, as already noted in Section III.C, it is convenient to distinguish two types of electron correlation: the internal or nondynamical, and the external or dynamical. The nondynamical correlation is related to rearrangements of the electrons by degenerate or nearly degenerate valence orbitals, and is therefore essential for the correct description of bond dissociation or bond formation. We shall use the abbreviation* EHF to imply either an Hartree–Fock energy or, whenever justified, an energy of the extended-Hartree–Fock type. The dynamical correlation (corr) is in turn associated with instantaneous nonspecific correlations of the electrons in their motion. Hence, it includes formally not only interfragment and intrafragment correlations, but also intra–inter coupling terms.

The extended-Hartree–Fock energy is essentially built up by the first-order exchange and electrostatic energy contributions, together with the second-order induction energy. Thus, it accounts for single excitations only in just one atom. In fact, the molecular wave function may, in principle, be written in terms of localized (atomic) orbitals, and according to Brillouin's theorem, single excitations do not mix with the Hartree–Fock fundamental configur-

*For clarity, this replaces the notation HF used in previous work.[17]

ation. In turn, the dynamical correlation includes all classes of double and multiple excitations in one of the atoms (intraatomic correlation) as well as single and multiple excitations in more than one atom (interatomic correlation and intra–inter coupling terms). This result is illustrated schematicaly in Fig. 11.

The partitioning of the potential energy into EHF and corr contributions proves to be most useful when used in conjunction with the many-body expansion in what has been called the double many-body expansion (DMBE).[17,21] The three first terms are

$$V_{\mathrm{A}}^{(1)} = V_{\mathrm{A,EHF}}^{(1)} + V_{\mathrm{A,corr}}^{(1)}, \tag{88a}$$

$$V_{\mathrm{AB}}^{(2)}(R_1) = V_{\mathrm{AB,EHF}}^{(2)}(R_1) + V_{\mathrm{AB,corr}}^{(2)}(R_1), \tag{88b}$$

$$V_{\mathrm{ABC}}^{(3)}(R_1, R_2, R_3) = V_{\mathrm{ABC,EHF}}^{(3)}(R_1, R_2, R_3) + V_{\mathrm{ABC,corr}}^{(3)}(R_1, R_2, R_3), \tag{88c}$$

where each term of the EHF and corr series expansions satisfies the appropriate asymptotic limits (Section IV.B). It is obvious that in the case of multivalued surfaces each Riemann sheet will have its own DMBE. Because such situations do not involve specific problems besides those already

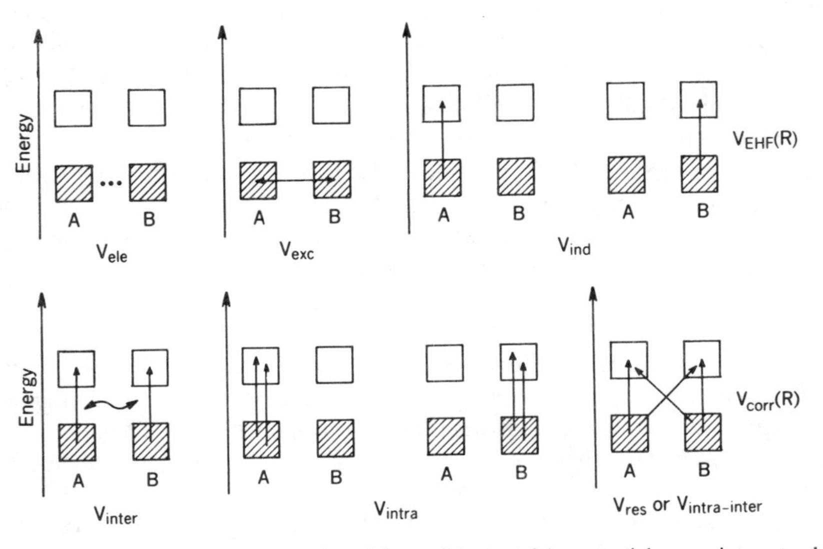

Figure 11. Schematic representation of the partitioning of the potential energy into extended Hartree–Fock (EHF), including the nondynamical correlation energy, and dynamical correlation (corr). See text.

mentioned in the previous section, we shall assume henceforward a single-valued surface.

If all the terms of the DMBE were to be calculated by exact ab initio methods, the potential-energy surface would be reproduced exactly. However, it is in the allowance for using reliable estimates of the various terms in the EHF and corr series expansions that the DMBE has its greatest value. Two conditions must be fulfilled in order that the DMBE may achieve this: (a) the EHF and corr series expansions converge fast enough that one may neglect for practical purposes the terms higher than some order; (b) there must be simple functional forms capable of describing analytically the n-body energy terms of that double expansion. Although it is not possible to make absolute statements about (a), it is at least reasonable to expect different convergence rates for the EHF and corr series expansions. Therefore, the terms of those expansions will contribute with different weights to the total energy, their individual importance depending on the region of the molecular configurational space. This, together with the fact that the EHF and corr energy terms have different functional forms, gives the DMBE method its greatest advantage over the MBE method. Moreover, even if convergence is slow, the DMBE formalism allows an estimate of the potential-energy surface in intermediate regions (where the dynamical correlation energy plays an important role, particularly for van der Waals interactions), once we know the asymptotic contributions of the interaction energy for small and large distances. On the other hand, the results from studies carried out for small polyatomic systems using the MBE method[16] allow us to conclude that the main topographical details of the potential surface are contained in the two-body and three-body energy terms, leaving the n-body terms ($n \geqslant 4$) for fine tuning for chemical accuracy. We must expect that this conclusion will not be drastically changed for the DMBE method. Therefore, in contrast to the two- and three-body energy terms, which need to be accurately represented, the higher-order terms should only need to be described in an approximate way. Accordingly, requirement (b) may be regarded as satisfied.

D. The Functional Form of the n-Body Energy Terms

As is already clear, a reliable functional representation of the potential-energy surface is of extreme importance. Such a functional form must be as simple as possible in order to facilitate solving the dynamical problem, but must contain sufficient complexity that the reliability of the representation will not be jeopardized.[105] In this spirit we have suggested models which have physical motivation and allow a simple, yet reliable, representation of the EHF and corr two-body[17,136,140] and three-body[17,21,141] energy terms. Since the EHF terms are, in principle, calculable by ab initio methods (and subsequently represented in analytic form) whereas the corr terms are calculated semiempirically, those models have been named EHFACE2 and EHFACE3; in general, EHFACEn is the abbreviation for extended-Hartree–Fock approximate correlation energy for n-body interactions. We concentrate in the remaining sections of this chapter on a brief analysis of these models.

1. *The* EHFACE2 *Model*

The extended-Hartree–Fock energy for an A $\cdots$ B interaction depends on the internuclear distance in a way similar to that of an overlap integral involving the outermost orbitals of A and B; see equation (48). It is therefore not surprising that the most popular representations are of the form

$$V^{(2)}_{\mathrm{AB,HF}}(R) = R^{\tilde{\alpha}} P_2(R) \exp[-P'_2(R)], \tag{89a}$$

where

$$P_2(R) = \sum_{i=0}^{m} a_i R^i, \tag{89b}$$

$$P'_2(R) = \sum_{i=1}^{m'} b_i R^i, \tag{89c}$$

and $\tilde{\alpha}$, a_i, and b_i are adjustable parameters. This includes the well-known simple Born–Mayer ($\tilde{\alpha} = 0, m = 0, m' = 1$) and modified Born–Mayer ($\tilde{\alpha} = 0$, $m = 0, m' = 2$) functions, as well as the corresponding screened-Coulomb-type ($\tilde{\alpha} = -1$, $m = 0$, $m' = 1, 2$) functions. Moreover, by defining the polynomials $P_2(R)$ and $P'_2(R)$ in terms of the displacement coordinate $R - R_m$, rather than R, where R_m is the equilibrium geometry of the total potential-energy curve, one gets, for $m = 3$ and $m' = 1$, the forms suggested in ref. 140.

In contrast with the EHF energy, which can generally be estimated by ab initio methods for many systems of practical interest (see ref. 140 for a semiempirical treatment), the interatomic correlation energy is most conveniently approximated from the dispersion energy by accounting semiempirically for orbital overlap and electron exchange effects. Thus the following mathematical form has been suggested[17,136].

$$V^{(2)}_{\mathrm{AB,corr}}(R) = -\sum_n C_n^{\mathrm{AB},*}(R) R^{-n}, \qquad n = 6, 8, 10, \ldots, \tag{90a}$$

where

$$C_n^{\mathrm{AB},*} = C_n^{\mathrm{AB}} \chi_n(R), \tag{90b}$$

$$\chi_n = \{1 - \exp[-A(n)x - B(n)x^2]\}^n, \tag{90c}$$

$$x = \frac{R}{\rho} = \frac{2R}{R_m + 2.5R_0}, \tag{90d}$$

$$R_0 = 2(\langle r_A^2 \rangle^{1/2} + \langle r_B^2 \rangle^{1/2}), \tag{90e}$$

$$A(n) = \alpha_0 n^{-\alpha_1}, \tag{90f}$$

$$B(n) = \beta_0 \exp(-\beta_1 n). \tag{90g}$$

Note the appearance of the AB equilibrium geometry (R_m, which can be obtained by a self-consistent procedure[136]) and Le Roy's parameter[142] [R_0, which represents the smallest value of the internuclear distance for which the asymptotic series of the dispersion energy is still a good representation of the damped series (49)] in the definition of the reduced coordinate x; $\langle r^2 \rangle$ represents the expectation value of the square of the radial coordinate for the outermost valence electrons, which is tabulated in the literature[143] for atoms with $1 \leqslant Z \leqslant 120$. Other important parameters in the dispersion damping functions are $A(n)$ and $B(n)$, which have been determined[17] from a fit to the ab initio perturbation results for the $H_2(b\,^3\Sigma_u^+)$ interaction: $\alpha_0 = 25.9528$, $\alpha_1 = 1.1868$, $\beta_0 = 15.7381$, $\beta_1 = 0.09729$. In turn, the factor ($=2.5$) which multiplies R_0 in the reduced coordinate x has been defined so that, using the average values of the dispersion coefficients[57] for Li_2 and the SCF energies of Olson and Konowalov,[144] the $Li_2(a\,^3\Sigma_u^+)$ potential of ref. 144 was well reproduced. Thus, it has been suggested[136] that by replacing the EHF energy and the parameters $\langle r^2 \rangle$ and C_n ($n = 6, 8, 10, \ldots$), which are specific for the A$\cdots$B interaction in equation (90), a reliable estimate of the interatomic correlation energy for that interaction may be obtained.

Concerning the interaction energy due to the intraatomic correlation, ab initio calculations for systems with few electrons (He_2,[145] HeH[146]) suggest (except for the strong-interaction region) a small but nonnegligible dependence on R. Such a dependence may be amplified in cases where there are a large number of effective valence electrons.[147] On the other hand, the correlation energy due to the intra–inter coupling terms also appears to be nonnegligible.[146–148] Although it is reasonable to expect that such energy terms should depend on R in a similar way to an overlap integral, such an overlap would mostly involve nonoccupied diffuse orbitals at centers A and B. Such expectations are well confirmed from the results of MC SCF calculations for the H–Rg (Rg = He, Ne, Ar) systems.[146] In fact, the total absolute value of those correlation energies, which is an order of magnitude smaller than that of the Hartree–Fock energy, is well represented by a curve parallel to it in a semilogarithmic plot.[136] Hence the residual part (res) of the dynamical correlation can be estimated as a fraction λ of the EHF curve according to

$$V_{\text{AB,res}}^{(2)}(R) = \lambda V_{\text{AB,EHF}}^{(2)}(R). \tag{91}$$

In principle, the value of λ may be determined from a fit to experimental data such as vibrational–rotational energies, second virial coefficients, viscosities, diffusion coefficients, and thermal conductivities.[149]

Finally, it should be noted that relativistic effects may be significant for interactions involving heavy atoms.[150] It is expected that such effects will show themselves through the exponential decay law of the EHF curve, since they result from a contraction of the electronic density due to the large kinetic energy of the core electrons. Preliminary studies for the Kr_2 and Xe_2 systems using a screened-Coulomb potential to represent the EHF curve suggest that the relativistic effects can be modeled by inclusion of an extra R^2 term in the exponential (thus, through a modified screened-Coulomb potential). Typically, the parameter b_2 for these interactions at the relativistic (non-relativistic) level is found to be 4–7% (3–5%) of b_1 for the corresponding screened-Coulomb fits.

For interactions in which both atoms are in non-S states, and may therefore have permanent electric moments (quadrupolar or higher order), the electrostatic energy assumes asymptotically the form of an inverse power series in R. As it is the dominant contribution at large distances, it should be treated separately from the remaining extended-Hartree–Fock energy. Due to the lack of ab initio calculations for the corresponding damping functions (which are representative of the orbital-overlap and electron-exchange effects), the use of equations (90b)–(90g) for extrapolation to the appropriate values of n has been suggested.[140]

For ion–atom interactions, and possibly for other interactions in which the induction energy plays an important role, it is also justified to study the latter separately from the remaining extended-Hartree–Fock energy. In relation to this, we note that by combining equations (40), (43) one obtains

$$C_n(l_a, l_b) = \binom{2l_a + 2l_b}{2l_a} \frac{\Delta\varepsilon_{l_a} \Delta\varepsilon_{l_b}}{\Delta\varepsilon_{l_a} + \Delta\varepsilon_{l_b}} C_n(l_a, 0) C_n(0, l_b), \tag{92}$$

where $n = 2(l_a + l_b + 1)$. It may be expected from equation (92) that a similar relationship can be found between the dispersion and induction damping functions. The results from ab initio calculations on H_2^+ ($\tilde{X}\,^2\Sigma_g^+$ and $1\,^2\Sigma_u^+$ states[69,70]) and H_2 ($\tilde{X}\,^1\Sigma_g^+$ and $b\,^3\Sigma_u^+$ states[69,73]) seem to corroborate this prediction through the relationship[151]

$$\chi_n(l_a, l_b; R) = \chi_n(l_a, 0; R)\chi_n(0, l_b; R) \tag{93}$$

provided the scaling factor ρ in the reduced coordinate x [and eventually also $\Delta\varepsilon_a$ and $\Delta\varepsilon_b$ in equation (92)], are suitably chosen. Thus, one may hope that equation (93) provides a consistent representation of the damping functions for the induction and dispersion coefficients. Note that the so-obtained term-to-term dispersion damping functions, $\chi_n(l_a, l_b; R)$, may vary for different combinations of l_a and l_b which correspond to the same value of $n = 2(l_a + l_b + 1)$. Of course, this problem does not exist for the leading terms $n = 6$ and 8, but is present for the case of $n = 10$, since in this case there are three possible values of (l_a, l_b), namely (1, 3), (3, 1), and (2, 2). It is unlikely, though, that the use of damping functions which depend only on the value of n will lead to errors of great practical significance.

2. *The* EHFACE3 *Model*

As for the EHFACE2 model, the three-body energy is partitioned into EHF and corr contributions. Of these, the three-body EHF is conveniently represented by the form

$$V^{(3)}_{\mathrm{ABC,EHF}}(R_1, R_2, R_3) = P_3(R_1, R_2, R_3)T_3(R_1, R_2, R_3), \tag{94}$$

where P_3 is a three-body polynomial in the interparticle coordinates,

$$P_3 = V^0\left(1 + \sum_i c_i Q_i + \sum_i \sum_{j \geqslant i} c_{ij} Q_i Q_j + \cdots\right), \tag{95}$$

and the three-body range-determining factor T_3 assumes either of the two alternative forms

$$T_3 = 1 - \tanh(\gamma_0 + \gamma_1 Q_1) \tag{96a}$$

or

$$T_3 = \prod_{i=1}^{3} [1 - \tanh(\gamma_0^{(i)} + \gamma_1^{(i)} R_i)]. \tag{96b}$$

In equations (95), (96), Q_i $(i = 1, 2, 3)$ are the D_{3h}-symmetry coordinates introduced in Section II.B, and c and γ are parameters.

In turn, the three-body dynamical correlation energy is semiempirically estimated from the dispersion coefficients for the various separate- and united-atom limits, and those also known for the equilibrium geometries of the subsystems. Two types of interactions can however be distinguished, depending on whether they can produce a chemical bond or are nonbonding (such as those involving closed-shell atoms only). To be more specific, in the first case, two of the atoms (e.g., B and C) interact strongly and even produce a stable chemical bond (BC); the corresponding united atom is represented by Ⓑⓒ. Significant orbital-overlap and electron-exchange effects (exc) therefore occur and play an important role. The problem is often treated as a pseudo two-body interaction in which one of the interacting species is A and the other is the molecule (BC);[77,152–156] similarly, the dispersion coefficients for this pseudo two-body interaction are $C_n^{A(BC)}$ (the corresponding dispersion coefficient involving A and the united atom BC will be denoted $C_n^{A\,\text{(BC)}}$). However, note that important relationships can be established between the total effective atom–atom dispersion coefficients for the pairs A···B and A···C and the corresponding dispersion coefficients for the interactions A···(BC) and A··· (BC). For example, one gets ($R_1, R_3 \to \infty$)

$$\lim_{R_{BC}\to 0} C_n^{AB,\text{eff}} + \lim_{R_{BC}\to 0} C_n^{AC,\text{eff}} = C_n^{A\,\text{(BC)}}, \tag{97a}$$

$$\lim_{R_{BC}\to R_{BC}^0} C_n^{AB,\text{eff}} + \lim_{R_{BC}\to R_{BC}^0} C_n^{AC,\text{eff}} = C_n^{A(BC)}, \tag{97b}$$

$$\lim_{R_{BC}\to \infty} C_n^{AB,\text{eff}} = C_n^{AB}, \qquad \lim_{R_{BC}\to \infty} C_n^{AC,\text{eff}} = C_n^{AC}, \tag{97c}$$

where R_{BC}^0 is a reference geometry (for example the expectation value of R_{BC} for the ground vibrational state, $\langle R_{BC}\rangle$), and (BC) and (BC) are considered to be in the electronic state consistent with the Wigner–Witmer spin–spatial correlation rules[116] (however, see below). In order to represent the three-body correlation energy in terms of the effective atom–atom dispersion coefficients, we must find a functional form which allows the interpolation amongst the limits defined by equations (97). For this purpose we have suggested[17] (e.g., for the AB pair)

$$C_n^{AB,\text{eff}} = \frac{C_n^{AB,*}}{2}(g_n^{BC}h_n^{AC} + g_n^{AC}h_n^{BC}), \tag{98a}$$

where

$$g_n^{BC} = 1 + k_n^{BC}\exp\left[-k_n'^{BC}(R_{BC} - R_{BC}^0)\right], \tag{98b}$$

$$h_n^{BC} = (\tanh \eta_n^{BC} R_{BC})^{\eta_n'^{BC}} \tag{98c}$$

The three-body correlation energy[17] is therefore specified by

$$V_{\text{corr,exc}}^{(3)} = \sum_{AB}\sum_{n} C_n^{AB,*}\left[1 - \tfrac{1}{2}(g_n^{BC}h_n^{AC} + g_n^{AC}h_n^{BC})\right]R_{AB}^{-n}, \qquad n = 6, 8, 10, \tag{99}$$

or, in a more compact notation, by

$$V_{\text{corr,exc}}^{(3)} = \sum_{i=1}\sum_{n} C_n\chi_n(R_i) \times \left\{1 - \tfrac{1}{2}\left[g_n(R_j)h_n(R_k) + g_n(R_k)h_n(R_j)\right]\right\}R_i^{-n}, \qquad n = 6, 8, 10, \tag{100}$$

where the indices j and k are defined by $j = i + 1$ (mod 3) and $k = i + 2$

(mod 3), respectively, with $i, j, k = i-1, j-1, k-1$ in an obvious correspondence. Note that this expression has 36 constants: $k_n, k'_n, \eta_n, \eta'_n$ ($n = 6, 8, 10$ for the AB, BC, and AC pairs). Of these, k'_n and η_n estimate the rate of decay of functions g_n and h_n respectively. Thus, it has been suggested[17] that k_n and k'_n should be determined for predefined values of η_n and η'_n by solving equations (97a, b). We note that there are still two pairs of equations similar to these which involve the B + AC and C + AB asymptotic limits. Hence, from equations (97), (98), one gets for a given pair of values (η_n, η'_n)

$$k_n^{\mathrm{BC}} = \frac{2C_n^{\mathrm{A(BC)}} - C_n^{\mathrm{AB}} - C_n^{\mathrm{AC}}}{C_n^{\mathrm{AB}} + C_n^{\mathrm{AC}}} - (\tanh \eta_n^{\mathrm{BC}} R_{\mathrm{BC}}^0)^{\eta_n'^{\mathrm{BC}}}, \tag{101a}$$

$$k_n'^{\mathrm{BC}} = \frac{1}{R_{\mathrm{BC}}^0} \ln\left(\frac{2C_n^{\mathrm{A(BC)}} - C_n^{\mathrm{AB}} - C_n^{\mathrm{AC}}}{(C_n^{\mathrm{AB}} + C_n^{\mathrm{AC}})\, k_n^{\mathrm{BC}}}\right). \tag{101b}$$

We note that the atom–diatom dispersion coefficients are usually expressed in the center-of-mass coordinate system by an expansion in terms of Legendre polynomials $P_n(\cos\theta)$. For example, for the A$\cdots$(BC) interaction, one gets

$$C_n^{\mathrm{A(BC)}}(R_2^0, \theta_2) = C_{n0}^{\mathrm{A(BC)}}(R_2^0) \times [1 + \tau_1(R_2^0)P_1(\cos\theta_2) + \tau_2(R_2^0)P_2(\cos\theta_2) + \cdots] \tag{102}$$

where $C_{n0}^{\mathrm{A(BC)}}$ represents the spherically symmetric component (averaged over all values of θ_2) of the atom–diatom dispersion coefficient, and τ_1 and τ_2 are the two first anisotropic components. The coefficients k_n and k'_n will, for a given R_2^0, turn out to be θ_2-dependent, thus ensuring an accurate description of the long-range potential anisotropy. However, in some of the applications considered in Section V.B this θ_2 dependence has been disregarded, since only the values of the spherically symmetric atom–diatom dispersion coefficients were used as input data for equations (101). In such applications, the anisotropy is predicted rather than built in from the available long-range data.

As regards η_n, the restriction $\eta_n \sim b_1$, where b_1 measures the exponential decay of the EHF curve, has proved to be useful.[126,141,157–159] As for η'_n, it has been suggested that it could be chosen so that the dispersion energy calculated from equation (99) would represent well the isotropic component of the asymptotic dispersion expansion for intermediate and large atom–diatom separations.[126]

In the case of interactions between closed-shell atoms and others which are formally identical, such as the interaction among three ^2S atoms with the same spin, the description of the correlation energy at intermediate and large distances is of great importance. In these regions of configuration space, the main three-body interatomic correlation energy should correspond to the nonadditive Axilrod–Teller–Muto dispersion energy. However, as for the other dispersion energy contributions, orbital overlap and exchange effects have to be taken into account. To approximately account for such effects, a modification of equation (50) has been suggested by the present author[160] which has the form

$$V^{(3)}_{\mathrm{corr,ddd}} = C_9[1 + 84.72e^{-10.84X} + 3\cos\alpha_1\cos\alpha_2\cos\alpha_3] \times \frac{\{1 - \exp\{-3.20(1 - 0.16\Upsilon_{123})X[1 + 1.51(1 + 0.69\Upsilon_{123})X]\}\}^9}{(R_1R_2R_3)^3} \tag{103a}$$

where

$$\Upsilon_{123} = \sum_{i=1}^{3} \sin\alpha_i, \tag{103b}$$

$$X^3 = \prod_{i=1}^{3} x_i, \tag{103c}$$

$$x_i = \frac{2R_i}{R_{m,i} + 2.5R_{0,i}}, \tag{103d}$$

and $R_{m,i}$ and $R_{0,i}$ represent the values of R_m and R_0 for the ith pair.

Similarly to the two-body problem, the numerical values of the constants in equation (103) have been determined from a fit to the perturbational results for the quartet state of H_3.[84] For this system, the orbital overlap effects represent a 5–10% reduction of the dispersion energy in the region of the van der Waals minimum. This percentage is higher in the case of more polarizable atoms such as the alkali metals, for which the above triple-dipole dispersion-energy term has been used in the calculation of the third virial coefficient for the corresponding vapors.[161] It should be noted that in the case of rare gases the triple-dipole dispersion-energy term, used in conjunction with good-quality two-body terms, leads to values of the third virial coefficients which are in good agreement with experiment.[17,81,162] A similar result has been found for other macroscopic properties.[81] However, this agreement may be partly due to fortuitous cancellations of the higher-order non-pair-additive dispersion energies, both among themselves and with the three-body first-order exchange energy (which differs from the three-body EHF energy only in the induction energy contribution[17] present in the latter[82,83]). This, in addition to the fact that there is not a clear-cut distinction between the exc and ddd components, suggests that the combined use of these energies may not be totally free of ambiguity.[17,21]

We conclude this section by addressing the case of an A ··· BC interaction where the electrostatic energy is the dominant contribution at long range, and hence should be treated independently of the remaining extended-Hartree–Fock energy. Since the electrostatic energy may be formally considered a three-body energy in so far as the leading quadrupole–dipole ($\Theta_A - \mu_{BC}$) component goes to zero whenever an atom is removed to infinity, the following representation has been suggested[141]:

$$V^{(3)}_{ABC,ele} = \frac{1}{2}\sum_{i=1}^{3}\sum_{n=4,5}[C_n G_n(R_j)h_n(R_k) + C'_n G_n(R_k)h_n(R_j)]\chi_n(R_i)R_i^{-n}, \tag{104a}$$

where

$$G_n(R) = K_n R^{\xi}\exp[-K'_n(R - R^0)], \tag{104b}$$

$h_n(R)$ is defined as in equation (98c), and the indices j and k have a definition similar to that given after equation (100). In equation (104a) C_n represents the long-range electrostatic coefficient for the atom–diatom interaction involving one of the atoms of the ith pair with the jth diatomic fragment, and C'_n has a similar meaning but referring to the remaining atom of the ith pair with the kth diatomic fragment. Note especially that the parameters in these equations are determined by specifying conditions formally similar to those of equations (97b, c) (e.g., for the AB and BC fragments, and $R_{AB}, R_{AC} \to \infty$):

$$\lim_{R_{BC}\to R^0_{BC}} C_n^{AB,eff} + \lim_{R_{BC}\to R^0_{BC}} C_n^{AC,eff} = C_n^{A(BC)}, \qquad n = 4, 5. \tag{105}$$

If, however, the united atom has a quadrupole (or a higher-order) moment, then the A ··· BC long-range electrostatic interaction energy may not vanish. In such cases, the values of the long-range electrostatic coefficients will depend on the particular electronic state of the A ··· (BC) interaction.[116,163] Yet, this term will not have an analog in R^{-4}, since the dipole moment vanishes for species with spherical symmetry. Accordingly, it seems reasonable to disregard the additional complication resulting from the requirement to satisfy those asymptotic limits. This may be achieved by forcing the three-body electrostatic energy to be zero whenever two atoms collapse to form the corresponding united atom. The parameter ξ in equation (104) accomplishes this; the value $\xi = 4$ has been found satisfactory in the case of O ··· OH.[141]

V. EHFACE POTENTIALS FOR VAN DER WAALS MOLECULES AND OTHER MORE STABLE SYSTEMS

A. Diatomic Systems

One may regard the problem of the functional representation of the diatomic potential curve as practically solved. Due to the realistic nature of the semiempirical models, it is nowadays possible to reproduce many experimental properties within the experimental error limits by using a small number of adjustable parameters in the potential function, a result which in the sixties could only be achieved through multiparametric[164] and seminumerical[165] forms. Moreover, the theoretical basis of these semiempirical models enables their calibration, in principle, from a single experimental property, as distinct from the multiproperty fit that is usually required with multiparametric potentials.[7–9] For example, the determination of the parameter λ in the EHFACE2 model [equation (91)] from a least-squares fit to second-virial-coefficient data of good quality leads, in many cases, to a potential function capable of reproducing other properties within the estimated experimental error limits. It should be pointed out that, although we are mainly concerned with our own model, there are others which have proved to be rather popular within the field of nonbonding interactions, namely the HFD model (Hartree–Fock plus damped dispersion energy[166]), the Tang–Toennies model,[152,167] and the XC (exchange-Coulomb) model.[168] For a historical survey and critical analysis of these semiempirical models see, e.g., refs. 8, 167a, and 169.

We shall now illustrate the application of the EHFACE2 model to the $Ar_2(\tilde{X}\,^1\Sigma_g^+)$ system. In this case a modified screened-Coulomb function has been used to represent the EHF energy contribution, and two possibilities have been considered in determining the parameters involved. For potential I, the EHF curve was chosen to fit single-configuration ab initio SCF calculations,[166b,170] while λ was determined from a least-squares fit to second-virial-coefficient data.[171] For potential II, a similar procedure was employed except for the extra flexibility introduced into the method by allowing the C_6, C_8, and C_{10} dispersion energy coefficients to be optimized within their reported[60] error bounds. Table I compares the observed vibrational spectrum of ground-state Ar_2 with that calculated from potentials I and II. Table II compares the argon viscosity (η^0) calculated from these potentials, at zero density and temperatures of 223 and 298.15 K, with the experimental results. Also included in this table are the results obtained from other models. There is good agreement between the experimental results and those calculated from potentials I and II, supporting the representation of the residual dynamical correlation energy (intraatomic and intra–inter) through equation (91). Also stressed from this comparison is the importance of optimizing C_6, C_8, and C_{10} for fine tuning for spectroscopic accuracy.

TABLE I Experimental and Calculated Spectroscopic Constants $G(v)$ for $Ar_2(\tilde{X}\,^1\Sigma_g^+)$[a]

	$G(v)(\mathrm{cm}^{-1})$		
v	EHFACE2 (I)[b]	EHFACE2 (II)[c]	Exp.[171c]
1	25.664	25.780	25.740 ± 0.1
2	46.095	46.312	46.149 ± 0.2
3	61.501	61.798	61.755 ± 0.2
4	72.264	72.616	72.661 ± 0.2
5	79.006	79.384	79.441 ± 0.25

[a]All values are relative to the ground-state constant $G(0)$. For further comparisons, see ref. 173.

[b]van der Waals parameters: distance of energy zero $\sigma = 6.355a_0$, equilibrium geometry $R_m = 7.119a_0$, and well depth $\varepsilon = 0.451mE_h$. Other coefficients [see equation (90)]: $\lambda = 0.8360$, $a_0 = 49.6983E_h a_0$, $b_1 = 0.8005a_0^{-1}$, $b_2 = 0.0829a_0^{-2}$, $R_0 = 7.2684a_0$. C_6, C_8, and C_{10} are the average values from ref. 60.

[c]$\sigma = 6.346a_0$, $R_m = 7.109a_0$, $\varepsilon = 0.453mE_h$; for the remaining coefficients see ref. 149b.

TABLE II Zero-Density Viscosity of Argon (η^0)[a]

		$\eta^0(\mu P)$	
Potential	Type	$T = 223$ K	298.15 K
BFW[174]	Multiparametric	174.60	225.05
MSVIII[175]	Multiparametric	177.84	229.08
CD[172]	Numerical	174.89	225.86
HFD-C[176]	3 parameters	174.89	225.19
KMA[173]	1 parameter	175.45	225.91
EHFACE2 (I)	1 parameter	174.90	225.20
EHFACE2 (II)[149b]	4 parameters	175.14	225.56
Exp.[177,178]		174.9 ± 0.5	226.0 ± 0.5
Exp.[179]		175.1 ± 0.5	

[a]Evaluated using various potentials, and compared with the experimental results of Hanley and Haynes[177] and Kestin et al.[178,179].

However, only the EHFACE2 model treats both bonding and nonbonding interactions in a unified way.[140] Note that even for molecular electronic states which become degenerate at $R \to \infty$, electron-exchange effects influence distinctly the damping functions χ_n for each of those states. This result is formally approximated by the EHFACE2 model through the inclusion of R_m in the reduced coordinate x. Note that R_m not only characterizes the balance of attractive and repulsive forces at the minimum, but also varies with the molecular electronic state.

Also important within the context of modeling bonding interactions semiempirically is the need to know the corresponding dispersion coefficients. Although there are available reliable estimates of C_n ($n = 6, 8, 10$) for many interactions involving H, alkali-metal, alkaline-earth, and rare-gas atoms,[56–60] the situation is markedly different for interactions which involve atoms with a nonzero angular momentum. In such cases, estimates of the dispersion coefficients are often limited[56] to C_6. In order to overcome this dificulty one usually turns to semiempirical formulae, among which one of the most popular is the generalized Starkshall–Gordon formula,[180,181]

$$\frac{C_8^{AB}}{C_6^{AB}} = \frac{3}{2}\left[\left(\frac{\langle r^4\rangle}{\langle r^2\rangle}\right)_A + \left(\frac{\langle r^4\rangle}{\langle r^2\rangle}\right)_B\right], \tag{106a}$$

$$\frac{C_{10}^{AB}}{C_6^{AB}} = 2\left(\frac{\langle r^6\rangle}{\langle r^2\rangle}\right)_A + 2\left(\frac{\langle r^6\rangle}{\langle r^2\rangle}\right)_B + \frac{21}{5}\left(\frac{\langle r^4\rangle}{\langle r^2\rangle}\right)_A\left(\frac{\langle r^4\rangle}{\langle r^2\rangle}\right)_B. \tag{106b}$$

Using the fact that the expectation values of the radial coordinate for the valence electrons of the abovementioned S atoms follow approximately (standard deviations of 9% and 23%, respectively) the relationships[140]

$$\langle r^4\rangle \sim 2.51(\langle r^2\rangle^{1/2})^{3.73}, \tag{107a}$$

$$\langle r^6\rangle \sim 12.40(\langle r^2\rangle^{1/2})^{5.30}, \tag{107b}$$

one gets from equations (106), (107), in the case of homonuclear interactions,

$$C_8^{AA}/C_6^{AA} = 7.53(\langle r^2\rangle^{1/2})^{1.73}, \tag{108a}$$

$$C_{10}^{AA}/C_6^{AA} = 49.58(\langle r^2\rangle^{1/2})^{3.30} + 26.48(\langle r^2\rangle^{1/2})^{3.46}. \tag{108b}$$

The generalized Starkshall–Gordon formulae (106) show average errors of 20% and 50% in relation to the estimates of C_8 and C_{10} (respectively) from refs. 56–58, and 38% and 77% in relation to those from ref. 60; note that the quoted uncertainties for C_8 and C_{10} are, respectively, about[57] 16 and 17% and[60] 26 and 25%. Despite the reasonably poor description of the available dispersion coefficients, the generalized Starkshall–Gordon formulae may be useful in providing theoretical guidance for deriving other expressions which best fit those dispersion coefficients. For example, the universal correlation suggested by the present author and Dias da Silva,[140]

$$C_n^{AB}/C_6^{AB} = \kappa_n R_0^{a(n-6)/2}, \qquad n = 8, 10, \tag{109}$$

with $a = 1.57$, $\kappa_8 = 1$, $\kappa_{10} = 1.13$, leads to

$$C_8^{AA}/C_6^{AA} = 8.82(\langle r^2\rangle^{1/2})^{1.57}, \tag{110a}$$

$$C_{10}^{AA}/C_6^{AA} = 88.59(\langle r^2\rangle^{1/2})^{3.14}. \tag{110b}$$

As shown in Fig. 12, the universal correlation (109) finds support in the

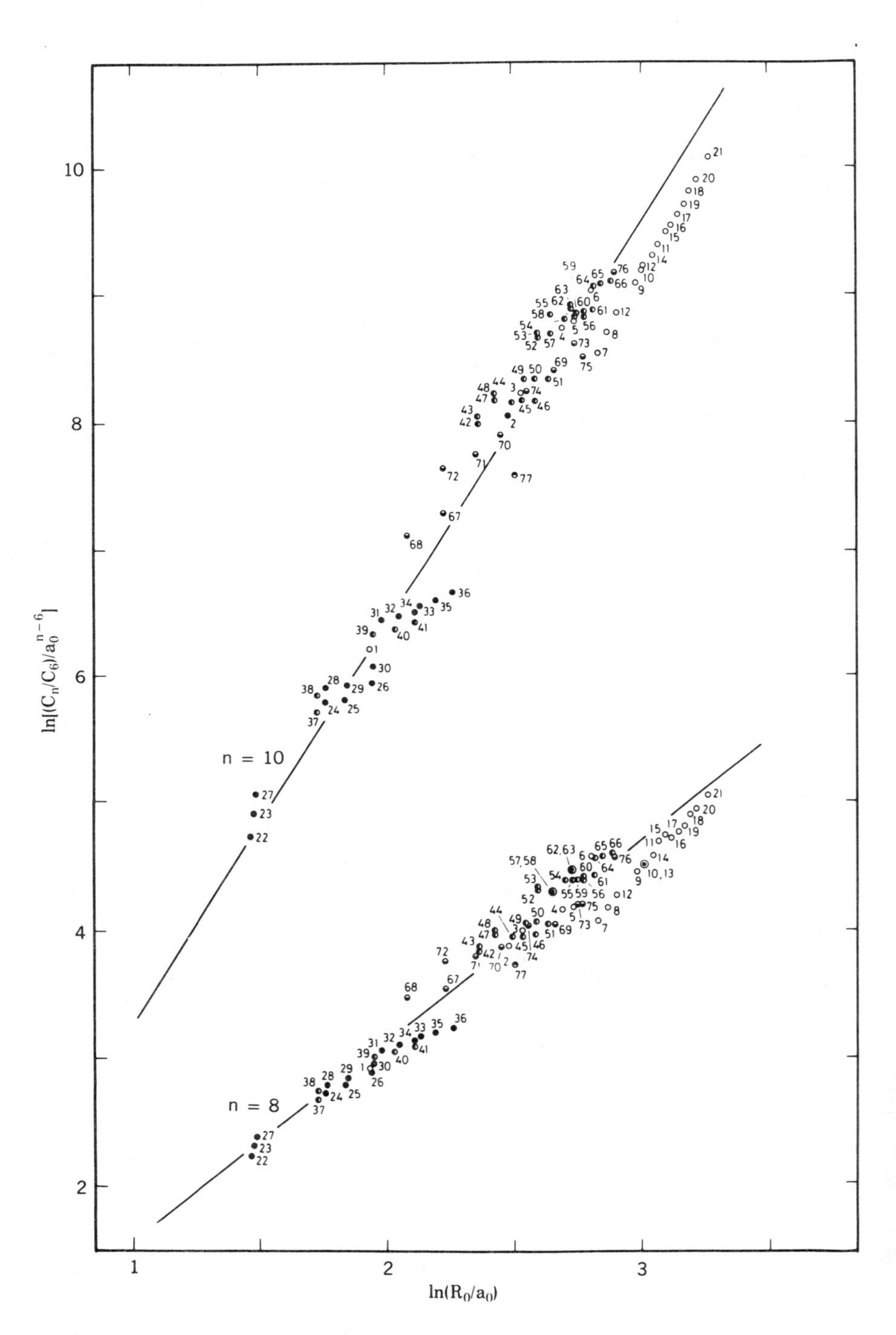

$\ln[(C_n/C_6)/a_0^{n-6}]$
$\ln(R_0/a_0)$
n = 10
n = 8

published values[57-59] of C_8 and C_{10}, and we have foreseen an error of 15% in C_8 and of 30% in C_{10} for the predictions based on it. Note, however, that the most recent compilation[60] of coefficients C_n ($n = 6, 8, 10$) was published after the above universal correlation was obtained using as calibration data the 77 coefficients reported in refs. 57–59. It should also be noted that ref. 60 provides additional estimates of C_n ($n = 6, 8, 10$) for two families of interactions which have not been considered before[57-59] and which involve alkali-metal and alkaline-earth atoms, these latter interacting with rare-gas atoms.

Using, as in the original work,[57] Padé approximants to provide bounds for the dynamic polarizabilities [see equation (41)], Standard and Certain[60] based their new values of the dispersion coefficients C_n on more recent estimates for the input data which are required in their semiempirical method, i.e., transition energies, sum rules, and static polarizabilities. Since the error bars for some of the dispersion coefficients have suffered considerable changes (particularly for C_8 and C_{10} for interactions involving heavy atoms) in relation to the values previously published,[57] we report here the values one obtains for the constants in the universal correlation (109) when these latest estimates of C_6, C_8, and C_{10} are used as calibrating data in the least-squares fitting procedure: $a = 1.54$, $\kappa_8 = 1$, $\kappa_{10} = 1.31$. Accordingly, the values of the constants in equation (110) become 8.46, 1.54, 93.67, and 3.08. The standard deviations for the new universal correlation are 18% and 33% for C_8 and C_{10}, respectively, and thus similar to those reported in ref. 140. We also note that the predicted values of C_8 and C_{10}, which are obtained from the original universal correlation[140] by taking the new values[60] of C_6 as input data, show standard deviations of 22% and 37%, respectively, for the 107 systems now being considered.

The EHFACE2 model (and others[166,167]) may use dispersion coefficients higher than C_{10}, though the contribution from the latter is typically less than 5% of the value of the total dispersion energy for values of R of the order of magnitude of the equilibrium distance in van der Waals interactions. In order to get such coefficients, one usually turns to semiempirical formulae similar to those suggested for the calculation of C_8 and C_{10} from C_6. For example Tang and Toennies[167] suggested

$$C_{n+4}^{AB}/C_{n-2}^{AB} = \kappa_n'(C_{n+2}/C_n)^3 \tag{111}$$

with $\kappa_n' = 1$ ($n = 8, 10, \ldots$), whereas Douketis et al.[166] used the same expression but with κ_n' depending on n, namely $\kappa_8' = 1.028$, $\kappa_{10}' = 0.975$. By repeatedly

---

Figure 12. Log–log representation of C_n/C_6 vs R_0 ($n = 8, 10$) for interactions which involve hydrogen (H), rare gas (Rg), alkali-metal (M), alkaline-earth (N) and mercury (Hg) atoms.[140] Open circles, H–H, H–M, and M–M′; solid circles, Rg–Rg; left-filled circles, H–Rg and M–Rg; bottom-filled circles, N–N; top-filled circles, Hg–Hg. For the numbers see ref. 140.

combining equations (109) and (111) it is now possible to obtain an estimate of higher-order dispersion coefficients from C_6.

In Table III we compare the vibrational spectrum calculated for the ground state of H_2 based on the EHFACE2 model with that observed experimentally. In this case the EHF curve was represented by equation (89) with $\tilde{\alpha} = 0$, $m = 3$, and $m' = 1$, with R_m [in equation (90d) and in the coordinate $r = R - R_m$ which replaces R in the two polynomials of equation (89)] and a_0 chosen to reproduce the experimental values of the equilibrium geometry and of the dissociation energy. The three remaining parameters (a_1 and b_1 are related to each other[140]) were determined from a least-squares fit to the experimental RKR (Rydberg–Klein–Rees) points.[182] In addition to the good quality of the resulting potential (which applies to all the studied systems[140]), one should note the good agreement between the semiempirical EHF curve thus obtained and the corresponding curve calculated by ab initio methods.[98] This result corroborates, once again, the realistic basis of the EHFACE2 model.

Finally, it should be mentioned that for $R \to 0$ the potential-energy curve must behave[5] like $V(R \to 0) = Z_A Z_B / R$, where Z_A and Z_B represent the nuclear charges of A and B. Despite the fact that the EHF curve (89) with $\tilde{\alpha} = -1$ ($m = 3$, $m' = 1$) used in ref. 140 shows this type of functional dependence, the correct behavior suggests that the condition of normalizing the kinetic field[183] should be observed. Preliminary results[184] obtained by forcing this

TABLE III Comparison of Spectroscopic Constant $G(v)$ Obtained from the EHFACE Model for $H_2(\tilde{X}\,^1\Sigma_g^+)$ with Experimental Results

v	$G(v)$ (cm^{-1})	
	EHFACE2	Exp.[182]
1	4 145.64	4 161.14
2	8 069.53	8 087.10
3	11 772.91	11 782.34
4	15 254.24	15 250.35
5	18 509.75	18 491.91
6	21 533.81	21 505.64
7	24 318.99	24 287.82
8	26 855.97	26 830.96
9	29 133.26	29 123.92
10	31 136.65	31 150.20
11	32 848.23	32 886.85
12	34 244.76	34 301.83
13	35 294.48	35 351.00
14	35 950.30	35 972.97

[a] $R_m = 1.4006a_0$, $\varepsilon = 38313.68$ cm^{-1}.

requirement appear promising. Therefore, we believe it possible to achieve a realistic, yet moderately simple, representation of the diatomic potential which is valid over the whole space of the coordinate R. This justifies the statement we opened this subsection with.

B. Triatomic Systems

1. *van der Waals Molecules*

Examples of the application of the DMBE method to van der Waals molecules currently include H_3,[126] HeH_2,[21,141] NeH_2,[157] and $HeLi_2$.[141,158] Although the ground electronic state of H_3 represents the simplest prototype of this class of molecules, its importance in chemical reactivity justifies a separate study in Section V.B.3. Here, we summarize the results obtained for other triatomic van der Waals molecules, examining in more detail the case of $HeLi_2$.

As already commented in Section IV.B, the application of the Wigner–Witmer rules to the fundamental electronic state of the RgX_2 systems (Rg = rare gas; X = H, Li, ..., Cs) suggests that the corresponding potential-energy surfaces can be represented by a single-valued function throughout the whole configuration space. For the case of $HeLi_2$, if one neglects spin–orbit coupling, the Wigner–Witmer rules allow one to write for the united atoms the possible asymptotic limits

$$\mathrm{HeLi_2}(\tilde{X}\,{}^1A') \rightarrow \begin{cases} \mathrm{He}({}^1S) + \mathrm{C}({}^1D), \\ \mathrm{Li}({}^2S) + \mathrm{B}({}^2P), \\ \mathrm{O}({}^1D), \end{cases} \tag{112}$$

where we have considered in all cases the lowest-energy atomic states compatible with spin conservation.[185] It should be noted that the correlation rules for the united atom lose meaning when both atoms involved in the interaction are polyelectronic, due to the fast oscillations in electronic energy over the short intervals of R where the mutual penetration of the core electrons takes place.[116] Nevertheless, as far as the DMBE method is concerned, such correlation rules are only of importance for determining the constants in the expression of the three-body correlation energy through the specification of the dispersion coefficients for the interactions between one of the atoms and the united atom which results from the other two (Section IV.D). It is possible that the definition of these coefficients may affect the calculation of the three-body correlation energy significantly only in regions of configuration space close to the united-atom limit (where that energy represents a very small fraction of the total interaction energy). For practical purposes, one may therefore consider the values of the dispersion energy as if the ground electronic state of the united atom were involved, for example He(^1S)–C(^3P) instead of He(^1S)–C(^1D).

A frequent problem in the representation of the three-body EHF energy emerges from the way in which the computational effort is generally distributed among the molecular geometries. For example, the single-configuration SCF calculations published for He(^1S)–Li$_2$($\tilde{X}\,^1\Sigma_g^+$) cover only structures with the Li$_2$ molecule fixed at the equilibrium geometry ($R_1 = R_m = 5.015a_0$) and angles θ_1 equal to[186] 0, 45, and 90°; for $\theta = 0$ and 90°, the energy was calculated for 13 different values of r_1 (the distance from the He atom to the Li$_2$ center of mass), whereas for $\theta_1 = 45°$ the calculations covered 12 values of r_1. By subtracting the single-configuration SCF interaction energies for the HeLi fragments[187,188] (represented by a modified Born–Mayer function) from the He–Li$_2$ single-configuration SCF interaction energies, it has been possible to calculate the three-body EHF interaction energy for those 38 structures. Still remaining, therefore, was the problem of the analytical continuation of the three-body EHF energy along the coordinate R_1. However, test studies for the He(^1S)–H$_2$($\tilde{X}\,^1\Sigma_g^+$) interaction, for which extensive studies including several values of R_1 are available by ab initio methods,[113] have suggested that this analytical continuation might be carried out to a good approximation by using for the three-body polynomial [see equation (95)] the form

$$\begin{aligned} P_3 = V^0(1 &+ c_1Q_1 + c_2Q_1^2 + c_3Q_2^2 + c_4Q_1^3 + c_5Q_1Q_2^2 + c_6Q_1^4 \\ &+ c_7Q_2^4 + c_8Q_1^2Q_2^2 + c_9Q_1^3Q_2^2 + c_{10}Q_1Q_2^4 + c_{11}Q_1^4Q_2^2 \\ &+ c_{12}Q_1^2Q_2^4 + c_{13}Q_1^3Q_2^4 + c_{14}Q_1^4Q_2^4). \end{aligned} \tag{113}$$

Note that equation (113) contains all powers (allowed by symmetry) obtained from the product of a fourth-order polynomial function $P(Q_1)$, dependent only on the molecular perimeter, with a function $F(Q_2, Q_3 = 0)$, also of fourth order, but only dependent on the shape of the triangle formed by the three atoms. This polynomial form was then combined with a three-body range-determining function [T_3 defined by equation (96a)], and the coefficients appearing in P, F, and T_3 calculated[141] from a least-squares fit to the HeLi$_2$ ab initio points.[186] More recent ab initio studies[158] of the HeLi$_2$ system, using the MONSTERGAUSS computer program[189] and the GTO basis set of functions of ref. 186 (which include values of $R_1 = 4.76$, 5.01, and $5.38a_0$ and $\theta_1 = 0$, 22.5, 45, 67.5, and 90°), have confirmed the usefulness of this procedure. Although the results are encouraging (particularly in view of the considerable anisotropy of the He–Li$_2$ interaction), the analytical continuation based on equation (113) may require further investigation on other closed-shell interactions.

The two-body and three-body correlation energies have been obtained from the method described in Section IV.D, using as input values for the He–Li_2 dispersion coefficients data estimated from correlated electronic-structure calculations using the CEPA method.[186] In the case of the EHFACE2 potential for the HeLi interaction it has not been found useful to take into account any intraatomic and intra–inter dynamical correlation energies [denoted by res in equation (91)], due to the considerable errors in the experimental values of ε and R_m. The $HeLi_2$ potential so obtained, which is free from adjustable parameters, has been called[141] DMBE I. Another potential, DMBE II, has also been derived by adding to potential I an energy term which is supposed to account for three-body dynamical correlation effects, and which (as in the two-body problem) was approximated by

$$V_{\text{res}}^{(3)} = (p_0 + p_2\xi^2 + p_4\xi^4)V_{\text{EHF}}^{(3)}, \tag{114}$$

where ξ represents the pseudoangular function

$$\xi = \frac{R_2 - R_3}{R_1}. \tag{115}$$

In equation (114), p_i $(i = 0, 2, 4)$ are adjustable parameters which may be determined by reproducing an ab initio point for each of the angles $\theta_1 = 0, 45$, and 90° or from a fit to experimental properties. In ref. 141 those coefficients were determined from a least-squares fit to the CEPA energies for $r \leqslant 8a_0$ at the angles $\theta = 0°$, 45°, and 90°.

Figure 13 shows the good agreement obtained for the He–Li_2 system, at $10 \leqslant r \leqslant 18a_0$, between the C_{2v} and $C_{\infty v}$ interaction curves obtained from the DMBE I and DMBE II potentials and the CEPA ab initio results. Table IV illustrates the performance of the DMBE method for HeH_2 by comparing the van der Waals parameters of the DMBE potentials so obtained[21,141] with the best ab initio estimates[113,114] and the experimental results.[190–194] Also included for comparison are the results obtained using other semiempirical models.[152,153] Note that for an interaction of the A–B_2 type the expansion of the potential in Legendre polynomials leads to

$$V(r, R, \theta) = V_0(r, R) + V_2(r, R)P_2(\cos\theta) + \cdots, \tag{116}$$

since the odd terms are zero for symmetry reasons; in this expansion, R has been fixed at the equilibrium distance ($R_m = 1.40a_0$) of the diatomic molecule when calculating the data for Table IV. As for $HeLi_2$, the agreement with the ab initio results is found to be quite satisfactory.

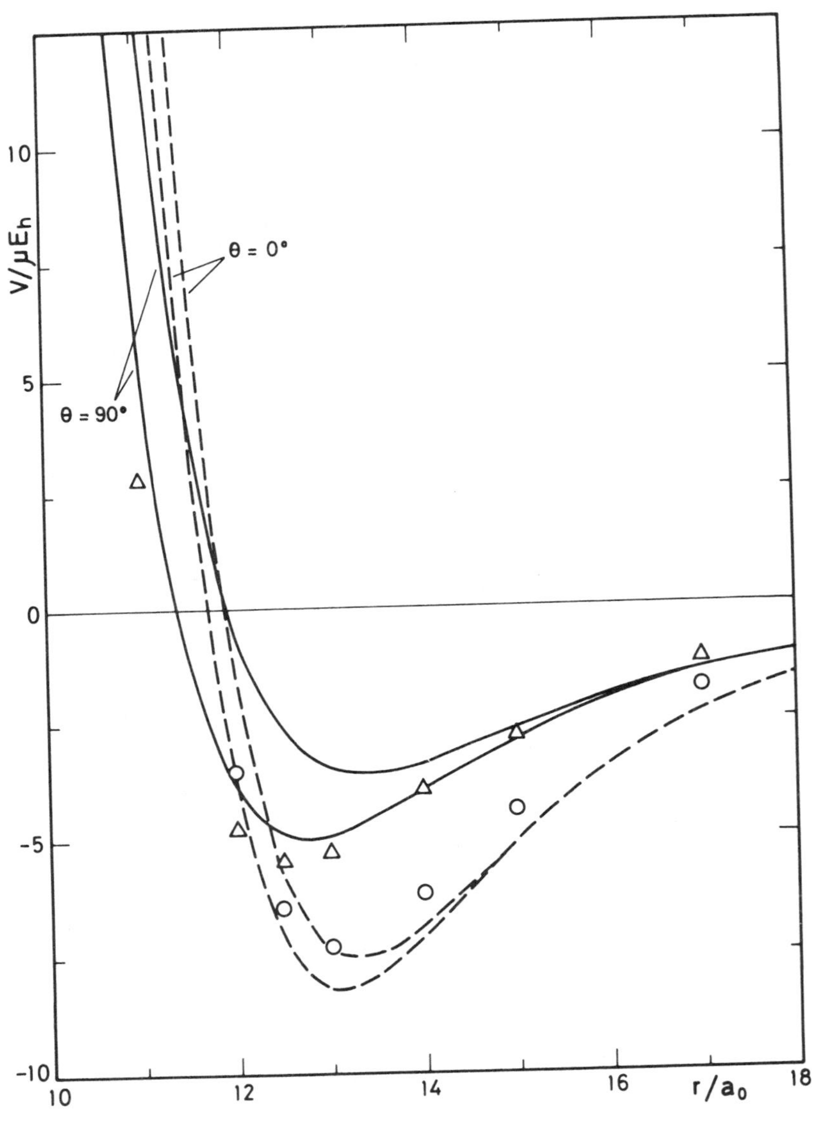

Figure 13. Van der Waals minimum for the He(^1S)–Li_2($\tilde{X}\ ^1\Sigma_g^+$) interaction. Key:—— and – – –, DMBE[141] for $\theta = 0$ and 90°, respectively (potential I, upper curves; II, lower curves); circles and triangles, CEPA.[186]

TABLE IV Comparison of van der Waals Parameters Obtained from Various Ab Initio Calculations and Semiempirical Models for $He(^1S)$–$H_2(\tilde{X}\,^1\Sigma_g^+)$ with Experimental Results

Potentials	V_0			V_2		
	σ/a_0	r_m/a_0	$\varepsilon/\mu E_h$	σ/a_0	r_m/a_0	$\varepsilon/\mu E_h$
		Ab initio[a]				
CEPA[113]	5.71	6.42	44.1	6.14	6.80	4.0
PNOCI–CEPA2[114]	5.67	6.46	39.8	6.19	6.85	4.4
		Semiempirical				
TT[152]	5.78	6.62	37.7	6.05	6.70	5.3
HFD (*A*)[153]	5.75	6.48	39.3	6.20	6.91	4.1
HFD (*B/C*)[153]	5.66	6.38	43.3	6.09	6.80	4.7
DMBE I[21]	5.66	6.40	44.3	5.93	6.58	5.3
DMBE I[141]	5.56	6.27	50.0	5.63	6.27	9.4
DMBE II[141]	5.67	6.40	43.8	5.99	6.66	5.8
		Experimental				
AM[190]	—	5.84	63	—	—	—
FR[191]	—	—	—	—	6.52	5.9
SG–GH[192,193]	5.71	6.39	49	6.60	7.28	2.9
Riehl et al.[194]	—	—	—	—	8.03	2.6

[a]Best estimate: $44.3\mu E_h$,[113] $42.0\mu E_h$.[114]

B. More Stable Molecules

The DMBE method has also been successfully applied to chemically stable triatomic molecules: so far $HO_2(\tilde{X}\,^2A'')$[141] and $O_3(\tilde{X}\,^1A')$.[159] In this subsection, we survey some of these results.

Four major reasons recommend the HO_2 potential-energy surface for application of the DMBE method: (a) its fundamental importance for dynamical calculations of the reaction $H + O_2 \rightarrow HO + O$ and its reverse, both of which are of considerable interest for understanding and modeling the chemistry of the atmosphere[195,196] and combustion processes[197]; (b) the important role[198] in the reaction dynamics played by the long-range electrostatic interactions involving the permanent quadrupole moment of the oxygen atom and the permanent electric multipoles (namely, the dipole and quadrupole moments) of the OH radical; (c) the fact that none of the existing potentials,[29,199,200] including that of Melius and Blint[200] based on MR CI ab initio calculations, has been found to be entirely satisfactory in dynamical studies using quasiclassical[201–207] and quantal[198] approaches and in calculations based on variational transition-state theory[208]; (d) simplicity in that the potential is single-valued.

To these motivations one may add that the HO_2 potential surface offers a good case study of how to use the DMBE method to semiempirically improve ab initio potentials. Indeed, these potentials often describe poorly the exothermicity of the reaction,[200,209] thus hindering its use for dynamical studies. Yet, it is reasonable to assume that the errors in the ab initio energy of the triatomic system persist in the study (at an equivalent level of sophistication to that of the basis set of functions and the type of ab initio approach) of the diatomic fragments which make up the dissociation limits. This means that the three-body energy, obtained by subtracting the sum of the ab initio energies for the diatomics from that for the triatomic, is likely to provide a good representation of the true three-body energy, particularly in regions of the potential where the ab initio calculations are supposed to be most reliable (generally, in regions of strong interactions). If we accept the above reasoning, it is only necessary to replace the two-body ab initio energies by the realistic EHFACE2 potentials to get an improved description of the triatomic potential-energy surface. This has been the methodology employed for the ground electronic state of the HO_2 molecule.

Also, the importance of the interaction energy at large distances,[198] and the evidence from the study of the $H + O_2$ [195] and $O + OH$ [208] reactions that those regions of the HO_2 potential surface are not satisfactorily described by the ab initio calculations of Melius and Blint,[200] suggest that the three-body electrostatic and correlation energy terms should be treated semiempirically in the way described in Section IV.B. Accordingly, the three-body EHF energy term has been obtained by removing the electrostatic and correlation energy contributions [equations (104), (99), respectively] from the three-body ab initio energy,[200] and represented analytically by equations (94)–(96); in equation (95), a complete polynomial expansion of fifth order has been employed, the linear and nonlinear coefficients being obtained from a least-squares fit to the ab initio points.[141] Note that the long-range electrostatic coefficients C_4 and C_5 depend on the atom–diatom orientation [equations (35–37)], though they vanish when averaged, for fixed R_1, over the angles θ and φ of Fig. 8, since species A and B are neutral. Note also that in the spirit of a recently proposed adiabatic theory,[198] the value of θ_b was fixed at $\theta_b = 0$ so as to give the lowest interaction energy for the quadrupole–dipole and quadrupole–quadrupole interactions. Thus, taking into account the definition[210] used[198] for the quadrupole moment, the two leading terms of the O···OH long-range electrostatic energy assume the form

$$V_{\text{ele}}^{\mu_{\text{O}}\theta_{\text{OH}}} = \frac{3}{2}\frac{\mu_{\text{OH}}\Theta_{\text{O}}\cos\theta_i}{r_i^4}, \qquad i = 2, 3, \tag{117}$$

$$V_{\text{ele}}^{\theta_{\text{OH}}\theta_{\text{O}}} = \frac{3}{4}\frac{(3\cos^2\theta_i - 1)\Theta_{\text{OH}}\Theta_{\text{O}}}{r_i^5}, \quad i = 2, 3. \tag{118}$$

These expressions are not, however, the most appropriate to define C_4 and C_5, due to the presence of $\cos\theta_i$, which is not defined for $r_i = 0$. The following approximations have therefore been suggested[141]:

$$C_4 \sim \tfrac{3}{2}\mu_{\text{OH}}\Theta_{\text{O}}\xi_i, \tag{119}$$

$$C_5 \sim \tfrac{3}{4}(3\xi_i^2 - 1)\Theta_{\text{OH}}\Theta_{\text{O}}, \tag{120}$$

where

$$\xi_i = \frac{R_j - R_k}{R_i}, \qquad i = 2, 3. \tag{121}$$

Note that the pseudoangular variable ξ_i has already been used in the previous subsection for the representation of the three-body residual dynamical correlation energy [equations (114)–(115)]. Figure 14 shows that for an

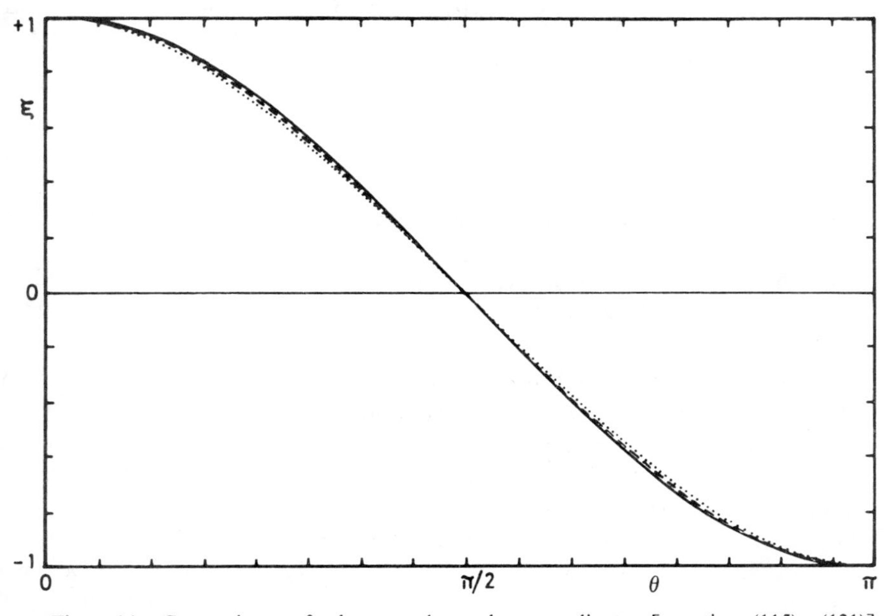

Figure 14. Comparison of the pseudoangular coordinate [equations (115), (121)] with $\cos\theta$ as a function of θ. Key: ——, $\cos\theta$;··· and ---, ξ for $r = 3a_0$ and $5a_0$ (respectively; $R = 2.3a_0$).

A–B_2-type interaction where the center of mass (usually chosen, though not obligatory, as origin) of the diatomic molecule coincides with the center of geometry, ξ_i is an excellent representation of $\cos\theta_i$ even for quite small values of r_i where the expansion in inverse powers of r_i itself loses meaning. In this case one gets

$$\cos\theta_i = \left[\frac{R_j + R_k}{2r_i}\right]\xi_i, \qquad i = 2, 3, \tag{122}$$

where the factor in square brackets can be shown to be approximately equal to unity in the regions of the configuration space which are of interest for the present analysis.

In the case of an interaction like O + OH, the center of geometry of the diatomic molecule may still be adopted as the reference provided one introduces the modifications connected with the change of origin. For example, if we define (r, θ) and (r', θ') as the coordinates of the center of mass and center of geometry, respectively, and if z' represents the distance between those centers, then the change from one coordinate system into the other will involve the following changes:

$$r^{-1} \to r'^{-1} = r^{-1} + z'\cos\theta\, r^{-2} + z'^2(\tfrac{3}{2}\cos^2\theta - \tfrac{1}{2})r^{-3} + \cdots, \tag{123a}$$

$$\cos\theta \to \cos\theta' = \cos\theta - z'r^{-1}\sin^2\theta - \tfrac{3}{2}z'^2r^{-2}\cos\theta\sin^2\theta + \cdots, \tag{123b}$$

$$\mu \to \mu' = \mu, \tag{123c}$$

$$\Theta \to \Theta' = \Theta - 2z'\mu, \text{ etc.} \tag{123d}$$

Although the change in the quadrupole moment of OH expressed by equation (123d) (the first nonzero permanent electric moment is always independent of the choice of origin[211]) was not taken into account in building up the HO_2 DMBE potential of ref. 141, it is likely that the associated error will have no practical significance.

Figure 15 shows a contour plot of the HO_2 potential-energy surface for the O atom moving around the OH molecule, which is kept fixed at its equilibrium geometry. Figure 16 shows a similar plot for the ad hoc functional form of Melius and Blint,[200] which has been obtained from a direct least-squares fit to the correlated ab initio HO_2 energies calculated by those authors. Note the somewhat better quality of the DMBE potential, as measured by the root-mean-square deviation, with respect to the direct least-squares fit[200] to the ab initio points: $0.0073E_h$ vs $0.0116E_h$ (the number of adjustable parameters is similar for the two potentials). Figures 17 and 18 show equipotential energy contours for a H atom moving around an O_2

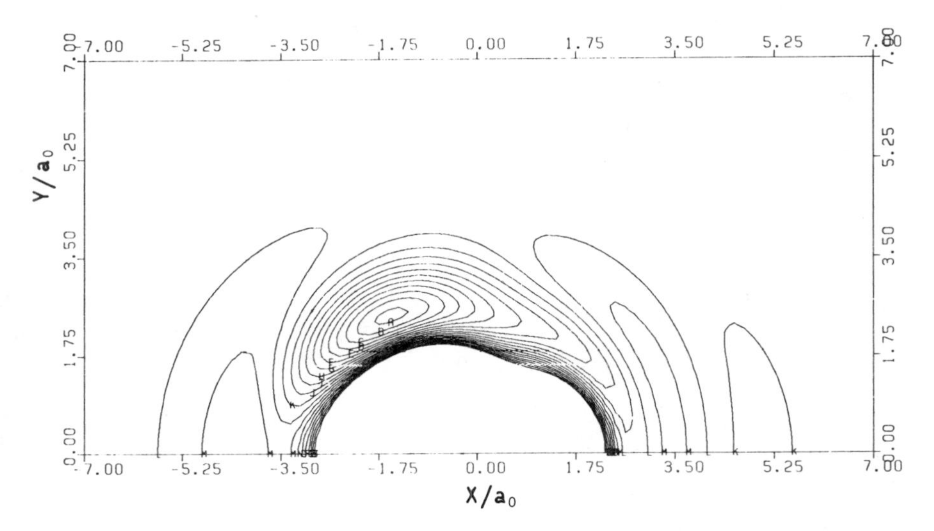

Figure 15. DMBE potential[141] for an $O(^3P)$ atom moving around a $OH(\tilde{X}\,^2\Pi)$ molecule fixed at the equilibrium geometry, with the center of the bond at the origin. Note the presence (contour K on the right) of the metastable minimum for the $C_{\infty v}$ hydrogen-bonded species $OH \cdots O$. The contours in this figure (and the following ones till Fig. 19) are equally spaced by $0.01E_h$, starting at $A = -0.277E_h$ (close to the equilibrium C_s geometry of the HO_2 molecule).

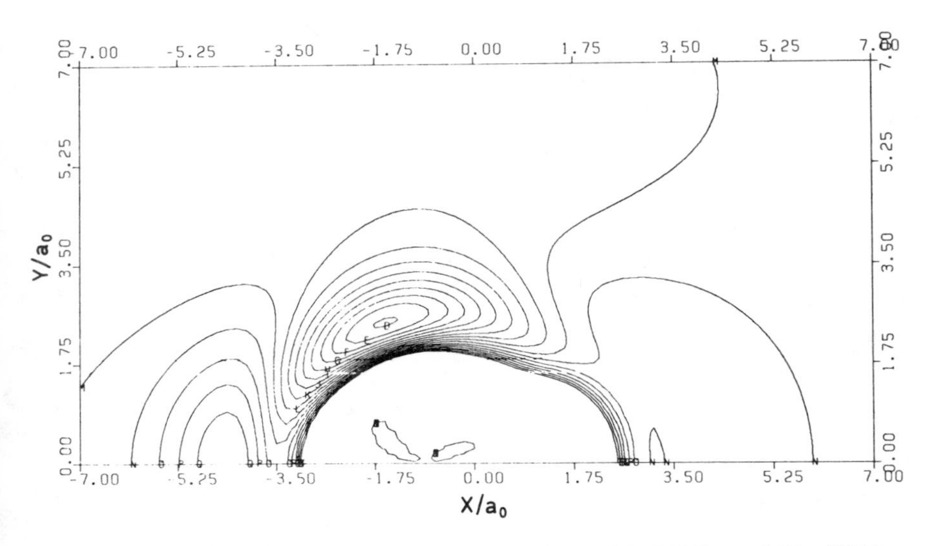

Figure 16. The same as Fig. 15, but for the ad hoc potential of Melius and Blint.[200] The asymptotic O + OH energy is now $-0.15876E_h$.

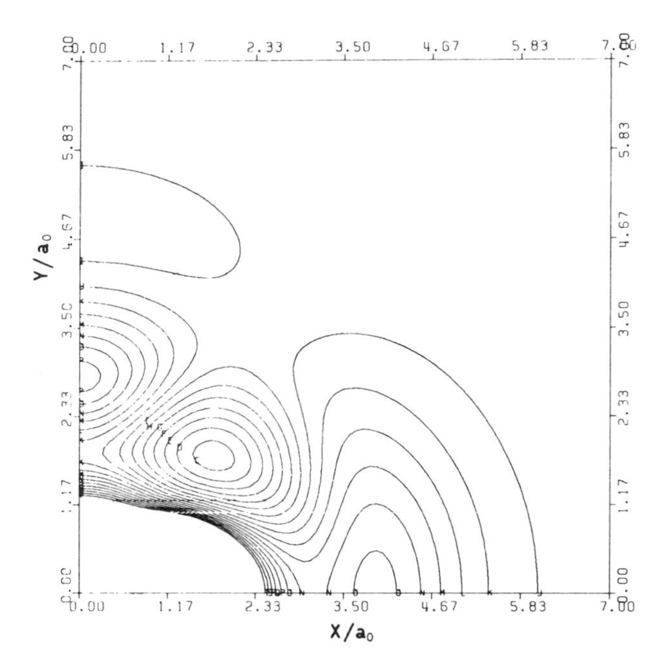

Figure 17. DMBE potential[141] for a $H(^2S)$ atom moving around a molecule of $O_2(\tilde{X}\,^3\Sigma_g^-)$ fixed at the equilibrium geometry, with the center of the bond fixed at the origin. Bound by contour H, is the metastable C_{2v} minimum of the T-shaped $H \cdots O_2$ van der Waals structure.

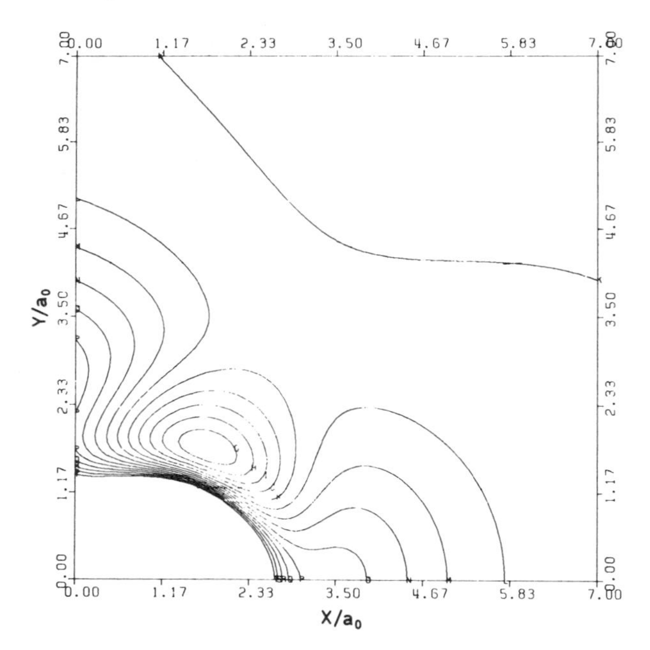

Figure 18. The same as Fig. 17, but for the ad hoc potential of Melius and Blint.[200] The asymptotic $H + O_2$ energy is now $-0.17787E_h$.

molecule which is fixed at its equilibrium geometry. It is apparent from Figs. 16 and 18 that the Melius–Blint potential overestimates the attraction at long-range atom–diatom separations, as first pointed out by Rai and Truhlar[208] in the case of the O + OH asymptotic channel, and by Cobos et al.[195] in the case of the $H + O_2$. Also, and in contrast with the original Melius–Blint potential surface, the ab initio potential corrected with basis in the DMBE method does not show any barrier for the $H + O_2$ reaction. This is in good agreement with the best available ab initio calculations[209] as well as with the fact that the activation energy for the three-body $H + O_2 + M$ reaction (where M is an inert gas) has a negative value.[212] In Fig. 19 we show a triangular plot of the DMBE potential surface for the equilibrium perimeter of the HO_2 molecule, providing evidence for the two C_s minima associated with the two permutation isomers. Finally, Table V summarizes the properties of the DMBE potential minima and their corresponding physical meaning. Also, for comparison, we include in this table the ab initio and experimental estimates. The good agreement with the results obtained from the DMBE potential should be noted. One should also note the lack of ab initio calculations over the regions of the configuration space where the metastable minima $H \cdots O_2(C_{2v})$ and $O \cdots HO(C_{\infty v})$ are predicted. However, the existence of such structures has some basis in chemical intuition, as do the orders of magnitude which are predicted for their geometries and energies[216].

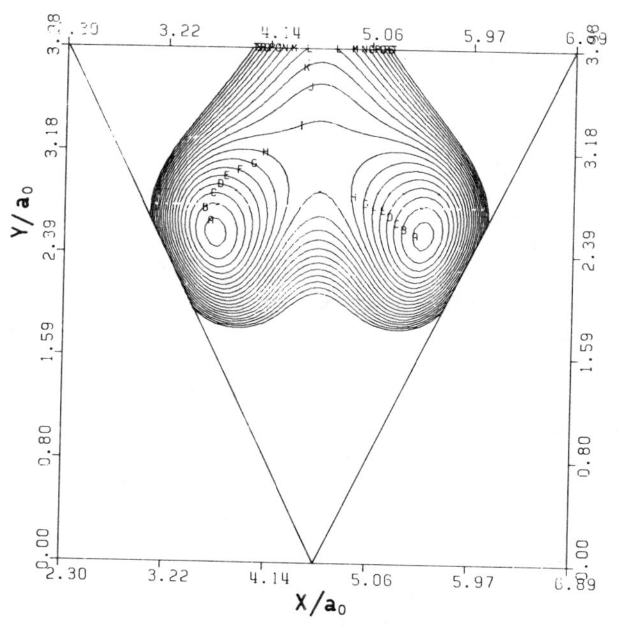

Figure 19. Triangular plot of the $HO_2(\tilde{X}^2A'')$ DMBE potential[141] with the molecular perimeter fixed at the equilibrium value ($\hbar = 7.56a_0$; see Fig. 3).

TABLE V Geometries and Energies of Various Minima on $HO_2(^2A'')$ Potential-Energy Surface[a]

Structure	DMBE	Ab initio[200]	Experiment Refs. 213, 214, 215	
Equilibrium (C_s)				
R_1/a_0	2.543	2.58	2.542	2.57
R_2/a_0	1.893	1.88	1.85	1.86
OOH angle (deg)	104.28	104.2	104.1	106
V/E_h	−0.2808	−0.2488	-0.2747 ± 0.003	—
van der Waals, $H \cdots O_2$ (C_{2v})				
R_1/a_0	2.243			
R_2/a_0	4.927			
OOH angle (deg)	76.84			
V/E_h	−0.2012[b]			
"H *bond*", $O \cdots HO$ $(C_{\infty v})$				
R_1/a_0	5.824			
R_2/a_0	1.875			
OHO angle (deg)	0			
V/E_h	−0.1796[c]			

[a]Energies, in units of E_h, are taken relative to the three isolated atoms. R_1 is the OO distance, and R_2 and R_3 the two OH distances.

[b]Dissociation energy of O_2, $\varepsilon = 0.19157E_h$.

[c]Dissociation energy of OH, $\varepsilon = 0.17020E_h$.

We complete this subsection by presenting some preliminary results[159] for the ground electronic state of the ozone molecule. Though this potential is in fact multivalued (Section IV.C), it is convenient for dynamical studies to have a single-valued function to describe the O_3 potential energy throughout the whole configuration space. This need, which results from the high computational costs when dealing with a matrix potential, is aggravated in the case of the double many-body expansion method by the fact that the analytical complexity of the potential function is inevitably greater. Thus, we have adopted a compromise single-valued representation similar to the one originally proposed in ref. 217, though based here on the DMBE method. Accordingly, the three-body EHF term has been represented by the energy forms P and G of ref. 217 parametrized so as to reproduce the spectroscopic force field at the equilibrium C_{2v} geometry of the O_3 molecule[218] as well as the geometry and the atomization energy of the cyclic D_{3h} ozone structure.[119] By further adding a new EHF three-body local term which vanishes at the C_{2v} and D_{3h}

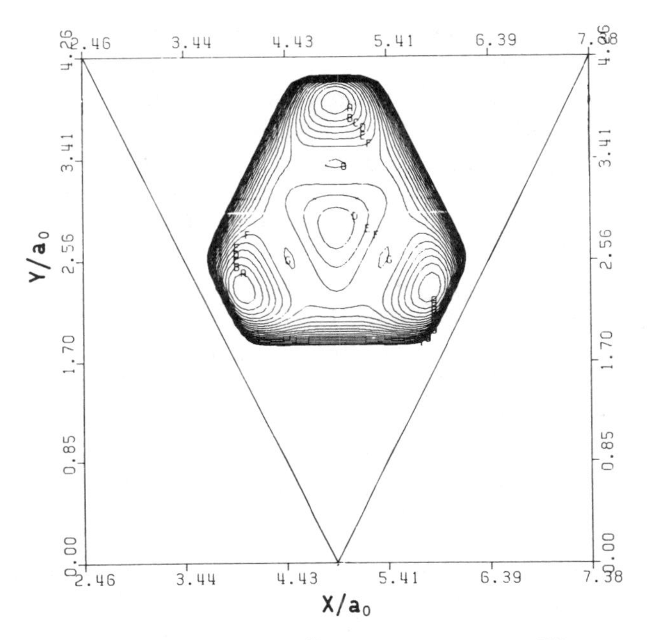

Figure 20. Triangular plot of the $O_3(\tilde{X}\,^1A_1)$ DMBE potential[159] with the molecular perimeter fixed at the average value of the perimeters for the C_{2v} equilibrium structure and the D_{3h} cyclic structure ($\mathscr{P} = 8.52a_0$). Contours are equally spaced by $0.007E_h$, starting at $A = -0.221E_h$.

geometries, it has also been possible to make the optimized C_{2v} bending of O_3 fit the best ab initio estimates[219] for this property. The long-range interaction energies include the dynamical correlation energy (80) and the electrostatic energy which arises from the interaction between the permanent quadrupoles of the O_2 molecule and the O atom, the latter being represented by expressions similar to the ones given for HO_2 in equations (90), (99), (104). Figure 20 shows a triangular plot of the O_3 DMBE potential for a perimeter equal to that of equilibrium O_3; the three equivalent C_{2v} minima and the D_{3h} metastable minima are clearly visible. The O_3 DMBE potential surface obtained from this procedure also shows the spherically symmetric component for the $O + O_2$ interaction, in good agreement with that obtained from incomplete total-scattering cross sections.[220] Moreover, the thermalized rate coefficient for the reaction $^{18}O + ^{16}O_2 \rightarrow ^{18}O\,^{16}O + ^{16}O$ has been found to be in good agreement with experiment.[221]

3. H_3: *A Well-Studied Jahn–Teller System*

The H_3 molecule in its ground electronic state may be classified as van der Waals, and as such been the object of many theoretical and experimental studies.[16,222] Nevertheless, it is the fact that it represents the simplest model of an elementary chemical reaction ($H + H_2 \rightarrow H_2 + H$), and so the favorite for ab initio studies, which makes it unique in the field of chemical reactivity.[223] Thus, it is an important test for the DMBE method. A brief summary of the results[126] obtained for this system will be presented in this subsection.

The ground-state potential-energy surface of the H_3 molecule was the subject of a very accurate (< 1 kcal/mol) CI-type ab initio study by Liu[107] in 1973 for linear geometries, and five years later by Siegbahn and Liu[108] for nonlinear geometries (for references to previous studies see refs. 16 and 222). These ab initio energies were subsequently fitted by Truhlar and Horowitz[132] to a potential (LSTH) of the LEPS type enlarged with three-body exponentially decaying terms; the root-mean-square deviation for the 267 fitted points is 0.17 kcal/mol with a maximum error of 0.55 kcal/mol. As a result of this accurate functional representation, the LSTH potential has been extensively used in dynamics studies. However, there are a few points in which the LSTH potential can be improved.

As we have already discussed in Section IV.B.1, the potential-energy surface for the ground state of H_3 is double-valued, with the two Riemann sheets showing an intersection of the Jahn–Teller type along the line of D_{3h} symmetry. This region of the potential corresponds, however, to high interaction energies, and as a result its topography has seldom been discussed.[16,30,124,132,134] However, a correct description of this region of the H_3 configuration space is important whenever nonadiabatic collisions might occur.[224] In particular, knowledge of the upper Riemann sheet is important in calculating nonadiabatic coupling integrals.[225,226]

Recently,[30,31] it has been shown that the lowest Riemann sheet of the two-valued potential-energy surface for a homonuclear triatomic system can be represented by a function which shows the correct analytical structure when expanded in terms of the D_{3h} coordinates (17), (18) near the conical intersection. Such a procedure has been suggested for the analytical continuation of the potential energy from the lower to the upper Riemann sheet.[30] By carrying out a similar expansion for the LSTH potential, it is easy to show that it contains improper terms in the sense of ref. 30, thus invalidating its use for the analytical continuation of the energy to the upper Riemann sheet of the H_3 surface.

Two other important points need to be corrected in the LSTH potential. First, a better ab initio estimate is available for the classical potential-energy barrier of the exchange reaction (9.67 ± 0.1 kcal/mol), which is 0.15 kcal/mol smaller than that of the LSTH potential. Second, it is important for the study of inelastic collisions to have a correct description of the $H{-}H_2$ interaction potential at large distances.

The DMBE method has been applied in a similar way to that described for the HO_2 system. Accordingly, the fitted three-body EHF-like ab initio energies have been obtained by removing the total two-body energy for the three $H_2(\tilde{X}\,^1\Sigma_g^+)$ fragments, in addition to the three-body correlation energy (100), from the CI energies corresponding to 316 molecular geometries. The latter include the 267 ab initio points used in ref. 132 to calibrate the LSTH function in addition to 31 points from ref. 112. The remaining 18 points have been calculated in ref. 126 for non-D_{3h} geometries near the conical intersection using the Liu–Siegbahn gaussian basis functions.[108] For these geometries, a CAS SCF calculation was first carried out which, together with three active orbitals, makes a total of eight configurations, followed by a CAS SCF CISD calculation based on this multireference space (in a total of 4895 configurations); all calculations were carried out with the COLUMBUS computer program.[227–229] This was followed by a semiempirical prediction of the remaining dynamical correlation energy using the SEC (scaled external correlation) method of Brown and Truhlar.[100] All points except those based on the SEC method were subjected to a small geometry-dependent correction so that the energy would be lower by 0.15 kcal/mol at the symmetrically collinear saddle point for the hydrogen-atom exchange reaction, and[134] by 0.25 kcal/mol at the minimum of the D_{3h} curve, i.e.,

$$V_{\text{corrected}} = V_{\text{ab initio}} - 0.15(1 - \tanh 1.2846Q_1), \qquad (124)$$

where $Q_1 = (R_1 + R_2 + R_3 - 7.028a_0)/\sqrt{3}$.

Due to the considerable success of the functional form used by Truhlar and Horowitz, this has been kept as much as possible to represent the EHF energy term of the H_3 potential after correcting for the formally unacceptable terms mentioned above. We note that in the Truhlar–Horowitz parametrization procedure the $H_2(b^3\Sigma_u^+)$ potential function is partly obtained from a least-squares fit to the $D_{\infty h}$ energies of H_3 (Section IV.B.1). Thus, in order to allow the upper Riemann sheet of the H_3 potential to dissociate into the correct triplet state of H_2, the following form has been used for the EHF triplet:

$$W = V^{\text{eff}}_{\text{HH,EHF}} f + [V^{(2)}_{\text{HH}}(b\,^3\Sigma_u^+) - V^{(2)}_{\text{HH,corr}}(\tilde{X}\,^1\Sigma_g^+)](1 - f), \qquad (125)$$

where $V^{(2)}_{\text{HH,corr}}$ is the two-body dynamical correlation energy (90),

$$V^{(2)}_{\text{HH}}(b^3\Sigma_u^+) = V^{(2)}_{\text{HH,EHF}}(b^3\Sigma_u^+) + V^{(2)}_{\text{HH,corr}}(b^3\Sigma_u^+) \qquad (126)$$

is the true diatomic potential for the triplet state of H_2, $V^{\text{eff}}_{\text{HH,EHF}}$ is an effective diatomic triplet-state function similar to equation (79), and f is a switching

TABLE VI Properties of Linear Symmetric ($D_{\infty h}$) Saddle Point for $H + H_2$ Reaction[a]

	DMBE[126]	LSTH[132]
$R^{\neq}/a_0$	1.755	1.757
$V^{\neq}$ (kcal/mol)	9.65	9.80
Harmonic frequencies (cm^{-1}):		
asymmetric stretch	1493i	1506i
symmetric stretch	2067	2059
bend	899	910

[a]As predicted from the DMBE and LSTH H_3 potential-energy surfaces.

function defined in terms of the hyperspherical coordinates of equations (17), (18) according to

$$f = \exp\left[-(1 + s^3 \cos 3\varphi)\left(\frac{q}{q_0}\right)^4\right], \tag{127}$$

where q_0 is a least-squares parameter; note that equation (90) for the $\tilde{X}\,{}^1\Sigma_g^+$ and $b\,{}^3\Sigma_u^+$ states of H_2 differ only in the value of R_m, respectively $1.40a_0$ and $7.82a_0$.

With a total of 27 parameters in the EHF term, the final potential shows a

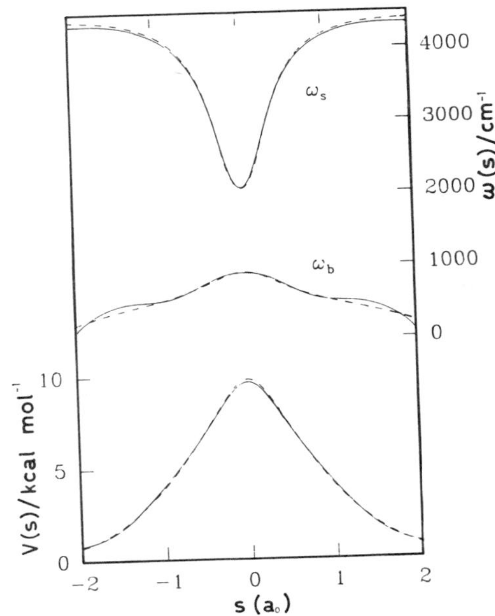

Figure 21. Comparison[230] between the classical potential-energy barrier and the harmonic frequencies for the stretching (ω_s) and bending (ω_b) degrees of freedom as a function of the reaction coordinate ($s = 0$ at the $D_{\infty h}$ saddle point) for the DMBE[126] and LSTH[132] potentials.

TABLE VII Comparison of van der Waals Parameters Obtained from Various $H(^2S)$–$H_2(\tilde{X}\,^1\Sigma_g^+)$ Potentials for Isotropic Interaction

Potential	σ/a_0	r_m/a_0	ε (cal/mol)
GHT[a] [231]	5.93	6.71	54
TD[232]	5.66	6.56	42
VT[c] [233]	5.73	6.58	49
H[b] [234]	5.65	6.48	47
LSTH[d] [132]	5.78	6.59	36
V[d] [134]	5.76	6.66	60
DMBE[d] [126]	5.78	6.45	48

[a]Born–Mayer spline + van der Waals $(-C_6r^{-6} - C_8r^{-8})$.
[b]Modified Born–Mayer, damped dispersion: $\chi_n = \exp[(-\rho r_m/r - 1)^2]$, $n = 6, 8, 10$.
[c]Modified Born–Mayer, damped dispersion: $\chi_n = \exp(h_1r^2 - h_2r^4)$, $n = 6, 8, 10$.
[d]Spherically symmetric term of the corresponding 3D potential, $V_0 = \frac{1}{2}\int_0^\pi V(R_1, R_2, R_3)\sin\theta\, d\theta$.

root-mean-square deviation (rmsd) of 0.24 kcal/mol for the 316 fitted points. This rmsd results from a rmsd of 0.175 kcal/mol for 153 collinear geometries and one of 0.291 kcal/mol for the remaining 163 nonlinear geometries.

Table VI summarizes the geometry of the saddle point ($D_{\infty h}$) for the $H + H_2$ exchange reaction, as well as the normal vibrational frequencies of the H_3 DMBE potential at that point.[230] Also included for comparison in this table are the corresponding attributes for the LSTH potential. Figure 21 compares the minimum-energy reaction path for these two potentials, as well as the variation of the harmonic frequencies for symmetric stretching (ω_s) and bending (ω_b) along the minimum-energy coordinate s. The agreement between the two potentials is excellent. Finally, Table VII compares the parameters representing the van der Waals minimum for the isotropic component of the H–H_2 interaction. In this case there is also good agreement with the best theoretical and experimental estimates.

VI. CONCLUDING REMARKS

In finishing the monograph of ref. 16 we emphasized the following principle of the scientific method: "one should start with the simple and introduce complexity in stages when the simple is shown to be inadequate." The same principle was employed here when showing that a small alteration of the diatomic potentials is enough to obtain a realistic description of the potential-energy curve from $R = 0$ to $R = \infty$. In particular, we only had to combine the extended Rydberg form with the long-range energy in order to introduce a significant improvement into the description of the diatomic potential. From the conceptual point of view, the alterations which have been introduced are important, since they have made it possible to rationalize the potential energy in terms of two well-established concepts: the extended Hartree–Fock energy and the dynamical correlation energy. With a stronger theoretical basis, the extension to triatomic (and larger) systems may be carried out according to the same principle of partitioning the potential energy into those two components, taking into account that the energy is not pairwise additive. Thus the introduction of the abbreviations DMBE for the method and EHFACEn for the models which have been used for describing the n-body terms of that double expansion.

From a more specific point of view one should ask what are the real advantages of introducing the DMBE–EHFACEn formalism. In the words of Kryachko and Koga,[5] referring to the diatomic problem: "the most important and pragmatic problem, even the cornerstone, of the theory of diatomic interactions consists in addressing the following question: how well does a given empirical potential function represent, provide, and predict the relevant observed data?" Concerning the diatomic problem, we have shown that with a small number of adjustable parameters the EHFACE2 model is capable of describing with quantitative accuracy not only bonding interactions but also nonbonding ones.

For the triatomic problem the answer is certainly less definite. As Herschbach saw in 1970, in a note that he found in a pile of computer paper, "the trouble with triatomic molecules is they have one atom too many!"[235] However, the parallelism with the history of the diatomic potential leads us to conclude that the problem will only be solved if a single function can be used to describe simultaneously the results of many experimental properties. This was the main aim of this work, in suggesting a general strategy which should be applied not only to weakly bound complexes, generally known as van der Waals molecules, but also to more stable (or "real") molecules. Such potentials, in contrast to most current "pseudo- two-body" models[152–156] that can only be applied to problems of rotational excitation, may be used for describing the dynamics of reactive and nonreactive collisions. In turn, these dynamical studies offer an important route for testing the accuracy of the final potentials and are thus an important area of our research interest.

A final word concerns the n-body problem with $n > 3$. Here the main problem is the rate of convergence of the EHF and corr series expansions, which we have discussed in Section IV.C. Although we may consider the knowledge of the potential-energy surface in its full dimensionality to be of fundamental importance in the case of four or five atoms, we suspect that the same may not be true for systems with a larger number of atoms, since the main role in the system chemical reactivity may then be attributed to three, four, or five atoms which define the active molecular center. Currently under way are studies for the HO_3 and O_4 systems, and we hope, by using the DMBE method to obtain 6D potentials for these systems, to understand and rationalize not only the aspects of reactivity in $H + O_3$ and $O + O_3$, but also the "bulk" properties for which nonreactive molecular collision processes play the dominant role. Nevertheless, "the trouble with tetraatomic molecules is they have two atoms too many!"

APPENDIX: UPDATE 1987

The main body of this article dates from March 1986, when, in a preliminary version, it was presented in Portuguese to the Academia de Ciências de Lisboa. This appendix, added in May 1987, provides additional references and remarks on subsequent developments.

Two books published in 1985 came to our attention only recently. One[236] of these books provides the proceedings of a symposium and reports advances in accurate ab initio methods that are capable of competing favorably with experiment. The other contains one chapter[237] on ab initio determination of potential-energy surfaces and another[238] devoted primarily to semiempirical potential surfaces for atom–diatom collisions. Another book of proceedings on the theory of chemical reaction dynamics was published during 1986, which has one chapter by Truhlar et al.[239] devoted to the representation of potential-energy surfaces in the wide vicinity of a reaction path for dynamical calculations. This topic is also discussed in a review article[240] that appeared in February 1987. In this article, Truhlar et al. focus on the general problem of constructing potential-energy surfaces for polyatomic reaction dynamics, with polyatomic being operationally defined to refer to a collision in which one collision partner or product is a triatomic or larger or to a unimolecular process involving four or more atoms. Though a reference is made to the H_3 potential-energy surface we have discussed in Section V and to the MBE method, the emphasis is on methods that had already been used to treat "polyatomic" reactions in the above sense. The reader is perhaps wondering at this point about our concluding remark in Section VI, as Truhlar et al.[240] discuss systems with up to ten atoms or more. To our knowledge, though, no single polyatomic ($n \geqslant 5$) potential-energy surface has yet been published that treats evenhandedly the dissociation limits into all possible subclusters, i.e., into diatomics, triatomics, etc. Thus, in most cases the illustrations presented by Truhlar et al.[240] refer to model "polyatomic" potential surfaces with groups of atoms taken as undissociative fragments. As already noted, a second section of the review by Truhlar et al.[240] (which has altogether 319 references) examines methods which are specifically designed to treat the potential surface in the region close to the minimum-energy reaction path (MRP methods) connecting a particular set of reactants and products and passing through the classical transition state. The proposed methodology is (1) to use variational transition-state theory to learn about regions of the surface that are more critical for determining thermal-rate data (including activation energies) and are responsible for the magnitudes or absence of delayed thresholds, and (2) to characterize those regions of the potential-energy surface. Note, of course, that the whole potential-energy surface is still required if full dynamical calculations are to be performed. The review of Truhlar et al.[240] does not discuss many-valued potential-energy surfaces or nonreactive systems.

In the field of intermolecular forces a book has been published by Kaplan[241] which provides a coverage of the theory from long-range forces (including retardation effects) to short-range forces and nonadditivity. The determination of molecular potentials from experimental data is also considered in one chapter of this book.

Several papers reporting research specially relevant to the current review should be mentioned. Knowles and Meath[242,243] have used the time-dependent coupled Hartree–Fock method to evaluate the second-order induction and dispersion energies and their associated damping functions. Results are presented for the dimer interactions arising from ground-state He, Ne, and HF, with the diatomic fixed at the corresponding equilibrium geometry. The calculated damping functions are found to be considerably more complicated than for atom–atom interactions. Given the success of semiempirical damping methods to describe the results from sophisticated ab initio calculations,[244] it will be interesting to see to what extent this comparison holds for atom–molecule and molecule–molecule interactions.

Fuchs and Toennies[245] reported integral scattering cross sections for collisions of He and Kr with Li_2 in selected rotational and vibrational states up to $v = 21$. In the highest excited states considered, the internal energy amounts to about 80% of the Li_2 dissociation energy. These measurements are particularly significant in that they represent the first experimental evidence to stress the importance of the intramolecular coordinate in closed-shell atom–diatom collisions at moderate collision energies. (For the very high-energy regime see ref. 246.) They also present a model potential for the He–Li_2 interaction which includes the dependence on R in addition to that on r and θ. This potential is written as a two-term Legendre series with $V_0(r, R)$ and $V_2(r, R)$ being determined from the potential-energy curves for the collinear and perpendicular attack of He on Li_2. Several assumptions were made in representing these $\theta = 0$ and 90° curves. Perhaps the more drastic one concerns the use of a simple Kihara (n, s) potential to represent the interaction energy; for $\theta = 90°$, a cutoff function was found necessary to model the interaction energy at short atom–diatom separations. The coefficient $C_6(R)$ (actually the only one considered to represent the long-range dispersion energy) has been estimated from the semiempirical London formula (43), and the assumption made that the R dependence could be obtained from that associated with the Li_2 static dipole polarizability $\alpha(R)$. Thus, it has been modeled from the $\alpha(R)$-vs-R curves computed by ab initio methods by Müller and Meyer,[247] which show a maximum at intermediate values of R [both for $\alpha_{\perp}(R)$ and $\alpha\ (R)$] before reaching the plateau associated with the sum of the atomic Li polarizabilities at $R = \infty$. Although this seems a reasonable approach, it is unknown whether the R dependence of the Li_2 polarizability is canceled or reinforced by that associated with the Li_2 ionization potentials (more generally, energy parameters), which also appear in the London formula. Another worrying point is that the model describes poorly the available ab initio data[158,186] on the repulsive wing of the He–Li_2 interaction potential. Despite these limitations, the Fuchs–Toennies model has been found to provide a rationalization of the reported scattering measurements. Unfortunately, no reference was made to our $HeLi_2$ DMBE potential-energy surface[141] that might guide us in assessing its capabilities and limitations.

During the past year some progress has also been made in our group. For NeH_2, a fully three-dimensional potential-energy surface has been obtained[157] which shows features not considered in previous applications of the DMBE theory. First, the three-body EHF term was obtained from a fit to Ne–H_2 SCF energies including those for $R = 0$, i.e., for the interaction involving Ne and He. Second, the three-body dynamical correlation energy was chosen to reproduce the known asymptotic behavior of V_0 and V_2 for the Ne–H_2 interaction as well as their associated R dependences. Also, a new HO_2 DMBE potential-energy surface has been completed.[248] [For a recent paper on the modeling of the HO_2 potential-energy surface with switching functions and extensions of the bond-encrgy-bond-order[249] (BEBO) concepts, which provides an extensive list of references to earlier work on this system, see ref. 250.] Although the HO_2 DMBE potential previously reported[141] shows some definite improvement over its predecessors, it has been found to describe poorly the experimental thermal-rate measurements for the reaction $O + OH \rightarrow O_2 + H$.[251,252] This has been attributed to a small potential barrier in the entrance channel for the $O + OH$ reaction which remained undetected during the numerical search for stationary points carried out in ref. 141. This artifact, which is not apparent from Fig. 16, has now been removed. In addition, a more satisfactory adiabatic theory (see Section V.B.2) has been suggested to describe the $O \cdots OH$ electrostatic interaction energy and used to model the HO_2 DMBE potential-energy surface at large atom–diatom separations. Essentially, one allows the orientation of the O-atom quadrupole to relax so as to give the lowest quadrupole–dipole and quadrupole–quadrupole electrostatic interaction energy at all atom–diatom angles of approach. Moreover, this improved HO_2 DMBE potential surface has been empirically corrected so as to fit the experimental[16,253] quadratic force field of the hydroperoxyl radical. Since it provides[254] in addition a good description of the above thermal-rate data for the reaction $O + OH \rightarrow O_2 + H$, there are grounds to believe that it is the most realistic representation currently available for the electronic energy of ground-state HO_2.

Because the graphical representation of potential-energy surfaces represents an important first step towards the interpretation of an elementary chemical reaction, it is of great interest to have a graphical display which reduces as much as possible the artifacts in the traditional contour plots. Note that for a triatomic system, such contour plots usually represent projections of the 3D surface onto a 2D space. Recently,[255] we have suggested an extension of the triangular plots discussed in Section II for which the hidden coordinate—the perimeter of the molecule or the sum of squares of the three bond distances—is allowed to relax in order to give the lowest potential energy. Such triangular plots, which preserve the full permutational symmetry of the problem, may also be very useful in visualizing the dynamics of atom–diatom molecular collisions.

Finally, we note the slightly different approach to the long-range terms in the double many-body expansion suggested by Murrell et al.[256] With a view to an easier extension to larger systems, these authors proposed to use only atom–atom contributions in developing the terms in the many-body expansion of the long-range interaction energy. Results are presented for the electrostatic interaction energy in HO_2 and H_2F_2. No attempts have so far been made to treat the dispersion and induction energies in a similar way. Nevertheless, either in our original form or the slightly different form proposed by Murrell et al.,[256] the DMBE method seems to be a promising route to developing the potential-energy surfaces of small polyatomic systems.

Acknowledgments

I wish to thank all my collaborators for their valuable contributions. It is also a pleasure to thank Professor S. J. Formosinho for his interest in this work, and Professor H. D. Burrows for reading the manuscript. The author also wishes to express his gratitude to the editors and publishers of all the Journals who have kindly given permission for reproduction of graphical material. This work has been sponsored by the Instituto Nacional de Investigação Científica (INIC), Lisbon.

References

1. R. Rydberg, *Z. Phys.* **73**, 376 (1932); O. Klein, *Ibid.* **76**, 226 (1932); A. L. G. Rees, *Proc. Phys. Soc.* **59**, 998 (1947); J. T. Vanderslice, E. A. Mason, W. G. Maisch, and E. R. Lippincott, *J. Mol. Spectros.* **5**, 83 (1960).
2. U. Buck, *Rev. Mod. Phys.* **46**, 369 (1974).
3. D. W. Gough, G. C. Maitland, and E. B. Smith, *Molec. Phys.* **24**, 151 (1972).
4. J. P. Toennies, in *Physical Chemistry, an Advanced Treatise, Vol. VIA, Kinetics of Gas Reactions*, H. Eyring, D. Henderson, and W. Jost, Eds., Academic Press, New York, 1974, Chapter 5, p. 227.
5. E. S. Kryachko and T. Koga, *Adv. Quant. Chem.* **17**, 97 (1985).
6. P. Claverie, in *Intermolecular Interactions from Diatomics to Biopolymers*, B. Pullman, Ed., Wiley, Chichester, 1978, Chapter 2, p. 69.

7. G. Scoles, *Ann. Rev. Phys. Chem.* **31**, 81 (1980).
8. G. C. Maitland, M. Rigby, E. B. Smith, and W. A. Wakeham, *Intermolecular Forces*, Clarendon Press, Oxford, 1981.
9. P. Hobza and R. Zahradnik, *Weak Intermolecular Interactions in Chemistry and Biology*, Elsevier, Amsterdam, 1980.
10. P. Arrighini, *Intermolecular Forces and Their Evaluation by Perturbation Theory*, Lecture Notes in Chemistry, Springer Verlag, Berlin, Vol. 25, 1981.
11. B. Jesiorski and W. Kolos, in *Molecular Interactions*, Vol. 3, H. Ratajczack and W. J. Orville-Thomas, Eds., Wiley, Chichester, 1982, p. 1.
12. H. Eyring and S. H. Lin, in *Physical Chemistry, An Advanced Treatise, Vol. VIA, Kinetics of Gas Reactions*, H. Eyring, D. Henderson, and W. Jost, Eds., Academic Press, New York, 1974, Chapter 3, p. 121.
13. H. F. Schaefer III, in *Atom–Molecule Collision Theory*, R. B. Bernstein, Ed., Plenum, New York, 1979, Chapter 2, p. 45.
14. D. G. Truhlar, Ed., *Potential Energy Surfaces and Dynamics Calculations*, Plenum, New York, 1981.
15. J. A. Pople, *Faraday Discuss. Chem. Soc.* **73**, 7 (1982).
16. J. N. Murrell, S. Carter, S. C. Farantos, P. Huxley, and A. J. C. Varandas, *Molecular Potential Energy Functions*, Wiley, Chichester, 1984.
17. A. J. C. Varandas, *J. Mol. Struct. (THEOCHEM)* **120**, 401 (1985).
18. N. Sathyamurthy, *Comp. Phys. Rep.* **3**, 1 (1985).
19. D. M. Hirst, *Potential Energy Surfaces*, Taylor & Francis, London, 1985.
20. F. London, *Z. Electrochem.* **35**, 552 (1929); H. Eyring and M. Polanyi, *Z. Phys. Chem. Abt. B* **12**, 279 (1931); S. Sato, *J. Chem. Phys.* **23**, 592, 2465 (1955).
21. A. J. C. Varandas, *Molec. Phys.* **53**, 1303 (1984).
22. M. Born and J. R. Oppenheimer, *Ann. Phys.* **84**, 457 (1927).
23. E. R. Cohen and B. N. Taylor, CODATA Bull., no. 63 (1986).
24. M. Sana, in *Structure and Dynamics of Molecular Systems*, R. Daudel, J. P. Korb, J. P. Lemaistre, and J. Maruani, Eds., Reidel, Dordrecht, 1985, p. 1.
25. K. S. Sorbie and J. N. Murrell, *Molec. Phys.* **29**, 1387 (1975).
26. E. R. Davidson, *J. Am. Chem. Soc.* **99**, 397 (1977).
27. A. J. C. Varandas and J. N. Murrell, *Chem. Phys. Lett.* **84**, 440 (1981).
28. H. Weyl, *The Classical Theory of Groups*, Princeton University Press, Princeton, New Jersey, 1946.
29. S. Farantos, E. C. Leisegang, J. N. Murrell, K. S. Sorbie, J. J. C. Teixeira Dias, and A. J. C. Varandas, *Molec. Phys.* **34**, 947 (1977).
30. C. A. Mead and D. G. Truhlar, *J. Chem. Phys.* **70**, 2284 (1979).
31. T. C. Thompson, G. Izmirlian, Jr., S. J. Lemon, D. G. Truhlar, and C. A. Mead, *J. Chem. Phys.* **82**, 5597 (1985); and references therein.
32. I. M. Mills, *Chem. Soc. Spec. Per. Rep., Theor. Chem.* **1**, 110 (1974).
33. J. N. Murrell and K. Laidler, *J. Chem. Soc. Faraday Trans.* **64**, 371 (1968).
34. J. v. Neumann and E. P. Wigner, *Phys. Z.* **30**, 467 (1927).
35. E. Teller, *J. Phys. Chem.* **41**, 109 (1937).
36. H. A. Jahn and E. Teller, *Proc. R. Soc. London Ser. A* **161**, 220 (1937).
37. G. Herzberg and H. C. Longuet-Higgins, *Faraday Discuss. Chem. Soc.* **35**, 77 (1963).

38. T. Carrington, *Faraday Discuss. Chem. Soc.* **53**, 27 (1972).
39. T. Carrington, *Acc. Chem. Res.* **7**, 20 (1974).
40. K. R. Naqvi and W. B. Brown, *Int. J. Quant. Chem.* **6**, 271 (1972).
41. K. R. Naqvi, *Chem. Phys. Lett.* **15**, 634 (1972).
42. G. J. Hoytink, *Chem. Phys. Lett.* **34**, 414 (1975).
43. H. C. Longuet-Higgins, *Proc. R. Soc. London Ser. A* **344**, 147 (1975).
44. A. J. C. Varandas, J. Tennyson, and J. N. Murrell, *Chem. Phys. Lett.* **61**, 431 (1979).
45. E. R. Davidson and W. T. Borden, *J. Phys. Chem.* **87**, 4783 (1983).
46. R. Renner, *Z. Phys.* **92**, 172 (1934).
47. S. Carter, I. M. Mills, and R. N. Dixon, *J. Mol. Spect.* **106**, 411 (1984).
48. R. N. Dixon and I. L. Robertson, *Chem. Soc. Spec. Per. Rep. Theor. Chem.* **3**, 100 (1978).
49. L. R. Kahn, P. J. Hay, and R. D. Cowan, *J. Chem. Phys.* **68**, 2386 (1978); Y. S. Lee, W. C. Ermler, and K. S. Pitzer, *J. Chem. Phys.* **67**, 5861 (1977).
50. E. R. Davidson, in *Algorithms for Chemical Computations*, R. E. Christoffersen, Ed. (ACS Symp. Series, Vol 46), American Chemical Society, Washington, DC, 1977, p. 21.
51. A. Szabo and N. E. Ostlund, *Modern Quantum Chemistry*, McMillan, 1982; H. F. Schaefer III, *The Electronic Structure of Atoms and Molecules*, Addison-Wesley, Reading, Massachusetts, 1972.
52. F. O. Ellison, *J. Am. Chem. Soc.* **85**, 3540 (1963).
53. P. J. Kuntz, in *Atom-Molecule Collision Theory*, R. B. Bernstein, Ed., Plenum, New York, 1979, Chapter 3, p. 79; J. C. Tully, *Adv. Chem. Phys.* **42**, 63 (1980).
54. J. N. L. Connor, *Comp. Phys. Comm.* **17**, 117 (1979).
55. K. T. Tang, *Phys. Rev. A* **177**, 108 (1969).
56. F. E. Cummings, *J. Chem. Phys.* **63**, 4960 (1975).
57. K. T. Tang, J. M. Norbeck, and P. R. Certain, *J. Chem. Phys.* **64**, 3063 (1976).
58. F. Maeder and W. Kutzelnigg, *Chem. Phys.* **42**, 95 (1979), and references therein.
59. G. Figari, G. F. Musso, and V. Magnasco, *Molec. Phys.* **50**, 1173 (1983).
60. J. M. Standard and P. R. Certain, *J. Chem. Phys.* **83**, 3002 (1985).
61. A. Conway and J. N. Murrell, *Molec. Phys.* **27**, 873 (1974).
62. J. N. Murrell and A. J. C. Varandas, *Molec. Phys.* **30**, 223 (1975).
63. A. J. C. Varandas, *Chem. Phys. Lett.* **69**, 222 (1980).
64. A. T. Amos and C. S. van der Berghe, *Molec. Phys.* **47**, 897 (1982).
65. W. A. Sokalski, P. C. Hariharan, and J. J. Kaufman, *J. Comp. Chem.* **4**, 506 (1983).
66. M. Gutowski, G. Chalasinski, and J. van Duijneveldt–van der Rijdt, *Int. J. Quant. Chem.* **26**, 971 (1984).
67. J. N. Murrell, M. Randić, and D. R. Williams, *Proc. Roy. Soc. London Ser. A* **284**, 566 (1965).
68. J. N. Murrell and G. Shaw, *J. Chem. Phys.* **49**, 4731 (1968).
69. H. Kreek and W. J. Meath, *J. Chem. Phys.* **50**, 2289 (1969).
70. T. R. Singh, H. Kreek, and W. J. Meath, *J. Chem. Phys.* **52**, 5565 (1970).
71. I. K. Snook and T. H. Spurling, *J. Chem. Soc. Faraday II* **71**, 852 (1975).
72. N. Jacobi and Gy. Csanak, *Chem. Phys. Lett.* **30**, 367 (1975).
73. A. Koide, W. J. Meath, and A. R. Allnatt, *Chem. Phys.* **58**, 105 (1981).
74. A. Koide, *J. Phys. B* **9**, 3173 (1976).
75. M. Battezzati and V. Magnasco, *J. Chem. Phys.* **67**, 2924 (1977).

76. M. Krauss and D. B. Neumann, *J. Chem. Phys.* **71**, 107 (1979); M. Krauss, D. B. Neumann, and W. J. Stevens, *Chem. Phys. Lett.* **66**, 29 (1979); M. Krauss, W. J. Stevens, and D. B. Neumann, *Chem. Phys. Lett.* **71**, 500 (1980); M. Krauss and W. J. Stevens, *Chem. Phys. Lett.* **85**, 423 (1982); M. E. Rosenkrantz, R. M. Reagan, and D. D. Konowalov, *J. Phys. Chem.* **89**, 2804 (1985).
77. M. E. Rosenkrantz and M. Krauss, *Phys. Rev. A* **32**, 1402 (1985).
78. B. M. Axilrod and E. Teller, *J. Chem. Phys.* **11**, 299 (1943); B. M. Axilrod, *J. Chem. Phys.* **19**, 719 (1951); Y. Muto, *Proc. Phys. Math Soc. (Japan)* **17**, 629 (1943).
79. R. J. Bell and I. J. Zucker, in *Rare Gas Solids*, M. L. Klein and J. A. Venables, Eds., Academic Press, London, 1976, Chapter 2, p. 122.
80. M. Diaz Peña, C. Pando, and J. A. R. Renuncio, *J. Chem. Phys.* **73**, 1750 (1980).
81. J. A. Barker, in *Rare Gas Solids*, M. L. Klein and J. A. Venables, Eds., Academic Press, London, 1976 Chapter 4, p. 212.
82. P. Schuster, A. Karpfen, and A. Beyer, in *Molecular Interactions*, Vol. 1, H. Ratajczack and W. J. Orville-Thomas, Eds., Wiley, Chichester, 1980, Chapter 5, p. 117.
83. W. J. Meath and R. A. Azis, *Molec. Phys.* **52**, 225 (1984).
84. S. F. O'Shea and W. J. Meath, *Molec. Phys.* **28**, 1431 (1974).
85. S. F. O'Shea and W. J. Meath, *Molec. Phys.* **31**, 515 (1976).
86. J. N. Murrell, A. J. C. Varandas, and M. F. Guest, *Molec. Phys.* **31**, 1129 (1976).
87. S. F. Boys, *Proc. Roy. Soc. London Ser. A* **200**, 542 (1950).
88. W. J. Hehre, R. F. Stewart, and J. A. Pople, *J. Chem. Phys.* **51**, 2657 (1969).
89. T. H. Dunning, Jr., *J. Chem. Phys.* **53**, 2823 (1970).
90. F. B. Duijneveldt, *Gaussian Basis Sets for Atoms H–Ne for use in Molecular Calculations*, IBM Research Laboratory, San José, California, 1971.
91. W. Kutzelnigg, *J. Chem. Soc. Faraday Discuss.* **62**, 111 (1976).
92. S. F. Boys and F. Bernardi, *Molec. Phys.* **19**, 553 (1970).
93. B. H. Wells and S. Wilson, *Chem. Phys. Lett.* **101**, 429 (1983).
94. K. Kitaura and K. Morokuma, *Int. J. Quant. Chem.* **10**, 325 (1976).
95. W. A. Sokalski, S. Roszack, P. C. Hariharan, and J. J. Kaufman, *Int. J. Quant Chem.* **23**, 847 (1983).
96. L. Brillouin, *Actualités Sci. Ind.* **71**, (1933).
97. (a) I. Shavitt, in *Advanced Theories and Computational Approaches to the Electronic Structure of Molecules*, C. E. Dykstra, Ed., Reidel, Dordrecht, 1984, p. 185. (b) B. O. Roos, P. R. Taylor and P. E. M. Siegbahn, *Chem. Phys.* **48**, 157 (1980); K. Ruedenberg, M. W. Schmidt, M. M. Gilbert, and S. T. Elbert, *Chem. Phys.* **71**, 41 (1982).
98. J. D. Bowman, Jr., J. O. Hirschfelder, and A. C. Wahl, *J. Chem. Phys.* **53**, 2743 (1970).
99. F. P. Billingsley II and M. Krauss, *J. Chem. Phys.* **60**, 4130 (1974).
100. F. B. Brown and D. G. Truhlar, *Chem. Phys. Lett.* **117**, 307 (1985).
101. A. Laforgue and A. J. C. Varandas, *C.R. Acad. Sci. Paris Sér. II* **302**, 395 (1986).
102. J. H. Ahlberg, E. N. Nilson, and J. L. Walsh, *The Theory of Splines and Their Applications*, Academic Press, New York, 1967.
103. A. R. Curtiss, in *Numerical Approximations to Functions and Data*, J. G. Hayes, Ed., Athlone, London, 1970, Chapter 4.
104. N. Sathyamurthy and L. M. Raff, *J. Chem. Phys.* **63**, 464 (1975).
105. S. K. Gray and J. S. Wright, *J. Chem. Phys.* **68**, 2002 (1978); S. Chapman, M. Dupuis, and S. Green, *Chem. Phys.* **78**, 93 (1983).

106. S. K. Gray, J. S. Wright, and X. Chapuisat, *Chem. Phys. Lett.* **48**, 155 (1977); J. S. Wright and S. K. Gray, *J. Chem. Phys.* **69**, 67 (1978).
107. B. Liu, *J. Chem. Phys.* **58**, 1925 (1973).
108. P. Siegbahn and B. Liu, *J. Chem. Phys.* **68**, 2457 (1978).
109. B. Liu, *J. Chem. Phys.* **80**, 581 (1984).
110. D. M. Ceperley and B. J. Alder, *J. Chem. Phys.* **81**, 5833 (1984).
111. R. N. Barnett, P. J. Reynolds, and W. A. Lester, Jr., *J. Chem. Phys.* **82**, 2700 (1985).
112. M. R. A. Blomberg and B. Liu, *J. Chem. Phys.* **82**, 1050 (1985).
113. W. Meyer, P. C. Hariharan, and W. Kutzelnigg, *J. Chem. Phys.* **73**, 1880 (1980); P. Hariharan and W. Kuttzellnigg, *Progress Report on the Calculation of the Potential Energy Surface for the* $He + H_2$ *Interaction in the Region of the van der Waals Minimum*, communicated to the author by J. P. Toennies.
114. U. E. Senff and P. G. Burton, *J. Phys. Chem.* **89**, 797 (1985).
115. B. T. Pickup, *Proc. Roy. Soc. London Ser. A* **333**, 69 (1973).
116. E. P. Wigner and E. E. Witmer, *Z. Phys.* **51**, 859 (1928); G. Herzberg, *Molecular Spectra and Molecular Structure: Electronic Spectra and Electronic Structure of Polyatomic Molecules*, Vol. III, Van Nostrand, New York, 1966; *Spectra of Diatomic Molecules*, Vol. II, Van Nostrand, New York, 1950.
117. J. N. Murrell, S. Carter, I. M. Mills, and M. F. Guest, *Molec. Phys.* **42**, 605 (1981).
118. S. Carter, I. M. Mills, J. N. Murrell, and A. J. C. Varandas, *Molec. Phys.* **45**, 1053 (1982).
119. P. G. Burton, *J. Chem. Phys.* **71**, 961 (1979).
120. R. O. Jones, *J. Chem. Phys.* **82**, 325 (1985).
121. J. N. Murrell, K. S. Sorbie, and A. J. C. Varandas, *Molec. Phys.* **32**, 1359 (1976).
122. P. J. Hay, T. H. Dunning Jr., and W. A. Goddard III, *J. Chem. Phys.* **62**, 3912 (1975).
123. J. N. Murrell and A. J. C. Varandas, *Molec. Phys.* **57**, 415 (1986).
124. R. N. Porter, R. M. Stevens, and M. Karplus, *J. Chem. Phys.* **49**, 5163 (1968).
125. A. J. C. Varandas and J. N. Murrell, *J. Chem. Soc. Faraday Discuss.* **62**, 92 (1977).
126. A. J. C. Varandas, F. B. Brown, A. C. Mead, D. G. Truhlar, and N. C. Blais, *J. Chem. Phys.*, **86**, 6258 (1987).
127. R. L. Martin and E. R. Davidson, *Molec. Phys.* **36**, 1713 (1978).
128. W. H. Gerber and E. Schumacher, *J. Chem. Phys.* **69**, 1692 (1978).
129. A. J. C. Varandas and V. M. F. Morais, *Molec. Phys.* **47**, 1241 (1982).
130. A. J. C. Varandas, V. M. F. Morais, and A. A. C. C. Pais, *Molec. Phys.* **58**, 285 (1986).
131. J. K. Cashion and D. R. Herschbach, *J. Chem. Phys.* **40**, 2358 (1964).
132. D. G. Truhlar and C. J. Horowitz, *J. Chem. Phys.* **68**, 2466 (1978).
133. W. Kolos and L. Wolniewicz, *J. Chem. Phys.* **49**, 404 (1968).
134. A. J. C. Varandas, *J. Chem. Phys.* **70**, 3786 (1979).
135. C. H. Wu, *J. Chem. Phys.* **65**, 3181 (1976).
136. A. J. C. Varandas and J. Brandão, *Molec. Phys.* **45**, 857 (1982).
137. A. J. C. Varandas, *Int. J. Quant. Chem.*, **32**, 563 (1987).
138. S. P. Keating and C. A. Mead, *J. Chem. Phys.* **82**, 5102 (1985).
139. B. H. Well and S. Wilson, *Molec. Phys.* **55**, 199 (1985); *ibid.* **57**, 21 (1986).
140. A. J. C. Varandas and J. Dias da Silva, *J. Chem. Soc. Faraday Trans. 2* **82**, 593 (1986).
141. A. J. C. Varandas and J. Brandão, *Molec. Phys.* **57**, 387 (1986).

142. R. J. Le Roy, *Chem. Soc. Spec. Per. Rep. Mol. Spect.* **1**, 113 (1973).
143. J. P. Descleaux, *Atom Data* **12**, 311 (1973).
144. M. L. Olson and D. D. Konowalov, *Chem. Phys.* **21**, 393 (1977).
145. P. J. Bertoncini and A. C. Wahl, *J. Chem. Phys.* **58**, 1259 (1973).
146. G. Das, A. F. Wagner, and A. C. Wahl, *J. Chem. Phys.* **68**, 4917 (1978).
147. W. J. Stevens and M. Krauss, *J. Chem. Phys.* **67**, 1977 (1977).
148. B. Liu and A. D. McLean, *J. Chem. Phys.* **59**, 4557 (1973).
149. (a) J. Brandão, J. Dias da Silva, and A. J. C. Varandas, *J. Mol. Struct.* (*THEOCHEM*), in press; (b) J. Dias da Silva, J. Brandão, and A. J. C. Varandas, to be published.
150. P. A. Christiansen, K. S. Pitzer, Y. S. Lee, J. H. Yates, W. C. Ermler, and N. W. Winter, *J. Chem. Phys.* **75**, 5410 (1981); J. Andzelm, S. Huzinaga, M. Klobukowski, and E. Radzio, *Molec. Phys.* **52**, 1495 (1984).
151. A. J. C. Varandas, *Molec. Phys.* **60**, 527 (1987).
152. K. T. Tang and J. P. Toennies, *J. Chem. Phys.* **68**, 5501 (1978); **74**, 1148 (1981); **76**, 2524 (1982).
153. W. R. Rodwell and G. Scoles, *J. Phys. Chem.* **86**, 1053 (1982).
154. G. Douketis, J. M. Hutson, B. J. Orr, and G. Scoles, *Molec. Phys.* **52**, 763 (1984).
155. R. R. Fuchs, F. R. W. McCourt, A. J. Thakkar, and F. Grein, *J. Phys. Chem.* **88**, 2036 (1984).
156. R. LeSar, *J. Phys. Chem.* **88**, 4272 (1984).
157. A. J. C. Varandas, C. A. Rocha, and M. Matias, to be published.
158. M. A. Matias and A. J. C. Varandas, *J. Comp. Chem.* **8**, 761 (1987); M. A. Matias, J. Brandão, L. M. Tel, and A. J. C. Varandas, to be published.
159. A. A. C. C. Pais and A. J. C. Varandas, *J. Mol. Struct.* (*THEOCHEM*), in press; A. J. C. Varandas and A. A. C. C. Pais, to be published.
160. A. J. C. Varandas, *Molec. Phys.* **49**, 817 (1983).
161. V. M. F. Morais and A. J. C. Varandas, *Chem. Phys. Lett.* **113**, 192 (1985).
162. A. J. C. Varandas, unpublished work.
163. T. Y. Chang, *Rev. Mod. Phys.* **39**, 911 (1967).
164. M. V. Bobetic and J. A. Barker, *Phys. Rev. B* **2**, 4169 (1970); J. A. Barker, M. V. Bobetic, and A. Pompe, *Molec. Phys.* **20**, 347 (1971); G. C. Maitland and F. B. Smith, *Molec. Phys.* **22**, 861 (1971).
165. J. M. Farrar and Y. T. Lee, *J. Chem. Phys.* **56**, 5801 (1972); J. M. Farrar, Y. T. Lee, V. V. Goldman, and M. L. Klein, *Chem. Phys. Lett.* **19**, 359 (1973).
166. (a) C. Douketis, G. Scoles, S. Marchetti, M. Zen, and A. J. Thakkar, *J. Chem. Phys.* **76**, 3057 (1982); (b) R. Ahlrichs, R. Penco, and G. Scoles, *Chem. Phys.* **19**, 119 (1977); (c) J. Hepburn, G. Scoles, and R. Penco, *Chem. Phys. Lett.* **36**, 451 (1975).
167. K. T. Tang and J. P. Toennies, *J. Chem. Phys.* **80**, 3726 (1984); *Ibid.* **66**, 1496 (1977).
168. K. C. Ng, W. J. Meath, and A. R. Allnatt, *Molec. Phys.* **37**, 237 (1979); *Chem. Phys.* **32**, 175 (1978).
169. R. Feltgen, *J. Chem. Phys.* **74**, 1186 (1981).
170. T. L. Gilbert and A. C. Wahl, *J. Chem. Phys.* **47**, 3425 (1967); W. J. Stevens and A. C. Wahl, unpublished work in T. L. Gilbert, O. C. Simpson and M. A. Williamson, *J. Chem. Phys.* **63**, 4061 (1975), ref. 26.
171. J. H. Dymond and E. B. Smith, *The Virial Coefficients of Pure Gases and Mixtures*, Oxford Science Research Papers, Oxford, 1980.
172. E. A. Colbourn and A. E. Douglas, *J. Chem. Phys.* **65**, 1741 (1976).

173. A. Koide, W. J. Meath, and A. R. Allnatt, *Molec. Phys.* **39**, 895 (1980).
174. J. A. Barker, R. A. Fischer, and R. O. Watts, *Molec. Phys.* **21**, 657 (1971).
175. J. M. Parson, P. E. Siska, and Y. T. Lee, *J. Chem. Phys.* **56**, 1511 (1972).
176. R. A. Aziz and H. H. Chen, *J. Chem. Phys.* **67**, 5719 (1977).
177. H. J. M. Hanley and W. M. Haynes, *J. Chem. Phys.* **63**, 358 (1975).
178. J. Kestin, S. T. Ro, and W. A. Wakeham, *J. Chem. Phys.* **56**, 4119 (1972).
179. J. J. de Groot, J. Kestin, and H. Sookiazian, *Physica* **75**, 454 (1974).
180. G. Starkschall and R. G. Gordon, *J. Chem. Phys.* **56**, 2801 (1972).
181. V. Staemmler and R. Jaquet, *Chem. Phys.* **92**, 141 (1985).
182. S. Weissman, J. T. Vanderslice, and R. Batino, *J. Chem. Phys.* **39**, 2226 (1963).
183. R. F. Nalewajski, *J. Phys. Chem.* **82**, 1439 (1978).
184. A. J. C. Varandas, M. C. A. Gomes, and J. Dias da Silva, to be published.
185. A. G. Gaydon, *Dissociation Energies*, Chapman and Hall, London, 1968.
186. U. Stahl, *Quantenchemische ab initio berechnung der van der Waals-Wechselwirkungen zwischen He im Grundzustand und Li_2 den Zustanden*, Diplomarbeit, Bochum, F. R. Germany; U. Stahl and V. Staemmler, to be published.
187. G. Das and A. C. Wahl, *Phys. Rev. A* **4**, 825 (1971).
188. S. B. Schneiderman and H. H. Michels, *J. Chem. Phys.* **42**, 3706 (1965).
189. M. R. Peterson and R. A. Poirier, MONSTERGAUSS *Program*, University of Toronto, Canada.
190. I. Amdur and A. P. Malinauskas, *J. Chem. Phys.* **42**, 3355 (1965).
191. K. R. Foster and J. H. Rugheimer, *J. Chem. Phys.* **56**, 2632 (1972).
192. R. Gengenbach and Ch. Hahn, *Chem. Phys. Lett.* **15**, 604 (1972).
193. R. Shafer and R. G. Gordon, *J. Chem. Phys.* **58**, 5422 (1973).
194. J. W. Riehl, J. L. Kinsey, J. S. Waugh, and J. H. Rugheimer, *J. Chem. Phys.* **49**, 5276 (1968); J. W. Riehl, C. J. Fisher, J. D. Baloga, and J. L. Kinsey, *J. Chem. Phys.* **58**, 4571 (1973).
195. C. J. Cobos, H. Hippler, and J. Troe, *J. Phys. Chem.* **89**, 342 (1985), and references therein.
196. M. Nicolet, *Adv. Chem. Phys.* **55**, 63 (1985).
197. G. L. Schott, *Combust. Flame* **21**, 357 (1973).
198. D. C. Clary and H. J. Werner, *Chem. Phys. Lett.* **112**, 346 (1984); D. C. Clary, *Molec. Phys.* **53**, 3 (1984).
199. A. Gauss Jr., *Chem. Phys. Lett.* **52**, 252 (1977).
200. C. F. Melius and R. J. Blint, *Chem. Phys. Lett.* **64**, 183 (1979).
201. A. Gauss Jr., *J. Chem. Phys.* **68**, 1689 (1978).
202. R. J. Blint, *J. Chem. Phys.* **73**, 765 (1980).
203. J. A. Miller, *J. Chem. Phys.* **74**, 5120 (1981); *Ibid.* **75**, 5349 (1981).
204. M. Bottomley, J. N. Bradley, and J. R. Gilbert, *Int. J. Chem. Kinet.* **13**, 957 (1981).
205. C. S. Gallucci and G. C. Schatz, *J. Phys. Chem.* **86**, 2352 (1982).
206. N. J. Brown and J. A. Miller, *J. Chem. Phys.* **80**, 5568 (1984).
207. K. Kleinermanns and R. Schinke, *J. Chem. Phys.* **80**, 1440 (1984).
208. S. N. Rai and D. G. Truhlar, *J. Chem. Phys.* **79**, 6046 (1983).
209. T. H. Dunning Jr., S. P. Walch, and A. C. Wagner, in *Potential Energy Surfaces and Dynamics Calculations*, D. G. Truhlar, Ed., Plenum, New York, 1981, p. 329; T. H. Dunning Jr., S. P. Walch, and M. M. Goodgame, *J. Chem. Phys.* **74**, 3482 (1981).

210. J. O. Hirschfelder, C. F. Curtiss, and R. B. Bird, *Molecular Theory of Gases and Liquids*, Wiley, New York, 1954; second printing, 1964.
211. A. D. Buckingham, *Adv. Chem. Phys.* **12**, 107 (1967).
212. M. J. Kurylo, *J. Phys. Chem.* **76**, 3518 (1972).
213. Y. Beers and C. J. Howard, *J. Chem. Phys.* **64**, 1541 (1976).
214. J. F. Ogilvie, *Can. J. Spectrosc.* **19**, 171 (1973).
215. S. N. Foner and R. L. Hudson, *J. Chem. Phys.* **36**, 2681 (1962).
216. P. A. Kollman, in *Modern Theoretical Chemistry, Applications of Electronic Structure Theory*, Vol. 4, H. F. Scheaffer III, Ed., Plenum, New York, 1977, Chapter 3, p. 109.
217. A. J. C. Varandas and J. N. Murrell, *Chem. Phys. Lett.* **88**, 1 (1982).
218. A. Barbe, C. Secroun, and P. Jouve, *J. Molec. Spectrosc.* **33**, 538 (1970).
219. S. Shih, R. J. Buenker, and S. D. Peyerimhoff, *Chem. Phys. Lett.* **28**, 463 (1974).
220. P. B. Foreman, A. B. Lees, and P. K. Rol. *Chem. Phys.* **12**, 213 (1976).
221. S. M. Anderson, F. S. Klein, and F. Kaufman, *J. Chem. Phys.* **83**, 1648 (1985).
222. D. G. Truhlar and R. E. Wyatt, *Adv. Chem. Phys.* **36**, 141 (1977).
223. D. G. Truhlar and R. E. Wyatt, *Ann. Rev. Phys. Chem.* **27**, 1 (1976).
224. B. C. Garrett and D. G. Truhlar, *Theor. Chem. Adv. Perspectives* **6A**, 215 (1981).
225. V. I. Osherov and V. G. Ushakov, *Dokl. Akad. Nouk. SSR* **236**, 68 (1977).
226. N. C. Blais and D. G. Truhlar, *J. Chem. Phys.* **78**, 2388 (1983).
227. R. M. Pitzer, *J. Chem. Phys.* **58**, 3111 (1973); M. Dupuis, J. Rys, and H. F. King, *J. Chem. Phys.* **65**, 111 (1976).
228. H. Lischka, R. Shepard, F. B. Brown, and I. Shavitt, *Int. J. Quantum Chem. Symp.* **15**, 91 (1981).
229. R. Shepard, I. Shavitt, and J. Simons, *J. Chem. Phys.* **76**, 543 (1982).
230. B. C. Garrett, D. G. Truhlar, A. J. C. Varandas, and N. C. Blais, *Int. J. Chem. Kinetics* **18**, 1065 (1986).
231. R. Gengenbach, Ch. Hahn, and J. P. Toennies, *J. Chem. Phys.* **62**, 3620 (1975).
232. F. Torello and M. G. Dondi, *J. Chem. Phys.* **70**, 1564 (1979).
233. A. J. C. Varandas and J. Tennyson, *Chem. Phys. Lett.* **77**, 151 (1981).
234. N. Hishinuma, *J. Chem. Phys.* **75**, 4960 (1981).
235. D. R. Herschbach, in *Proceedings of the Conference on Potential Energy Surfaces in Chemistry*, W. A. Lester, Ed., IBM Research Laboratory, San Jose, California, 1971.
236. R. J. Bartlett, Ed., *Comparison of Ab Initio Quantum Chemistry with Experiment for Small Molecules*, Reidel, Dordrecht, 1985.
237. T. H. Dunning Jr. and L. B. Harding, in *Theory of Chemical Reaction Dynamics*, M. Baer, Ed., CRC, Boca Raton, FL, 1985, Vol. 1, p. 1.
238. P. J. Kuntz, in *Theory of Chemical Reaction Dynamics*, M. Baer, Ed., CRC, Boca Raton, FL, 1985, Vol. 1, p. 71.
239. D. G. Truhlar, F. B. Brown, R. Steckler, and A. D. Isaacson, in *The Theory of Chemical Reaction Dynamics*, D. C. Clary, Ed., Reidel, Dordrecht, 1986, p. 285.
240. D. G. Truhlar, R. Steckler, and M. S. Gordon, *Chem. Rev.* **87**, 217 (1987).
241. I. G. Kaplan, *Theory of Molecular Interactions*, Elsevier, Amsterdam, 1986.
242. P. J. Knowles and W. J. Meath, *Molec. Phys.* **59**, 965 (1986).

243. P. J. Knowles and W. J. Meath, *Molec. Phys.* **60**, 1143 (1987).
244. M. Gutowski, J. Verbeek, J. H. Van Lenthe, and G. Chalasinski, *Chem. Phys.* **111**, 271 (1987).
245. M. Fuchs and J. P. Toennies, *J. Chem. Phys.* **85**, 7062 (1986).
246. A. P. Kalinin and V. B. Leonas, *Chem. Phys. Lett.* **114**, 557 (1985).
247. W. Müller and W. Meyer, *J. Chem. Phys.* **85**, 953 (1986).
248. A. J. C. Varandas, J. Brandão, and L. A. M. Quintales, *J. Phys. Chem.*, in press.
249. H. S. Johnson, *Gas Phase Reaction Rate Theory*, Ronald, New York, 1966, p. 55.
250. W. J. Lemon and W. L. Hase, *J. Phys. Chem.* **91**, 1596 (1987).
251. J. Troe, *J. Phys. Chem.* **90**, 3485 (1986); see also ref. 195.
252. For a critical evaluation of the experimental data see N. Cohen and K. R. Westberg, *J. Phys. Chem. Ref. Data* **12**, 531 (1983).
253. R. P. Tuckett, P. A. Freedman, and W. J. Jones, *Molec. Phys.* **37**, 403 (1979); **37**, 379 (1979).
254. L. A. M. Quintales, A. J. C. Varandas, and J. M. Alvariño, *J. Phys. Chem.*, in press.
255. A. J. C. Varandas, *Chem. Phys. Lett.*, **138**, 455 (1987).
256. J. N. Murrell, N. M. R. Hassani, and B. Hudson, *Molec. Phys.* **60**, 1343 (1987).

AUTOWAVE MODES OF CONVERSION IN LOW-TEMPERATURE CHEMICAL REACTIONS IN SOLIDS

V. V. BARELKO, I. M. BARKALOV, V. I. GOLDANSKII,
D. P. KIRYUKHIN, AND A. M. ZANIN

Institute of Chemical Physics of the
Academy of Sciences of the USSR
Chernogolovka Branch
Chernogolovka, Moscow District, USSR

CONTENTS

I. INTRODUCTION

This chapter generalizes the results of the studies of traveling waves of chemical conversion observed by the authors in a wide variety of solids at liquid-nitrogen and -helium temperatures.

As recently as in the early 1970s the very possibility of chemical reaction in a solid† at absolute zero seemed most doubtful. The event that has led to a revision of the generally accepted views was the discovery of low-temperature limits for the velocities of chemical reactions, first in the case of the growth of chains during the irradiation-induced polymerization of formaldehyde[2] and then for a number of other chain and nonchain reactions (see ref. 3 and the references therein). Systems were shown to exist in which reactions can run even at liquid-helium temperatures with quite measurable (though low) and temperature-independent velocity. To explain the phenomenon of a low-temperature limit in cryochemical conversions, the concept of molecular tunneling was proposed, and was further developed to include low-temperature intermolecular vibrations (see refs. 3–7 and the references therein).

A most important aspect of the tunneling mechanism as applied to low-temperature irradiation-induced chain conversions in a solid is the assumption of energy equilibrium in a reacting solid-state system. However, the universality of kinetic models of the above processes based on the Arrhenius equilibrium law has neven been considered an axiom—especially in studies of chemical reactions stimulated externally by ionizing radiation, that is, under conditions where temperature was no longer the only parameter characterizing the energy state of the system.

It is quite natural that alternative mechanisms were developed in which the leading part was played by the nonequilibrium stages in low-temperature solid-state conversions of the above-specified type. As far back as 1960, Semenov proposed a mechanism of energy chains to describe the irradiation-induced solid-state polymerization, according to which the energy released in each elementary stage in the growth of a polymeric chain serves to overcome the activation barrier in the next such stage.[8] By this mechanism, the growth of the polymeric chain may have a velocity equal to that of a propagating exciton ($\sim 10^5$ cm/sec). A comprehensive theoretical treatment of an exciton chain model was given in ref. 9. It was concluded that rather high rates of solid-state conversion are possible even at very low temperatures due to the chain energy transfer along a frozen matrix.

Reference 10 describes a combined (in a sense) scheme which permits the action of the tunneling mechanism at the initial stage of a reaction but assumes very rapid subsequent development of the chain process owing to recouping of the energy of elementary stages. In addition to such a chain-tunneling mechanism, another example of the combined type should be mentioned. This

†A reaction is defined, for our purposes, as a conversion accompanied by a change in the lengths and angles of interatomic valence bonds and by the corresponding rearrangement of atoms. The reactions of electron transfer[1] in a solid at low temperatures are not considered in this chapter.

is a chain-thermal mechanism suggested and developed in refs. 11–13 to explain the limiting yield of radicals accumulated during irradiation in frozen samples of various organic compounds.

However, we would like to point here not to the differences between the equilibrium tunneling mechanism and the above examples of mechanisms of the nonequilibrium type in low-temperature chemical conversions, but, on the contrary, to a simplifying assumption which relates them but which has to be rejected in a number of cases—and that is the subject matter of this chapter. In the above models the solid matrix itself was considered, in essence, from a special point of view, namely, as an ideal system, devoid of defects, which is in mechanical equilibrium. In other words, the fact that the systems in question are significantly out of equilibrium with respect to their mechanoenergetic state was ignored. This property of the experimentally studied samples was the result of both their preparation conditions and the ionizing radiation.

It is well known that an irradiated solid shows a considerable distortion in molecular arrangement (possibly even a transition from the polycrystalline to the amorphous state), greatly altered fields of internal tension, and markedly lowered strength. Changing over to a postradiation regime in experiments does not remove the question about the role of the mechanical disequilibrium of a sample in solid-state conversions, because it remains appreciable even after switching off the γ-ray energy flow, due to the suppression at low temperatures of all kinds of relaxation processes. Besides, the chemical reaction itself, initiated in a solid matrix, must lead to an increase in the lattice distortions, that is, to further accumulation in it of the mechanical potential energy.

The above considerations were a starting point in the formulation of the problems in the series of investigations discussed in this chapter.[14–24] The motive for elucidating the mechanochemical aspect of the radiation cryochemistry of solids was the discovery[14] of the effect of excitation of a reaction in the irradiated, low-temperature-stabilized samples of reactants in response to a local fracture.

The main questions raised at the very beginning of the studies can be briefly formulated as follows. Can the potential energy accumulated in the mechanically nonequilibrium matrix of a frozen sample of reactants be released by its brittle fracture? Will the accumulated mechanical energy transform into chemical energy, and will a chemical conversion take place in this case?

The answer yes was suggested by the general statements of the theory of the strength of solids, from which it follows that the mechanical energy of elastic deformation at the instant of fracturing is concentrated in the region immediately adjacent to the surface being formed. It is known[25] that the process of energy transformation is accompanied by a significant change in the physicochemical properties of a substance on the newly formed surface from those in the bulk: brittle fracture is connected with such phenomena as electron and ion emission, luminescence, radical formation, and dramatic acceleration of the transfer processes on fracture surfaces.

In other words, it seemed probable that switching over the process from the homogeneous to the essentially heterogeneous state would switch on the nonequilibrium mechanism of energy transfer to active centers prefrozen in a three-dimensional matrix and would thereby cause a chemical conversion at such low temperatures. It should be added that in the processes of traditional mechanochemistry brittle fracture (realized under conditions of forced dispersion of a sample) was always assigned a prominent role (see ref. 26 and the references therein).

Besides the above questions, there was one more important point in the definition of the objectives of the present series of works. In earlier studies, not only the kinetic models of solid-state radiation-induced chemical conversions but also the experimental procedures were based on the assumption of homogeneity of the system and of the chemical reaction in it. This was quite natural, because it was difficult to examine the spatial peculiarities of a chemical conversion with the then existing calorimetric and spectral techniques operating with the integrated signals. Note that the permissibility of zero-dimensional kinetic models in the radiation chemistry of the solid was always doubted, since it is well known that homogeneous samples do no exist: in the course of radiolysis there appear highly active regions—tracks and spurs—and the radicals and ions being stabilized are nonuniformly distributed over the volume. Besides disregarding the actual inhomogeneity of a sample, the homogeneous approach had the disadvantage of not allowing one to exceed the limits of linear (or almost linear) models and making it difficult to bring to light the action of essentially nonlinear (feedback) mechanisms in the reactions considered. Their existence, however, in such complex chemical systems as a reactionable solid is most probable, particularly as far as the feedback between the mechanical state of a system and its chemical activity is concerned. These considerations have led to study of the sensitivity of the solid-state processes to all kinds of artificially introduced or spontaneously arising local disturbances.

As shown by experiments, the approach employed has proved its value, the initial premises have been found to be right, and the study has led to the discovery of highly active autowave regimes of low-temperature solid-state conversion, which became the subject of the entire series of studies.

II. OBJECTS OF THE STUDY

In conformity with the problems formulated, we used as objects systems which were known to be incapable of producing spontaneously chains of conversion at radiolysis temperatures of 77 and 4.2 K. In the course of γ irradiation of these systems, nonchain radiation-induced chemical conversions took place which led to the stabilization of active centers and the formation of various kinds of defects. The radiation-induced chain reactions in such systems were earlier studied in sufficient detail at higher temperatures. A brief characterization of the systems employed is given below.

1. Chlorination of saturated hydrocarbons:

$$\begin{aligned} &RH \rightsquigarrow R^{\cdot} + H, \\ &R^{\cdot} + Cl_2 \rightarrow RCl + Cl^{\cdot}, \\ &RH + Cl \rightarrow HCl + R^{\cdot}, \\ &\left.\begin{array}{l} R^{\cdot} \rightarrow \\ Cl^{\cdot} \rightarrow \end{array}\right\} \text{termination.} \end{aligned}$$

2. Hydrobromination of olefines:

$$\begin{aligned} &RH \rightsquigarrow R^{\cdot} + H^{\cdot}, \\ &R^{\cdot} + HBr \rightarrow RH + Br^{\cdot}, \\ &Br + \rangle C{=}C\langle \rightarrow \rangle\dot{C}{-}C(Br)\langle, \\ &\rangle\dot{C}{-}C(Br){-} + HBr \rightarrow {-}C(H){-}C(Br){-} + Br^{\cdot}, \\ &\left.\begin{array}{l} Br \rightarrow \\ R \rightarrow \end{array}\right\} \text{termination.} \end{aligned}$$

3. Polymerization:

$$\begin{aligned} &M \rightsquigarrow R^{\cdot} + H^{\cdot}, \\ &R^{\cdot} + M \rightarrow RM^{\cdot}, \\ &RM^{\cdot} \rightarrow \text{termination.} \end{aligned}$$

We shall briefly describe some concrete examples.

Butyl chloride (BC) + Cl_2: On being cooled to 77 K, a solution of Cl_2 in BC (mole ratio 1:3) completely vitrifies ($T_g \sim 100$ K). The irradiation of the system by ^{60}Co γ rays leads to the accumulation of stabilized active centers. The chain chlorination reaction, not observed during radiolysis, is initiated only on heating up the irradiated system, starting from the devitrification region. The heat of reaction is 138 ± 12 kJ/mole.[27]

Methylcyclohexane (MCH) + Cl_2: When frozen, a solution of Cl_2 in MCH (1:3) vitrifies ($T_g \sim 93$ K). The chain chlorination does not occur until the system is heated into the devitrification region. The heat of reaction is 130 kJ/mole.[14]

Ethylene + HBr: An equimolar reactant mixture was prepared at $T \leqslant 178$ K. Cooling the mixture results in the formation of the crystalline complex $C_2H_4 \cdot HBr$, whose melting point is 110 K. The phase transitions for this complex and the low-temperature chain hydrobromination kinetics at $T > 30$ K were described in refs. 28, 29. The chain hydrobromination was not observed below 30 K.

2-Methylbutadiene-1, 3 (MB) + SO_2. A solution of SO_2 in MB (1:1) was prepared at $T \sim 170$ K. On freezing, such a solution vitrifies. The thermoactivated copolymerization in the system, containing active centers stabilized during the low-temperature radiolysis, also begins in the devitrification temperature region at ~ 100 K.[17]

III. RECORDING OF THERMAL SIGNALS ACCOMPANYING THE FRACTURE OF A REACTIONABLE SOLID SAMPLE IN A MICROCALORIMETRIC CELL

In order to produce brittle fracture in a sample, we used initially the method of thermoelastic stresses, ensuring that the rate of temperature change was large enough to cause the sample to crack. For recording this rate and the thermal signals accompanying the process, a differential scanning calorimeter was employed. In the course of heating or cooling the sample, the temperature on its surface and the thermal effects of the processes occurring in it (the formation of a branched network of cracks or chemical conversion) were registered simultaneously. For details of the method, see ref. 30.

As shown in ref. 14, for the vitrified MCH + Cl_2 system, slow heating of the sample from 4.2 to 77 K did not produce any cracking and the calorimeter registered no thermal effects (Fig. 1*A*, *B*, solid lines). Fast heating led to the fracture of the sample, and the concomitant thermal effect was registered by the calorimeter (Fig. 1*C*, *D*, solid lines). Note that the formation of cracks in the sample could also be registered visually and acoustically. After determining the rate of heating that results in brittle fracture of the sample, analogous experiments were performed with samples preirradiated with a certain dose of ^{60}Co γ rays to produce stabilized active centers in the system. In the course of slow heating of the irradiated sample, unable to cause any fracture, no heat production due to the chemical reaction could be detected (Fig. 1*A*, *B*, dashed lines). On the contrary, at the instant of brittle fracture during fast heating, the reaction did take place (Fig. 1*C*, *D*, dashed lines). The concentration of radicals accumulated during the irradiation was 5×10^{18} cm^{-3}, and the degree of Cl_2 conversion was experimentally found to be 10–15%.

Brittle fracture of the sample could also be produced during fast cooling. Figure 1*E*, *F* presents data for a sample which was cooled rapidly enough to cause cracking. As expected, the fracture of the irradiated sample was

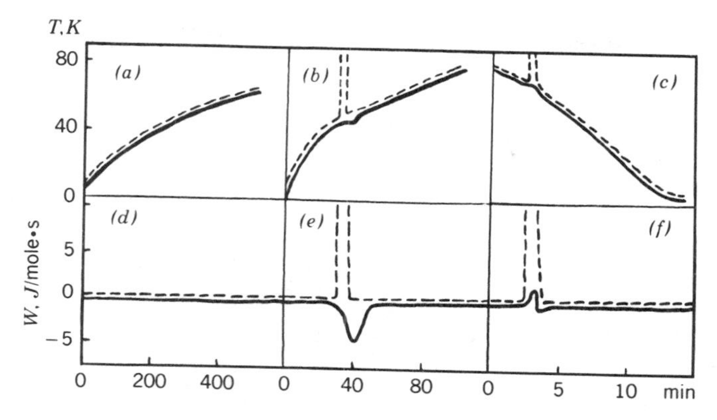

Figure 1. (*A*, *B*, *C*) Time dependence of the temperature of the sample (Cl_2 + MCH, molar ratio 1:3) and (*D*, *E*, *F*) thermal effects; solid lines: nonirradiated samples; dashed lines: samples irradiated by γ rays from ^{60}Co at 77 K, dose 27 kGy.

accompanied by the fast chemical conversion, whereas no such conversion was observed in the nonirradiated sample.

To prove the decisive role of fracturing in the initiation of chemical conversion in a sample, experiments were carried out in which the sample was subjected to mechanical fracture by an external force. The brittle fracture of the sample containing stabilized active centers (γ radiolysis) was accomplished at a constant thermostat temperature of 4.2 K by turning a frozen-in thin metallic rod (see Fig. 2). At the instant of the disturbance, a rapid (explosive) chemical conversion occurred. Being initiated locally, the reaction then spread over the sample—the dark color due to γ radiolysis disappeared. Such mechanically induced chemical conversions were observed both in vitreous (BC + Cl_2, MCH + Cl_2, MB + SO_2) and in polycrystalline (C_2H_4HBr [15]) systems with different types of reactions studied (chlorination, hydrobromination, polymerization).[†]

It should be noted that in the initiation of the cryochemical reactions observed, what is important is the very fact of formation of a new crack (or zone cut by such cracks), and not the increase in the specific internal surface of the sample. Indeed, slow heating of a monolithic sample, or of a sample

[†]The explosive character of the photoinduced solid-state chlorination reaction of MCH was first described in ref. 31, the phenomenon being interpreted on the assumption of a decrease in the chain-growth activation energy due to the thermoelastic stresses induced in the sample. A possible role of brittle fracture was not considered in that case. However, it would be of interest also to take account of that effect under the conditions used in ref. 31, the more so in that the evaluated values of stresses required to reduce the activation energy markedly are far above the thresholds of brittle fracture of the corresponding matrices (for details, see Section XII).

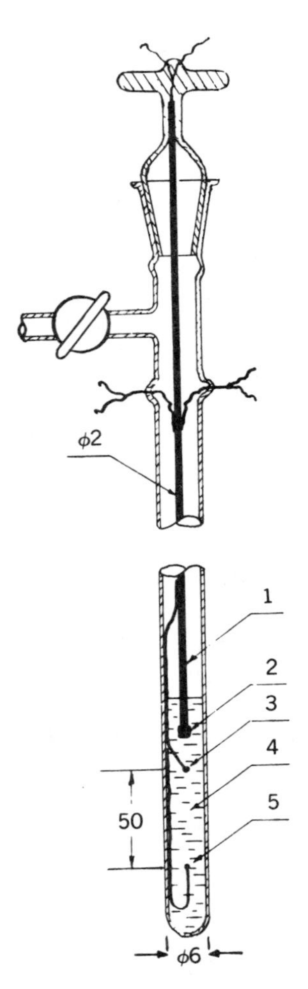

Figure 2. Schematic of the device for investigating the reaction dynamics in a sample: 1, iron rod; 2, heater; 3 and 5, thermocouples; 4, sample.

fractured artificially before the formation of active centers in it (i.e., before the radiolysis), does not result in a reaction burst. Consequently, it is at the instant of fracture that the conditions crucial for the initiation of the reaction arise.

These results led us to two conclusions of principal importance. First, in the systems studied the brittle fracture of the sample produces a burst of chemical conversion. Second, the process initiated by the fracture is self-accelerated and spreads over the entire sample. Therefore, the formation of a primary crack acts as a trigger switching on a certain positive feedback mechanism, which leads to self-acceleration of the process. To shed light on this mechanism, an experimental investigation into the dynamics of the reaction response to an external disturbance was performed.

IV. THERMOGRAPHICAL ANALYSIS OF THE DYNAMICS OF THE REACTION BURST INDUCED BY LOCAL FRACTURE OF A SOLID SAMPLE OF REACTANTS. THRESHOLD EFFECTS

Figure 2 schematizes a device used in the study of the reaction dynamics in a sample. A constantan microheater of 1-mm diameter, through which a capacitor could be discharged, was embedded into the upper portion of the sample during freezing. The energy released during the pulse heating varied. The initiating pulse energy was such that the maximum temperatures in the initiation region would not exceed the temperature of devitrification (melting); that is, the conditions required (according to ref. 27) for the thermoactivated reaction to proceed at an appreciable rate were not reached. Using this technique, it was possible to obtain the temperature gradients required for fracturing the sample in the vicinity of the heater.

The dynamics of heating in the initiation region was studied with the help of a copper–constantan thermocouple mounted on the heater surface (not shown in Fig. 2). The results of the measurements are presented in Fig. 3 (the arrow indicates the instant of switching on the pulse heater). Whatever the pulse energy, no reaction burst is observed in nonirradiated samples after switching on the heater (Fig. 3, solid curve). In the preirradiated $CB + Cl_2$ system,

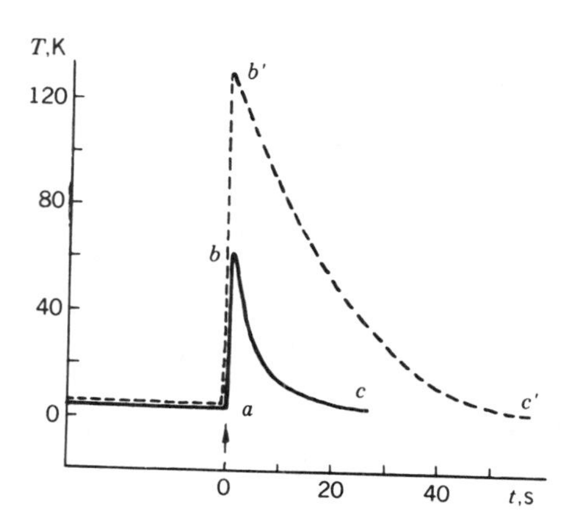

Figure 3. Dynamics of heating in the initiation zone of frozen samples, nonirradiated (solid line) and irradiated by γ rays from ^{60}Co, dose 27 kGy (dashed line), for the $BC + Cl_2$ (3:1) reaction. The arrow indicates the instant of application of the initiating thermal pulse ($q = 5$ J).

however, the pulse does initiate the reaction in the sample after having reached an overthreshold value (dashed curve).

Figure 4 shows the results of measuring the threshold (critical) values of the energy released during the capacitor discharge (q_{cr}) which must be reached for the reaction to occur. It is seen that at $q < q_{cr}$, triggering of the reaction does not occur, but at $q \geqslant q_{cr}$ the system responds to a pulse with initiation (ignition) of the reaction. Note that no qualitative changes in the reaction dynamics were observed even when the temperature in the region near the heater was above T_g ($q > 16$ J). An oscillographic display of the reaction burst (stage *ab* in Fig. 3) made it possible to measure the characteristic time of this stage as a function of the pulse energy (Fig. 4*B*).

These experiments led to the following conclusions: below the ignition threshold ($q < q_{cr}$) the sample is reactionless and the preburst heating is unrecordable (Fig. 4*A*); the reaction burst is practically synchronous with the application of the initiating pulse (Fig. 3); the burst of the reaction is a rapid stage, proceeding without any induction period and with the characteristic

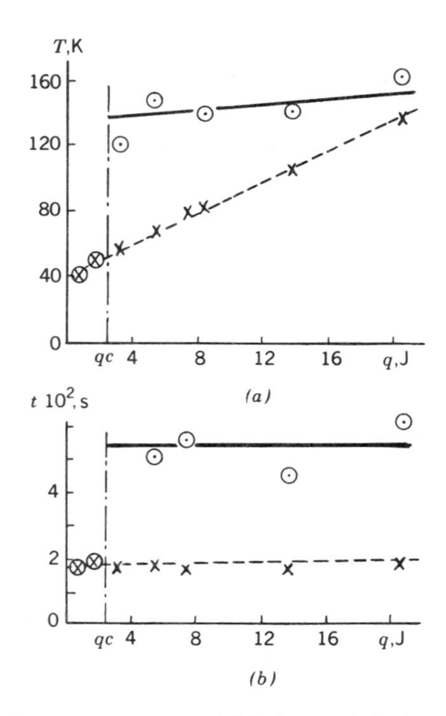

Figure 4. (*A*) Maximum temperature and (*B*) characteristic time (that required to reach $0.5T_m$) for the $BC + Cl_2$ (3:1) reaction as functions of pulse energy q. ×, nonirradiated sample; ·, sample irradiated by γ rays from ^{60}Co, dose 27 kGy.

time ~ 0.05 s (for comparison, thermal lagging of the sample amounts of tens of seconds), and is independent of the pulse energy (Fig. 4*B*).

These features of the reaction burst dynamics evidence convincingly that the classical thermal mechanism of feedback does not play any decisive role in this phenomenon. The discovered crucial significance of newly formed surfaces in fracturing the sample suggests the following nonthermal mechanisms of self-acceleration in the systems studied: the chemical reaction occurring on a newly formed surface and in its adjacent regions generates discontinuities, for example as a result of the temperature or density gradients arising in the course of the reaction and producing stresses which lead to the sample fracture.

Based on these considerations about the possible nature of the effects observed, let us construct a phenomenological model of the process, not going deeply for the time being into the reason for the high chemical activity of the surfaces formed during the fracture.

According to the hypothesis, chemical conversions in a solid occur on the surface of a new crack (or in the layer adjacent to it), that is, the reaction rate in such a reacting system is a certain function of the specific surface area S (active surface area per unit volume). As noted above, the positive feedback in this model manifests itself in the fact that the rate of formation of cracks (i.e., the active surface growth rate in the sample volume) is proportional to the reaction velocity. Therefore, the equation describing the formation of a new surface can be written in a form analogous to that of a branched-chain process:

$$\frac{dS}{dt} = F(S) - G(S), \tag{1}$$

where $F(S)$ is the generation rate of an active surface and $G(S)$ is the rate of its deactivation. If $F(S)$ has a higher order with respect to S than $G(S)$, then the model will qualitatively describe the observed experimental facts: (1) the critical phenomena of the generation and self-accelerating development of the reaction; (2) the possibility of initiating the reaction burst by creating at the initial instant a certain primary network of cracks, either mechanically or thermally (in the absence of such disturbances the reaction rate is immeasurably small).

As $G(S)$ must be, from the general considerations, a linear function (analogously to a unimolecular process), it is sufficient for the above condition to be reached that $F(S)$ have an effective order higher than the first with respect to S.

Consider the question of degeneration of the critical phenomena. It follows from (1) that it is possible to suppress the burst of the process by decreasing the generation rate of an active surface. This can be easily done by decreasing

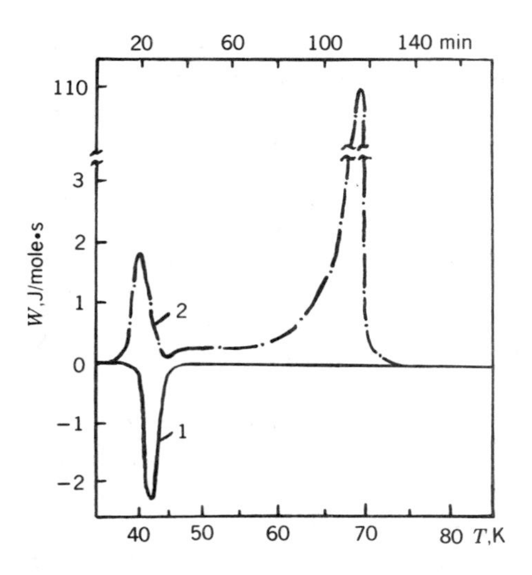

Figure 5. Calorimetric curves of rapid thawing of the equimolar $C_2H_4 \cdot HBr$ complex, nonirradiated (1) and irradiated by γ rays from ^{60}Co at 4.2 K, dose 4 kGy (2).

concentrations of the reactants. For instance, we may reduce the concentration of active centers stabilized during the radiolysis. This conclusion is confirmed by experiments on the $C_2H_4 \cdot HBr$ system.[15] Figure 5 presents results obtained with this system preirradiated at 4.2 K with a small dose of 4 kGy. Under these conditions the sample fracture provoked by the high rate of heating does not lead to a burst of the process. At this instant of time the calorimeter registers only a weak exothermal signal (the first exothermal peak on curve 2 of Fig. 5; the endothermal peak in this figure corresponds to the fracture of the nonirradiated sample), which reflects, apparently, the chemical conversion only in the artificially fractured region. With the small irradiation dose used, the process does not reach the self-sustaining state; no progressively growing increase in the number of cracks and no burst of the process take place. This conclusion is convincingly confirmed by further heating of the sample: at higher temperatures during the melting the microcalorimeter registers the onset of the hydrobromination reaction, which proceeds in conformity with the regularities described earlier.[28,29]

V. AUTOWAVE PROPAGATION OF THE REACTION OVER THE SAMPLE

The principal feature of systems described by nonlinear models of the type (1) is their ability to generate autowave phenomena. The autowave regimes of conversion must occur in samples of sufficient extent in response to a local disturbance. These considerations have determined the experimental problem at this stage of research, namely searching for autowave phenomena, ascertaining the fact of their existence, and studying the dynamic characteristics and structure properties of the chemical-reaction wave front propagating over a solid sample.

Already in the first experiments[14] with the MCH + Cl_2 system it was

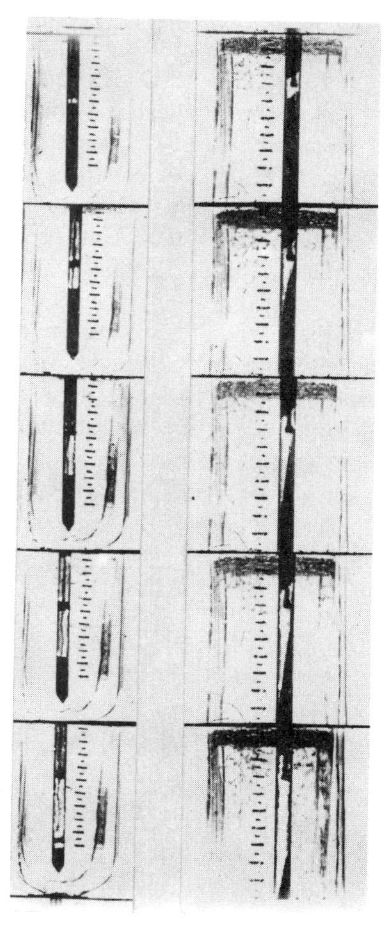

Figure 6. Cinegrams of reaction-front propagation in a BC + Cl_2 (3:1) system for different directions of the main crack ($T = 77$ K): (*A*) normal to the sample axis; (*B*) at an angle to the sample axis. The times in seconds, from top to bottom, are (*A*) 0.06, 0.47, 0.88, 1.09, 1.66; (*B*) 0.03, 0.09, 0.13, 0.16, 0.22. γ-irradiation dose 27 kGy.

possible to initiate spontaneous propagation of a reaction wave by a local disturbance produced either mechanically or thermally (pulse heater) in a limited area of the solid matrix. Similar processes were observed in all the systems studied. Below are described the typical results obtained for the BC $+ Cl_2$ system in experiments with cylindrical samples of frozen reactant mixture 0.5–1.0 cm in diameter and 5–20 cm in length (Fig. 2). As the reaction propagation over a sample was accompanied by the change in its color, the method of cinegram could be used to measure the front propagation velocity. The front structure was studied thermographically with the help of thermocouples embedded in frozen samples.

Figure 6*A* presents a typical cinegram of the autowave process in a sample immersed in a transparent Dewar flask filled with liquid nitrogen. The reaction was switched on either by a local mechanical disturbance (by turning the rod 1 in the upper portion of the sample, Fig. 2) or by pulse heating (discharge of a capacitor through the heater 2, Fig. 2). As seen from the cinegram, a flat front was rapidly formed after switching on the reaction, which propagated parallel to itself along the reaction-tube axis. Figure 7 shows the change with time of the reaction-front coordinate. The reaction wave moved with practically constant velocity equal to ~ 2.5 cm/s.

According to the hypothesis, the direction of the propagating front must correspond to that of the initiating primary crack. For instance, if the initiating crack, which cuts the sample along its entire length, makes an acute angle with the sample axis, the reaction front is expected to move parallel to the crack without any change throughout its motion. This situation could be realized by selecting the initiating force applied to the frozen-in rod. The cinegram of Fig. 6*B* shows the result of such an experiment: one can see the initiating crack

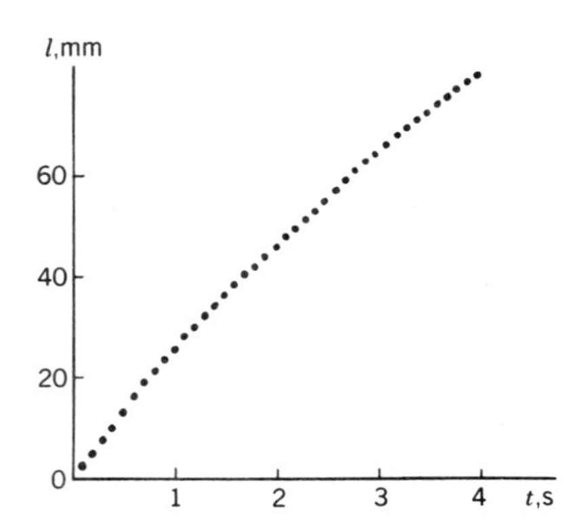

Figure 7. Time dependence of the reaction-front coordinate in the $BC + Cl_2$ system, obtained from the cinegram of Fig. 6*A*.

(*ab*) and the reaction front moving parallel to it. Thus, further evidence for the significance of autodispersing as a chemically activating factor has been obtained, now in experiments with propagating reaction waves.

To study the wave-front structure (its temperature profile), the autowave process was registered thermographically in a series of experiments. The propagation velocity was measured by the time required for the wave to travel a distance (3–5 cm) between two thermocouples (Fig. 2).

Figure 8 presents typical profiles of the traveling temperature wave front displayed both in time and in space at initial temperatures of 77 and 4.2 K. As seen, the front profile remains qualitatively the same on changing the temperature of the thermostat. It includes three stages: *ab*, inert heating (at 77 K this stage was not registered, being practically isothermic); *bc*, the jumplike onset and ignition of the reaction (the characteristic break point *b*); *cd*, the stage of cooling. Stage *bc* is a rapid one. Its duration was found oscillographically to be ~0.1 s at 77 K and 0.3 s at 4.2 K, so that at the wave velocity $U = 2$ cm/s the width of the zone of the traveling wave front was $\delta \sim 0.2$ cm, and at $U = 1$ cm/s, $\delta \sim 0.3$ cm.

This structure of the traveling front of a low-temperature reaction exhibits features utterly atypical of classical thermal self-propagation. They are: (1) a weak or nonexistent stage of inert preburst heating; (2) a jumplike switching on and off of the reaction—characteristic break points *b* and *c* (Fig. 8), the temperature where the reaction switches on at point *b* being far below that for

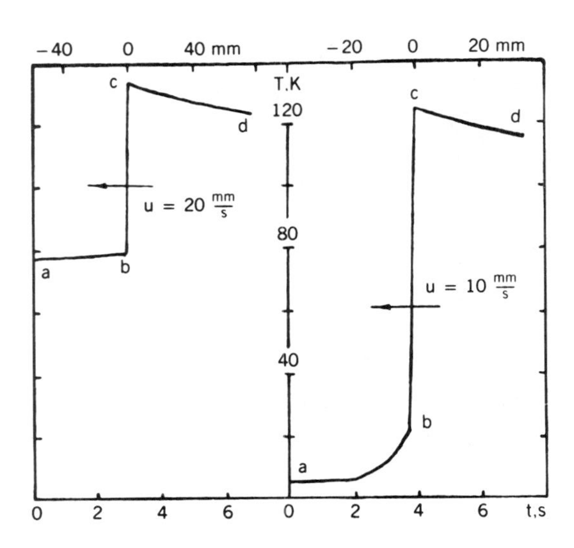

Figure 8. Temporal and spatial display of typical temperature profiles of a BC + Cl_2 reaction wave at (*A*) 77 K and (*B*) 4.2 K. Preirradiation dose 27 kGy.

the initiation of a thermoactivated reaction ($T_g \sim 100$ K for the BC + Cl_2 system[7]); (3) similarity of the dynamical patterns of the process development in the case of spontaneous initiation of a reaction by a traveling wave and in the case of forced reaction ignition by the accelerated heating (see the initiation-stage dynamics in Fig. 3); (4) weak alteration in the wave velocity on changing the temperature in the thermostat from 77 to 4.2 K, accompanied by strong alteration in the heat release.

It should be noted that the temperature gradient in the front at 4.2 K is about 2 times that at 77 K (see Fig. 8), that is, the degree of chemical conversion increases with decreasing initial temperature. This is also atypical of the thermal mechanism based on the Arrhenius law, but can be explained by the above hypothesis, considering that the lifetime of a new surface in the active state increases with decreasing temperature, which results in the enhanced conversion.

Using the experimental values for the width of the traveling wave front (portion *bc*, Fig. 8), let us estimate the propagation velocity for the case of a thermal mechanism based on the Arrhenius law of heat evolution from the known relationship $U = a/\delta$, where $a \sim 10^{-2}$ cm^2/s is the thermal conductivity determined by the conventional technique. We obtain 5×10^{-2} and 3×10^{-2} cm/s for 77 and 4.2 K, respectively, which are below the experimental values by about 1.5–2 orders of magnitude. This result is further definite evidence for the nonthermal nature of the propagation mechanism of a low-temperature reaction initiated by brittle fracture of the irradiated reactant sample.

All the reaction systems considered, despite being greatly different chemically, have been found to have similar dynamic characteristics of the autowave processes occurring therein. Particularly, the linear velocities of the wave-front propagation are in the range of 1–4 cm/s for all systems. All of them have a certain critical irradiation dose below which the excitation of an autowave process becomes impossible and the system responds to a local disturbance only with local conversion incapable of self-propagating (the situation discussed above and illustrated by Fig. 5).

Essentially different from the reactions studied is the MB + SO_2 system.[17] Whereas in the BC + Cl_2 system the degree of chemical conversion did not exceed 10–15%, in the MB + SO_2 copolymerization reaction the conversion was practically complete. The small degree of conversion in the BC + Cl_2 system is likely to be connected with the heating up of the wave front to the devitrification temperature, which leads to the effective decay of active centers as a result of their recombination. Perhaps this reason, along with the others discussed earlier, may be called to account for the increase in the degree of conversion in the wave front on passing from the nitrogen to helium temperature. In other words, in the reactions of chlorination and hydrobromination the heating above the temperature of devitrification and the melting of a solid matrix are factors that suppress rather than accelerate the conversion, whereas in the copolymerization reaction the limited molecular mobility of active centers in the matrix of a high-molecular-weight product favors reaching a high degree of conversion in the wave front.

The next important step in the elucidation of the role of the thermal factor in the mechanism of the phenomena in question was studying the effect of the sample size on the characteristics of the autowave process. Can the self-sustained wave regime of conversion be made impossible by intensification of heat release at the expense of a decrease in the diameter of a cylindrical sample containing the reactant mixtures? By analogy with combustion physics, the question of a critical sample size has been raised.

It has been established, in experiments performed in capillaries of 0.5–1.0-mm diameter (which corresponds to more than a tenfold increase in the parameter *Bi* characterizing the intensity of heat release), that none of the

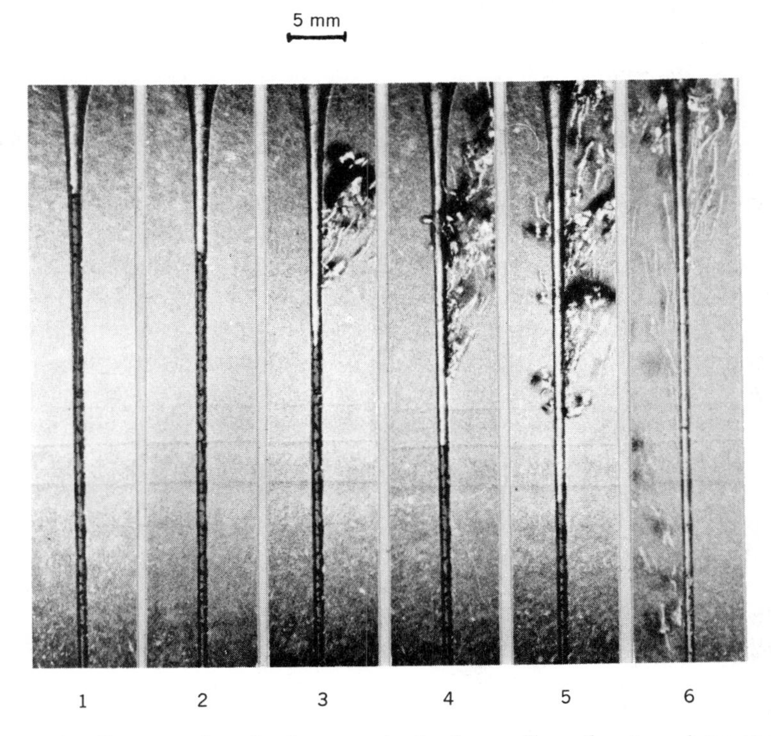

Figure 9. Cinegram of reaction-front propagation in a capillary, diameter ~ 1 mm, at 4.2 K. Time intervals between frames 1 and 2: 0.14 s; 2 and 3: 0.06 s; 3 and 4: 0.14 s; 4 and 5: 5 and 6: 0.72 s. Dose 45 kGy.

systems studied displays a degeneration of the autowave process. Moreover, the characteristic values for the reaction self-propagation do not undergo any noticeable changes under these conditions. Figure 9 shows a cinegram of the reaction-front propagation ($BC + Cl_2$) in a capillary of 1-mm diameter immersed in liquid helium (irradiation dose 45 kGy, wave-front propagation velocity ~ 2.5 cm/s). This series of experiments has thus provided additional evidence against the classical thermal mechanism being involved in the phenomena considered. It should be added that even the use of film samples in the subsequent experiments to avoid heat screening by the reaction-tube wall did not lead to the disappearance of the autowave regimes of low-temperature solid-state conversions (for details, see Section VIII of this chapter).

VI. THEORETICAL TREATMENT OF AUTOWAVE PROCESSES IN SOLID-STATE CRYOCHEMICAL CONVERSIONS (THE SIMPLEST MODEL)

The experimental results systematized in the previous section have made it possible to further develop the hypothesis of the mechanochemical nature of the self-activation of solid-state conversion and to describe qualitatively the possible pattern of the autowave process of chemical interaction in the systems considered. It can be pictured in the form of a narrow region propagating over a solid sample and cut all over with an intricate network of newly formed cracks. The chemical reaction occurs exactly in this region, on the surface of or near the cracks, and in its turn creates the conditions for continuation of the process of fracture, which plays an activating role, in the neighboring layer of the solid matrix. This dispersing of the sample, which proceeds layer by layer, is caused by the propagating field of stresses. The appearance of the field may be due to, first, the difference in the densities of the initial and final products (this mechanism will be called *isothermal*) and, second, the steep temperature gradients resulting from exothermicity of the reaction (nonisothermal mechanism). The two factors can, naturally, act simultaneously as well. Based on this qualitative picture, let us construct the simplest phenomenological models of the process.

Models of the isothermal mechanism can be constructed using a balance equation (1) for the area of active surface per unit volume of a solid sample, with a term added which describes the propagation of this surface into the nonfractured matrix. The term requires that a certain effective transfer coefficient (analogous to the diffusion coefficient) should be introduced. To a first approximation, it can be written as $D = vl \equiv v^2\tau$, where v is the velocity of sound in the sample, l is the length of the "free run" of a crack for the time τ, and τ is the time of mechanical unloading (or the characteristic relaxation time of stresses in the real solid matrix of a reactant sample). It seems impossible to operate in the framework of the above model with any real values of parameters and variables and to perform any sound comparison with experiment before accomplishing detailed studies of the mechanical and strength characteristics of the samples as well as of fracturegrams of the solid matrix ruptured by the reaction. For that reason, we restrict ourselves at this stage of research to analysis of the nonisothermal mechanism under the conditions realized in massive cylindrical samples, in which, as shown above, heating due to the reaction is significant and the role of thermal stresses in the sample dispersion must be most pronounced.

The model proposed in ref. 18 describes a sample of infinite length without regard to heat exchange with the environment, that is, it is assumed that the characteristic time of the reaction heat release is much below that of heat transfer.[†] This assumption makes it possible to perform the analysis in terms of a one-dimensional model. In accordance with the hypothesis, we believe that switching on of the reaction at a given cross section of the sample occurs when that cross section suffers a stress exceeding the ultimate strength of the solid matrix. Since we speak about thermal stresses and are concerned with elastic deformation and brittle fracture, we can easily change over, as can be shown, from stresses to temperature gradients, assuming a single-valued dependence between them. Such a substitution of variables greatly simplifies the model and allows elimination of the mechanical equations and reduction of the analysis to only one equation of thermal balance, in which we introduce a certain critical value of the temperature gradient $dT/dx = (dT/dx)^*$ as parameter.

With the above assumptions, the equation describing the autowave process in the systems in question takes a form which looks similar to the fundamental combustion equation[32]:

$$\lambda \frac{d^2T}{dx^2} - U\,c\rho \frac{\alpha T}{dx} + Q = 0. \tag{2}$$

Here T is the temperature; x is the coordinate; λ is the thermal conductivity; c and ρ are, respectively, the heat capacity and density of the solid mixture of reactants; Q is the rate of reaction heat release, and U is the propagation velocity of the temperature wave front.

The principal difference between the autowave and combustion processes is that in the former case Q is a function of the temperature gradient but not the temperature itself. Let Q change jumpwise from 0 to Q^* at $dT/dx = (dT/dx)^*$, and let it retain this value for a certain time interval τ. This implies physically

[†]The assumption is quite justified: a study of the dynamics of the reaction burst (Fig. 4) gave a value for the time of reaction $\sim 10^{-1}$ s, whereas the time of thermal relaxation is of the order of 10 s.

that the heat release in a reaction is switched on only in response to brittle fracture of the sample produced by thermal stresses equal to the ultimate strength of the material. The employment of the reaction time τ as parameter reflects that part of the hypothesis according to which the time period of chemical activity in the course of fracture formation is limited by the deactivation process (for instance, by recombination of active particles on fracture surfaces). In terms of the stationary model (2) considered here, the parameter τ contains information on the reaction zone (zone of dispersivity) at the propagating front, whose size is equal to U_τ. At each point within this zone the reaction proceeds at a rate Q^*, and outside it $Q = 0$.

The boundary conditions can be written as

$$T|_{x=-\infty} = T_0, \qquad T|_{x=+\infty} = T_0 + \frac{Q\tau}{c\rho} \equiv T_m,$$

where T_0 is the initial temperature of the sample and T_m is the maximum temperature of its adiabatic heating due to the reaction heat. The model is considered quasiharmonic, that is, it is assumed that the characteristic size of a grain in the zone of dispersion is much less than the width of the temperature wave front.

Equation (2) with the heat source as a function of the type indicated and with the above boundary conditions is integrated analytically and has a solution in the form of a traveling wave. The principal attention will be paid to

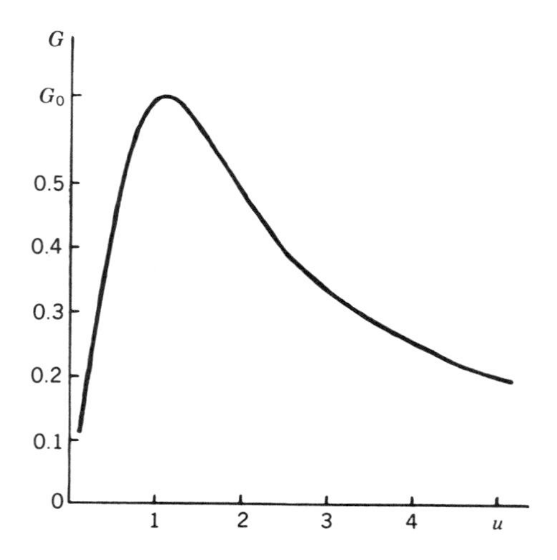

Figure 10. Critical temperature gradient as a function of reaction-wave propagation velocity.

analysis of the dependence of the wave propagation velocity on the parameters. This dependence, shown in Figure 10, has the form

$$g(u) \equiv \{1 - \exp(-u^2)\}u^{-1} = G. \tag{3}$$

Here $u = U(\tau/a)^{0.5}$ is the dimensionless velocity of a propagating wave (the parameter to be determined), $G = (dT/dx)^* a^{0.5}/(T_m - T_0)$ is a given dimensionless critical temperature gradient, and $a = \lambda/c\rho$ is the coefficient of thermal conductivity. The expression for the temperature-wave profile is

$$\Theta = u^{-2}\{1 - \exp[-u^2(1-\xi)]\} + \xi,$$

where $\Theta = (T - T_0)/(T_m - T_0)$ is the dimensionless temperature and $\xi = x/U\tau$ is the dimensionless coordinate.

As seen from Figure 10, the autowave solution of equation (2) exists only at $G < G_0 \sim 0.64$. Physically this implies that in a system described by this model the autowave mode of the reaction propagation over a solid reactant mixture becomes impossible at a definite increase in the strength of the sample, decrease in the thermal effect and reaction velocity, and increase in the thermal conductivity.

The essential feature of equation (2) is the nonuniqueness of its solution. In the range $G < G_0$, two different values of the steady-state velocity of wave propagation correspond to the same value of G. Let us compare the characteristics of the two autowave regimes. For simplicity, we shall consider the range $G \ll G_0$, in which equation (3) is soluble for U. The expression for U corresponding to the smaller velocity mode has the form

$$U = \frac{\alpha(dT/dx)^*}{T_m - T_0} \quad \text{or} \quad U = \frac{\alpha(dT/dx)^*}{Q\tau/c\rho}, \tag{4}$$

and that for the greater velocity mode

$$U = \frac{T_m - T_0}{(dT/dx)^*\tau} \quad \text{or} \quad U = \frac{Q^*/c\rho}{(\alpha T/dx)^*}. \tag{5}$$

The slower autowave process is similar in some respects to classical combustion, despite the differences in their physical nature. The wave velocity shows the same dependence on thermal conductivity as in the case of flame propagation. Analogously to combustion, the reaction zone is near the maximum temperature T_m [it is near T_m that the critical gradient $(dT/dx)^*$ switching on the reaction is realized], whereas the greater part of the front temperature profile corresponds to inert heating of the sample (the "Michelson zone" in combustion theory).

The faster process is significantly different. The most conspicuous feature of (5) is that the velocity does not depend on the thermal conductivity. This result, unexpected for a problem with conductive heat transfer, is connected with the peculiarities of the front structure. The temperature profile of the wave front is considerably steeper than that of the slower wave, and the coordinate of the reaction switching on is in its fore part, near T_0. It is this that explains the Q independence of the wave process, because the temperature gradients are high, approaching a critical value even in the unheated part of the front. These considerations also explain the different effects of the critical temperature gradient on the front velocity in the two modes of the autowave process: with strengthening of the reacting sample [with increase in $(dT/dx)^*$] the velocity of the faster wave decreases and that of the slower wave increases.

The problem of the stability of the two modes of steady-state propagation of the temperature wave over the reaction sample needs a separate study. From qualitative considerations it follows that the faster process is less sensitive to disturbances (both mechanical and thermal) than the slower one. It is evident that in the case of slow motion any kind of inhomogeneities in the sample [e.g., a local reduction in the strength, leading to a decrease in $(dT/dx)^*$] may cause a displacement of the reaction-onset coordinate to the fore part of the front and thereby induce a spontaneous transformation of the slower wave into the faster one.

Let us consider the experimental data of the previous section from the standpoint of the theoretical results presented here. The faster wave regime is closer to the experimental conditions in two respects. The first one concerns the initial conditions, that is, the mode of initiation (ignition) of the reaction. To initiate the slower wave, the heating up of a local area of the solid sample must proceed at a rate such that the time of this process is comparable to the characteristic time of thermal relaxation of the system. Indeed, it is this mode of heating that produces a temperature profile of the slower wave in which the critical gradient approaches the maximum temperature. The faster wave is initiated at a high rate of heating that triggers at the initial instant the reaction heat release in the unheated zone of the sample as a result of its dispersing. These conditions of wave initiation were realized at the first stage of the experimental study.

The second resemblance is in the structure of a traveling wave. As indicated above, the reaction zone of the faster wave front starts at temperatures close to the initial one. This peculiarity was revealed in experiments carried out in liquid helium or nitrogen (Figure 8).

Having obtained some qualitative agreement, we shall try to make a quantitative comparison as well. In order to calculate the wave velocity from

(5), we need numerical data on the temperature difference in the wave front, critical gradient, and reaction duration. The values for these parameters were obtained in experiments on the reaction-burst dynamics and traveling waves in the $BC + Cl_2$ system. In particular, in liquid nitrogen $T_m - T_0 = 60\,K$, $\tau \sim 0.1\,s$, and $(dT/dx)^* \sim 300\,K/cm$. The calculated value $U \sim 2\,cm/s$ is very close to the experimental one.

To test the fit of the theoretical mechanism to the experimentally observed phenomena, it seemed principally important to try to realize experimentally the second mode of the autowave process. Its initiation was performed, in accordance with the theory, not by pulse heating but with the help of a heater whose temperature could be raised slowly. Under such conditions the slower wave could not be excited either in liquid helium or nitrogen, that is, there was only one mode of wave propagation. This was possibly connected with the fact that under conditions of intense heat release into liquid media, high (close to critical) transverse temperature gradients occurred in the samples, which might be a source of severe disturbances impeding the realization of the slower wave mode.

To eliminate this factor, in specially designed experiments the sample was placed in a cryostat in the vapor of boiling nitrogen. In that case we could

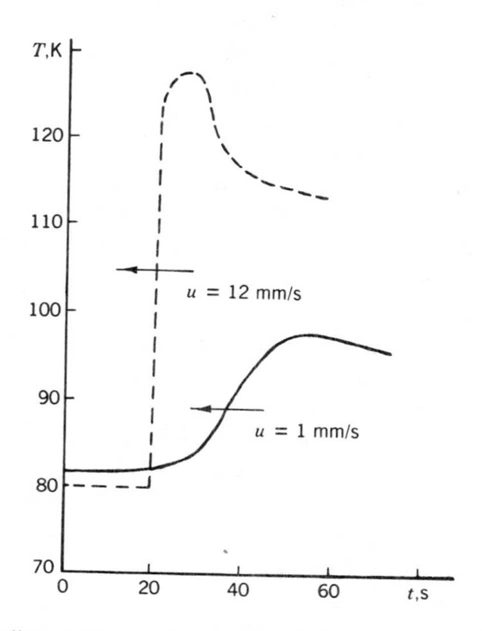

Figure 11. Time display of temperature profiles of $BC + Cl_2$ reaction waves. The system was placed in the vapor of liquid nitrogen. The solid curve resulted from initiation by slow heating, and the dashed curve by rapid heating (discharge of capacitor through heater).

observe, along with the faster wave (having the same velocity as that described earlier), also the slower wave (see Figure 11). As seen, the velocity of the slower wave is about one order of magnitude less than that of the faster wave. Its value calculated from (4) is $U = 5 \times 10^{-2}$ cm/s [$a \sim 10^{-2}$ cm^2/s; the values of $T_m - T_0$ and $(dT/dx)^*$ are given above], which is close to the experimental value.

The realization of the slower wave involved certain difficulties. Not in every experiment could it be initiated by the slow local heating. Frequently the faster wave was initiated as well. There were cases when the well-developed slower wave transformed into the faster one in the course of its propagatioon, so that the first thermocouple in its path registered a front profile similar to that of the solid curve in Figure 11, and the second, a profile similar to the dashed curve. The higher sensitivity of the slower propagation mode to various disturbances has already been noted in the theoretical treatment of the autowave process.

VII. AUTOWAVE PROCESSES UNDER CONDITIONS OF UNIFORM COMPRESSION OF THE SAMPLE

From the theoretical analysis it follows that acting on the strength characteristics of the solid matrix of a frozen reactant mixture may be an effective means for testing the concepts developed. This was an impetus for a study of the effect of high pressures on the dynamic characteristics of the autowave regimes of chemical conversion,[19] since it is known that uniform compression of solid materials results in significant strengthening.

The study was performed on hydrocarbon chlorination in a vitreous solution of Cl_2 in butyl chloride, using a specially designed fixed-pressure chamber. The solution of Cl_2 in BC (mole ratio 1:3) in Ar atmosphere at 100–120 K was placed in the pressure chamber, and the whole system was then cooled to the temperature of liquid nitrogen. Two copper–constantan thermocouples were positioned in the inner channel of the chamber to measure the propagation velocity of the reaction wave and its temperature profile. The chamber also contained a microheater. The discharge of a capacitor through the microheater caused fracture of the sample in the adjacent zone, which led to the initiation of a reaction wave. At 77 K, the sample in the chamber was compressed to a definite pressure by a hydraulic press with the help of a piston system. To achieve the overthreshold concentration of active centers, the compressed sample was irradiated at 77 K by ^{60}Co γ rays with a dose of 27 kGy.

Figure 12 shows the most typical results. As predicted by the theory, the increase in the pressure was accompanied by a decrease in the reaction wave-front velocity. It is seen that at 6×10^8 Pa the velocity is significantly reduced—approximately by a factor of 2.5–3 below the uncompressed

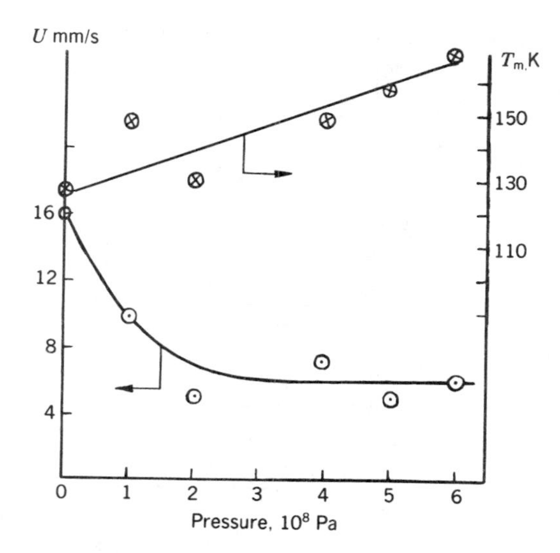

Figure 12. Wave propagation velocity U and maximum temperature T_m of reaction wave front as functions of the applied pressure in $BC + Cl_2$ system. Dose 27 kGy.

sample. Thus, the achievement by the matrix of a more homogeneous and ordered structure due to the compression, together with its simultaneous strengthening and the impeded fracture formation, slows down the propagation of a traveling dispersion–conversion wave. This can be naturally connected, on the basis of the relationship (5), with the growth of the critical value of the temperature gradient $(dT/dx)^*$ causing brittle fracture of the sample.

The wave structure also undergoes typical changes in complete agreement with the theoretical results (Figure 13): the front steepness decreases with increasing pressure; the maximum gradient of the temperature profile is displaced from the fore part of the front to the coordinate of maximum heating; in the low-temperature zone there appear "tails" of inert heating, which become pronounced on loading. In other words, the strengthening of a solid mixture of reactants is accompanied, in accordance with the solution (3) (see Figure 10), by the development in the front structure of the characteristic properties of the slower wave.

It is interesting to note that in spite of the decrease in the wave propagation velocity, the maximum heating at the front increases markedly (Fig. 12), that is, the degree of conversion grows. This is another trait that distinguishes the phenomena under study from those of the combustion type: the increase in the heat flow does not compensate the effect of mechanical strength on the characteristics of a traveling wave.

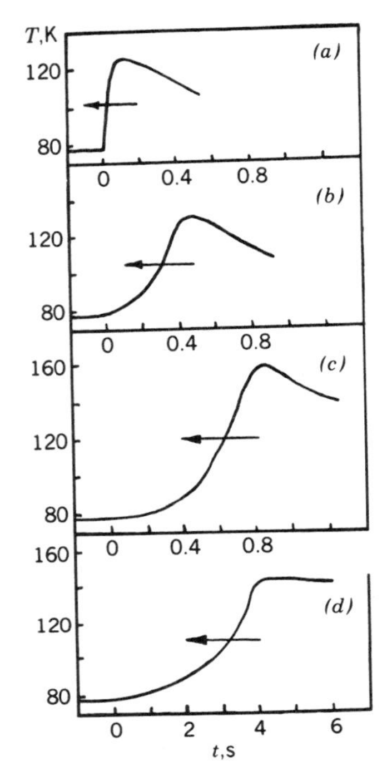

Figure 13. Temperature profiles of reaction wave front at external pressure (*a*) 10^5 Pa, (*b*) 2×10^8 Pa, and (*c*) 5×10^8 Pa, and (*d*) in the burnt sample at 10^5 Pa. $BC + Cl_2$ system, dose 27 kGy.

The dependence of heating on pressure itself may be due to a decrease, with increasing pressure, of the characteristic size of dispersion grains. There is also an alternative explanation, connected with increasing devitrification temperature of a compressed matrix: this factor causes a displacement of the temperature region of the intense recombination of active centers and creates the conditions for a higher degree of conversion at the front.

The strengthening of a sample can be achieved by other means than pressure treatment. An important factor responsible for the degree of homogeneity of a matrix and hence for its strength parameters is the condition of freezing. In all the above experiments the freezing of a reactant mixture was performed by immersing the reaction tube directly in liquid nitrogen. Experiments with the $BC + Cl_2$ (3:1) system were carried out in which vitrification was produced under conditions of slow cooling with long exposure to a temperature close to T_g (100 K). In such samples a higher degree of order and regularity could be observed even visually: they contained many fewer actual cracks. Indeed, the wave propagation velocities in the samples thus ordered, just as in the case of treatment with pressure, were found to be much lower (0.2–0.5 cm/s instead of 1.0–2.0 cm/s), and the wave-front structure acquired features typical of the slower wave (see Figure 13*d*).

VIII. AUTOWAVE PROCESSES IN FILM SAMPLES OF REACTANTS

The next important step in the study of the regularities of the autowave modes of cryochemical conversion was to perform a series of experiments with thin-film samples of reactants. The changeover to such objects, characterized by the most intense heat absorption, allowed the realization of quasi-isothermal conditions of the process development and thus favored the manifestation of the abovementioned isothermal mechanism of wave excitation, which involves autodispersing the sample layer by layer due to the density difference between the initial and final reaction products. The new conditions not only not suppressed the phenomenon, but made it possible to reveal some details of the traveling-wave-front structure, which will be discussed here and also in Section X.

Experiments were performed with films of reactants frozen on a flat substrate and immersed in liquid nitrogen, that is, in the absence of any heat screen between the free film surface and the pool of a cooling agent. To obtain film samples, a liquid reactant mixture, free of O_2 and cooled to 150 K, was measured out into a flat horizontally positioned cylindrical cuvette. With adequate moistening, the mixture spread over the cuvette bottom to yield a film 50–150 μm thick, depending on the mixture dose. The cuvette with the film was then placed in liquid nitrogen, and the frozen reactant layer was subjected to γ radiolysis from a ^{60}Co source directly in the cryostat. The dynamics of the process in a film sample was registered by microfilming at a magnification up to 20 ×.

Comparison of the experiments carried out in glass and metallic cuvettes did not show any appreciable effect of the substrate material on the characteristics of the autowave processes in solid-state conversions. The critical doses required for realization of the self-sustained regimes of conversion in films turned out to be noticeably greater than in massive samples.

The best-studied were two film systems: the chlorination reaction in a solid solution of Cl_2 in BC and the copolymerization reaction of sulfur dioxide with isoprene. Under the above-described conditions of cooling, the liquid films in both systems vitrified.

The autowave chemical conversion in a film was initiated by a local mechanical disturbance in the form of a puncture or scratch. Figure 14 presents the results of measuring the averaged value of the wave-front propagation velocity as a function of the irradiation dose. It is seen that for

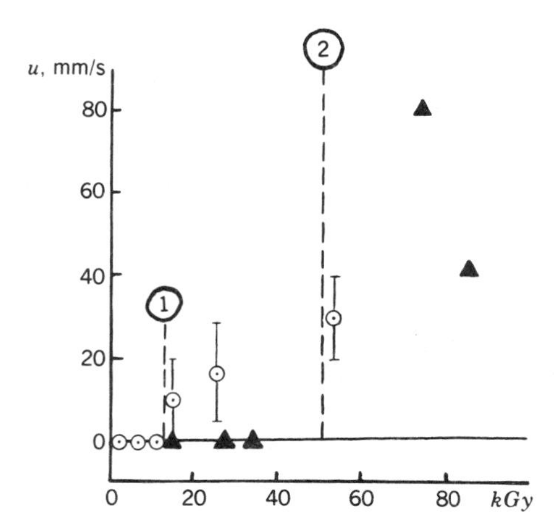

Figure 14. Velocity of $BC + Cl_2$-reaction wave-front propagation as a function of γ irradiation dose at 77 K for massive cylindrical samples of diameter 5–6 nm (circles) and thin-film samples $\sim 100\,\mu m$ thick (triangles). Critical doses of γ irradiation for massive (1) and film (2) samples are shown by dashed lines.

films the critical irradiation dose (below which the process of mechanical self-activation is not realized) is greater by a factor of 3–3.5 than for massive samples. It may be assumed that the velocity of the reaction wave switched on at the front and hence the velocity of the reaction heat release grow linearly with the concentration of active centers (i.e., with the irradiation dose). If this is the case, then the above increase in the threshold dose fails to compensate the intensification of heat transfer from film samples, by more than two orders of magnitude (as evaluated by the change in the parameter Bi calculated for the thermophysical conditions in the massive cylindrical and film samples). This may be considered as definite evidence for the isothermal mechanism being involved in the autowave processes in the experiments described. It cannot even be excluded that the growth of the threshold dose may have no relation at all to the intensification of heat transfer in a film and may be determined only by the fact that film samples have a lower volume-averaged coefficient of irreversible absorption of the energy of formation of stabilized particles than massive samples have. This feature of film samples may be due to their greater transparency (the difference in thickness is about two orders of magnitude) and to the ratio of the free surface of a sample to its volume during the stabilization of active particles formed in the matrix as a result of radiolysis. Besides, the increase in the threshold dose may be linked with the increased role of mechanical imperfection of films in comparison with massive samples: the actual cracks, by breaking the thermal and mechanical contacts between separate grains of a film, severely suppress the movement of a reaction wave.

This is evidenced by the character of the process displayed in the cinegram of Fig. 15, and even more clearly in the cinegram of Fig. 16, obtained with oblique illumination. It is seen that the wave process develops essentially asymmetrically, slowing down or even coming to a halt in some places. Figure 16 (frames 5 and 6) shows that when the reaction front reaches the crack AB, its propagation in the direction perpendicular to AB ceases. The front moves to the periphery and, bypassing this zone, continues its propagation (frame 6, right side). Hence it follows that the true velocity of the wave front in a defectless sample must be far above the averaged experimental value, whereas the limit dose must be below the measured one.

Analogous results have been obtained with thin films of equimolar SO_2 solution in isoprene. When the surface of such a film immersed in liquid nitrogen was damaged with a thin needle, a copolymerization wave occurred which spread over the sample. The preirradiation dose required to excite the autowave process in this system was ~ 200 kGy, which is also well above the critical values for cylindrical samples.

The key role of the density mechanism of the autowave process in film samples is also evidenced by the fact that in these systems (as well as in capillaries), under no initiation conditions could the slower wave be excited:

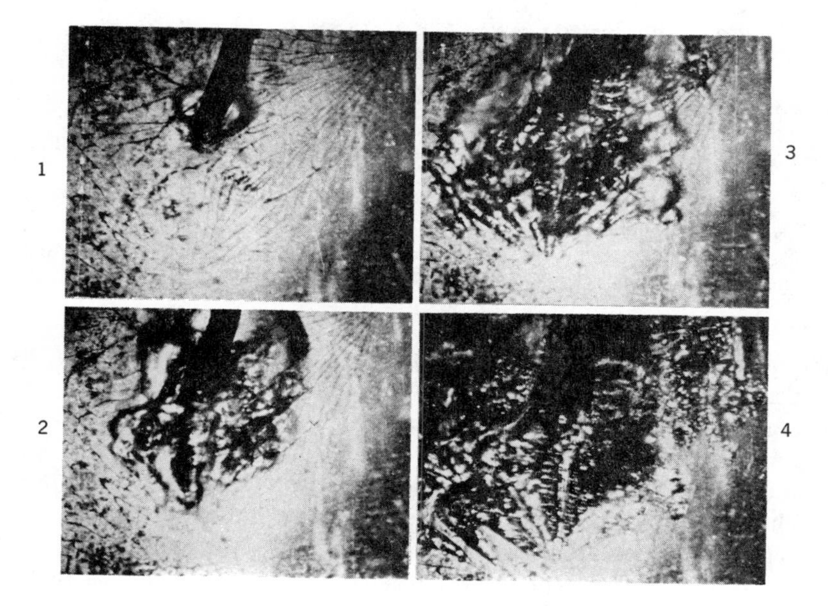

Figure 15. Cinegram of reaction-wave propagation on scratching the surface of a film produced from solution of Cl_2 in BC. The film was immersed in liquid nitrogen. Sample thickness $\sim 80\ \mu m$, dose 72.5 kGy.

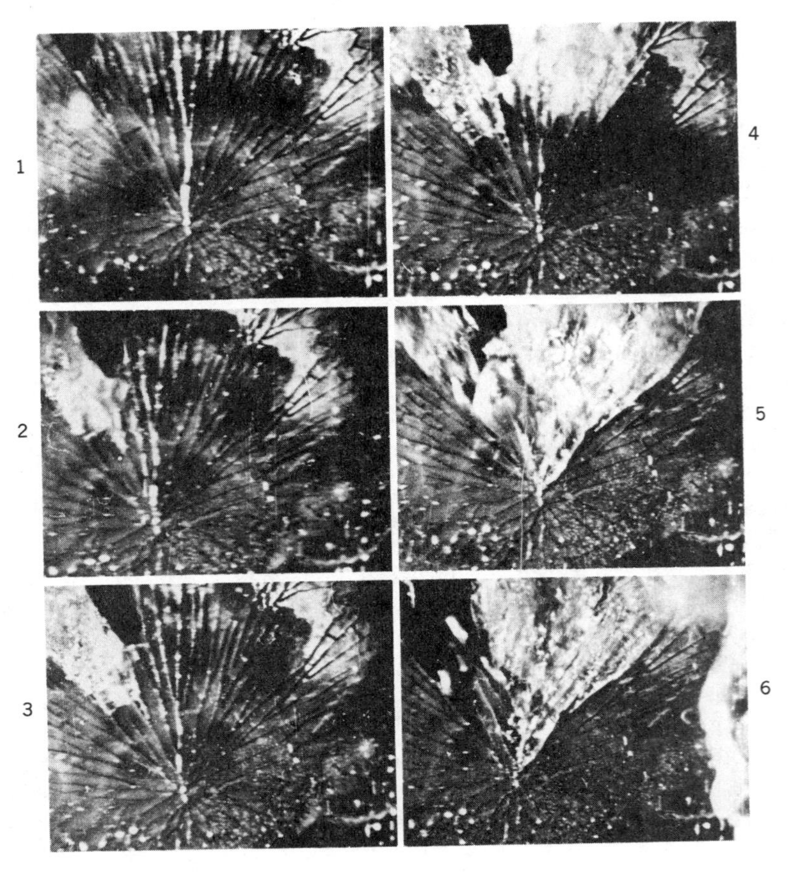

Figure 16. Cinegram of reaction-wave propagation on scratching the surface of a film prepared from solution of Cl_2 in BC. The film was immersed in liquid nitrogen. Sample thickness $\sim 100\ \mu m$, dose 87.5 kGy; oblique illumination.

the enhancement of heat transfer suppressed this phenomenon, characteristic of the nonisothermal autodispersion–conversion mechanism.

An accurate study of the temperature profile structure in film and capillary samples involves considerable technical difficulties, which accounts for the lack of direct information on the role of the isothermal and nonisothermal mechanisms in the systems considered. However, some features of the structure are evident from the cinegram of Fig. 9. It shows that the wave front traveling in a capillary is noticeably ahead of the zone of intense reaction-heat release, marked by violent boiling of liquid helium in the cryostat. This observation allows the conclusion that here the fore part of the wave front is located in the not yet heated portion of the sample; that is, small degrees of conversion seem to be sufficient to produce fracture of the matrix due to the density gradient and to switch on the reaction in the neighboring layer of the initial solid sample of reactants. A separate study on the relative position of the temperature and dispersion fronts will allow evaluation of the contributions of the heat and density factors to the autowave modes of conversion under different experimental conditions.

IX. THE ROLE OF MECHANICAL LOADING DYNAMICS AND FRACTURE MODE IN THE INTIATION OF AUTOWAVE PROCESSES

The concepts developed in the series of works presented place emphasis on the decisive role of brittle fracture of a solid matrix of reactants both at the initiation stage and in the course of autowave conversion. In this connection, the question of the possible activating effect of plastic deformations on the phenomena under investigation seemed to be of principal importance, especially as this type of deformation was traditionally assigned a prominent part in the mechanochemistry of the solid.[25,33] To elucidate this question, the effect of the dynamics of loading and of the mode of fracturing a sample on the reaction initiation has been studied.[19] First, it was necessary to find out whether the solid systems in question are capable of plastic deformation or are absolutely fragile at such low temperatures. The data obtained in the previous experiments were insufficient to answer this question, because the destruction of a sample both at the stage of the autowave process initiation by a local fracture and during the spontaneous propagation of the wave front is characterized by high shock rates of loading, under which destruction typically occurs even in the region of elastic deformations. Evaluating the time scale of the sample loadings at the initiation stage during the discharge through the microheater and at the front of a traveling wave by their dynamical characteristics, we get values of the order of 10^{-1}–10^{-2} s. This confirmation of the existence of plastic deformations required a change to regimes of slow loading.

The first series of experiments was performed with film samples 50–100 μm thick, which were prepared from a chlorine + butyl chloride mixture (mole ratio 1:3) by the method described in the previous section.

The microphotographs of Fig. 17*a* and *b* demonstrate the brittle fracture of a film occurring in response to an impact at the point *A* (the initial state—frame *a*) with a needle, developing a pressure of $\sim 10^8$ Pa for the time ~ 0.1 s. Frame *b* shows that the impact produces a network of cracks in the grain affected. An analogous effect on a film preirradiated with a supercritical dose of γ-rays initiates a traveling wave of chemical conversion (Fig. 17*e*) similar to those described in Section VIII.

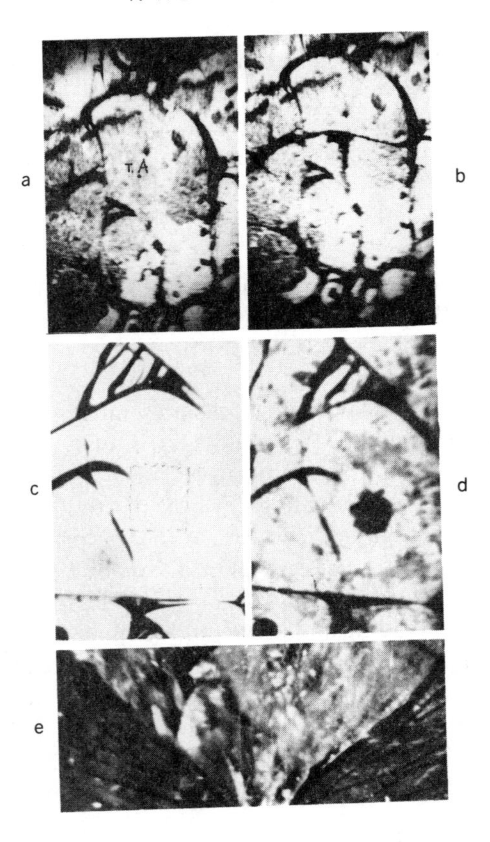

Figure 17. (*a*) Nonirradiated initial sample; (*b*) the same fragment after shock loading of the sample at point *A*; (*c*) initial sample irradiated by γ rays from ^{60}Co, dose 90 kGy; (*d*) the same sample irradiated by γ rays after slow loading on the area shown by the dashed line; (*e*) wave propagation (the boundary between light and dark regions) along the γ-irradiated sample (dose 90 kGy) after rapid loading in the regime of *b*.

A local disturbance with a needle (diameter 0.6 mm) in the regime of slow loading (approximately 10^3 Pa/s) of the film subjected to the same dose of γ-radiation did not lead to any autowave propagation of chemical reaction, although it left a distinct trace as a pressed-in, plastically deformed area having the form of the needle cross section (in Fig. 17, frame *c* shows the initial state of the film, frame *d* its final state: there are no signs of brittle fracture).

The series of experiments allows the conclusion that plastic deformation by itself (of which the systems under study are, in principle, capable) cannot initiate the development of the autowave process without the occurrence of brittle fracture. Therefore, creation of high but plastically deforming pressures is not a sufficient condition for the initiation of autowave processes. In order to excite a traveling reaction wave in a sample with a supercritical concentration of active centers, it is necessary that not only the absolute threshold value of a mechanical load but also a sufficiently high rate of its change be ensured to cause brittle fracture of the solid matrix.

This conclusion was confirmed also by a different series of experiments elucidating the role of plastic deformations. The experiments were performed in the regime of practically uniform rather than local loading. To this end we employed the procedure developed to study the initiation and development of autowave processes under conditions of uniform compression (see Section VII). But whereas previously what were subjected to γ radiation were massive samples under conditions of high static pressure (i.e., the stage of accumulation of active centers in the sample was preceded by plastic deformation during compression), in this work the experimental procedure was modified to fit the task formulated above.

The reactant mixture of the above indicated composition was frozen at 77 K and at atmospheric pressure and was then pressed at $(3-4) \times 10^8$ Pa. The pressing caused 6% shrinkage, which was probably due to compression of the sample material itself and curing of pores, blisters, and cracks formed during freezing. The sample was then unloaded to atmospheric pressure and subjected to γ irradiation to produce an overthreshold concentration of active centers. The irradiated sample, capable of chemical conversion, was again subjected to uniform compression to $(3-4) \times 10^8$ Pa in the regime of smooth loading at a rate 6×10^5 Pa/s with 2-min exposure to a fixed pressure every 10^7 Pa. No switching on of chemical reaction was observed, either in the course of loading or on its completion, in spite of high static pressures and significant plastic deformations evidenced by shrinkage of the irradiated sample during its compression. The occurrence of local brittle fracture did initiate the autowave propagation of chemical conversion with characteristics analogous to those described above.

Therefore, neither the appreciable plastic deformation (both in the case of uniform compression and of local fracture) of the solid reaction systems studied nor their static state of high stress is a factor conditioning the critical phenomena and autowave processes observed during the chemical conversion in the systems. In other words, this series of experiments has provided another telling argument for the decisive role of brittle fracture in the mechanism of the phenomena considered.

X. ON THE STABILITY OF THE STEADY-STATE PROPAGATION OF A SOLID-STATE CONVERSION WAVE

In Section VI we have touched upon the subject of the stability of steady-state wave propagation and pointed out the signs of a monotonic instability in the low-velocity autowave process. Here we shall consider qualitatively another possible physical mechanism for the oscillatory instability in the propagation of the reaction zone over a frozen sample.

The model considered above contained implicitly a simplifying assumption that the reaction front follows closely the fracture front in which the temperature gradient reaches a critical value; that is, it was assumed that the two fronts coincide spatially. But the fact is that the fracture zone always has its final dimensions, and the boundary of this zone, which occurs at $(dT/dx)^*$, moves forward jumpwise (with a rate of the order of the sound velocity) into the region of lower temperature gradient, to a depth comparable with the characteristic size of a fracture grain. It is not until this process is completed that the heating before the fracture front starts, and when the temperature reaches its critical value $(dT/dx)^*$, a second spatial jump of the reaction zone occurs. Thus, it might be expected that the propagation of the reaction along the sample would be discrete rather than continuous.

Indeed, the pulsating character of chemical-reaction propagation has been demonstrated[20] both in time (analysis of acoustical effects which accompany the sample fracture in the reaction front) and in space (investigation of the fine structure of the front by microfilming).

A phonogram of reaction wave propagation was recorded on magnetic tape with the help of an MD-47 microphone (frequency pass band 10^2–10^4 Hz). Spectral analysis of the acoustic signals was not performed; only the integral characteristics of the sound energy released were recorded in the experiments. Analysis of the phonogram has shown that the sound power is radiated in the form of successive pulses (clicks). This result is evidence for the discreteness of the fracture process, that is, for the pulsating character of the reaction wave-front propagation. The average wave velocity calculated from the total duration of sound pulses generated by the reacting sample, which was taken as the time required for the wave to travel along the entire sample length, is in good agreement with the values obtained from thermographic and optical measurements. Further development of the acoustic method of observation of wave processes (spectral analysis of a signal, in particular) is expected to give additional information on the character of the material breakup during reaction, the dispersivity of a fractured layer, and the mechanical properties of the system.

Experiments on the observation of the spatial reaction front structure by microfilming were performed using flat quartz tubes with slits ~ 1 mm wide. The reaction was initiated, as usual, by a pulsed microheater positioned in the upper part of the tube. The wave of chemical conversion thus propagated downwards along the sample, which is immersed in liquid nitrogen. Microfilming was performed in transmitted light with weak oblique illumination. Figure 18 presents a cinegram of the autowave process. The initial sample (frame 1) is characterized by a coarse-grained structure (grain size 1–2 mm)

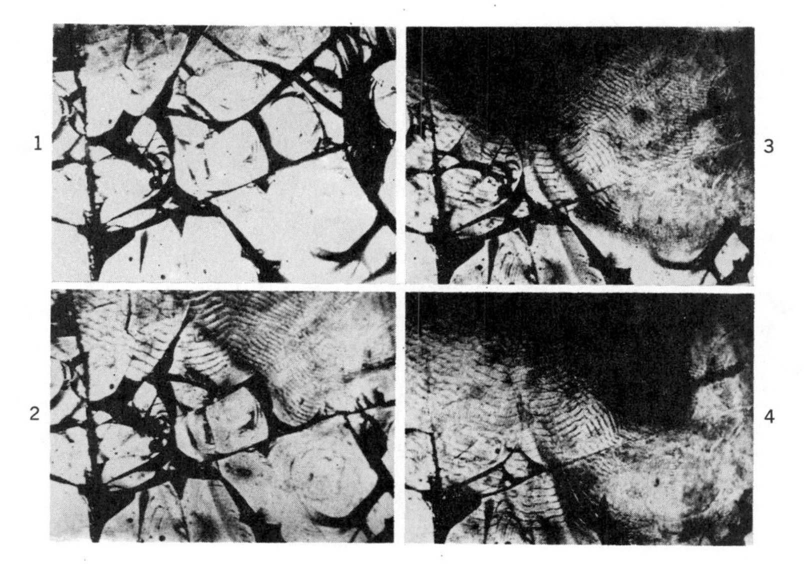

Figure 18. Cinegram of a traveling front of chemical conversion in BC + Cl_2 (3:1) at 77 K. Time intervals between frames 1 and 2: 0.11 s; 2 and 3: 0.06 s; 3 and 4: 0.03 s; dose 27 kGy.

produced during preirradiation freezing. The next frames show the process of wave propagation through the investigated fragment of the sample. It is seen that the traveling wave leaves a trace in the form of grooves parallel to wave front and separated by 50–150 μm. Such a trace is indicative of pulsatory, rather than smooth, wave-front propagation. The trace remains unchanged even on completion of the reaction, whereas the primary network of cracks becomes less pronounced and even disappears altogether in some places. The latter observation may be connected with the fact that the transparency and the scattering power of the sample greatly vary due to its fine-grained breakup during the passage of the wave front.

The form of the grooves suggests curvature of the reaction front, which appears flat on macrocinegrams (see Section V). In all likelihood, this is due to either inhomogeneity of mechanical properties of the sample (of its strength, in particular) or the loss of stability of the flat front and the formation of spatial inhomogeneities of the type of dissipative structures.[34] It is for these reasons that zones of accelerated motion and delay zones (characteristic size of inhomogeneities is ~0.5 mm) appear on the initially flat front. As the front propagates, the character of its curvature alters.

It is interesting to note that the striped "imprints" of chemical conversion observed in our experiments bear a striking resemblance to the pattern produced by the fatigue destruction of metals[35] and polymers.[36] A deeper analogy may underlie this resemblance—the pulsatory character of the fracture due to a chemical reaction is connected, as in the case of fatigue destruction, with the cyclicity of loading.

To conclude this section, we note that a theoretical treatment of the phenomena described can be conducted on the basis of the earlier model by introducing into it an additional parameter, the characteristic depth of the layer (grain size) of dispersion. This parameter seems to be close to the average length of a crack, whose physical meaning was discussed in Section VI.

This effect may also cause an oscillatory instability of the front propagation in the case of the isothermal mechanism of the autowave process. However, in this case a jumplike onset of dispersion in the next layer of the solid matrix will be connected not with the critical temperature gradient of thermal fracture, but with the critical concentration of the final product accumulated at the boundary of this layer (fracture due to a local alteration of the density).

XI. EXTENSION OF THE RANGE OF PROBLEMS

The series of studies performed testifies to a wide occurrence of autowave phenomena in cryochemical conversions of the solid and to the generality of their physical mechanism for a variety of chemical systems. Traveling waves were observed in the reactions of chlorination of saturated hydrocarbons and in the hydrobromination of olefins, as well as in the copolymerization of isoprene with sulfur dioxide, the phenomenon being realized both in vitreous and in polycrystalline matrices. These results impelled us to search for new examples of solid-state systems reacting in the traveling-wave condition. In this respect, the reaction of postpolymerization of solid acetaldehyde subjected to γ radiolysis seemed to be of great interest. Its kinetics was studied previously by the integral calorimetric method.[37] However, the development of the concepts of autowave processes stimulated an analysis of the properties of this reaction at a qualitatively new level. The system was considered particularly interesting in two respects. First, the discovery of an autowave regime in acetaldehyde polymerization would provide a first example of cryochemical waves in a one-component system. Second, the thermal effect of this reaction is almost one order of magnitude less than in the previously investigated systems. This allowed a deeper insight into the role of isothermal and nonisothermal mechanisms and their interaction in the autowave process. It is also interesting that as long as 20 years ago a study of this reaction revealed a spontaneously propagating wave of conversion at relatively high temperatures, near the melting point (108–123 K).[38] The authors of ref. 38 explained the occurrence of such a wave on the basis of the classical thermal mechanism of combustion. In this connection, the question of the continuation of the wave regime in the passage to the low-temperature region

($\leqslant 77$ K) and to the conditions of film samples, as well as of the role of pulsatory dispersion as a factor governing the traveling wave of polymerization, was of particular importance for this system.

Without going into details, we shall note only some important results obtained in the study of the system.

1. Polymerization of the radiolyzed acetaldehyde at 77 K is initiated by local brittle fracture and then spreads over the solid matrix layer by layer in the form of a traveling wave.
2. Autowave conversion is realized both in massive samples and in thin films, and both in crystalline and in vitreous states.[†]
3. Propagation of the reaction front causes intense dispersion of the frozen matrix of the monomer, and the temperature profile of the wave preserves all qualitative features characteristic of the mechanochemical wave of a solid-state conversion.

In other words, evidence has been obtained that in this system also, the autowave self-activation of the reactant is due to the positive feedback between the chemical reaction and its brittle fracture.

Publications[39,40] that appeared in 1982 contained experimental data on the explosion-type reaction in condensates of atomic magnesium with halohydrocarbons and on the polymerization of cyclopentadiene with $TiCl_4$ initiated at 77 K by an impact with a metallic needle. The authors of these works have pointed to the explosive character of chemical conversion in the systems, leaving open, however, the question of the dynamics of the processes and of the pattern of their development in time and space. It seemed natural to suppose that these publications describe a new type of autowave cryochemical reaction realized under qualitatively different conditions, which did not require preactivation of the solid matrix with γ radiation. In addition, the authors[40] took notice of the wave character of the polymerization process in a film and evaluated the front velocity as ~ 20 cm/s. We have discovered analogous phenomena in experiments with nonirradiated thin-film cocondensates of ethylene + Cl_2, propylene + Cl_2, and vinyl chloride + Cl_2.[23] It has been firmly established[23] that (1) the chlorination of these hydrocarbonds at 77 K proceeds not in the form of a homogeneous reaction burst, but in the form of an autowave process induced by pricking or scratching the film, and (2) the process is characterized by strongest dispersion in the front of the self-propagating reaction, which showed itself, particularly, in the phenomenon of "breaking out" of fine-grained fragments of the reacting matrix. Considering the latter fact and also the high velocities of front movement (tens of

[†]When frozen, pure acetaldehyde yields polycrystals. Small additions of the polymer, BC, or chloroform promote vitrification.

centimeters and even meters per second), we have good reason to agree with the authors of ref. 4 on the possibility of a different (mechanochemical) type of low-temperature wave in the solid, which however resembles closely the detonation phenomena, showing signs of degeneration (infrasonic velocities). This problem is a complicated one and calls for the development of fundamentally new theoretical and experimental approaches to the elucidation of the mechanisms of energy transfer during low-temperature detonation in solid-state systems.

In concluding this section, we shall consider one more, apparently very broad, spectrum of solid reactions at low temperatures in the form of a traveling wave. We mean the recent experiments on local mechanical initiation of a reaction in photoactivated solid matrices. It has been found that pretreatment of the solids previously studied (e.g., BC + Cl_2 at 77 K) with light also leads to the accumulation in the matrix of potential energy that may be used for the subsequent initiation of a predetermined chemical conversion by local mechanical fracture. As in the case of γ activation, the process takes the form of a traveling wave of dispersion, which invades layer after layer and is accompanied by chemical conversion. Already in the first experiments with photolysed matrices, some new properties of the autowave process have been revealed, along with the general features of self-propagating reactions, in particular, the possibility of rather high velocities of the reaction front. There are a great number of questions raised in connection with the effect of the spectral characteristics of the activating light on the dynamics of the wave process, but these are beyond the scope of the present review and require a separate study.

XII. CONCLUSION: ON THE HISTORY OF THE PROBLEM AND PERSPECTIVES FOR ITS DEVELOPMENT

In order to place the phenomena considered among the diverse processes of self-propagation of chemical conversions, we shall start with an attempt to introduce a classification of chemical systems exhibiting autowave properties. Not going into details, well shall distinguish two large phenomenological groups which include practically all the known systems of this kind: processes of the combustion type (with thermal nonlinearity as factor determining the existence of the traveling wave), and processes of the branched-chain type (the so-called "chain combustion", based on kinetic nonlinearity of an autocatalytic character). It is evident that the self-activation mechanism of the frozen reactionable system, which determines the regularities of the autowave phenomena studied in this series of works, is close in its essence to the branched-chain mechanism.

The branched-chain concept per se is not new in the chemistry of the solid and has been discussed in the literature for many decades.[41,42] It is based on the idea that the active centers ("embryos") in a reaction are various kinds of defects distributed in the solid matrix of reactants and progressively increasing in number under the influences induced in the solid by the reaction itself. The action of this mechanism has received the best study in a large series of works [43-47] concerned with the thermal decomposition of amminium perchlorate at temperatures near 500 K. By observing the alteration in the structure of a decomposing crystalline sample, Rayevskii et al. have established that (1) the conversion of a solid reactant proceeds under the strong influence of mechanical stresses due to the decomposition reaction and consists of several stages—the formation and growth of embryos of the reaction, their reproduction, increase in density, and confluence; and (2) the confluence of separately developing reaction zones may result in the formation of a single large-scale reaction front propagating over the sample as a cloud of reproducing embryos with small but quite definite velocities (of the order of 10^{-5} cm/s). The critical phenomena which, as is known, are attributes of the branched-chain and autowave processes had not been observed in refs. 43–47. Most likely, this is because the thermal mechanism plays a leading part in the reaction of perchlorate decomposition. This is evidenced by the strong exponential dependence of the total rate of the salt decomposition (as well as of the front propagation velocity) on temperature: the reaction slows down rapidly with cooling and practically comes to a halt at temperatures of 460–470 K.

It should be added that the decomposition of perchlorate is not a solid-state reaction in a strict sense, because the conversion is accompanied by an alteration of the phase state. It is this fact (the formation of the gas products of decomposition) that is considered by the authors of refs. 43–47 to be the cause of mechanical deformation of the solid reactant matrix and of the development in it of the reaction-activating imperfections.

A few years ago the concept considered was introduced also in the low-temperature chemistry of the solid.[31] Benderskii et al. have employed the idea of self-activation of a matrix due to the feedback between the chemical reaction and the state of stress in the frozen sample to explain the so called "explosion during cooling" observed by them in the photolyzed MCH + Cl_2 system. The model proposed in refs. 31, 48, 49 is unfortunately not quite concrete, because it includes an abstract quantity called by the authors "the excess free energy of internal stresses." No means of measuring this quantity or estimating its numerical values are proposed. Neither do the authors discuss the connection between this characteristic and the imperfections of a solid matrix. Moreover, they have to introduce into the model a heat-balance equation to specify the feedback, although they proceed from the nonthermal mechanisms of self-activation of reactants at low temperatures. Nevertheless, the essence of their concept is clear and can be formulated phenomenologically as follows: the mechanical energy accumulated in the matrix of reactants directly facilitates the chemical solid-state conversion at the expense of a decrease in the reaction activation barrier by a certain equivalent value. In other words, the model of low-temperature reactions in the solid proposed in refs. 31, 48, 49, unlike the mechanism of molecular tunneling,[4] is wholly developed within the framework of classical Arrhenius kinetics, supplemented only with some elements of the kinetic theory of strength.[33]

Let us try to evaluate the validity of the above statement in the light of our own data obtained in the studies of cryochemical solid-state reactions carried out under conditions of the artificially created state of high stress in the sample (see Sections VII and IX). For definiteness, we shall consider the same reaction of chlorination of hydrocarbons that was investigated in refs. 31, 48, 49, with the sole difference that preactivation was accomplished not with light but with γ radiation. The experiments have shown that the reaction not only is not facilitated but is even hindered by loading the sample to pressures $P = 6 \times 10^8$ Pa in the regime of uniform compression: the state of high stress of the matrix does not cause any chemical reaction, and the velocity of the autowave process in the compressed sample decreases markedly. Evaluation of the specific energy supplied to the sample under the pressure realized in the experiment gives a value $\Delta U \sim 6$–12 kJ/mol [$\Delta U = P^2/E$, where E is the elasticity modulus; according to ref. 50, for the investigated vitreous systems at low temperatures $E = (3$–$6) \times 10^9$ Pa]. This value of U is in good agreement with the previously measured[27] values of the activation energy for the conversions in question (~ 20 kJ/mol) and might be expected to lead to a considerable reduction in the activation barrier and to acceleration rather than suppression of the reaction. Certainly, the above considerations are far from being sufficient to reject the concept developed in refs. 31, 48, and 49 but they are forceful arguments questioning the very idea of the possible direct involvement of the mechanical energy of the deformed solid matrix in the process of its chemical activation.

Our own ideas of the nature of highly active autowave regimes of chemical conversion in cryochemistry are based on a mechanism that may be called, by analogy with the previously mentioned mechanism,[8] the *energetic chain.* The data presented in this review justify this parallelism because the mechanism suggested consists of a chain of stages connected by feedback and accompanied by transformation of the energy of chemical conversion in time and in space. It is assumed that during the reaction, part of the energy released is accumulated in a solid matrix in the form of elastic deformation; that is, at this stage part of the sample that did not react changes over to a state of mechanical nonequilibrium—in a sense, an "excited" state. The potential energy of this metastable state is then transformed into the chemical form, not immediately but through the stage of brittle fracture. It is at this stage that the conditions are created for the realization of acts of chemical conversion on the newly formed energy-concentrating fracture surfaces or near them. The action of this mechanism provides realization of the self-sustained regime of conversion induced by autodispersing of the solid sample. The concept of the branched-chain character of the process finds its picturesque reflection in the network of cracks, continuously branching and moving over the reacting matrix. The appearance of such a mechanism of recuperation of the energy of chemical conversion leading to self-activation of the solid matrix is greatly favored by sufficiently low temperatures. It is just under these conditions, under which the dissipation of the energy stored in the mechanically nonequilibrium matrix slows down (due to the extremely limited molecular mobility and almost complete loss of plasticity), that the accumulation of large amounts of the lattice is possible. This energy, released during fracture, enables further development of the self-sustained conversion of reactants.

In a most interesting series of studies on solid-state chemical conversions under conditions of high pressure combined with shifting stresses, a superhigh molecular mobility has been discovered.[51] What appears most striking for a variety of physicochemical processes, under these specific experimental conditions, is the rather large values of the diffusion coefficients, greater by 10–15 orders of magnitude than the ordinary ones. The autowave concepts accounting for the autodispersion of the reacting matrix will presumably be useful in interpreting phenomena of this kind.

The concepts developed can rightfully be called on the account also for the "cold" evolution of matter in the universe. In particular, one can imagine the formation, from the frozen mixture of elements, of compounds such as ammonia and methane that are found in appreciable amounts as solids on the cold planets of the solar system. Daily temperature variations and the concomitant thermal stresses in the solid cover of a planet might serve as a mechanism of its continuous destruction, resulting in the chemical binding of the components, which might be activated by solar radiation as well. A next stage of evolution, at which the solid ammonia and hydrocarbons dispersed in the "solar mill" might transform into the simplest amines, also seems most probable. A direct experimental check of the chain of cryosyntheses described above is expected to be an important step in understanding many of the mysteries in space chemistry.

The data on the autowave phenomena considered in this chapter may turn out to be significant for the general chemistry of solids outside the low-temperature region. Probably, it is autodispersion of the solid matrix of reactants (the process that brings a chemical reaction from a volume to a highly active surface of fracture) that accounts for the extremely high velocities of conversion not infrequently observed experimentally in systems of diverse chemical nature. They may be comparable with or even be far above the velocities of the corresponding processes in liquid media. Therefore, special attention should be paid to the action in solid systems of non-thermal, nonequilibrium mechanisms of matrix activation conditioned by the possibility of incorporation into a chemical process of the energy accumulated in the lattice. An example illustrating this general statement is furnished in ref. 52, which considers the self-activating effect of dispersion of the refractory component particles in the formation of intermetallides from powders, which occurs in quite a different temperature region. In this system, the dispersing of the refractory component is due to the moistening of particles with liquid reaction products (a kind of mechanochemical autocatalysis).

In the introduction, we touched upon the question of the validity of the quasi-homogeneous approach widely used in solid-state chemical kinetics and based on models disregarding the processes of matter and energy transfer in a solid matrix. The concepts developed in the abovementioned work on the autowave dynamics of solid reactant conversion, the sensitivity of the process to local disturbances, and inhomogeneity of the reaction development in a sample strengthen the long-existing doubts as to the permissibility of the above classical approach to solid-state reaction kinetics. Nonlinear mechanisms characteristic of the chemistry of solids, provide conditions which make spatially homogeneous reaction highly in probable. Because of this, the integrated signal of chemical conversion (and it is just this type of signal that is recorded in most works on solid-state reaction kinetics) is a superposition of signals generated at various points of the sample and differing both in rate and in phase. It is evident that such a signal cannot carry reliable information on the true kinetics of the process, and attempts to treat it in terms of kinetic models based on systems of ordinary differential equations would be erroneous.

Let us take as an example the well-known case of "polychronous" kinetics. To explain the phenomenon of kinetic stopping of a reaction recorded by integral methods, a spectrum of kinetic constants is introduced in the description of the process, which is connected with inhomogeneity in the energy states of the active centers distributed in the matrix. The autowave approach makes it possible to explain the phenomenon without recourse to the polychronous kinetic scheme: the process can be easily pictured as separate, locally induced zones of conversion propagating over the sample as a wave front and stopping at some inhomogeneities or macrodefects of the matrix. Also quite probable is a situation in which the chemical process simply cannot develop homogeneously because of spatial instability, and the reaction is accompanied by the formation of the so-called "dissipative structures"[54] (analogs of the domain phenomena widely occurring in the physics of solids.[†]

[†]Similar phenomena in heterogeneous catalysis were described in refs. 55–57 for a field related to the chemistry of solids.

Thus in each concrete case quasi-homogeneous models (in particular, models of the "polychronous" kinetics) need to be confirmed by special experiments eliminating the effects of autowave processes on the reaction dynamics.

The ideas based on the activating effect of self-destruction may turn out to be useful in a number of adjacent scientific fields concerned with solids. Probably it is this phenomenon that contains the clue to the high sensitivity of explosives to friction and impact. Up to the present time, the general theory of the sensitivity of explosives to mechanical influences has been based entirely on thermal concepts, according to which the energy supplied to a sample must first be transformed into thermal energy to produce a heated area, and only after that can an explosion be excited by the nidus mechanism. The initiation of the explosion-type conversion is likely to be due not to the transformation of mechanical energy into thermal but to the appearance in the sample of chemically highly active fracture surfaces resulting from impact or friction.

The extensively studied (especially during the recent years) transitions of solids from the metastable amorphous state to the polycrystalline state (see ref. 58 and the references therein) are of autowave character and resemble very much the above regimes of solid-state cryochemical reactions. The action of autodispersion, which facilitates phase transition by allowing it to proceed on the surface of a fracture instead of in the glass volume, cannot be excluded in the case of those processes either. Actually, the two classes of processes are similar in their physical nature: both are connected with rearrangement of the solid matrix and are of exothermic character, differing only in the extent of the thermal effect. It should be added that fracturing and autodispersion of the sample are very typical of the autowave destruction of amorphous states and can be seen even by the unaided eye.

Another example belongs to a field very remote, as it may seem, from the chemistry of solids—the breakup of transparent media by a high-power flux of radiant energy (radiation stress).[59] These processes also have an autowave character but are explained on the basis of a theory that is very close to the theory of combustion. The only difference is that the exponential temperature dependence of chemical heat release in the case of combustion is replaced by the exponential absorption of the radiant energy by a solid medium in the breakup problems. Analysis of the literature on the optical strength of glass has led us to suggest that the mechanism of autoacceleration conceived in application to chemical reactions in the solid may be used to describe the dynamics of radiative destruction in glass. Using the concepts considered, the destruction of a sample by a light flux can be pictured as follows. The light-absorbing admixtures of submicrometer dimensions, which are always present in the glass, are heated up by the light flux. The temperature reached is not dangerous for the material, because there is intense conductive heat dissipation from such small midi and the zone surrounding an admixture is as small as the absorbing center itself. However, the resulting temperature gradients are very high because of the great curvature of a spherical nidus of such a size. It is this fact that is destructive of the material strength, leading to fracture and exfoliation of the admixture from the matrix. The nonequilibrium states occurring at the instant of fracture are characterized by intense energy absorption, which results in the heating up of a newly formed fracture surface and in the subsequent crumbling of the next layer of the irradiated transparent medium. The initiation of breakup in a mixture and the subsequent wave propagation of destruction, dispersing the sample layer by layer, may occur in just this fashion. This process may be treated in the framework of a phenomenological model completely identical to that considered in Section VI of this chapter. The two experimentally modes of glass breakup,[59] dependent for their realization on the initial excitation conditions, are qualitatively predicted by the model. That these considerations on the mechanism of the optical destruction of transparent media are plausible is indicated by their acceptance by known experts in the field.[60]

To conclude our review of the studies on traveling waves of cryochemical conversions in the solid, we note that what has been accomplished so far is mainly phenomenological. The next stage of investigation faces a number of fundamental problems connected with the elementary steps of the process. The major ones involve elucidating the nature of the active states occurring on fracture surfaces and their lifetimes, determining the mechanism of intensification of the transfer processes on newly formed surfaces, and searching for electrical signals related to the emission of charged particles in the course of fracturing and also for signals of the trivoluminescence type originating from the front of a traveling wave of dispersion and chemical conversion.

References

1. K. I. Zamaraev, P. F. Khairutdinov, and V. P. Zhdanov, *Tunnelling of Electrons in Chemistry*, Nauka, Novosibirsk, 1985.
2. D. P. Kiryukhin, A. M. Kaplan, I. M. Barkalov, and V. I. Goldanskii, *Vysokomol. Soyedin.* **14(A)**, 2115–2119 (1972).
3. V. I. Goldanskii, L. I. Trakhtenberg, and V. N. Phlerov, *The Tunnel Phenomena in Chemical Physics*, Nauka, Moscow, 1986.
4. V. I. Goldanskii, *Khimii* **44**, 2121–2149 (1975).
5. M. Ya. Ovchinnikova, *Chem. Phys.* **36**, 85–95 (1979).
6. V. A. Benderskii, V. I. Goldanskii, and A. A. Ovchinnikov, *Chem. Phys. Lett.* **73**, 492–495 (1980).
7. L. I. Trakhtenberg, V. L. Klochikhin, and S. Ya. Pshezhetsky, *Chem. Phys.* **59**, 191–198 (1981); **69**, 121–134 (1982).
8. N. N. Semenov, *Chem. and Technol. Polymers*, **7–8**, 196 (1960).
9. E. I. Adirovich, *Dokl. AN SSSR* **136**, 117 (1961).
10. M. V. Basilevskii, P. N. Gerasimov, L. N. Kitrosskii, C. I. Petrochenko, and V. A. Tikhomirov, *Kinetika i Kataliz* **22**, 1134–1147 (1981).

11. J. L. Jackson, *J. Chem. Phys.* **31**, 154, 722 (1959).
12. G. K. Vasilyev and V. L. Tal'rose, *Kinetika i Kataliz* **4**, 497 (1963).
13. E. B. Gordon, L. P. Mezhov-Deglin, O. F. Pugachev, and V. V. Khmelenko, *ZhETF* **73**, 952–960 (1977) (in Russian).
14. A. M. Zanin, D. P. Kiryukhin, I. M. Barkalov, and V. I. Goldanskii, *Pisma ZhETF* **33**, 336–339 (1981) (in Russian).
15. A. M. Zanin, D. P. Kiryukhin, I. M. Barkalov, and V. I. Goldanskii, *Dokl. AN SSSR* **260**, 1171–1173 (1981).
16. A. M. Zanin, D. P. Kiryukhin, V. V. Barelko, I. M. Barkalov, and V. I. Goldanskii, *Khimicheskaya Fizika* **1**, 265–275 (1982).
17. A. M. Zanin, D. P. Kiryukhin, I. M. Barkalov, and V. I. Goldanskii, *Vysokomol. Soyedin.* **24(B)**, 243–244 (1982).
18. V. V. Barelko, I. M. Barkalov, D. A. Vaganov, A. M. Zanin, and D. P. Kiryukhin, *Khimicheskaya Fizika* **2**, 980–984 (1983).
19. A. M. Zanin, D. P. Kiryukhin, V. S. Nikolskii, I. M. Barkalov, and V. I. Goldanskii, *Izv. AN SSSR Ser. Chim.* **6**, 1228–1231 (1983).
20. A. M. Zanin, D. P. Kiryukhin, V. V. Barelko, I. M. Barkalov, and V. I. Goldanskii, *Dokl. AN SSSR* **268**, 1146–1150 (1983).
21. I. M. Barkalov, V. I. Goldanskii, D. P. Kiryukhin, and A. M. Zanin, *Int. Rev. Phys. Chem.* **3**, 247–262 (1983).
22. I. M. Barkalov, V. V. Barelko, V. I. Goldanskii, D. P. Kiryukhin, and A. M. Zanin, "Threshold Phenomena and Autowave Processes in Low-Temperature Solid-State Chemical Reactions", preprint, Inst. Chem. Physics AN SSSR, Chernogolovka (1983) (in Russian).
23. A. M. Zanin, D. P. Kiryukhin, V. V. Barelko, and I. M. Barkalov, *Dokl. AN SSSR* **281**, 372–374 (1985).
24. D. P. Kiryukhin, A. M. Zanin, V. V. Barelko, I. M. Barkalov, and V. I. Goldanskii, *Dokl. AN SSSR* **288**, 406–409 (1986).
25. N. K. Baramboim, *Mechanochemistry of High-Molecular-Weight Compounds*, Khimiya, Moscow (in Russian).
26. N. Z. Lyakhov and V. V. Boldyrev, *Izv. SO AN SSSR Ser. Chim.* **12**, 3–8 (1983).
27. D. P. Kiryukhin, I. M. Barkalov, and V. I. Goldanskii, *Khimiya Vysohikh Energii* **11**, 438–442 (1977).
28. D. P. Kiryukhin, I. M. Barkalov, and V. I. Goldanskii, *Dokl. AN SSSR* **238**, 388–391 (1978).
29. D. P. Kiryukhin, I. M. Barkalov, and V. I. Goldanskii, *J. Chim. Phys.* **76**, 1013–1015 (1979).
30. I. M. Barkalov and D. P. Kiryukhin, *Vysokomol. Soyedin.* **224**, 723–737 (1980).
31. V. A. Benderskii, E. Ya. Misochko, A. A. Ovchinnikov, and P. G. Philippov, *Pisma ZhETF* **32**, 429–432 (1980).
32. Ya. B. Zeldovich, G. I. Barenblatt, V. B. Librovich, and G. M. Makhviladze, *Mathematical Theory of Combustion and Explosion*, Nauka, Moscow, 1980.
33. V. P. Regel, A. I. Slutsker, and E. E. Tomashevskii, *Kinetic Nature of Strength of Solids*, Nauka, Moscow, 1974.
34. N. Glensdorf and I. Progogine, *Thermodynamic Theory of Structure, Stability and Fluctuations*, Wiley-Interscience, New York, 1971.
35. C. Lair and G. C. Smith, *Philos. Mag.* **7**, 847–857 (1962).
36. W. Döll, *J. Mater. Sci.* **10**, 935–942 (1975).

37. D. P. Kiryukhin, I. M. Barkalov, and V. I. Goldanskii, *Europ. Polymer J.* **10**, 309–313 (1974).
38. V. C. Pshezhetskii and V. I. Tupikov, *Heterochain High-Molecular-Weight Compounds*, 1964, pp. 220–226 (in Russian).
39. G. B. Sergeyev, V. V. Zatorskii, and A. M. Kosolapov, *Khimicheskaya Fizika* **1**, 1719–1721 (1982).
40. V. A. Kabanov, V. G. Sergeyev, G. M. Lukovkin, and V. Yu. Baranovskii, *Dokl. AN SSSR* **266**, 1410–1414 (1982).
41. W. E. Harner and H. R. Hailes, *Proc. Roy. Soc. London Ser. A* **139(A)**, 576 (1933).
42. B. Delmon, *Kinetika Geterogennykh Reaktsii*, Moscow, MIR, 1972.
43. A. V. Rayevskii and G. B. Manelis, *Dokl. AN SSSR* **155**, 886 (1963).
44. A. V. Rayevskii, G. B. Manelis, V. V. Boldyrev, and L. A. Votinova, *Dokl. AN SSSR* **160**, 1136 (1965).
45. G. B. Manelis, Yu. I. Rubtsov, and A. V. Rayevskii, *Fizika Goreniya i Vzryva* **6**, 3 (1970).
46. G. B. Manelis, *Problems of Chemical Kinetics*, MIR, Moscow, 1979, p. 266.
47. A. V. Rayevskii, in *Mechanisms of Thermal Decomposition of Ammonium Perchlorate*, Inst. Chem. Phys. Acad. Sci. USSR, Chernogolovka Branch, 1981, p. 30.
48. V. A. Benderskii, E. Ya. Misochko, A. A. Ovchinnikov, and P. G. Philippov, *Khimicheskaya Fizika* **1**, 685–691 (1982).
49. V. A. Benderskii, E. Ya. Misochko, A. A. Ovchinnikov, and P. G. Philippov, *Zh. Fizicheskoi Khimii* **57**, 1079–1090 (1983).
50. V. A. Benderskii, V. M. Beskrovnyi, E. Ya. Misochko, and P. G. Philippov, *Khimicheskaya Fizika* **3**, 1172–1183 (1984).
51. N. S. Enikolopyan, *Dokl. AN SSSR* **283**, 897–899 (1985).
52. V. N. Doronin, V. I. Itin, and V. V. Barelko, *Dokl. AN SSSR* **286**, 1155 (1986).
53. A. I. Mikhailov, A. I. Bolshakov, Ya. S. Lebedev, and V. I. Goldanskii, *Fizika Tverdogo Tela* **14**, 1172–1179 (1972).
54. G. Nicolis and I. Prigogine, *Self-Organization in Non-equilibrium Systems*, Wiley-Interscience, New York, 1977.
55. V. V. Barelko, *Problems of Kinetics and Catalysis* Moscow, Nauka, 1981, p. 61 (in Russian).
56. S. A. Zhukov and V. V. Barelko, *Khimicheskaya Fizika* **1**, 516 (1982).
57. Yu. E. Volodin, V. V. Barelko, and A. G. Merzhanov, *Khimicheskaya Fizika* **1**, 670 (1982).
58. V. A. Shklovskii, *Pisma ZhETF* **82**, 536 (1982).
59. Ya. A. Imas, "Optical Break-up of Transparent Dielectric", Preprint no. 13, ITMO AN BSSR, Minsk, 1982.
60. N. M. Bityurin, V. N. Gepkin, and M. Yu. Mylnikov, *Pisma ZhETF* **8**, 1395 (1982).

AUTHOR INDEX

Numbers in parentheses are reference numbers and indicate that the author's work is referred to although his name is not mentioned in the text. Numbers in *italics* show the pages on which the complete references are listed.

SUBJECT INDEX